MW01101012

Application of Nonlinear Systems in Nanomechanics and Nanofluids

Application of Nonlinear Systems in Nanomechanics and Nanofluids

Analytical Methods and Applications

Davood Domairry Ganji
Professor in Department of Mechanical Engineering,
Babol Noshirvani University of Technology,
Babol, Iran

Sayyid Habibollah Hashemi Kachapi
PhD in Department of Mechanical Engineering,
Babol Noshirvani University of Technology,
Babol, Iran

AMSTERDAM • BOSTON • HEIDELBERG • LONDON • NEW YORK • OXFORD
PARIS • SAN DIEGO • SAN FRANCISCO • SINGAPORE • SYDNEY • TOKYO
William Andrew is an imprint of Elsevier

William Andrew is an imprint of Elsevier
The Boulevard, Langford Lane, Kidlington, Oxford, OX5 1GB, UK
225 Wyman Street, Waltham, MA 02451, USA

British Library Cataloguing-in-Publication Data
A catalogue record for this book is available from the British Library

Library of Congress Cataloging-in-Publication Data
A catalog record for this book is available from the Library of Congress

ISBN: 978-0-323-35237-6

For information on all William Andrew publications
visit our website at http://store.elsevier.com/

This book is dedicated to
our dear mother, father, honorable family, and all the saints of Allah

We sincerely hope that the final outcome of this book helps the students, researcher and other user in developing an appreciation for the topic of nonlinear differential equations and analytical methods.

his page intentionally left blank

Contents

Preface xiii
Acknowledgments xv

CHAPTER 1 Introduction to Nanotechnology, Nanomechanics, Micromechanics, and Nanofluid 1

1.1 Nanotechnology 1
1.1.1 Introduction to Nanotechnology 1
1.1.2 Origins 3
1.1.3 Fundamental Concepts 4
1.1.4 Nanomaterials 4
1.2 Nanomechanics 5
1.3 Micromechanics 6
1.4 Nanofluid 8
1.4.1 Introduction 8
1.4.2 Synthesis of Nanofluids 9
1.4.3 Smart Cooling Nanofluids 9
1.4.4 Response Stimuli Nanofluids for Sensing Applications 9
1.4.5 Applications 10
References 10

CHAPTER 2 Semi Nonlinear Analysis in Carbon Nanotube 13

2.1 Introduction of Carbon Nanotube 14
2.1.1 Single-Wall Nanotubes 15
2.1.2 Multiwall Nanotubes 15
2.1.3 Double-Wall Nanotubes 15
2.2 Single SWCNT over a Bundle of Nanotube 17
2.2.1 Introduction 17
2.2.2 Formulations 18
2.2.3 Results 24
2.2.4 Conclusion 27
2.3 Cantilevered SWCNT as a Nanomechanical Sensor 28
2.3.1 Introduction 30
2.3.2 Analysis of the Problem 31
2.3.3 Numerical Results 33
2.3.4 Mass Sensor Mode Comparison 33
2.3.5 Conclusion 35
2.4 Nonlinear Vibration for Embedded CNT 36
2.4.1 Introduction 36

2.4.2 Basic Equations 37
2.4.3 Solution Methodology 40
2.4.4 Numerical Results and Discussion 41
2.4.5 Conclusion 45
2.5 Curved SWCNT 45
2.5.1 Introduction 45
2.5.2 Vibrational Model 46
2.5.3 Solution Methodology 48
2.5.4 Numerical Results and Discussion 49
2.5.5 Conclusion 53
2.6 CNT with Rippling Deformations 54
2.6.1 Introduction 54
2.6.2 Vibration Model 55
2.6.3 Results and Discussion 62
2.6.4 Conclusion 67
References 68

CHAPTER 3 Physical Relationships Between Nanoparticle and Nanofluid Flow 71
3.1 Turbulent Natural Convection Using Cu/Water Nanofluid 71
3.1.1 Introduction 72
3.1.2 Numerical Method 73
3.1.3 Code Validation and Mesh Results 77
3.1.4 Result and Discussion 78
3.1.5 Conclusions 81
3.2 Heat Transfer of Cu-Water Nanofluid Flow Between Parallel Plates 83
3.2.1 Introduction 83
3.2.2 Governing Equations 84
3.2.3 Analysis of the HPM 86
3.2.4 Implementation of the Method 87
3.2.5 Results and Discussion 88
3.2.6 Conclusion 91
3.3 Slip Effects on Unsteady Stagnation Point Flow of a Nanofluid over a Stretching Sheet 92
3.3.1 Introduction 94
3.3.2 Governing Equations 94
3.3.3 Result and Discussion 97
3.3.4 Conclusion 106
References 106

CHAPTER 4 Heat Transfer in Nanofluid **109**
4.1 Boundary-Layer Flow of Nanofluids Over a Moving Surface in a Flowing Fluid 110
4.1.1 Introduction 110
4.1.2 Mathematical Model 111
4.1.3 Analytical Solution by Homotopy Analysis Method 113
4.1.4 Convergence of the HAM Solution 115
4.1.5 Results and Discussion 115
4.1.6 Conclusions 127
4.2 Heat Transfer in a Liquid Film of Nanofluid on an Unsteady Stretching Sheet 127
4.2.1 Introduction 127
4.2.2 Problem Formulation and Governing Equation 128
4.2.3 Numerical Procedure and Validation 131
4.2.4 Results and Discussion 132
4.2.5 Conclusions 137
4.3 Investigation of Squeezing Unsteady Nanofluid Flow Using ADM 138
4.3.1 Introduction 138
4.3.2 Governing Equations 138
4.3.3 Fundamentals of Adomian Decomposition Method (ADM) 141
4.3.4 Solution with Adomian Decomposition Method 142
4.3.5 Results and Discussion 143
4.3.6 Conclusion 149
4.4 Investigation on Entropy Generation of Nanofluid Over a Flat Plate 149
4.4.1 Introduction 150
4.4.2 Governing Equation 150
4.4.3 Entropy Generation 153
4.4.4 Results and Discussion 154
4.4.5 Conclusion 161
4.5 Viscous Flow and Heat Transfer of Nanofluid Over Nonlinearly Stretching Sheet 162
4.5.1 Introduction 162
4.5.2 Basic Concepts of HPM 163
4.5.3 Formulation of Problem 164
4.5.4 Homotopy Perturbation Solution 166
4.5.5 Padé Approximation 169
4.5.6 Results and Discussion 169
4.5.7 Conclusions 176
References 178

CHAPTER 5 Thermal Properties of Nanoparticles 181
5.1 Effects of Adding Nanoparticles to Water and Enhancement in Thermal Properties 181
5.1.1 Introduction 182
5.1.2 Governing Equations 182
5.1.3 Basic Idea of HAM 185
5.1.4 Application of HAM to Falkner-Skan Problem 186
5.1.5 Convergence of HAM Solution 188
5.1.6 Results and Discussion 189
5.1.7 Conclusion 194
5.2 Temperature Variation Analysis for Micro and Nanoparticle's Combustion 194
5.2.1 Introduction 194
5.2.2 Problem Description 195
5.2.3 Applied Analytical Methods 196
5.2.4 Results and Discussion 200
5.2.5 Conclusion 201
References 202

CHAPTER 6 Natural, Mixed, and Forced Convection in Nanofluid 205
6.1 Natural Convection Flow of Nanofluid in a Concentric Annulus 206
6.1.1 Introduction 206
6.1.2 Problem Definition and Mathematical Model 207
6.1.3 Boundary Conditions 209
6.1.4 The Lattice Boltzmann Model for Nanofluid 210
6.1.5 Grid Testing and Code Validation 211
6.1.6 Results and Discussion 212
6.1.7 Conclusions 218
6.2 Mixed Convection Flow of a Nanofluid in a Horizontal Channel 219
6.2.1 Introduction 220
6.2.2 Describe Problem and Mathematical Formulation 220
6.2.3 Homotopy Perturbation Method Applied to the Problem 223
6.2.4 Results and Discussion 224
6.2.5 Conclusion 226
6.3 The Effect of Nanofluid on the Forced Convection Heat Transfer 229
6.3.1 Introduction 229
6.3.2 Governing Equations 230
6.3.3 Solution Using the HAM 233
6.3.4 Results and Discussion 235
6.3.5 Conclusion 241

6.4 Heat Transfer in Slip-Flow Boundary Condition of a Nanofluid in Microchannel ... 241
6.4.1 Introduction ... 242
6.4.2 Problem Statement and Governing Equation ... 243
6.4.3 Numerical Procedure and Validation ... 246
6.4.4 Results and Discussion ... 248
6.4.5 Conclusion ... 250
6.5 Forced Convection Analysis for Magnetohydrodynamics (MHD) Al_2O_3-Water Nanofluid Flow ... 253
6.5.1 Introduction ... 254
6.5.2 Description of the Problem ... 254
6.5.3 Basic Idea of HAM ... 257
6.5.4 Numerical Method ... 260
6.5.5 Results and Discussion ... 260
6.5.6 Conclusion ... 266
References ... 266

CHAPTER 7 Nanofluid Flow in Porous Medium ... 271
7.1 Introduction of Porous Medium ... 272
7.2 Stagnation Point Flow of Nanofluids in a Porous Medium ... 272
7.2.1 Introduction ... 272
7.2.2 Mathematical Formulation ... 273
7.2.3 Numerical Procedure and Validation ... 276
7.2.4 Results and Discussions ... 277
7.2.5 Conclusion ... 284
7.3 Flow and Heat Transfer of Nanofluids in a Porous Medium ... 286
7.3.1 Introduction ... 286
7.3.2 Problem Statement ... 287
7.3.3 Flow Analysis and Mathematical Formulation ... 288
7.3.4 The HAM Solution of the Problem ... 291
7.3.5 Convergence of the HAM Solution ... 293
7.3.6 Results and Discussions ... 293
7.3.7 Conclusions ... 300
7.4 Natural Convection in a Non-Darcy Porous Medium of Nanofluids ... 301
7.4.1 Introduction ... 303
7.4.2 Governing Equations ... 303
7.4.3 Solution Using HAM ... 306
7.4.4 Convergence of HAM Solution ... 307
7.4.5 Results and Discussions ... 308
7.4.6 Conclusion ... 314
References ... 314

CHAPTER 8 Nanofluid Flow in Magnetic Field 317
8.1 MHD Nanofluid flow Analysis in Divergent and Convergent Channels.......... 318
8.1.1 Introduction.......... 318
8.1.2 Problem Description.......... 319
8.1.3 Weighted Residual Methods 320
8.1.4 Results and Discussions 324
8.1.5 Conclusion 327
8.2 MHD Stagnation-Point Flow of a Nanofluid and Heat Flux 330
8.2.1 Introduction.......... 330
8.2.2 Mathematical Model.......... 331
8.2.3 Methods of Solution 333
8.2.4 Analytical Solution Statement.......... 334
8.2.5 Results and Discussion.......... 335
8.2.6 Conclusions.......... 347
8.3 Jeffery-Hamel Flow with High Magnetic Field and Nanoparticle.......... 349
8.3.1 Introduction.......... 350
8.3.2 Governing Equations 351
8.3.3 Fundamentals of ADM.......... 353
8.3.4 Application.......... 354
8.3.5 Results and Discussion.......... 355
8.3.6 Conclusion 359
8.4 The Transverse Magnetic Field on Jeffery-Hamel Problem with Cu-Water Nanofluid 359
8.4.1 Introduction.......... 360
8.4.2 Problem Statement and Mathematical Formulation 360
8.4.3 Application of HAM on MHD Jeffery-Hamel Flow 362
8.4.4 Convergence of the HAM Solution 363
8.4.5 Results and Discussions 365
8.4.6 Conclusion 369
8.5 Investigation of MHD Nanofluid Flow in a Semiporous Channel.......... 369
8.5.1 Introduction.......... 370
8.5.2 Problem Description.......... 370
8.5.3 Weighted Residual Methods 373
8.5.4 Results and Discussions 376
8.5.6 Conclusion 379
References.......... 385

Index 389

Notation and Units
Both the SI and the US/English system of units have been used throughout the book.

Preface

INTRODUCTION

This book deals with recent progress in nonlinear science and its application in systems with nature of nanotechnology field to all ramifications of science and engineering, highlighting the most important advances and challenging application of nonlinear science. This book strives to have chapters readable by a broad audience of the wider communities in varies fields of nanotechnology science, such as mathematics, physics, engineers, material science, and others. The book seeks to tackle a range of new problems of practical and theoretical interest especially in nanofluids, nanomechanics, and dynamical systems.

The book developed from problems on nonlinear differential equations in nanotechnology given over several years in nonlinear dynamics team in the Mechanical Department of Babol Noshirvani University of Technology that have been introduced to journal paper and academic lectures. It presents an introduction to dynamical, nanoparticle, and nanofluids systems with nanotechnology field in the context of nonlinear differential equations and is intended for students of engineering, mathematics, and the sciences, and workers in these areas who are mainly interested in the more direct applications of the subject.

To develop engineering and applied sciences, it is necessary to carefully study analytical and numerical methods for solving of all available problems in case of linear and nonlinear equations. It is of great importance to study nonlinearity in nanotechnology science because almost all applied processes act nonlinearly, and on the other hand, nonlinear analysis of complex systems is one of the most important and complicated tasks, especially in engineering and applied sciences problems.

None of the books in this area have completely studied and analyzed all applied processes in both linear and nonlinear forms for important nonlinear differential equations in this field, so that the user can solve the problems without the need of studying too many different references. Thereby in this book, by the use of the latest analytic methods and using more than 100 references like books, papers, and the researches done by the authors and by considering almost all possible processes and situation, new theories have been proposed to encounter applied problems in engineering and applied sciences. In this way, the user (bachelor's, master's, and PhD students, university teachers, and even in research centers in different fields) can encounter such systems in confidently. In the different chapters of the book, not only are the linear and nonlinear problems broadly discussed, but also applied examples are practically solved by the proposed methodology.

The users of this collection can achieve very strong capabilities in the area, especially in linear nonlinear phenomena, such as follows:

- A complete understanding of the dynamical, nanomechanical, and nanofluids systems with nanotechnology field in the context of nonlinear differential equations.
- A complete study of mathematical problems, analytic methods.
- Complete familiarity with specialized processes and applications in different areas of the nanodynamically system, studying them, elimination of complexities and controlling them, and also applying them in real-life application cases.
- The ability to encounter, model, and interpret an application process or system and to solve the related complexities.

AUDIENCE

This book is a comprehensive and complete text on nonlinear differential equations in nanotechnology field and application in engineering and applied sciences. It is self-contained and the subject matter is presented in an organized and systematic manner.

This book is quite appropriate for several groups of people including:

- Graduate students taking the course and research in nonlinear differential equation and analytical methods in dynamical, nanomechanical, and nanofluids systems with nanotechnology field.
- This book can be adapted for a short professional course on the subject matter.
- Design and research engineers and applied science such as mathematical and physics will be able to draw upon the book in selecting and developing mathematical models for analytical and design purposes in applied conditions.
- Generally, the users are bachelor's, master's, and PhD students, university teachers, and even in research centers in different fields of engineering and applied sciences, etc.

Because this book is aimed at a wider audience, the level of mathematics is kept intentionally low. All the principles presented in the book are illustrated by numerous worked examples. This book draws a balance between theory and practice.

Acknowledgments

We are grateful to all those who have had a direct impact on this work. Many people working in the general areas of dynamical systems with nanotechnology field, analytical methods, and nonlinear phenomena especially nonlinear differential equations, mathematical and physical problems, and design have influenced the format of this book.

The authors are very thankful to Babol Noshirvani University of Technology, Iran, and National Elite Foundation of Iran (Bonyad Melli Nokhbeghan), Mazandaran Province, Sari, Iran and especially nonlinear dynamics team in Mechanical Engineering Department, especially, M. Sheikholeslami, M. Hatami, I. Mehdipour, Y. Rostamiyan, A. Fereidoon, M. Taeibi-Rahni, M. Gorji, A. Barari, F. Hedayati, M. Khaki, A. Farshidianfar, R. Nouri, H.R. Ashorynejad, A. Malvandi, S. Tavakoli, M. Abdollahzadeh, A.R. Yousefi, A. Rafiei, A. Kimiaeifar, H. Jahani, A. Rasekh, P. Soltani, H. Sajjadi, G.H.R. Kefayati, A.M. Babaee, M. Hassani, M. Mohammad Tabar, A. Mohammad Tabar, M. Toomaj, M. Esmaeilpour, S.M.J. Hashemi, S.M. Hamidi, K. Boubaker, M. Fakour, A. Bakhshi, A. Vahabzadeh, A. Vosoughi, S. Naeejee, N. Ghadimi, M. Ramzannezhad, B. Haghighi, M. Hosseini, R.A. Talarposhti, S. Iman Pourmousavi.k, B. Jalilpour, S. Jafarmadar, A.B. Shotorban, Houman B. Rokni, I. Rahimi Petroudi, M. Khazayi Nejad, J. Rahimi, E. Rahimi, A. Rahimifar, and all the professors and students of all Iranian universities who helped them develop research skill, editing the electronic text, and gave useful consultations and also precious guidance. The references they provided for the authors, especially applied examples, were used in different chapters.

We are indebted to many colleagues and to numerous authors who have made contributions to the literature in this field. In addition, I greatly owe my indebtedness to all the authors of the articles listed in the bibliography of this book. I also like to thank the reviewers for their efforts and for the comments and suggestions, which have well served to compile the best possible book for the intent and targeted audience.

A special thanks is extended to the publisher and also the chief editor of the publication for his excellent revision of the English language of the book and for editing the electronic text.

Finally, we would very much like to acknowledge the devotion, encouragement, patience, and support provided by our family members.

I would appreciate being informed of errors or receiving other comments about the book. Please write to the authors at the Babol Noshirvani University of Technology address or send e-mail to ddg_davood@yahoo.com (Davood Domairry Ganji) and sha.hashemi.kachapi@gmail.com (Sayyid Habibollah Hashemi Kachapi).

This page intentionally left blank

CHAPTER

INTRODUCTION TO NANOTECHNOLOGY, NANOMECHANICS, MICROMECHANICS, AND NANOFLUID

1

CHAPTER CONTENTS

1.1 Nanotechnology 1
1.1.1 Introduction to Nanotechnology 1
1.1.2 Origins 3
1.1.3 Fundamental Concepts 4
1.1.4 Nanomaterials 4
1.2 Nanomechanics 5
1.3 Micromechanics 6
1.4 Nanofluid 8
1.4.1 Introduction 8
1.4.2 Synthesis of Nanofluids 9
1.4.3 Smart Cooling Nanofluids 9
1.4.4 Response Stimuli Nanofluids for Sensing Applications 9
1.4.5 Applications 10
References 10

1.1 NANOTECHNOLOGY

1.1.1 INTRODUCTION TO NANOTECHNOLOGY

Nanotechnology (sometimes shortened to "nanotech") is the manipulation of matter on an atomic, molecular, and supramolecular scale. The earliest, widespread description of nanotechnology referred to the particular technological goal of precisely manipulating atoms and molecules for fabrication of macroscale products, also now referred to as molecular nanotechnology. A more generalized description of nanotechnology was subsequently established by the National Nanotechnology Initiative, which defines nanotechnology as the manipulation of matter with at least one dimension sized from 1 to 100 nm. This definition reflects the fact that quantum mechanical effects are important at this quantum-realm scale, and so the definition shifted from a particular technological goal to a research

Application of Nonlinear Systems in Nanomechanics and Nanofluids. http://dx.doi.org/10.1016/B978-0-323-35237-6.00001-7

category inclusive of all types of research and technologies that deal with the special properties of matter that occur below the given size threshold. It is therefore common to see the plural form "nanotechnologies" as well as "nanoscale technologies" to refer to the broad range of research and applications whose common trait is size. Because of the variety of potential applications (including industrial and military), governments have invested billions of dollars in nanotechnology research. Through its National Nanotechnology Initiative, the United States has invested 3.7 billion dollars. The European Union has invested 1.2 billion and Japan 750 million dollars (see wikipedia.org/wiki/Nanotechnology).

Nanotechnology as defined by size is naturally very broad, including fields of science as diverse as surface science, organic chemistry, molecular biology, semiconductor physics, microfabrication, etc. (see Figure 1.1). The associated research and applications are equally diverse, ranging from extensions

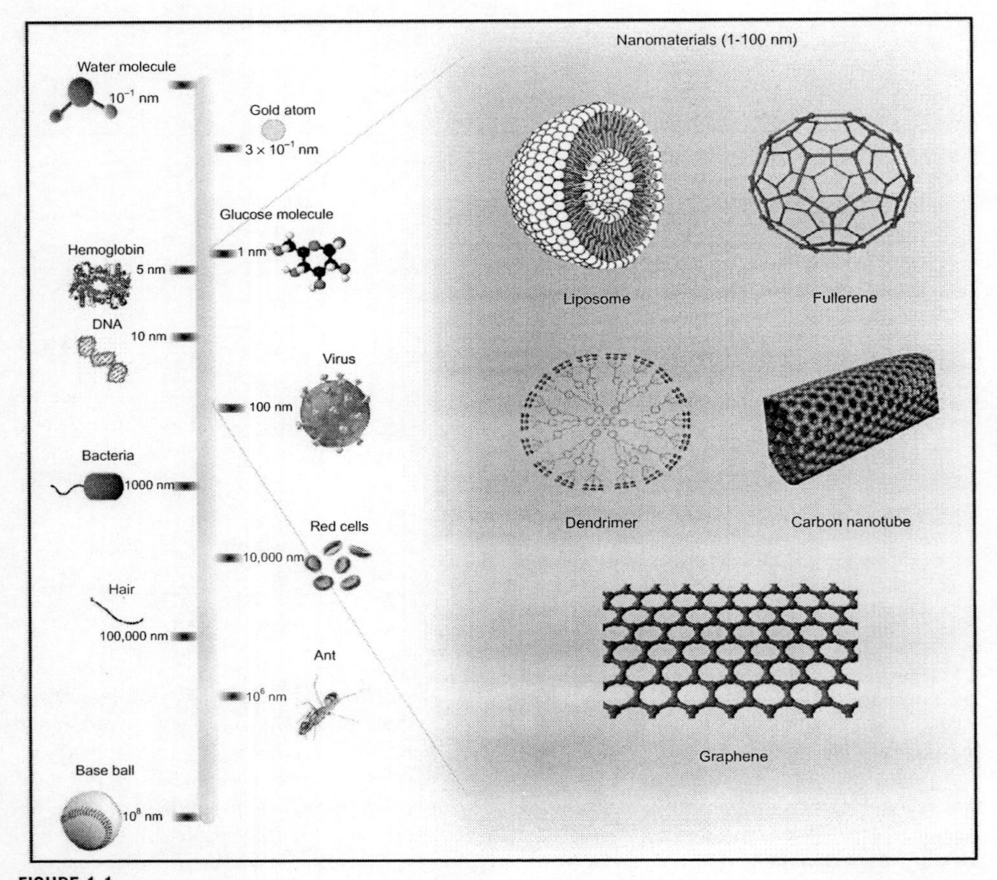

FIGURE 1.1

Comparison of nanomaterial sizes.

of conventional device physics to completely new approaches based upon molecular self-assembly, from developing new materials with dimensions on the nanoscale to direct control of matter on the atomic scale.

Scientists currently debate the future implications of nanotechnology. Nanotechnology may be able to create many new materials and devices with a vast range of applications, such as in medicine, electronics, biomaterials, and energy production. On the other hand, nanotechnology raises many of the same issues as any new technology, including concerns about the toxicity and environmental impact of nanomaterials, and their potential effects on global economics, as well as speculation about various doomsday scenarios. These concerns have led to a debate among advocacy groups and governments on whether special regulation of nanotechnology is warranted. (see nanotechweb, n.d; NASA, n.d; Drexler, 1986,1992; Bhushan, 2007; Buzea et al., 2007; Saini et al., 2010; Nanotechnology Information Center, 2011; Sattler, 2011; Minkowycz et al., 2013)

1.1.2 ORIGINS

The concepts that seeded nanotechnology were first discussed in 1959 by renowned physicist Richard Feynman in his talk "There's Plenty of Room at the Bottom," in which he described the possibility of synthesis via direct manipulation of atoms. The term "nano-technology" was first used by Norio Taniguchi in 1974, though it was not widely known.

Inspired by Feynman's concepts, K. Eric Drexler used the term "nanotechnology" in his 1986 book "Engines of Creation: The Coming Era of Nanotechnology," which proposed the idea of a nanoscale "assembler," which would be able to build a copy of itself and of other items of arbitrary complexity with atomic control. Also in 1986, Drexler cofounded The Foresight Institute (with which he is no longer affiliated) to help increase public awareness and understanding of nanotechnology concepts and implications.

Thus, emergence of nanotechnology as a field in the 1980s occurred through convergence of Drexler's theoretical and public work, which developed and popularized a conceptual framework for nanotechnology, and high-visibility experimental advances that drew additional wide-scale attention to the prospects of atomic control of matter. In 1980s, two major breakthroughs incepted the growth of nanotechnology in modern era.

First, the invention of the scanning tunneling microscope in 1981 provided unprecedented visualization of individual atoms and bonds and was successfully used to manipulate individual atoms in 1989. The microscope's developers Gerd Binnig and Heinrich Rohrer at IBM Zurich Research Laboratory received a Nobel Prize in Physics in 1986. Binnig, Quate, and Gerber also invented the analogous atomic force microscope that year.

Second, fullerenes were discovered in 1985 by Harry Kroto, Richard Smalley, and Robert Curl, who together won the 1996 Nobel Prize in Chemistry. C_{60} was not initially described as nanotechnology; the term was used regarding subsequent work with related graphene tubes (called carbon nanotubes and sometimes called Bucky tubes) which suggested potential applications for nanoscale electronics and devices.

In the early 2000s, the field garnered increased scientific, political, and commercial attention that led to both controversy and progress. Controversies emerged regarding the definitions and potential implications of nanotechnologies, exemplified by the Royal Society's report on nanotechnology. Challenges were raised regarding the feasibility of applications envisioned by advocates of molecular nanotechnology, which culminated in a public debate between Drexler and Smalley in 2001 and 2003.

Meanwhile, commercialization of products based on advancements in nanoscale technologies began emerging. These products are limited to bulk applications of nanomaterials and do not involve atomic control of matter. Some examples include the Silver Nano platform for using silver nanoparticles as an antibacterial agent, nanoparticle-based transparent sunscreens, and carbon nanotubes (CNTs) for stain-resistant textiles.

Governments moved to promote and fund research into nanotechnology, beginning in the United States with the National Nanotechnology Initiative, which formalized a size-based definition of nanotechnology and established funding for research on the nanoscale.

By the mid-2000s, new and serious scientific attention began to flourish. Projects emerged to produce nanotechnology roadmaps which center on atomically precise manipulation of matter and discuss existing and projected capabilities, goals, and applications.

1.1.3 FUNDAMENTAL CONCEPTS

Nanotechnology is the engineering of functional systems at the molecular scale. This covers both current work and concepts that are more advanced. In its original sense, nanotechnology refers to the projected ability to construct items from the bottom up, using techniques and tools being developed today to make complete, high performance products.

One nanometer is one billionth, or 10^{-9}, of a meter. By comparison, typical carbon-carbon bond lengths, or the spacing between these atoms in a molecule, are in the range of 0.12-0.15 nm, and a DNA double helix has a diameter around 2 nm. On the other hand, the smallest cellular life forms, the bacteria of the genus *Mycoplasma*, are around 200 nm in length. By convention, nanotechnology is taken as the scale range of 1-100 nm following the definition used by the National Nanotechnology Initiative in the United States. The lower limit is set by the size of atoms (hydrogen has the smallest atoms, which are approximately a quarter of a nanometer diameter) since nanotechnology must build its devices from atoms and molecules. The upper limit is more or less arbitrary but is around the size that phenomena not observed in larger structures start to become apparent and can be made use of in the nano device. These new phenomena make nanotechnology distinct from devices which are merely miniaturized versions of an equivalent macroscopic device; such devices are on a larger scale and come under the description of microtechnology.

To put that scale in another context, the comparative size of a nanometer to a meter is the same as that of a marble to the size of the earth. Or another way of putting it: a nanometer is the amount an average man's beard grows in the time it takes him to raise the razor to his face.

Two main approaches are used in nanotechnology. In the "bottom-up" approach, materials and devices are built from molecular components which assemble themselves chemically by principles of molecular recognition. In the "top-down" approach, nano-objects are constructed from larger entities without atomic-level control.

Areas of physics such as nanoelectronics, nanomechanics, nanophotonics, and nanoionics have evolved during the last few decades to provide a basic scientific foundation of nanotechnology.

1.1.4 NANOMATERIALS

The nanomaterials field includes subfields which develop or study materials having unique properties arising from their nanoscale dimensions.

- Interface and colloid science has given rise to many materials which may be useful in nanotechnology, such as CNTs and other fullerenes, and various nanoparticles and nanorods. Nanomaterials with fast ion transport are related also to nanoionics and nanoelectronics.
- Nanoscale materials can also be used for bulk applications; most present commercial applications of nanotechnology are of this flavor.
- Progress has been made in using these materials for medical applications (see Saini et al., 2010).
- Nanoscale materials such as nanopillars are sometimes used in solar cells which combat the cost of traditional silicon solar cells.
- Development of applications incorporating semiconductor nanoparticles to be used in the next generation of products, such as display technology, lighting, solar cells, and biological imaging.

1.2 NANOMECHANICS

Nanomechanics is a branch of *nanoscience* studying fundamental *mechanical* (elastic, thermal, and kinetic) properties of physical systems at the nanometer scale. Nanomechanics has emerged on the crossroads of classical mechanics, solid-state physics, statistical mechanics, materials science, and quantum chemistry. As an area of nanoscience, nanomechanics provides a scientific foundation of nanotechnology (see wikipedia.org/wiki/Nanomechanics; Cleland, 2003; Liu et al., 2006).

Nanomechanics is that branch of nanoscience, which deals with the study and application of fundamental mechanical properties of physical systems at the nanoscale, like elastic, thermal, and kinetic.

Often, nanomechanics is viewed as a *branch* of nanotechnology, i.e., an applied area with a focus on the mechanical properties of *engineered* nanostructures and nanosystems (systems with nanoscale components of importance). Examples of the latter include nanoparticles; nanopowders; nanowires; nanorods; nanoribbons; nanotubes, including CNTs and boron nitride nanotubes; nanoshells; nanomembranes; nanocoatings; nanocomposite/nanostructured materials (fluids with dispersed nanoparticles); nanomotors, etc.

Some of the well-established *fields of nanomechanics* are nanomaterials, nanotribology (friction, wear and contact mechanics at the nanoscale), nanoelectromechanical systems, and nanofluidics.

As a fundamental science, nanomechanics is based on some empirical principles (basic observations): (1) general mechanics principles; (2) specific principles arising from the smallness of physical sizes of the object of study or research.

General mechanics principles include:

- Energy and momentum conservation principles
- Variational Hamilton's principle
- Symmetry principles

Due to smallness of the studied object, nanomechanics also accounts for:

- Discreteness of the object, whose size is comparable with the interatomic distances
- Plurality, but finiteness, of degrees of freedom in the object
- Importance of thermal fluctuations
- Importance of entropic effects
- Importance of quantum effects

These principles serve to provide a basic insight into novel mechanical properties of nanometer objects. Novelty is understood in the sense that these properties are not present in similar macroscale objects or much different from the properties of those (e.g., nanorods vs. usual macroscopic beam structures). In particular, smallness of the subject itself gives rise to various surface effects determined by higher surface-to-volume ratio of nanostructures, and thus affects mechanoenergetic and thermal properties (melting point, heat capacitance, etc.) of nanostructures. Discreteness serves a fundamental reason, for instance, for the dispersion of mechanical waves in solids, and some special behavior of basic elastomechanics solutions at small scales. Plurality of degrees of freedom and the rise of thermal fluctuations are the reasons for thermal tunneling of nanoparticles through potential barriers, as well as for the cross-diffusion of liquids and solids. Smallness and thermal fluctuations provide the basic reasons of the Brownian motion of nanoparticles. Increased importance of thermal fluctuations and configuration entropy at the nanoscale gives rise to superelasticity, entropic elasticity (entropic forces), and other exotic types of elasticity of nanostructures. Aspects of configuration entropy are also of great interest in the context of self-organization and cooperative behavior of open nanosystems.

Quantum effects determine *forces of interaction* between individual atoms in physical objects, which are introduced in nanomechanics by means of some averaged mathematical models called *interatomic potentials.*

Subsequent utilization of the interatomic potentials within the classical multibody dynamics provides deterministic mechanical models of nanostructures and systems at the atomic scale/resolution. Numerical methods of solution of these models are called molecular dynamics (MD), and sometimes molecular mechanics (especially, in relation to statically equilibrated (still) models). Nondeterministic numerical approaches include Monte-Carlo, kinetic Monte-Carlo, and other methods. Contemporary numerical tools include also hybrid multiscale approaches allowing concurrent or sequential utilization of the atomistic scale methods (usually, MD) with the continuum (macro) scale methods (usually, field emission microscopy) within a single mathematical model. Development of these complex methods is a separate subject of applied mechanics research.

Quantum effects also determine novel electrical, optical, and chemical properties of nanostructures, and therefore they find even greater attention in adjacent areas of nanoscience and nanotechnology, such as nanoelectronics, advanced energy systems, and nanobiotechnology.

1.3 MICROMECHANICS

Generally speaking, micromechanics is a scientific discipline that studies: (1) mechanical, electrical, and, in general, thermodynamical behaviors of a material with microstruture or (2) materials' behaviors at micro (nano) or mesoscale (see Li, Lecture Note).

In recent years, micromechanics has become an indispensible part of theoretical foundation for many engineering fields and emerging technologies such as nanotechnology and biomedical technology.

The term "micromechanics" has become a truly interdiscipline jargon. It has been used with different meanings in different contexts. Traditionally, in the area of applied mechanics, micromechanics is referred to as a hierarchical mechanics paradigm that deals the effective material properties that are statistical averages of a nested two-level structure: microscopic and macroscopic structures. A material point at a macrolevel can be viewed as an ensemble microscope material space. The physical laws at macrolevel or the material behaviors at macrolevel are derived from the ensemble average of massive micro objects governed by the physical laws at microlevel. For instance, the effective material properties at macrolevel are the average of material properties of microstructures at fine scale. In general, the two-level paradigm is a special mathematical abstraction that is not associated with any fixed length scale. When studying material properties of a metal, 1 mm may be viewed as macroscale, and the length scale at microlevel may range from $\overset{0}{A}$ to nanometer; whereas, studying the deformation of a dam, the macroscale could be up to 103 m, and the length scale at microlevel may be around 10^{-2} m. In this sense, traditional micromechanics is essentially a particular (in some sense classical) averaging theory that takes into account the overall effects of microstructures.

In practice, it deals with subjects of a broad spectrum: material properties of composite/synthetic materials, e.g., composite structures, cementitious materials, geotechnical materials, and phase transformations; material properties of biomaterials, e.g., constitutive modeling of bone, muscle, blood flow; environmental problems, e.g., air pollutions, ground water transport and diffusion, oil spill in the ocean, etc.

In condensed matter physics and today in applied mechanics as well, the term micromechanics is used to describe a three-level physics realm: micromechanics at molecular or atomic level ($\overset{0}{A}$), mesomechanics at nanometer length scale, and macroscopic phenomenological theory at millimeter level or up.

The main task of contemporary micromechanics, or nanomechanics, is to seek unknown physical laws or mechanic regulations at the nanoscale. Different from traditional micromechanics, a salient feature of nanomechanics is its multiscale and multiphysics character. It includes some features that are present in quantum mechanics, or quantum statistical mechanics, a manifestation of the effects at atomic or subatomic level; on the other hand, it also shares with many features from the description of continuum mechanics, because of the size statistical ensemble.

The impetus for contemporary micromechanics or nanomechanics is primarily due to the emergence of nanoscience and biomedical technology. It appears that physics alone is not sufficient to deal with the many problems that are appearing from today's nanotechnologies and nanoengineering. There is a call for nanomechanics and nanocomputational mechanics to serve as the infrastructure of these emerging engineering fields. For instance, much attention has been focused on material properties of thin film, manufacturing devices, and components of a microelectromechanical system, e.g., submicro-size sensors, motors, the mechanics of nanotube and nanowire, computer-aided material design, and microbiophysics/biochemistry systems, e.g., protein/DNA interaction in biomolecular simulation, etc.

From the perspective of higher learning and intellectual advancement, micromechanics has developed into a rigorous mathematical theory, philosophical methodology, and beautiful computational realization. Forty years ago, micro elasticity started with simple definitions of eigenstrain and inclusion, came along with Eshelby's elegant equivalent homogenization theory and Hashin and Shtrikman's variational principle; it is now the foundation of an entire composite material industry.

Less than 10 years ago, Lattice Boltzmann method first debuted as a numerical emulation of continuous Boltzmann equation in statistical physics. Today, Lattice Boltzmann method has become a bona fide computational mesomechanics paradigm, and it has been used to solve problems such as turbulence flow, combustion, and flow pass through porous media and even cooling of packed flowers; In later 1980s, Clementi and his coworkers [1988] initiated the idea of multiscale modeling, or multiscale simulation, i.e., using supercomputers to conduct large-scale computations that combine *ab initio* modeling, classical molecular dynamic modeling, and phenomenological modeling in a single simulation.

The unified macroscopic, atomistic, *ab initio* dynamics description brings all three descriptions together into a seamless union, embracing all the size scales, from the very small to the very big.

The simplest and earliest multiscale modeling notion is the so-called Cauchy-Born rule. By combining this concept with the finite element methods, the so-called quasicontinuum method was developed by Tadmor, Ortiz, and Phillips and their coworkers (Tadmor et al., 1996). The Cauchy-Born rule is essentially a simplistic "homogenization postulation" in lattice kinematics and it serves as passage to link between the MD and continuum mechanics.

The Born rule assumes that the continuum energy density W can be computed using an atomic potential, with the link to the continuum being the deformation gradient F. To briefly review continuum mechanics, the deformation gradient F maps an undeformed line segment $\mathrm{d}X$ in the reference configuration onto a deformed line segment $\mathrm{d}x$ in the current configuration,

$$\mathrm{d}x = F\mathrm{d}X \tag{1.1}$$

In general, F can be written as

$$F = I + \frac{\mathrm{d}u}{\mathrm{d}x} \tag{1.2}$$

where u is the displacement vector. If there is no displacement in the continuum, the deformation gradient is equal to unity.

The major restriction and implication of the Cauchy-Born rule is that the continuum deformation must be homogeneous. This results from the fact that the underlying atomic system is forced to deform according to the continuum deformation gradient F. By using the Born rule, one may be able to derive a continuum stress tensor and tangent stiffness directly from the interatomic potential, which allowed the usage of the standard nonlinear finite element method. This procedure is now called as the so-called quasicontinuum method.

Apparently, the contemporary micromechanics or nanomechanics is only at its infancy. There are many unknown approaches to be explored and many new phenomena to be studied. In this lecture notes, we are attempting to synthesize the most recent research results in the forefront of nanomechanics while presenting traditional micromechanics in a coherent fashion. By doing so, we hope that it may serve as a stepping stone for us to reach a new height in the quest for a multiscale nanomechanics of our time.

1.4 NANOFLUID

1.4.1 INTRODUCTION

A nanofluid is a fluid containing nanometer-sized particles, called nanoparticles. These fluids are engineered colloidal suspensions of nanoparticles in a base fluid. The nanoparticles used in nanofluids are typically made of metals, oxides, carbides, or CNTs. Common base fluids include water, ethylene glycol, and oil (see wikipedia.org/wiki/Nanofluid; nanofluid, n.d).

Nanofluids have novel properties that make them potentially useful in many applications in heat transfer, including microelectronics, fuel cells, pharmaceutical processes, and hybrid-powered engines, engine cooling/vehicle thermal management, domestic refrigerator, chiller, heat exchanger, in grinding, machining and in boiler flue gas temperature reduction. They exhibit enhanced thermal conductivity and the convective heat transfer coefficient compared to the base fluid. Knowledge of the rheological behavior of nanofluids is found to be very critical in deciding their suitability for convective heat transfer applications.

In analysis such as computational fluid dynamics, nanofluids can be assumed to be single-phase fluids. However, almost all of new academic papers use two-phase assumption. Classical theory of single-phase fluids can be applied, where physical properties of nanofluid are taken as a function of properties of both constituents and their concentrations. An alternative approach simulates nanofluids using a two-component model see (Maiga et al., 2005; Buongiorno, 2006; Philip et al., 2006; Das et al., 2007; Chen et al., 2009; Kakaç et al., 2009; Kuznetsovet al., 2010; Mahendran et al., 2012, 2013; Taylor et al., 2013).

1.4.2 SYNTHESIS OF NANOFLUIDS

Nanofluids are supplied by two methods called the one-step and two-step methods. Several liquids including water, ethylene glycol, and oils have been used as base fluids. Nanomaterials used so far in nanofluid synthesis include metallic particles, oxide particles, CNTs, graphene nano-flakes, and ceramic particles.

1.4.3 SMART COOLING NANOFLUIDS

Realizing the modest thermal conductivity enhancement in conventional nanofluids, a team of researchers at Indira Gandhi Centre for Atomic Research Centre, Kalpakkam developed a new class of magnetically polarizable nanofluids where the thermal conductivity enhancement up to 300% of base fluids is demonstrated. Fatty acid-capped magnetite nanoparticles of different sizes (3-10 nm) have been synthesized for this purpose. It has been shown that both the thermal and the rheological properties of such magnetic nanofluids are tunable by varying the magnetic field strength and orientation with respect to the direction of heat flow. Such response stimuli fluids are reversibly switchable and have applications in miniature devices such as micro- and nanoelectromechanical systems. Recently, Azizian et al. (2014) considered effect of an external magnetic field on the convective heat transfer coefficient of water-based magnetite nanofluid experimentally under laminar flow regime. Up to 300% enhancement obtained at $Re = 745$ and magnetic field gradient of 32.5 mT/mm. The effect of the magnetic field on the pressure drop was not as significant.

1.4.4 RESPONSE STIMULI NANOFLUIDS FOR SENSING APPLICATIONS

Researchers have invented a nanofluid-based ultrasensitive optical sensor that changes its color on exposure to extremely low concentrations of toxic cations. The sensor is useful in detecting minute traces of cations in industrial and environmental samples. Existing techniques for monitoring cation levels in industrial and environmental samples are expensive, complex, and time consuming. The sensor is designed with a magnetic nanofluid that consists of nanodroplets with magnetic grains suspended in water. At a fixed magnetic field, a light source illuminates the nanofluid where the color of the nanofluid changes depending on the cation concentration. This color change occurs within a second after exposure to cations, much faster than other existing cation-sensing methods.

Such response stimulus nanofluids are also used to detect and image defects in ferromagnetic components. The photonic eye, as it has been called, is based on a magnetically polarizable nanoemulsion that changes color when it comes into contact with a defective region in a sample. The device might be used to monitor structures such as rail tracks and pipelines.

1.4.5 APPLICATIONS

Nanofluids are primarily used as coolant in heat transfer equipment such as heat exchangers, electronic cooling system (such as flat plate), and radiators. Heat transfer over flat plate has been analyzed by a lot of researchers. Graphene-based nanofluid has been found to enhance polymerase chain reaction efficiency. Nanofluids in solar collectors are another application where nanofluids are employed for their tunable optical properties.

REFERENCES

Azizian, R., Doroodchi, E., McKrell, T., Buongiorno, J., Hu, L.W., Moghtaderi, B., 2014. Effect of magnetic field on laminar convective heat transfer of magnetite nanofluids. Int. J. Heat Mass 68, 94–109.

Bhushan, B. (Ed.), 2007. Springer Handbook of Nanotechnology, second ed. Springer, New York.

Binnig, G., Rohrer, H., 1986. Scanning tunneling microscopy. IBM J. Res. Dev. 30, 4.

Buongiorno, J., 2006. Convective transport in nanofluids. J. Heat Transf. (Am. Soc. Mech. Eng.) 128 (3 (March)), 240. http://dx.doi.org/10.1115/1.2150834, Retrieved 27 March 2010.

Buzea, C., Pacheco, I., Robbie, K., 2007. Nanomaterials and nanoparticles: sources and toxicity. Biointerphases 2 (4), MR17–71. http://dx.doi.org/10.1116/1.2815690, PMID 20419892.

Chen, H., Witharana, S., et al., 2009. Predicting thermal conductivity of liquid suspensions of nanoparticles (nanofluids) based on rheology. Particuology 7, 151–157. http://dx.doi.org/10.1016/j.partic.2009.01.005.

Cleland, A.N., 2003. Foundations of Nanomechanics. Springer, Berlin.

Clementi, E., Reddaway, S.F., 1988. Global scientific and engineering simulations on scalar, vector and parallel LCAP-type supercomputers. Phil. Trans. R. Soc. Lond. A Math. Phys. Eng. Sci. 326, 445–470.

Das, S.K., Stephen, U.S.C., Yu, W., Pradeep, T., 2007. Nanofluids: Science and Technology. Wiley InterScience, USA, p. 397. Retrieved 27 March 2010.

Drexler, K.E., 1986. Engines of Creation: The Coming Era of Nanotechnology. Doubleday, New York. ISBN: 0-385-19973-2.

Drexler, K.E., 1992. Nanosystems: Molecular Machinery, Manufacturing, and Computation. John Wiley & Sons, New York. ISBN: 0-471-57547-X.

Kakaç, S., Pramuanjaroenkij, A., 2009. Review of convective heat transfer enhancement with nanofluids. Int. J. Heat Mass Transf. (Elsevier) 52, 3187–3196. http://dx.doi.org/10.1016/j.ijheatmasstransfer.2009.02.006, Retrieved 27 March 2010.

Kroto, H.W., Heath, J.R., O'Brien, S.C., Curl, R.F., Smalley, R.E., 1985. C60: buckminsterfullerene. Nature 318 (6042), 162–163. http://dx.doi.org/10.1038/318162a0, Bibcode:1985Natur.318.162K.

Kuznetsov, A.V., Nield, D.A., 2010. Natural convective boundary-layer flow of a nanofluid past a vertical plate. Int. J. Therm. Sci. 49 (2), 243–247. http://dx.doi.org/10.1016/j.ijthermalsci.2009.07.015.

Li, S., Wang, G., 2008. Introduction to Micromechanics and Nanomechanics. World Scientific Pub. ISBN: 10 9812814132.

Liu, W.K., Karpov, E.G., Park, H.S., 2006. Nano Mechanics and Materials: Theory, Multiscale Methods and Applications. In: Wiley, West Sussex.
Mahendran, V., Philip, J., 2012. Nanofluid based optical sensor for rapid visual inspection of defects in ferromagnetic materials. Appl. Phys. Lett. 100, 073104. http://dx.doi.org/10.1063/1.3684969.
Mahendran, V., Philip, J., 2013. Spectral response of magnetic nanofluid to toxic cations. Appl. Phys. Lett. 102, 163109. http://dx.doi.org/10.1063/1.4802899.
Maiga, S.E.B., Palm, S.J., Nguyen, C.T., Roy, G., Galanis, N., 2005. Heat transfer enhancement by using nanofluids in forced convection flows. Int. J. Heat Fluid Flow 26 (June), 530–546. http://dx.doi.org/10.1016/j.ijheatfluidflow.2005.02.004.
Minkowycz, W., et al., 2013. Nanoparticle Heat Transfer and Fluid Flow. CRC Press, Taylor & Francis, USA.
http://nanofluid.ir.
Nanotechnology Information Center: properties, applications, research, and safety guidelines. American Elements. Retrieved 13 May 2011.
http://nanotechweb.org/cws/article/tech/48783.
NASA Draft Nanotechnology Roadmap.
Philip, J., Shima, P.D., Raj, B., 2006. Nanofluid with tunable thermal properties. Appl. Phys. Lett. 92, 043108. http://dx.doi.org/10.1063/1.2838304.
Saini, R., Saini, S., Sharma, S., 2010. Nanotechnology: the future medicine. J. Cutan. Aesthet. Surg. 3 (1), 32–33. http://dx.doi.org/10.4103/0974-2077.63301, PMC 2890134. PMID 20606992.
Sattler, K.D., 2011. Handbook of Nanophysics. In: Principles and Methods, vol. 1. CRC Press, Boca Raton, FL.
Tadmor, E.B., Ortiz, M., Phillips, R., 1996. Quasicontinuum analysis of defects in solids. Phil. Mag. A 73, 1529–1563.
Taylor, R.A., et al., 2013. Small particles, big impacts: a review of the diverse applications of nanofluids. J. Appl. Phys. 113 (1), 011301–011301-19. http://jap.aip.org/resource/1/japiau/v113/i1/p011301_s1?bypassSSO=1.
http://en.wikipedia.org/wiki/Nanotechnology.
http://en.wikipedia.org/wiki/Nanomechanics.
http://en.wikipedia.org/wiki/Nanofluid.

his page intentionally left blank

CHAPTER

SEMI NONLINEAR ANALYSIS IN CARBON NANOTUBE

2

CHAPTER CONTENTS

2.1 Introduction of Carbon Nanotube 14
2.1.1 Single-Wall Nanotubes 15
2.1.2 Multiwall Nanotubes 15
2.1.3 Double-Wall Nanotubes 15
2.2 Single SWCNT Over a Bundle of Nanotube 17
2.2.1 Introduction 17
2.2.2 Formulations 18
2.2.2.1 Schematic of Problem 18
2.2.2.2 Modeling the Individual SWCNT as a Beam 19
2.2.2.3 Differential Quadrature and Solution Procedure 20
2.2.2.4 Finite Element Method 22
2.2.3 Results 24
2.2.3.1 Mesh Point Number Effect 24
2.2.3.2 Length Effect 25
2.2.3.3 Validation of GDQ Approach 25
2.2.4 Conclusion 27
2.3 Cantilevered SWCNT as a Nanomechanical Sensor 28
2.3.1 Introduction 30
2.3.2 Analysis of the Problem 31
2.3.2.1 Basic Bending Vibration and Resonant Frequencies of SWCNT with Attached Mass 31
2.3.2.2 Resonant Frequency of Cantilevered SWCNT Where the Mass is Rigidly Attached to the Tip 31
2.3.3 Numerical Results 33
2.3.3.1 Vibration Mode Analysis 33
2.3.4 Mass Sensor Mode Comparison 33
2.3.5 Conclusion 35
2.4 Nonlinear Vibration for Embedded CNT 36
2.4.1 Introduction 36
2.4.2 Basic Equations 37

Application of Nonlinear Systems in Nanomechanics and Nanofluids. http://dx.doi.org/10.1016/B978-0-323-35237-6.00002-9

2.4.3 Solution Methodology 40
2.4.4 Numerical Results and Discussion 41
2.4.5 Conclusion 45
2.5 Curved SWCNT 45
2.5.1 Introduction 45
2.5.2 Vibrational Model 46
2.5.3 Solution Methodology 48
2.5.4 Numerical Results and Discussion 49
2.5.5 Conclusion 53
2.6 CNT with Rippling Deformations 54
2.6.1 Introduction 54
2.6.2 Vibration Model 55
2.6.2.1 Boundary Conditions 55
2.6.2.2 Nonlinear Vibration Model 55
2.6.2.3 Nonlinear Analysis 57
2.6.3 Results and Discussion 62
2.6.4 Conclusion 67
References 68

2.1 INTRODUCTION OF CARBON NANOTUBE

Since their initial discovery by Iijima (1991), carbon nanotubes (CNTs) have come under ever-increasing scientific scrutiny.

A CNT is a tube-shaped material, made of carbon, having a diameter measuring on the nanometer scale. A nanometer is one-billionth of a meter, or about one ten-thousandth of the thickness of a human hair. The graphite layer appears somewhat like a rolled-up chicken wire with a continuous unbroken hexagonal mesh and carbon molecules at the apexes of the hexagons.

CNTs have many structures, differing in length, thickness, and in the type of helicity and number of layers. CNTs possess excellent mechanical properties, such as extremely high strength, stiffness, and resilience. These points, together with other distinctive physical properties, result in many prospective applications, such as strong, light, and high toughness fibers for nanocomposite structures, parts of nanodevices, hydrogen storage (high frequency) micromechanical oscillators, etc. (see www.nanocyl.com/en/CNT-Expertise-Centre/Carbon-Nanotubes).

In fact, CNTs are unique nanostructured materials. The extraordinary mechanical and physical properties in addition to the large aspect ratio and low density have made CNTs ideal components of nanodevices.

Although they are formed from essentially the same graphite sheet, their electrical characteristics differ depending on these variations, acting either as metals or as semiconductors.

As a group, CNTs typically have diameters ranging from <1 up to 50 nm. Their lengths are typically several microns, but recent advancements have made the nanotubes much longer, and measured in centimeters.

CNTs can be categorized by their structures:

1. Single-wall nanotubes (SWNT)
2. Multiwall nanotubes (MWNT)
3. Double-wall nanotubes (DWNT)

2.1.1 SINGLE-WALL NANOTUBES

SWNT are tubes of graphite that are normally capped at the ends. They have a single cylindrical wall. The structure of a SWNT can be visualized as a layer of graphite, a single atom thick, called graphene, which is rolled into a seamless cylinder.

Most SWNT typically have a diameter of close to 1 nm. The tube length, however, can be many thousands of times longer.

SWNT are more pliable yet harder to make than MWNT. They can be twisted, flattened, and bent into small circles or around sharp bends without breaking.

SWNT have unique electronic and mechanical properties which can be used in numerous applications, such as field-emission displays, nanocomposite materials, nanosensors, and logic elements. These materials are on the leading-edge of electronic fabrication, and are expected to play a major role in the next generation of miniaturized electronics.

2.1.2 MULTIWALL NANOTUBES

MWNT can appear either in the form of a coaxial assembly of SWNT similar to a coaxial cable, or as a single sheet of graphite rolled into the shape of a scroll.

The diameters of MWNT are typically in the range of 5-50 nm. The interlayer distance in MWNT is close to the distance between graphene layers in graphite.

MWNT are easier to produce in high volume quantities than SWNT. However, the structure of WNT is less well understood because of its greater complexity and variety. Regions of structural imperfection may diminish its desirable material properties.

The challenge in producing SWNT on a large scale as compared to MWNT is reflected in the prices of SWNT, which currently remain higher than MWNT.

SWNT, however, have a performance of up to 10 times better, and are outstanding for very specific applications.

2.1.3 DOUBLE-WALL NANOTUBES

DWNT are an important sub-segment of MWNT.

These materials combine similar morphology and other properties of SWNT, while significantly improving their resistance to chemicals. This property is especially important when functionality is required to add new properties to the nanotube.

Since DWNT are a synthetic blend of both SWNT and MWNT, they exhibit the electrical and thermal stability of the latter and the flexibility of the former.

Because they are developed for highly specific applications, SWNT that have been functionalized are more susceptible to breakage. Creating any structural imperfections can modify their mechanical and electrical properties.

However, with DWNT, only the outer wall is modified, thereby preserving the intrinsic properties.

Also, research has shown that DWNT have better thermal and chemical stability than SWNT. DWNT can be applied to gas sensors and dielectrics, and to technically demanding applications like field-emission displays, nanocomposite materials, and nanosensors.

The intrinsic mechanical and transport properties of CNTs make them the ultimate carbon fibers. The following Tables 2.1 and 2.2 compare these properties to other engineering materials.

Overall, CNTs show a unique combination of stiffness, strength, and tenacity compared to other fiber materials which usually lack one or more of these properties. Thermal and electrical conductivities are also very high, and comparable to other conductive materials.

Carbon nanotube technology can be used for a wide range of new and existing applications:

- Conductive plastics
- Structural composite materials
- Flat-panel displays
- Gas storage
- Antifouling paint
- Micro- and nano-electronics
- Radar-absorbing coating
- Technical textiles

Table 2.1 Mechanical Properties of Engineering Fibers

Fiber Material	Specific Density	*E* (TPa)	Strength (GPa)	Strain at Break (%)
Carbon nanotube	1.3-2	1	10-60	10
HS steel	7.8	0.2	4.1	<10
Carbon fiber—PAN	1.7-2	0.2-0.6	1.7-5	0.3-2.4
Carbon fiber—pitch	2-2.2	0.4-0.96	2.2-3.3	0.27-0.6
E/*S*—glass	2.5	0.07/0.08	2.4/4.5	4.8
Kevlar* 49	1.4	0.13	3.6-4.1	2.8

Kevlar is a registered trademark of DuPont.

Table 2.2 Transport Properties of Conductive Materials

Material	Thermal Conductivity (W/m k)	Electrical Conductivity
Carbon nanotubes	>3000	10^6-107
Copper	400	6 × 107
Carbon Fiber—pitch	1000	2-8.5 × 10^6
Carbon Fiber—PAN	8-105	6.5-14 × 10^6

- Ultra-capacitors
- Atomic force microscope tips
- Batteries with improved lifetime
- Biosensors for harmful gases
- Extra strong fibers

2.2 SINGLE SWCNT OVER A BUNDLE OF NANOTUBE

The deformation of an individual single-walled carbon nanotube (SWCNT) over a bundle of nanotubes has been studied using the generalized differential quadrature (GDQ) method. The effects of length, diameter, and minimum value of Lennard-Jones experimental potential have been considered in governing equation which is derived based on the GDQ and the issues related to the implementation of the boundary and compatibility conditions were addressed. The explanation of reliability and flexibility of the GDQ is occurred by solving several selected examples which are evaluated by comparing them with the existent exact or approximate solutions which is done before by finite element approach (this section has worked by A. Fereidoon, A.M. Babaee, Y. Rostamiyan, and D.D. Ganji in nonlinear dynamics team in Mechanical Engineering Department, 2011-2012).

2.2.1 INTRODUCTION

The tubular carbon structures were observed by Iijima (1991) for the first time. The nanotubes consisted of up to several tens of graphitic shells (so-called multiwalled carbon nanotubes, MWCNTs) with adjacent shell separation of 0.34 nm, diameters of 1 nm, and large length/diameter ratio. Recently, Iijima and Ichihashi (1993) and Bethune et al. (1993) synthesized SWCNTs. The synthesized nanotube samples are characterized by means of Raman, electronic, and optical spectroscopies. Important information is derived by mechanical, electrical, and thermal measurements. Along with the improvement of the production and characterization techniques for nanotubes, progress is being made in their application. The estimated high Young's modulus and tensile strength of the nanotubes has led to speculations for their possible use in composite materials with improved mechanical properties (Overney et al., 1993).

The configuration of current problem consists of a single SWCNT over a bundle of nanotubes which is assumed to be rigid due to relative mechanical properties. The van der Waals interaction plays an important role on the performance of SWCNT structures, such as mechanical properties (Fereidoon et al., 2011; Saether et al., 2003). The van der Waals nonlinear interaction plays the most important role in variation of separation distance between individual SWCNT and its substrate. Wong et al. (1997) implied cantilevered beam model in the research in which a microscopic point force bent a single MWCNT. Salvetat et al. (1999) used the simply supported beam model to simulate the deflections of the MWCNTs and of some different species of SWCNT ropes. In this current section, the effects of physical SWCNT properties on the deflection of an individual SWCNT in different substrate curvature is investigated by modeling the SWCNT as an Euler-Bernoulli beam model. GDQ is implemented as a practical numerical method in solving higher order differential equation (Wu and Liu, 1999), to solve the governing equation of nanotube simplification as a beam model. The proposed GDQ employs the same number of independent variables as that of the conditions at any discrete point. Therefore, the

GDQ can deal with the differential equations, which may be constrained by multiple conditions at any discrete point. Several recent publications have reviewed the modeling and simulation of mechanical properties of CNTs and nanocomposites (Sammalkorpi et al., 2005; Spitalskya et al., 2010). But these review articles do not cover separation an individual SWCNT from a substrate of its own kind. Therefore the purpose of this study is the explanation of reliability and flexibility of the GDQ by solving several selected examples which are evaluated by comparing them with the existent exact or approximate solutions, which is done before by finite element approach.

2.2.2 FORMULATIONS

2.2.2.1 Schematic of problem

As shown in Figure 2.1 the problem configuration consist of two main parts, one is an individual SWCNT and the other is a bundle of nanotube which plays a role as a substrate to fixed situation of coordinate axis. The substrate curvature is assumed to be a parabolic which is shown in Figure 2.1.

The analytical equation of above-mentioned curvature is

$$y = \frac{m}{2}x^2 \tag{2.1}$$

From Kudin et al. (2001) the stiffness of nanotube and its diameter has the following relationship:

$$EI = \pi C d^3 \tag{2.2}$$

where EI is the bending stiffness and C is 2152.8 eV/nm^2 which is computed for in-plane stiffness based on *ab initio* calculations, and d is the diameter of the tube. From Figure 2.2 it is easily understood that bending stiffness of the bundle is much bigger than the individual SWCNT (Qian et al., 2003). This means that assuming substrate as a rigid body will be acceptable.

In Figure 2.2, the diameter of the bundle is about 4 times bigger than each individual SWCNT diameter thus it's bending stiffness is 4^3 times bigger than each individual SWCNT (Equation (2.2)).

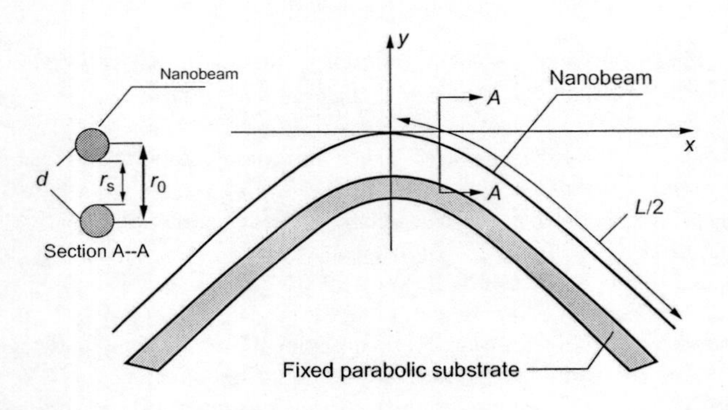

FIGURE 2.1

Initial condition of the nanobeam model.

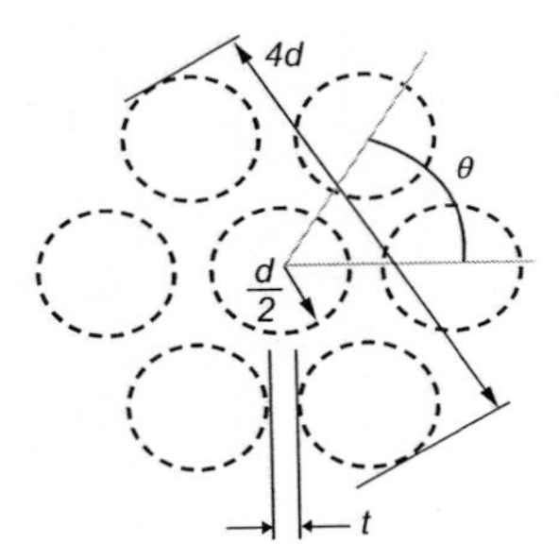

FIGURE 2.2

Schematic of a bundle of nanotubes consist of 7 SWCNTs.

2.2.2.2 Modeling the individual SWCNT as a beam

Based on aspect ratio of SWCNTs and their mechanical properties, the simulation of SWCNT as a straight inextensible beam is accepted, the only interaction force between nanotubes is van der Waals force (Li et al., 2004) which is determined in many papers from experimental Lennard-Jones potential (Coffin et al., 2006; Li et al., 2004). Using above information and Euler beam theory for single SWCNT will determine the governing equation which explains deflection of the individual tube.

From the Euler beam theory the governing equation of the one-dimensional inextensible beam is (Boresi and Schmidt, 2002):

$$\frac{d^2}{dX^2}\left(EI\frac{d^2y}{dX^2}\right)+F_{\text{substrate}}=-F_{\text{external}} \tag{2.3}$$

Figure 2.1 shows the symmetry about y-axis therefore the analysis of half part of SWCNT is sufficient to obtain the results and the only force that acts is van der Waals force which is expressed in per unit length as:

$$F(r(X))=17.81U_0\left[-\left(\frac{3.41}{3.13\frac{r(X)+r_0-d}{r_0-d}+0.28}\right)^{11}+\left(\frac{3.41}{3.13\frac{r(X)+r_0-d}{r_0-d}+0.28}\right)^{5}\right] \tag{2.4}$$

where, U_0 is the minimum energy in the Lennard-Jones energy potential, r_s is the distance between the surfaces of nanobeam and substrate when van der Waals force is zero (Figure 2.1), r_0 is the distance which is measured by $r_0=r_s+d$ which d is the diameter of the nanobeam, and $r(X)$ is offset distance between the surfaces of nanobeam and substrate during deformation. From Equation (2.1) the curvature of individual SWCNT has the following form:

$$y(X)=\frac{m}{2}X^2+r(X) \tag{2.5}$$

The boundary condition due to symmetry and configuration of problem will be

$$\begin{aligned} y^{(1)}(0) &= 0; \\ y^{(3)}(0) &= 0; \\ y^{(2)}(L) &= 0; \\ y^{(3)}(L) &= 0; \end{aligned} \tag{2.6}$$

considering Equation (2.5), the governing equation and the B.Cs in Equation (2.6) will be

$$EI\frac{d^4r(X)}{dX^4} + F(r(X)) = 0 \tag{2.7}$$

with B.C's:

$$\begin{aligned} r^{(1)}(0) &= 0 \\ r^{(3)}(0) &= 0 \\ r^{(2)}(L) &= -m \\ r^{(3)}(L) &= 0 \end{aligned} \tag{2.8}$$

By applying Taylor series expansion to the van der Waals force and truncating higher order terms one can simplify high-order nonlinearities to practical form as follows:

$$F_{\text{van der Waals}} \approx -1.10^{-8} + 274.5242066r(X) + O\left(r^2(X)\right) \tag{2.9}$$

2.2.2.3 Differential quadrature and solution procedure

The GDQ method has been proved to be an efficient higher order numerical technique for the solution of initial and boundary value problems. The GDQ technique has been widely reported to yield successful solutions for various dynamic and stability problems (Mohammad et al., 2007; Pradhan and Murmu, 2009). The essence of GDQ method is that a derivative of a function F is approximated as a weighted linear sum of all functional values within the computational domain.

$$\left.\frac{d^nF}{dX^n}\right|_{X=X_i} = \sum_{j=1}^{N} c_{ij}^{n} F(X_i) \tag{2.10}$$

where

$$c_{ij}^{1} = \frac{\pi(X_i)}{(X_i - X_j)(\pi(X_i))}; \quad i,j = 1,2,\ldots,N; \quad i \neq j \tag{2.11}$$

where $\pi\ (X_i)$ defined as:

$$\pi(X_i) = \prod_{j=1}^{N} (X_i - X_j); \quad i \neq j \tag{2.12}$$

and when $i=j$:

$$c_{ij}^{1}=c_{ii}^{1}=-\sum_{k=1}^{N}c_{ik}^{(1)},\quad i=1,2,\ldots,N;\ i\neq k;\ i=j \tag{2.13}$$

where N is the number of grid points along x direction. The weighting coefficients for the second, third, and fourth derivatives are determined via following formula:

$$c_{ij}^{(m)}=m\left(c_{ij}^{1}c_{ii}^{(m-1)}-\frac{c_{ij}^{(m-1)}}{(X_j-X_i)}\right);\ i,j=1,2,\ldots,N;\ i\neq j;\ m=2,3,\ldots,N-1 \tag{2.14}$$

$$c_{ii}^{(m)}=-\sum_{\substack{j=1\\ j\neq i}}^{N}c_{ij}^{(m)};\ i=1,2,\ldots,N;\ i\neq k;\ i=j \tag{2.15}$$

One of the most accurate meshes in GDQ formulation is Chebyshev nodes which are defined by following inverse node numbering:

$$X_i=X_1+\frac{1}{2}\left(1-\cos\frac{i-1}{N-1}\pi\right)(X_N-X_1);\ i=1,2,\ldots,N \tag{2.16}$$

For convenience and generality the following nondimensional variables are introduced in the present analysis:

$$x=\frac{X}{L} \tag{2.17}$$

where x is the nondimensional variable which varies between 0 and 1, L is the length of nanotube, and X represent vertical axis before nondimensionalisation. By this assumption the derivatives will have the following form:

$$\begin{aligned}\frac{\mathrm{d}r}{\mathrm{d}X}&=\frac{\mathrm{d}r}{L\mathrm{d}x}\\ \frac{\mathrm{d}^3r}{\mathrm{d}X^3}&=\frac{1}{L^3}\frac{\mathrm{d}^3r}{\mathrm{d}x^3}\\ \frac{\mathrm{d}^4r}{\mathrm{d}X^4}&=\frac{1}{L^4}\frac{\mathrm{d}^4r}{\mathrm{d}x^4}\end{aligned} \tag{2.18}$$

After nondimensionalisation the governing equation will be

$$EI\frac{1}{L^4}\frac{\mathrm{d}^4r(x)}{\mathrm{d}x^4}+F(r(x))=0 \tag{2.19}$$

with B.Cs:

$$r^{(1)}(0)=0,\ r^{(3)}(0)=0,\ r^{(2)}(1)=-mL^2,\ r^{(3)}(1)=0 \tag{2.20}$$

By applying Equations (2.10–2.20)

$$\frac{EI}{L^4}\begin{bmatrix} c_{11}^{(4)} & c_{12}^{(4)} & \cdots & c_{1(n-1)}^{(4)} & c_{1n}^{(4)} \\ c_{21}^{(4)} & c_{22}^{(4)} & & c_{2(n-1)}^{(4)} & c_{2n}^{(4)} \\ \vdots & & & & \vdots \\ c_{(n-1)1}^{(4)} & c_{(n-1)2}^{(4)} & & c_{(n-1)(n-1)}^{(4)} & c_{(n-1)n}^{(4)} \\ c_{n1}^{(4)} & c_{n2}^{(4)} & \cdots & c_{n(n-1)}^{(4)} & c_{nn}^{(4)} \end{bmatrix}\begin{bmatrix} r_1 \\ r_2 \\ \vdots \\ r_{(n-1)} \\ r_n \end{bmatrix}+\begin{bmatrix} 274.5242066r_1 \\ 274.5242066r_2 \\ \vdots \\ 274.5242066r_{(n-1)} \\ 274.5242066r_n \end{bmatrix}=\begin{bmatrix} 1\cdot10^{-8} \\ 1\cdot10^{-8} \\ \vdots \\ 1\cdot10^{-8} \\ 1\cdot10^{-8} \end{bmatrix} \tag{2.21}$$

Because of high instability in borders, for applying B.Cs, above equations should be substituted into first, second $(n-1)$th, and last line of Equation (2.21):

$$\begin{bmatrix} c_{11}^{(1)} & c_{12}^{(2)} & \cdots & c_{1(n-1)}^{(1)} & c_{1n}^{(1)} \\ c_{11}^{(3)} & c_{12}^{(3)} & \cdots & c_{1(n-1)}^{(3)} & c_{1n}^{(3)} \\ \frac{EI}{L^4}c_{31}^{(4)} & \frac{EI}{L^4}c_{32}^{(4)} & \cdots & \frac{EI}{L^4}c_{3(n-1)}^{(4)} & \frac{EI}{L^4}c_{3n}^{(4)} \\ \frac{EI}{L^4}c_{41}^{(4)} & \frac{EI}{L^4}c_{42}^{(4)} & \cdots & \frac{EI}{L^4}c_{4(n-1)}^{(4)} & \frac{EI}{L^4}c_{4n}^{(4)} \\ \vdots & \vdots & \cdots & \vdots & \vdots \\ \frac{EI}{L^4}c_{(n-3)1}^{(4)} & \frac{EI}{L^4}c_{(n-3)2}^{(4)} & \cdots & \frac{EI}{L^4}c_{(n-3)(n-1)}^{(4)} & \frac{EI}{L^4}c_{(n-3)n}^{(4)} \\ \frac{EI}{L^4}c_{(n-2)1}^{(4)} & \frac{EI}{L^4}c_{(n-2)2}^{(4)} & \cdots & \frac{EI}{L^4}c_{(n-2)(n-1)}^{(4)} & \frac{EI}{L^4}c_{(n-2)n}^{(4)} \\ c_{n1}^{(2)} & c_{n2}^{(2)} & \cdots & c_{n(n-1)}^{(2)} & c_{nn}^{(2)} \\ c_{n1}^{(3)} & c_{n2}^{(3)} & \cdots & c_{n(n-1)}^{(3)} & c_{nn}^{(3)} \end{bmatrix}\begin{bmatrix} r_1 \\ r_2 \\ \vdots \\ r_{(n-1)} \\ r_n \end{bmatrix}+\begin{bmatrix} 0 \\ 0 \\ 274.5242066r_3 \\ 274.5242066r_4 \\ \vdots \\ 274.5242066r_{(n-3)} \\ 274.5242066r_{(n-2)} \\ mL^2 \\ 0 \end{bmatrix}=\begin{bmatrix} 0 \\ 0 \\ 1\cdot10^{-8} \\ \vdots \\ 1\cdot10^{-8} \\ 0 \\ 0 \end{bmatrix} \tag{2.22}$$

Clearly, above set of equations can be solved using several known methods.

2.2.2.4 Finite element method

Finite element method (FEM) is applied to the governing equation in order to verification and corresponding results show the great agreement between analytical and FEM solutions. Expressing Equation (2.2) in Galerkin weak form:

$$\int_{x=0}^{L}\left(EI\frac{\mathrm{d}^2w(x)}{\mathrm{d}x^2}\frac{\mathrm{d}^2r}{\mathrm{d}x^2}+w(x)F(r(x))\right)\mathrm{d}x=-mEI\left.\frac{\mathrm{d}w(x)}{\mathrm{d}x}\right|_{x=L} \tag{2.23}$$

where $w(x)$ is acceptable test function. In force term, using Newton-Raphson method, yields

$$F(r)=F(\bar{r}_0)+\frac{\mathrm{d}}{\mathrm{d}r}F(\bar{r}_0)\Delta r+O\left(\Delta r^2\right) \tag{2.24}$$

w is assumed to be:

$$w = \sum_{A\in\eta_g} C_A \Phi_A \tag{2.25}$$

and Δr will be determined by:

$$r = \sum_{B\in\eta} (d_B + \Delta d_B)\Phi_B \tag{2.26}$$

$$\Delta r = \sum_{B\in\eta} \Delta d_B \Phi_B \tag{2.27}$$

which Φ is shape function, computed values of d_B are used to evaluate the Δd_B which is unknown variable, the set of all unknown degrees of freedom at points grid in the FEM mesh is η_g, and the total number of points grid multiplied by the degrees of freedom at each point is called η. Equation (2.24) is substituted into Equation (2.23) and then Newton-Raphson approach is applied to derive Equation (2.28).

$$\begin{aligned}&\sum_{A\in\eta_g} C_A\left[\sum_{B\in\eta}\left(\int_{x=0}^{L} EI\frac{\mathrm{d}^2\Phi_A}{\mathrm{d}x^2}\frac{\mathrm{d}^2\Phi_B}{\mathrm{d}x^2}\mathrm{d}x + \int_{x=0}^{L}\Phi_A\frac{\mathrm{d}F(\bar{r}_0)}{\mathrm{d}r}\Phi_B\mathrm{d}x\right)\Delta d_B\right]\\ &= -mEI\frac{\mathrm{d}w(x)}{\mathrm{d}x}\bigg|_{x=L} - \sum_{A\in\eta_g} C_A\left[\int_{x=0}^{L}\Phi_A F(\bar{r}_0)\mathrm{d}x + \sum_{B\in\eta_g}\left(\int_{x=0}^{L} EI\frac{\mathrm{d}^2\Phi_A}{\mathrm{d}x^2}\frac{\mathrm{d}^2\Phi_B}{\mathrm{d}x^2}\mathrm{d}x\right)d_B\right]\end{aligned} \tag{2.28}$$

explaining

$$K_{AB} = \int_{x=0}^{L} EI\frac{\mathrm{d}^2\Phi_A}{\mathrm{d}x^2}\frac{\mathrm{d}^2\Phi_B}{\mathrm{d}x^2}\mathrm{d}x,\quad K^*_{AB} = \int_{x=L}^{L}\Phi_A\frac{\mathrm{d}F(\bar{r}_0)}{\mathrm{d}r}\Phi_B\mathrm{d}x,\quad F_A = \int_{x=0}^{L}\Phi_A F(\bar{r}_0)\mathrm{d}x \tag{2.29}$$

and rewriting Equation (2.28):

$$\sum_{A\in\eta_g} C_A\left[\sum_{B\in\eta}(K_{AB}+K^*_{AB})\Delta d_B\right] = -mEI\frac{\mathrm{d}w(x)}{\mathrm{d}x}\bigg|_{x=L} - \sum_{A\in\eta} C_A\left[\sum_{B\in\eta} K_{AB}d_B + F_A\right] \tag{2.30}$$

Hermite interpolation polynomials which are used should be at least in order of 3, because every element has 4 unknown.

$$\Phi^e_1(x) = 1 - 3s^2 + 2s^3,\quad \Phi^e_2(x) = l^e s(s-1)^2,\quad \Phi^e_3 = s^2(3-2s),\quad \Phi^e_4(x) = l^e s^2(s-1) \tag{2.31}$$

l^e represents the length of one element, s is determined by $(x-x_1)/(x_2-x_1)$ which x_1 and x_2 are the left and right values of coordinates axis of the element. The first step to guess a value for d_B is important to converge the Newton-Raphson approach. Because the Equation (2.30) is high nonlinear equation, an incremental load is used. At the beginning the curvature of substrate is assumed to be zero and resultant solutions are taken as initial guess. In the next step the larger curvature of substrate is used, for the initial guess in this step the solution of the previous step is used therefore the convergence of approach is determined. Above steps are repeated until required m is reached.

2.2.3 RESULTS

2.2.3.1 Mesh point number effect

Because of high importance of the number of points in grid to converge the approach, it should be examined. As a simple show, Figures 2.3 and 2.4 belongs to 100 points grid and continuity of results are visible. Deformation behavior and interaction force are depicted in Figures 2.3 and 2.4, respectively.

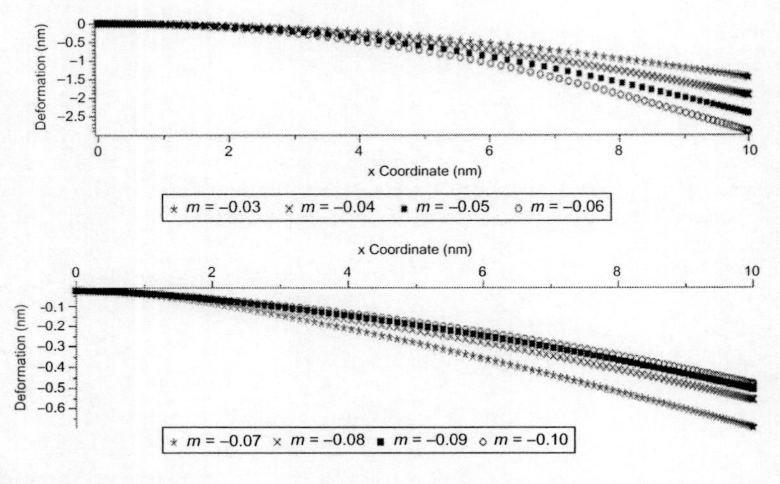

FIGURE 2.3

Deformation of single SWCNT ($2L=20$, $d=0.4$) the mesh grid has 100 points (GDQ).

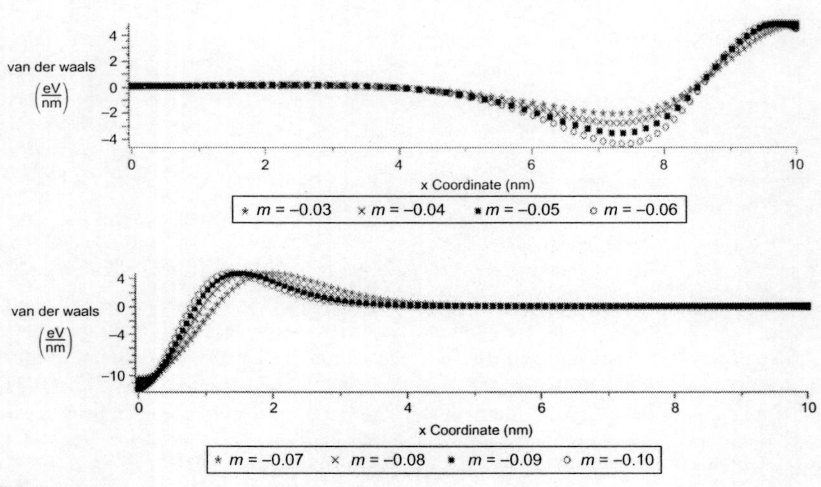

FIGURE 2.4

Interacting force of single SWCNT ($2L=20$, $d=0.4$) the mesh grid has 100 points (GDQ).

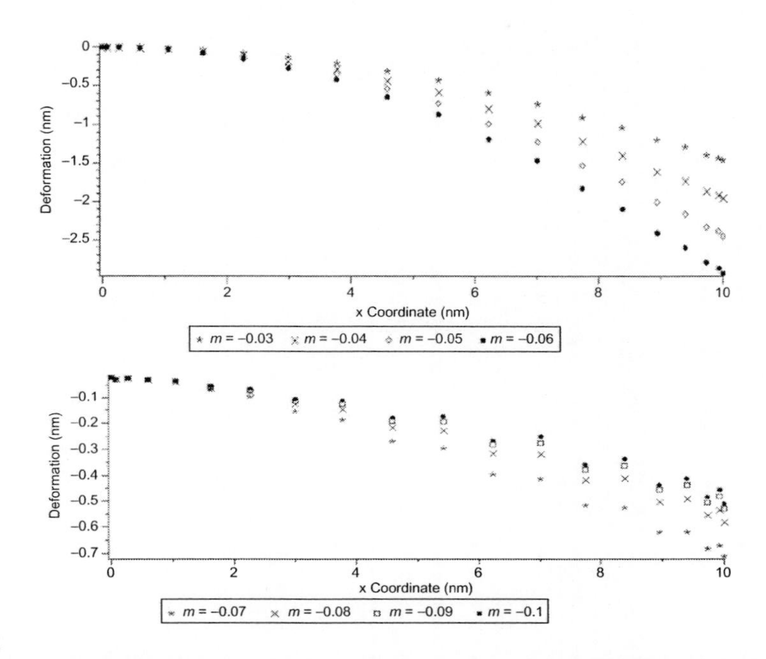

FIGURE 2.5

Interacting force of single SWCNT ($2L=20$, $d=0.4$) and mesh grid has 20 points (GDQ).

By substituting the deformation results into the Equation (2.4), van der Waals interaction will be shown in Figure 2.4, which has adequate accuracy in engineering point of view.

Figure 2.5 shows the results for the same SWCNT which uses 20 points. Deformation results respond emphasize to instability in this mesh grid responds.

2.2.3.2 Length effect

Herein the length of SWCNT is the parameter which is studied in GDQ results and each case is solved for 100 points grid. In Figures 2.6 and 2.7 the length of the individual SWCNT is $2L=40$ (nm) and the diameter is 1.4 (nm) and these figures are about deformation and interaction, respectively. The observed behaviors are similar to those which were seen in Figure 2.3.

Using last results and Equation (2.4), the interactions in above case are plotted in Figure 2.6 again reasonable values are obtained.

Figures 2.8 and 2.9 show the results related to SWCNT which has 200 (nm) length and 1.4 (nm) diameter. Figure 2.8 is related to deformation manner.

After obtaining deformation values, interaction will be achieved by Equation (2.4), the related responds are displayed in Figures 2.9.

2.2.3.3 Validation of GDQ approach

To validate the present GDQ approach for deformation solutions of the SWCNT, comparisons have been carried out with the results of Li et al. (2004). In their analysis, FEM was used to describe the deformation of the individual nanotube.

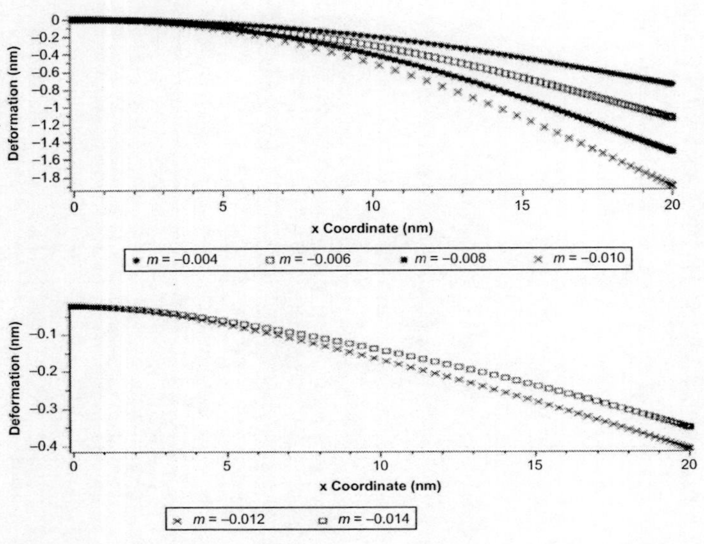

FIGURE 2.6

Deformation of single SWCNT ($2L = 40$, $d = 1.4$), mesh grid is 100 points (GDQ).

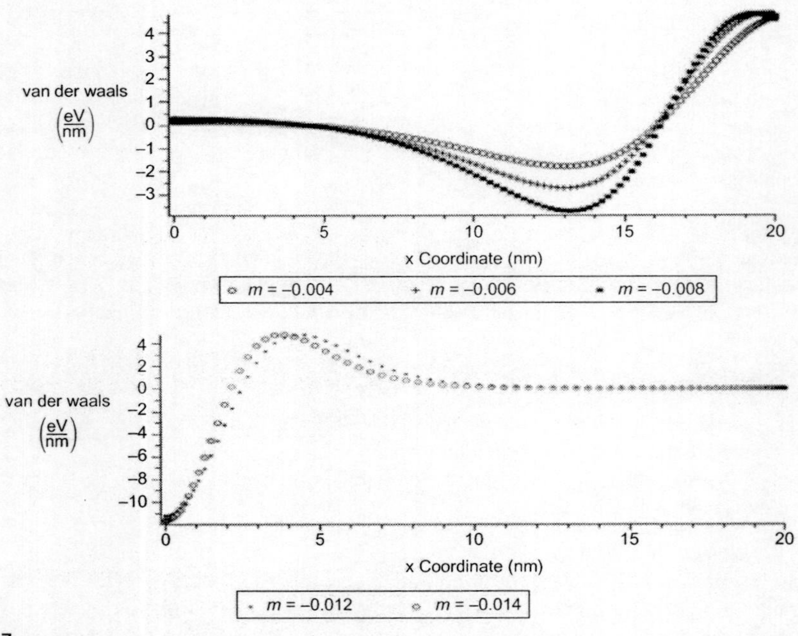

FIGURE 2.7

Interacting force of single SWCNT ($2L = 40$, $d = 1.4$), mesh grid is 100 points (GDQ).

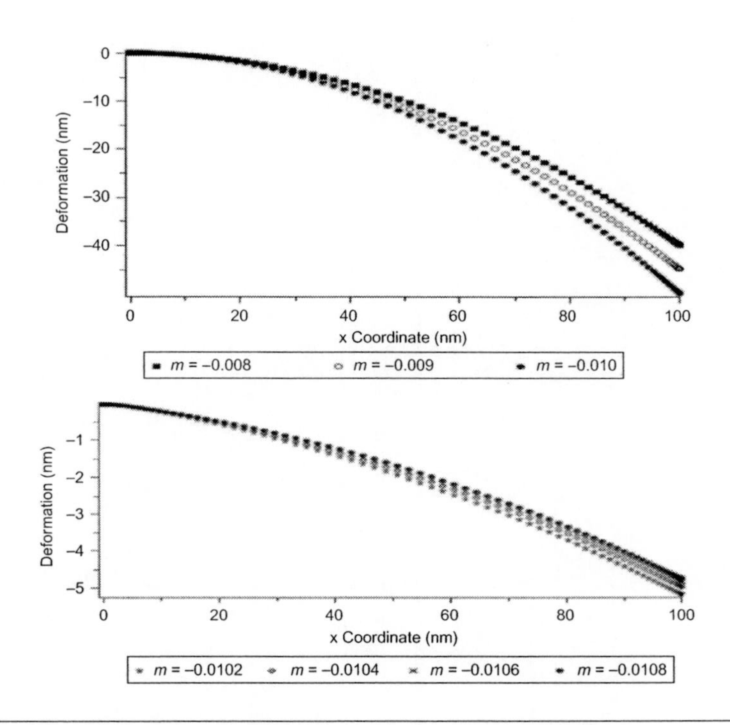

FIGURE 2.8

Deformation of single SWCNT ($2L=200$, $d=1.4$) by mesh grid with 100 points (GDQ).

The research was performed in two cases, the first one belongs to a single tube which has 20 (nm) length and 1.4 (nm) diameter. Corresponded responds are depicted in Figures 2.10.

In the second case, the nanotube has 40 (nm) length and 1.4 (nm) diameter. Computations related to this case are depicted in Figure 2.11.

2.2.4 CONCLUSION

The deformation of individual SWCNT located over a bundle of nanotube is analyzed based on small deformation theory and the Euler beam theory. The van der Waals forces are significant in SWCNTs which are located closely. Influence of the length and diameter of the individual SWCNT beside the curvature of substrate on deformation behavior of the single SWCNTs is shown. The governing equation and the boundary conditions for the SWCNT as an Euler beam are solved using the GDQ method. From the GDQ solutions, it can be clearly seen that the length of single SWCNT has a great effect of deformation behavior related to the same curvature. With the same SWCNT the number of mesh points

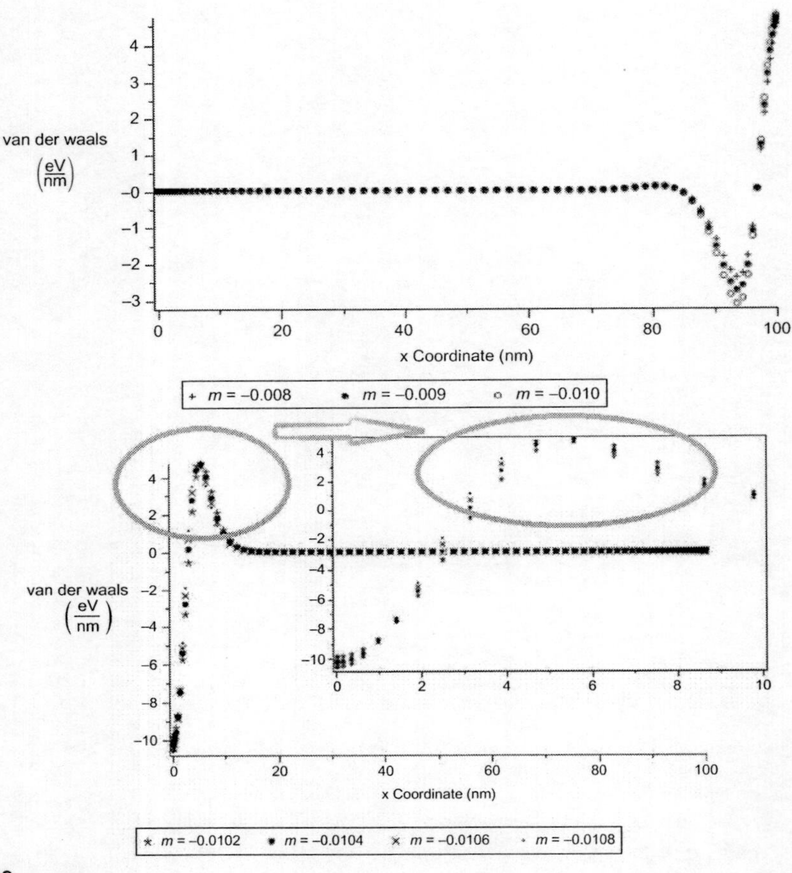

FIGURE 2.9

Interacting force of single SWCNT ($2L=200$, $d=1.4$) the mesh grid has 100 points (GDQ).

is important especially in distribution of interacting forces. As a parameter the aspect ratio is important and it's obvious that in small aspect ratio the individual SWCNT can deform more significantly and be near the substrate in high curvature value of substrate.

2.3 CANTILEVERED SWCNT AS A NANOMECHANICAL SENSOR

In this section, the continuum mechanics method and a bending model are applied to obtain the resonant frequency of the fixed-free SWCNT where the mass is rigidly attached to the tip. This method used the Euler-Bernoulli theory with cantilevered boundary conditions where the effect of attached mass on

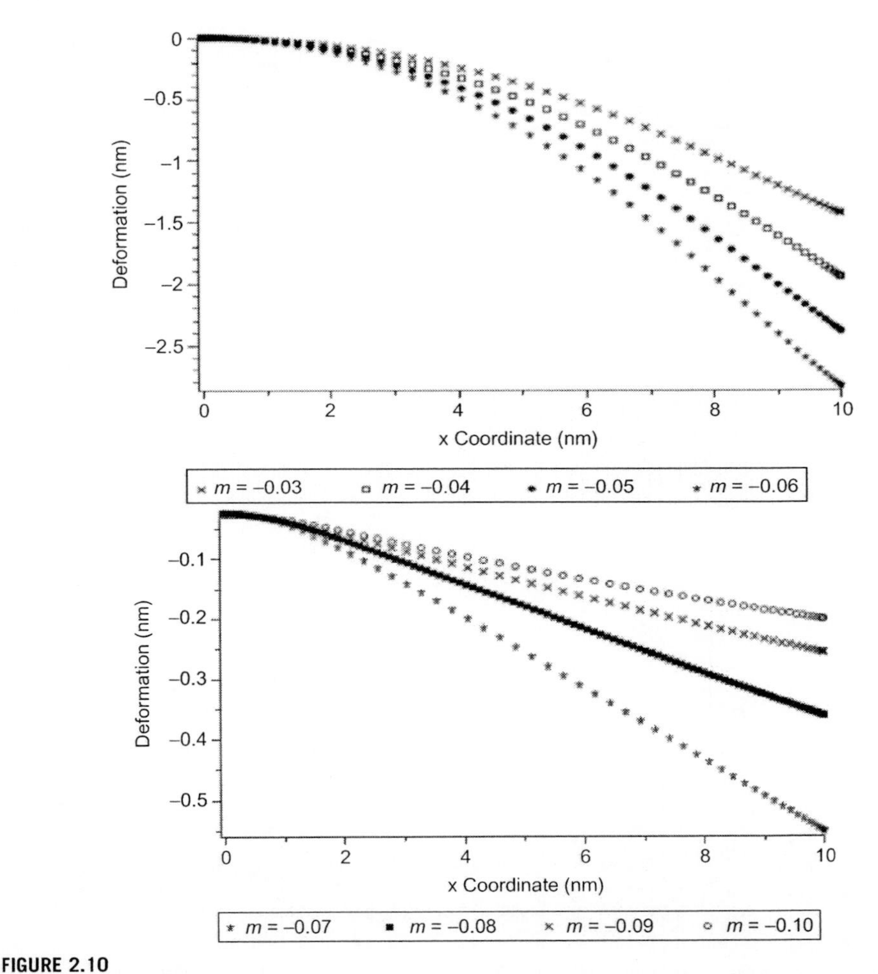

FIGURE 2.10

Deformation of single SWCNT ($2L=20$, $d=0.4$) the mesh grid has 100 points (FEM).

resonant frequency, is added at the free end condition. The resonant frequencies of the fixed-free SWCNT have been investigated. The results showed the sensitivity of the SWCNTs to different masses. The results indicate that by increasing the value of attached mass, the values of resonant frequency are decreased. The validity and the accuracy of these formulas are examined with other sensor equations in the literatures. The results indicate that the new sensor equations can be used for CNT like CNT-based biosensors with reasonable accuracy (*this section has been worked by I. Mehdipour, A. Barari, and G Domairry (D.D. Ganji) in nonlinear dynamics team in Mechanical Engineering Department, 2010-2011*).

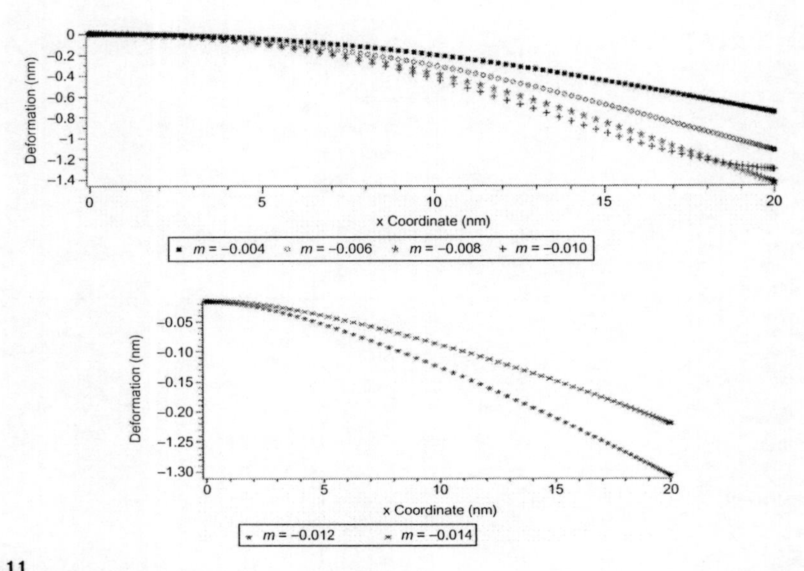

FIGURE 2.11

Deformation of single SWCNT ($2L=40$, $d=1.4$) the mesh grid has 100 points (FEM).

2.3.1 INTRODUCTION

CNTs are unique nanostructured materials. The extraordinary mechanical and physical properties in addition to the large aspect ratio and low density have made CNTs ideal components of nanodevices. A wide range of applications of CNTs have been reported in the literature, one of these applications is in biological applications (Baughman et al., 1999; Mattson et al., 2000), especially in medical technology (Lu et al., 2009) and sensors (Gu et al., 2005) which can be broadly classified into two categories (Yogeswaran et al., 2008): chemical sensors (Wang and Musameh, 2003) and biosensors (Allen et al., 2007).

Resonance based sensors recommend the deeper potential of achieving the high-performance requirement of many sensing applications. The principle of mass detection using resonators is based on the fact that the resonant frequency is sensitive to the resonator mass, which includes the self-mass of the resonator and the attached mass. The change of the attached mass on the resonator causes a shift to the resonant frequency. The key issue of mass detection is in quantifying the change in the resonant frequency due to the attached mass.

Recently, finite element analysis has been used to perform the rapid computation of the mechanical properties of nanostructures (Lau et al., 2004). Several studies have investigated the use of a CNT as a mass sensor. In previous study (Joshi et al., 2010), the current authors used the continuum mechanics method is combined with commercial FEM software to simulate the mechanical responses of individual CNTs treated as cylindrical beams or thin shells with thickness. The resonant frequency of the fixed-free and the bridged SWCNT is investigated using a bending model, and the frequency shift calculated when a small mass is attached to the tip. The use of SWCNT as bio sensors has been recently examined by Chowdhury et al. (2009) using continuum mechanics approach and equations are derived to detect the mass of biological objects.

In the present section, we used the continuum mechanics method and a bending model (Rossit and Laura, 2001; Rossit et al., 1998), to obtain the resonant frequency of the fixed-free SWCNT where the mass is rigidly attached to the tip, as shown in Figure 1. In addition, the validity and the accuracy of these results are examined by the new sensor equations for CNT-based biosensors (Chowdhury et al., 2009).

2.3.2 ANALYSIS OF THE PROBLEM

Recently, the continuum mechanics method has been successfully applied to analyze the dynamic responses of individual CNTs (Chowdhury et al., 2009; Joshi et al., 2010; Wu et al., 2006).

2.3.2.1 Basic bending vibration and resonant frequencies of SWCNT with attached mass

The continuum models based on beam have been used extensively for SWCNT and MWCNT. In order to obtain simple analytical expressions of the mass of attached biochemical entities, Based on the Euler-Bernoulli beam model (Timoshenko and Gere, 1961), it is well known that the equation of motion of a free vibration rod in the limit of small amplitude is governed by the fourth-order wave equation, i.e.:

$$EI\frac{\partial^4 w(x,t)}{\partial x^4}+\rho A\frac{\partial^2 w(x,t)}{\partial t^2}=0 \tag{2.32}$$

where E is the Young's modulus, I is the second moment of the cross-sectional area A, and ρ is the density of the beam material. In previous studies (Chowdhury et al., 2009; Joshi et al., 2010; Wu et al., 2006) considered the cases of a fixed-free and a bridge SWCNT with a nanoscale particle attached to its tip. The operation of a SWCNT based mass sensor is based on the fact that mass addition to the tip causes a measurable shift in the resonant frequency of the beam. The minimum detectable mass change of the sensor can be approximated by the formula:

$$f_n=\frac{1}{2\pi}\sqrt{\frac{k_{eq}}{m_{eq}}} \tag{2.33}$$

Here k_{eq} and m_{eq} are, respectively, equivalent stiffness and mass of SWCNT with attached mass in the first mode of vibration. The equivalent spring stiffness of the cantilever beam can be expressed in the form (Shigley, 1988):

$$k_{eq}=\frac{3EI}{L^3} \tag{2.34}$$

where I is the moment of inertia, L is the length of the beam, and E is the Young's modulus.

2.3.2.2 Resonant frequency of cantilevered SWCNT where the mass is rigidly attached to the tip

The present study considers the case of a fixed-free SWCNT with a nanoscale particle that is rigidly attached to its tip, as shown in Figure 2.12.

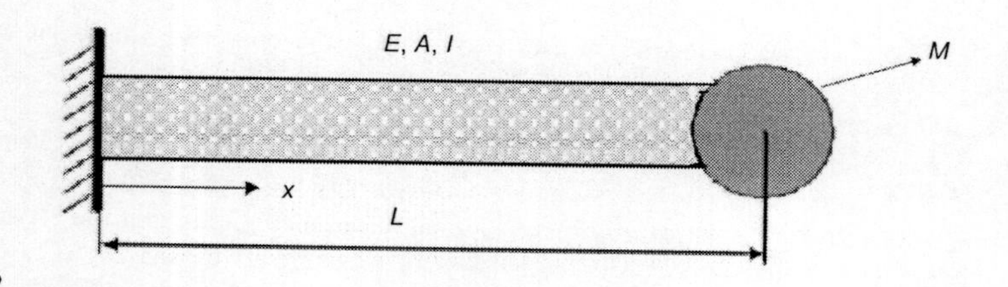

FIGURE 2.12

Cantilevered carbon nanotube resonator with an attached mass at the tip.

It is assumed that the problem under consideration is governed by the classical Euler-Bernoulli beam theory. Accordingly the differential system is described by the partial differential equation:

$$EI\frac{\partial^4 w(x,t)}{\partial x^4}+\rho A\frac{\partial^2 w(x,t)}{\partial t^2}=0 \tag{2.35}$$

This in the case of normal modes becomes

$$\frac{d^4 W(x)}{dx^4}-\frac{\rho A}{EI}\omega^2 W(x)=0 \tag{2.36}$$

where $w(x,t)=W(x)e^{i\omega t}$

The boundary conditions are given by:

$$W(0)=\frac{\partial W}{\partial x}(0)=0,\quad \frac{\partial^2 W}{\partial x^2}(L)=0,\quad -EI\frac{\partial^3 W}{\partial x^3}(L)=M\omega^2 W(L) \tag{2.37}$$

The solution of Equation (2.36) is

$$W(x)=A\cosh\beta\bar{x}+B\sinh\beta\bar{x}+C\cos\beta\bar{x}+D\sin\beta\bar{x} \tag{2.38}$$

where

$$\bar{x}=\frac{x}{L},\quad \beta^2=\sqrt{\frac{\rho A}{EI}}\omega L^2 \tag{2.39}$$

By substituting Equation (2.38) into the governing boundary conditions Equation (2.37), it can be obtained:

$$\begin{bmatrix} 1 & 0 & 1 & 0 \\ 0 & 1 & 0 & 1 \\ \cosh\beta & \sinh\beta & -\cos\beta & -\sin\beta \\ -\sinh\beta-m\beta\cosh\beta & -\cosh\beta-m\beta\sinh\beta & -\sin\beta-m\beta\cos\beta & \cos\beta-m\beta\sin\beta \end{bmatrix}\begin{Bmatrix} A \\ B \\ C \\ D \end{Bmatrix}=\begin{Bmatrix} 0 \\ 0 \\ 0 \\ 0 \end{Bmatrix} \tag{2.40}$$

where

$$m = \frac{M}{\rho AL} \tag{2.41}$$

A solution to this homogeneous system of equations exists only if the determinant of the coefficient's matrix set equal to zero; hence, Expansion of the determinant of Equation (2.40) leads to the following transcendental expression in the frequency coefficients:

$$\cosh\beta \ \cos\beta - m\beta \cosh\beta \ \sin\beta + m\beta \sinh\beta \ \cos\beta + 1 = 0 \tag{2.42}$$

2.3.3 NUMERICAL RESULTS

2.3.3.1 Vibration mode analysis

In this study, the single-walled fixed-free SWCNT is modeled as cantilevered beam. Table 2.3 summarizes the four first natural frequencies of the fixed-free SWCNT for different attached mass to the tip. The dimensions of the SWCNT are as follows: inner diameter 18.8 nm, outer diameter 33 nm, and length 5.5 μm. At the zero mass ($M=0$), the four first natural frequencies of the fixed-free SWCNT is equal the four first natural frequencies of the cantilevered beam under transverse vibration. From Table 2.3, it can be seen that by increasing the value of the added mass at the tip, the four first natural frequencies are decreasing. Figure 2.13 plots the first mode shape for the different value of the attached mass to the tip. At the first mode shape, Figure 2.13 shows that by increasing the value of the added mass at the tip, the first natural frequency is decreased.

Modal shapes of four first natural frequencies in the fixed-free SWCNT boundary conditions for the different values of the attached mass to the tip are depicted in Figure 3. For case (a), $M=0$, it can be seen that the beam motion is largely amplified in correspondence with the first mode when compared with case (b), $M=20$ fg. The effects of the attached mass to the tip on the other natural frequency and other modal shapes are obviously shown in Figure 2.14.

2.3.4 MASS SENSOR MODE COMPARISON

Chowdhury et al. (2009) developed the simple analytical formulas for CNT-based nanoresonators with attached mass. They gave a virtual force at the location of the mass so that the deflection under the mass becomes unity. For this case it can be shown that $k_{eq} = 3EI/L^3$. According to the deflection shape along

Table 2.3 Values of β_i ($i=1,\ldots,4$) for Different Attached Mass (M)

M	0	20 fg	30 fg	40 fg	50 fg
β_1	1.875104	0.881493	0.799706	0.745715	0.706114
β_2	4.694091	3.951260	3.943212	3.939125	3.936652
β_3	7.854757	7.083313	7.078458	7.076010	7.074535
β_4	10.995541	10.220406	10.217022	10.215320	10.214296

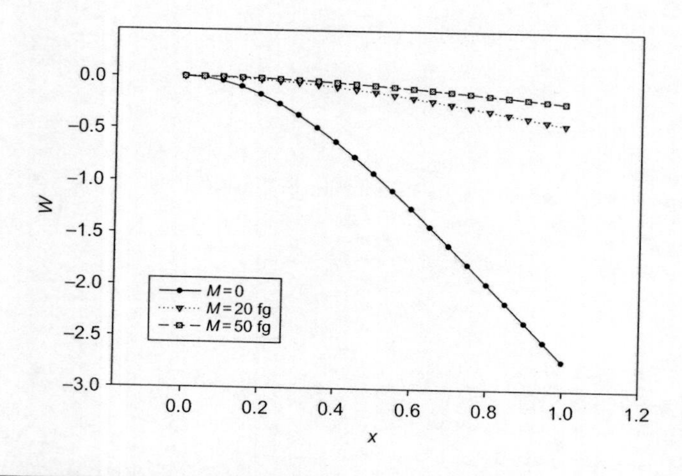

FIGURE 2.13

First mode shapes of the fixed-free SWCNT for the different value of the attached mass to the tip.

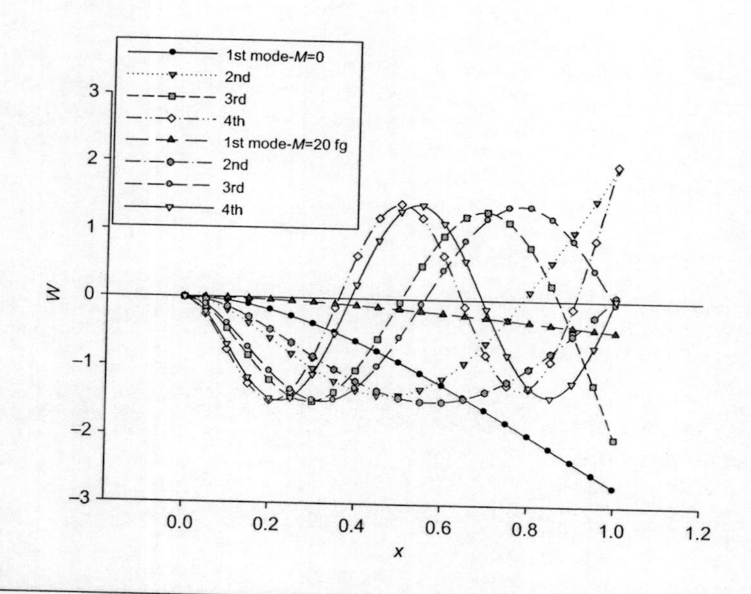

FIGURE 2.14

Four first mode shapes of the fixed-free SWCNT for the different value of the attached mass to the tip.

the length of the SWCNT for this case, a closed-form expression derived to detect the mass of biological objects from the frequency-shift and obtained the resonant frequency of cantilevered SWCNT with mass at the tip by the formula:

$$f_n = \frac{1}{2\pi}\sqrt{\frac{k_{eq}}{m_{eq}}} = \frac{1}{2\pi}\sqrt{\frac{3EI/L^3}{(33/140)\rho AL + M}} \tag{2.43}$$

where, M is the value of the added mass at the tip.

In this study, we obtained β by using Equation (2.42) with numerical method and by substituting β in Equation (2.39) can be obtained the resonant frequency as:

$$f_n = \frac{1}{2\pi}\frac{\beta^2}{L^2}\sqrt{\frac{EI}{\rho A}} \tag{2.44}$$

Table 2.4 depicts the comparison of resonant frequency for fixed-free SWCNT with different mass additions between Equations (2.43) and (2.44). The dimensions of the SWCNT are as follows: inner diameter 18.8 nm, outer diameter 33 nm, and length 5.5 μm. In addition, the Young's module is 32 GPa. From Table 2.4, it can be seen that by increasing the value of attached mass, the value of resonant frequency in both theories of Equations (2.43) and (2.44) are decreasing and also the error percentages are decreased. So, in the big values of attached mass, the theoretical values of Equations (2.43) and (2.44) are approximately the same.

2.3.5 CONCLUSION

This chapter has developed a new mass sensor equation for modeling the vibration behavior of a fixed-free SWCNT with attached mass at the tip. The resonant frequencies of the fixed-free SWCNT have been investigated. In addition, the validity and the accuracy of these formulas are examined with other sensor equations in the literatures. The result indicate that

Table 2.4 Comparison of Resonant Frequency for Fixed-Free SWCNT with Different Mass Additions

Attached Mass (fg)	Theoretical Values, Equation (2.43) (Hz)	Theoretical Values, Equation (2.44) (Hz)	%Error
20	190407.982	190401.785	0.003255
22	181939.662	181934.726	0.002713
24	174509.213	174505.207	0.002296
26	167920.601	167917.297	0.001968
28	162025.999	162023.235	0.001706
30	156711.541	156709.208	0.001489
35	145420.889	145419.280	0.001106
40	136264.666	136263.504	0.000853
50	122176.044	122175.371	0.000551

(1) By increasing the values of the added mass at the tip, the natural frequencies are decreased.
(2) In the big values of attached mass, the theoretical values of Equations (2.43) and (2.44) are approximately the same.
(3) The results indicate that the new sensor equations can be used for CNT-based mass sensors with reasonable accuracy. So, a SWCNT can be employed as a high-precision mass sensor.

2.4 NONLINEAR VIBRATION FOR EMBEDDED CNT

In this study, based on continuum mechanics and an elastic beam model, a nonlinear free vibration analysis of embedded SWCNT considering the effects of rippling deformation and midplane stretching on nonlinear frequency is investigated. By utilizing He's energy balance method (HEBM), the relationship of nonlinear amplitude and frequency for the SWCNT is expressed. The amplitude frequency response curves of the nonlinear free vibration for the SWCNT are obtained and the effects of rippling deformation, midplane stretching and surrounding elastic medium on the amplitude frequency response characteristics are discussed. In addition, the rippling instability of CNTs and the effective parameters on their behavior are briefly discussed (*this section has been worked by I. Mehdipour, A. Barari, and G. Domairry (D.D. Ganji) in nonlinear dynamics team in Mechanical Engineering Department, 2012-2013*).

2.4.1 INTRODUCTION

It is important to have accurate theoretical models for vibrational behavior of CNTs for several reasons. For instance, natural frequencies of CNTs play an important role on nanomechanical resonators using them. In addition, the effective young modulus of a nanotube may be determined indirectly from its measured natural frequencies or mode shapes if a sufficiently precise theoretical model is used. Molecular dynamics (MD) method simulates CNTs accurately. However, MD simulation is limited to systems with small number of atoms (say less than 10^{16}) and remains time consuming and expensive (Gibson et al., 2007; Yaghmaei and Rafii-Tabar, 2009; Zhang et al., 2009). For large-scale systems, continuum mechanics approach has widely and successfully modeled mechanical and vibrational characteristics of CNTs (Gibson et al., 2007; Ranjbartoreh et al., 2007). The continuum modeling approach is much less computational effort and much cheaper than the MD simulations and experimental verification, respectively.

The high elastic modulus of CNT (higher than 1 TPa) and remarkable bending flexibility (up to 20%) without breaking, exhibit new phenomena in bending vibration of nanotubes called rippling. The rippling affects directly on the resonant frequencies and elastic modulus of CNTs; for example, Wang et al. (2005) constructed a three-dimensional finite element model based on orthotropic theory of finite elasticity deformation, to estimate the effective bending modulus of CNT with rippling effect. The nonlinear bending moment-curvature relationship has been derived using FEM and expressed as a ninth-order polynomial equation. A nonlinear vibration analysis method has been used to calculate the effective modulus of a CNT in rippling mode. The results declare that effective bending modulus of CNTs decreases substantially with increasing diameter.

Beams resting on elastic foundations have wide application in modern engineering and pose great technical problems in structural design. As a result, numerous research reports involving the calculation and analysis approach for beams on elastic foundation have been presented (see for example Gibson et al., 2007; Ranjbartoreh et al., 2007).

In the present work, a nonlinear elastic beam model with midplane stretching is developed for transverse vibration of a SWCNT on elastic foundation on the basis of rippling deformation. Following this model, the nonlinear fundamental frequency is derived by using the HEBM (He, 2002) for the case of the influences of the stiffness of the foundation, are discussed. In addition, the instability caused by rippling deformation and midplane stretching is introduced and the conditions causing this kind of instability are determined.

2.4.2 BASIC EQUATIONS

Schematic diagram, of a CNT embedded in an elastic medium, is considered as a hollow cylindrical tube of length L, cross-sectional area A, cross-sectional inertia moment I, Young's modulus E, and density ρ, as shown in Figure 2.15. Assume that the transverse displacement is $w(x, t)$ in terms of the spatial coordinate x and the time variable t.

The free vibration equation of embedded CNT considering the midplane stretching of the structure is (Fu et al., 2006).

$$EI\frac{\partial^4 w}{\partial x^4}+\rho A\frac{\partial^2 w}{\partial t^2}+kw=\left[\frac{EA}{2l}\int_0^l\left(\frac{\partial w}{\partial x}\right)\mathrm{d}x\right]\frac{\partial^2 w}{\partial x^2} \tag{2.45}$$

According to the classical analysis of beam $M(x,t)=EI(\partial^2 w/\partial x^2)$, so Equation (2.45) can be written in the following form:

$$M''(x,t)+\rho A\frac{\partial^2 w}{\partial t^2}+kw=\left[\frac{EA}{2l}\int_0^l\left(\frac{\partial w}{\partial x}\right)\mathrm{d}x\right]\frac{M(x,t)}{EI} \tag{2.46}$$

where $M(x, t)$ expresses the bending moment and $M''(x,t)$ the partial derivatives $(\partial^2 M(x,t))/\partial x^2$. The corresponding boundary conditions for a simply supported beam are expressed as:

$$w(0,t)=\frac{\partial^2 w(0,t)}{\partial x^2}=0,\quad w(l,t)=\frac{\partial^2 w(l,t)}{\partial x^2}=0 \tag{2.47}$$

When CNT bends, the rippling formation occurs specially for the relatively and locally large deformation. Because the bending curvature $\kappa(x, t)$ is a function of the beam deflection $w(x, t)$, it is necessary to get a nonlinear relation between $M(x, t)$ and $\kappa(x, t)$ from the full three-dimensional theory of finite deformation. Thus, substituting the nonlinear relation between $M(x, t)$ and $\kappa(x, t)$ into Equation (2.46) yields a nonlinear differential equation governing the deflection function $w(x, t)$.

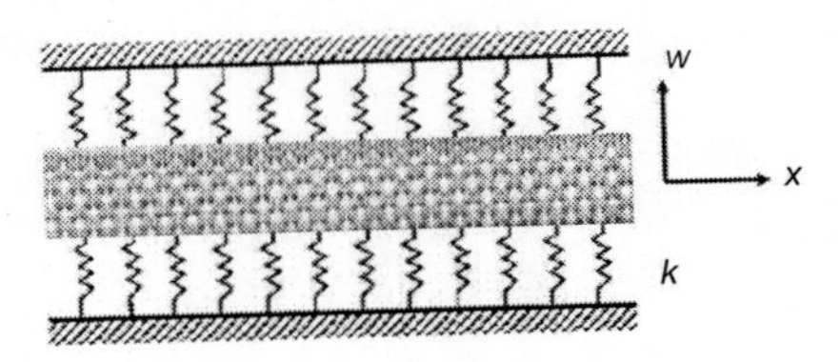

FIGURE 2.15

Model of an embedded carbon nanotube in an elastic medium of constant k.

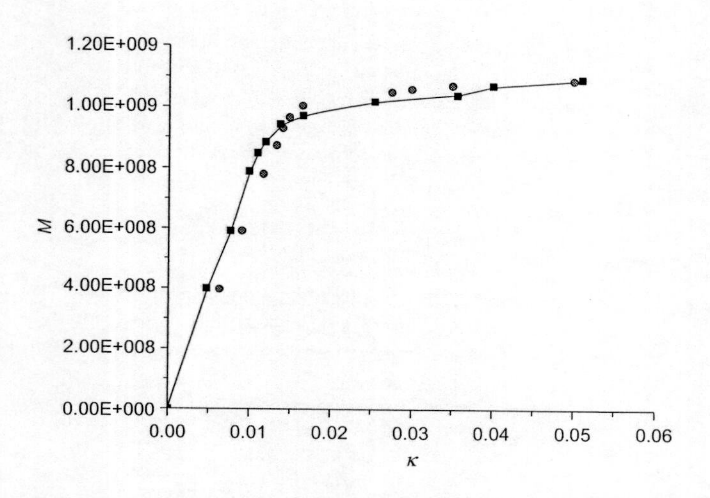

FIGURE 2.16

Two results derived from different ratios to *L/D*, where ■ represents *L/D*=10, • represents *L/D*=20.

The nonlinear effect of rippling mode for a bent CNT has been calculated using FEM simulation (Liu et al., 2003). Figure 2.16 shows the bending moment M versus the bending curvature κ with the length-to-diameter ratios $L/D=10$, and 20 from FEM simulating result (Wang et al., 2005).

According to the beam bending theory, we know that $M(x, t)$ should be an odd function $\kappa(x, t)$. Using a polynomial up to the ninth order, the discrete points in Figure 2.2 are fitted as:

$$M(x,t)=EI\kappa\left(1-a_3D^2\kappa^2+a_5D^4\kappa^4-a_7D^6\kappa^6+a_9D^8\kappa^8\right) \tag{2.48}$$

where $\kappa=\kappa(x,t)$, $a_3=1.755\times10^3$, $a_5=2.0122\times10^6$, $a_7=1.115\times10^9$, and $a_9=2.266\times10^{11}$

The two orders of partial derivatives to x variable in Equation (2.48) gives

$$\begin{aligned}M''(x,t)=EI\big[&\kappa''\left(1-3a_3D^2\kappa^2+5a_5D^4\kappa^4-7a_7D^6\kappa^6+9a_9D^8\kappa^8\right)\\&+(\kappa')^2\left(-6a_3D^2\kappa+20a_5D^4\kappa^3-42a_7D^6\kappa^6-72a_9D^8\kappa^7\right)\big]\end{aligned} \tag{2.49}$$

Based on the large deflection deformation, the relation between the bending curvature $\kappa(x,t)$ and the beam deflection $w(x,t)$ can be expressed as:

$$\begin{aligned}\kappa(x,t)&=\frac{w''(x,t)}{\left[1+w'(x,t)^2\right]^{3/2}}\\&=w''\left[1-\frac{3}{2}(w')^2+\frac{15}{8}(w')^4-\cdots\right]\\&\cong w''(x,t)\left[1-\frac{3}{2}(w')^2\right]\end{aligned} \tag{2.50}$$

Substituting Equation (2.50) into Equations (2.48) and (2.49) yields

$$M(x,t) = EI\left\{w'' - 3a_3D^2w''^3 - \frac{3}{2}w'^2w''\right\} \tag{2.51}$$

and

$$M''(x,t) = EI\left\{w'''' - 3a_3D^2\left[2w''(w''')^2 + (w'')^2w''''\right] - \frac{3}{2}\left[2(w'')^3 + 6w'w''w''' + (w')^2w''''\right]\right\} \tag{2.52}$$

The nonlinear vibration equation will be obtained by substituting Equations (2.51) and (2.52) into Equation (2.46).

$$\begin{aligned} EI\left(w'''' - 3a_3D^2\left[2w''(w''')^2 + (w'')^2w''''\right] - \frac{3}{2}\left[2(w'')^3 + 6w'w''w''' + (w')^2w''''\right]\right) + \rho A\frac{\partial^2 w}{\partial t^2} + kw \\ = \left[\frac{EA}{2l}\int_0^l \left(\frac{\partial w}{\partial x}\right)\mathrm{d}x\right]\left(w'' - 3a_3D^2w''^3 - \frac{3}{2}w'^2w''\right) \end{aligned} \tag{2.53}$$

According to the boundary conditions, the unknown function $w(x, t)$ may be given as:

$$w(x,t) = W(t)\sin\frac{\pi x}{l} \tag{2.54}$$

It satisfies the boundary conditions.

By substituting Equation (2.54) into Equation (2.53), the nonlinear differential equation for the time function $W(t)$ can be obtained as follows:

$$\frac{\mathrm{d}^2W}{\mathrm{d}t^2} + \left(\frac{\pi^4EI}{l^4\rho A} + \frac{k}{\rho A}\right)W + (\alpha_1 + \alpha_2)W^3 + \alpha_3W^5 = 0 \tag{2.55}$$

where

$$\alpha_1 = 2\left(-\frac{0.375\pi^8EIa_3D^2}{l^8\rho A} - \frac{180.260474EI}{l^6\rho A}\right) \tag{2.56}$$

$$\alpha_2 = \frac{0.25\pi^4E}{l^4\rho} \tag{2.57}$$

and

$$\alpha_3 = 2\left(-\frac{0.28125\pi^8Ea_3D^2}{l^8\rho} - \frac{45.06511845E}{l^6\rho}\right) \tag{2.58}$$

In Equation (2.55), parameters, α_1 and α_3 are considered to analysis the effect of rippling deformation and also α_2 to simulate the midplane stretching on vibration equation of motion for CNT embedded in an elastic medium.

2.4.3 SOLUTION METHODOLOGY

Recently, considerable attention has been directed towards to analytical solutions for nonlinear equations without possible small parameters. The traditional perturbation methods have many shortcomings, and they are not useful to strongly nonlinear equations. In present analysis, a very effective and convenient method, the HEBM (He, 2002) is employed to seek the analytical solution of Equation (2.55).

HEBM can be applied to nonlinear vibrations and oscillations, while this method does not require linearization or small parameter and finally the results will reveal the simplicity of the method as well.

According to Equation (2.55), its variational principle can be easily established using the semi-inverse method (He, 2004, 2007)

$$J(W)=\int_0^t\left(\frac{1}{2}\dot{W}^2-\left(\frac{\pi^4EI}{l^4\rho A}+\frac{k}{\rho A}\right)\frac{W^2}{2}+\frac{(\alpha_1+\alpha_2)}{4}W^4+\frac{\alpha_3}{6}W^6\right)dt \tag{2.59}$$

In the functional (2.59), $(1/2)\dot{W}^2$ is kinetic energy (K) and $\left(\frac{\pi^4EI}{l^4\rho A}+\frac{k}{\rho A}\right)\frac{W^2}{2}+\frac{(\alpha_1+\alpha_2)}{4}W^4+\frac{\alpha_3}{6}W^6$ is potential energy (V), so the functional (2.59) is the least Lagrangian action ($L=K-V$), from which we can immediately obtain its Hamiltonian ($H=K+V=\text{constant}=H_0$), which reads

$$H=\frac{1}{2}\dot{W}^2+\left(\frac{\pi^4EI}{l^4\rho A}+\frac{k}{\rho A}\right)\frac{W^2}{2}+\frac{(\alpha_1+\alpha_2)}{4}W^4+\frac{\alpha_3}{6}W^6=\left(\frac{\pi^4EI}{l^4\rho A}+\frac{k}{\rho A}\right)\frac{a^2}{2}+\frac{(\alpha_1+\alpha_2)}{4}a^4+\frac{\alpha_3}{6}a^6 \tag{2.60}$$

or

$$\frac{1}{2}\dot{W}^2+\left(\frac{\pi^4EI}{l^4\rho A}+\frac{k}{\rho A}\right)\frac{W^2}{2}+\frac{(\alpha_1+\alpha_2)}{4}W^4+\frac{\alpha_3}{6}W^6-\left(\frac{\pi^4EI}{l^4\rho A}+\frac{k}{\rho A}\right)\frac{a^2}{2}-\frac{(\alpha_1+\alpha_2)}{4}a^4-\frac{\alpha_3}{6}a^6=0 \tag{2.61}$$

Equation (2.60) implies that the total energy keeps unchanged during the conservative oscillation.

Oscillatory systems contain two important physical parameters, i.e., the frequency ω and the amplitude of oscillation, a. So let us consider such initial conditions:

$$W(0)=a,\quad \dot{W}(0)=0 \tag{2.62}$$

Assume that its initial approximate guess can be expressed as:

$$W(t)=a\cos\tilde{\omega}t \tag{2.63}$$

By substituting Equation (2.63) into Equation (2.61), obtains a residual

$$R(t)=\frac{1}{2}a^2\tilde{\omega}^2(\sin\tilde{\omega}t)^2+\left(\frac{\pi^4EI}{l^4\rho A}+\frac{k}{\rho A}\right)\frac{(\cos\tilde{\omega}t)^2}{2}+\frac{(\alpha_1+\alpha_2)}{4}(\cos\tilde{\omega}t)^4+\frac{\alpha_3}{6}(\cos\tilde{\omega}t)^6-\left(\frac{\pi^4EI}{l^4\rho A}+\frac{k}{\rho A}\right)\frac{a^2}{2}$$
$$-\frac{(\alpha_1+\alpha_2)}{4}a^4-\frac{\alpha_3}{6}a^6=0 \tag{2.64}$$

If, by any chance, the exact solution had been chosen as the trial function, then it would be possible to make R zero for all values of t by appropriate choice of x. Since (2.62) is only an approximation to the exact solution, R cannot be made zero everywhere. By locating at some a special point, i.e., $\widetilde{\omega}t=(\pi/4)$, and setting $R(t=(\pi/4\widetilde{\omega}))=0)$ we can obtain an approximate frequency-amplitude relationship of the studied nonlinear oscillator

$$\widetilde{\omega}=\frac{1}{6}\sqrt{\left(\frac{36EI\pi^4+36kl^4}{\rho Al^4}+21\alpha_3a^4+27(\alpha_1+\alpha_2)a^2\right)} \tag{2.65}$$

By substituting Equations (2.60) and (2.61) into Equation (2.65), can be obtained:

$$\widetilde{\omega}=\frac{1}{6}\sqrt{\left(\frac{36EI\pi^4+36kl^4}{\rho Al^4}+42\left(-\frac{0.28125\pi^8Ea_3D^2}{l^8\rho}-\frac{45.06511845E}{l^6\rho}\right)a^4+54\left(-\frac{0.375\pi^8EIa_3D^2}{l^8\rho A}-\frac{180.260474EI}{l^6\rho A}+\frac{0.125\pi^4E}{l^4\rho}\right)a^2\right)} \tag{2.66}$$

2.4.4 NUMERICAL RESULTS AND DISCUSSION

Assume the linear free vibration frequency to be $\overline{\omega}$ in Equations (2.49) and (2.59), and $\overline{\omega}^2=(\pi^4EI/l^4\rho A)+k/\rho A$. The parameters of material and geometry are taken as $E=1.1$ TPa, $\rho=1.3\times10^3$ kg/m^3, $l=45$ nm, the outside diameter $D=d_o=3$ nm, and the inside diameter is $d_i=2.32$ nm (Fu et al., 2006).

According to Equation (2.65), when parameters α_1 and α_3 are zero, it can be obtained the nonlinear frequency due to only the effect of midplane stretching. Figure 2.17 shows the ratio of nonlinear frequency

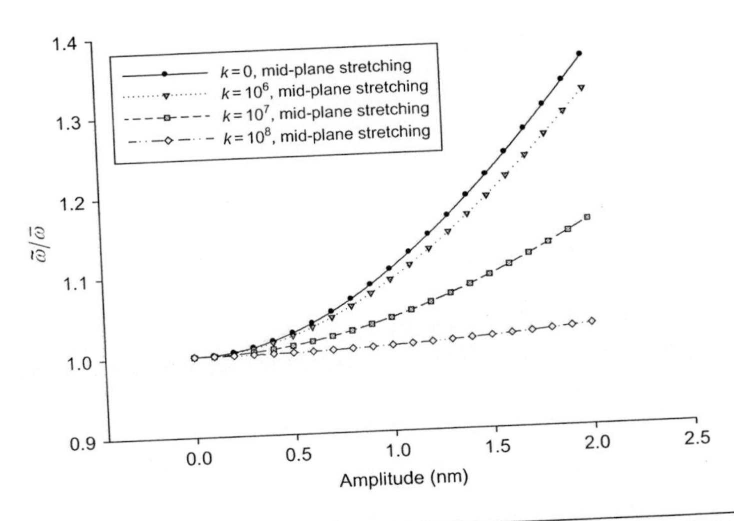

FIGURE 2.17

The effects of surrounding stiffness k and midplane stretching on the nonlinear amplitude frequency response curves of SWCNT when $\alpha_1=\alpha_3=0$.

resonant frequency to linear frequency caused by midplane stretching configuration of a SWCNT. In Figure 2.17 the nonlinear frequency ratio to linear frequency is presented as a function of amplitude frequency for different values of surrounding stiffness, k, with simply supported boundary conditions. It can be seen when the surrounding elastic medium parameter remains constant the nonlinear frequency ratio to linear frequency can increase as the amplitude frequency increases. Moreover, for a CNT with a given value of amplitude frequency, as the stiffness of medium increases, the nonlinear effects of midplane stretching decreases and the nonlinear frequency ratio to linear frequency approaches to 1.

The effect of rippling deformation without midplane stretching ($\alpha_2 = 0$) on the nonlinear frequency ratio to linear frequency of simply supported SWCNT is shown in Figure 2.18. The effect of rippling deformation on the frequency is obvious, especially at large amplitude frequency. On the contrary, the effect is zero for large surrounding elastic medium parameter. It can be found that increasing the amplitude frequency increases the effect of rippling deformation on nonlinear frequency of SWCNT. When the amplitude frequency remains constant, it can be found that increasing the surrounding stiffness (k) decreases the effect of rippling deformation on nonlinear frequency of SWCNT.

Figure 2.19 shows the effects of rippling deformation and midplane stretching on the nonlinear frequency ratio to linear frequency of simply supported SWCNT with different Winkler constant (k) and amplitude frequency. In this figure, it can be seen that by increasing the amplitude frequency, the rippling formation, and midplane stretching affects on the vibration of the SWCNT increases but when the

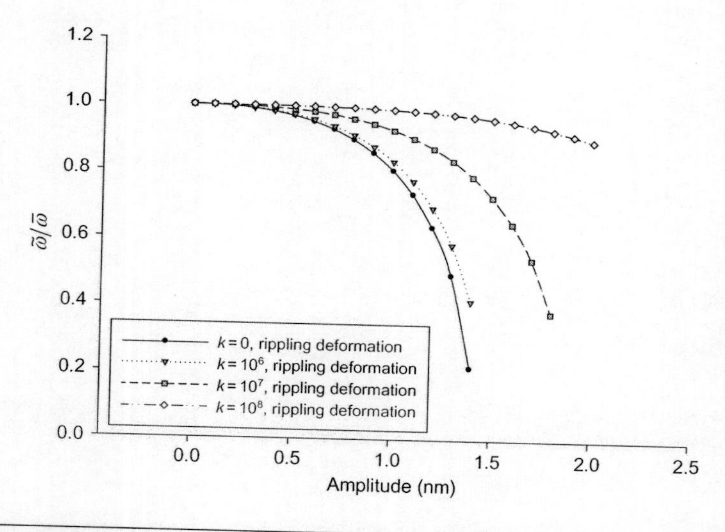

FIGURE 2.18

The effects of surrounding stiffness k on the nonlinear amplitude frequency response curves of SWCNT base rippling deformation.

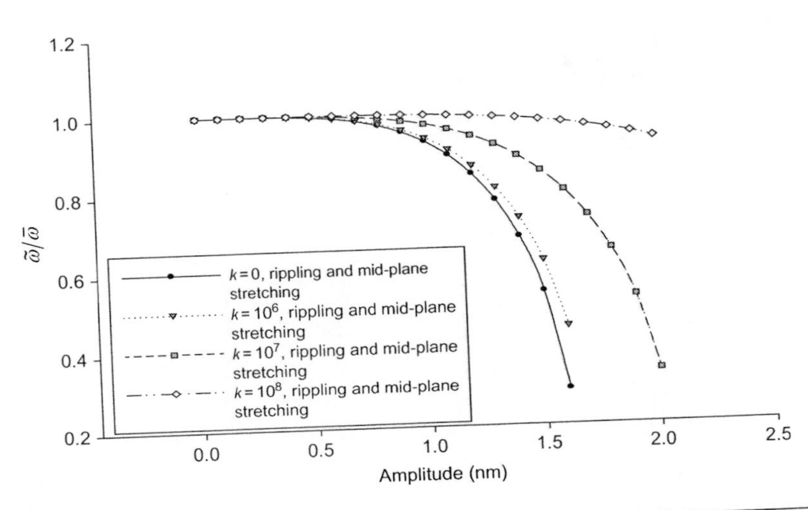

FIGURE 2.19

The effects of surrounding stiffness k on the nonlinear amplitude frequency response curves of SWCNT base rippling deformation and midplane stretching.

amplitude frequency remains constant, it can be found that by increasing the surrounding stiffness (k), the rippling formation and midplane stretching affects on the nonlinear frequency decreases.

The effect of midplane stretching nonlinearity and surrounding stiffness (k) on the nonlinear frequency ratio to linear frequency of simply supported SWCNT is shown in Figure 2.20. It is seen that by decreasing the amplitude frequency and increasing the surrounding stiffness (k), the effect of rippling deformation on nonlinear frequency decreases. In addition, in the same amplitude frequency and surrounding stiffness (k), the midplane stretching nonlinearity can decrease the effect of rippling deformation of the nonlinear frequency.

Rippling formation may cause instability on CNT vibration under certain conditions, especially in large deformations. When the nonlinear frequency takes nonpositive values, the model shows rippling instability thus the instability threshold will be $\widetilde{\omega}=0$. Consequently, the effect of variables influences on rippling the instability phenomenon can be calculated from Equation (2.66). In the instability threshold ($\widetilde{\omega}=0$), the variation of amplitude frequency against surrounding stiffness is shown in Figure 2.21. From Figure 2.21, it can be found that rippling instability occurs in larger values of amplitude frequency as the surrounding stiffness rises. In other words, as the foundation of vibrating CNT becomes stiffer, the instability due to rippling happens in CNTs with larger amplitude frequency. In this figure, we discussed the effect of midplane stretching nonlinearity on rippling instability. When we considered the midplane stretching nonlinearity ($\alpha_2 \neq 0$) in Equation (2.65), it can be seen that rippling instability occurs in larger values of amplitude frequency in the same surrounding stiffness. So, the midplane stretching nonlinearity can increase the instability threshold ($\widetilde{\omega}=0$).

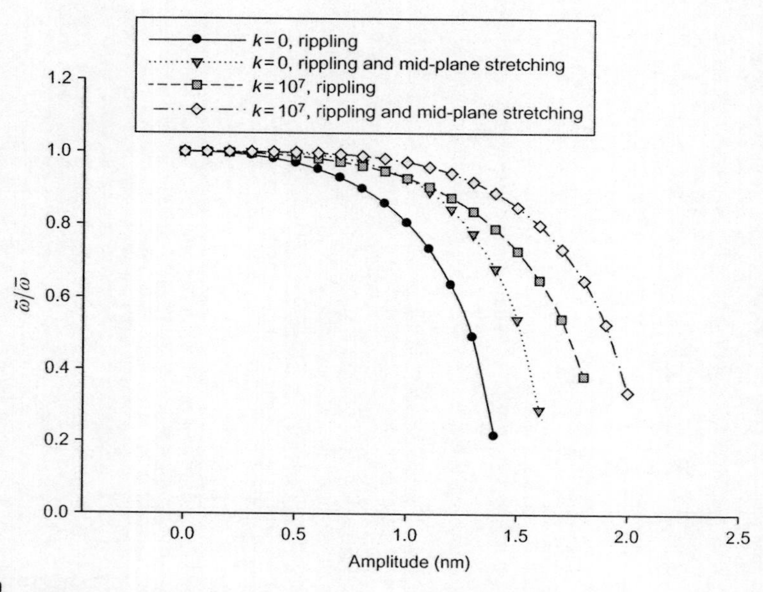

FIGURE 2.20

The effects of surrounding stiffness k and midplane stretching on the nonlinear amplitude frequency response curves of SWCNT base rippling deformation.

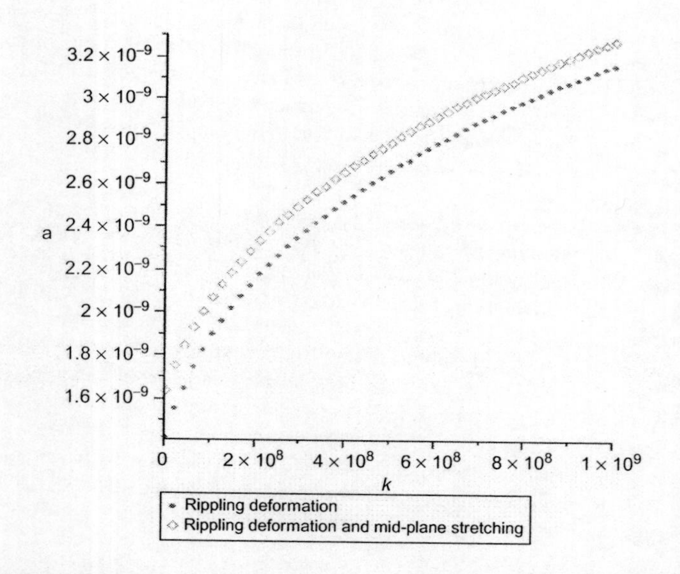

FIGURE 2.21

The effects of surrounding stiffness k and midplane stretching on the stability threshold ($\tilde{\omega} = 0$) of CNT due to rippling deformation.

2.4.5 CONCLUSION

Based on rippling deformation and midplane stretching, a nonlinear elastic beam model was presented for transverse vibration of a SWCNT embedded in an elastic foundation. Based on the analysis, it was observed that the influences of midplane stretching nonlinearity, elastic constant, and amplitude frequency on nonlinear frequency of the simply supported SWCNT with rippling deformation is significant. In addition, the rippling instability has been introduced as a geometrical instability. The results indicate that

(1) By increasing the amplitude frequency, the effect of rippling deformation on nonlinear frequency of SWCNT increases.
(2) By increasing the surrounding stiffness (k), the effect of rippling deformation and midplane stretching on nonlinear frequency of SWCNT decreases.
(3) Rippling instability occurs in larger values of amplitude frequency as the surrounding stiffness rises.
(4) The midplane stretching nonlinearity increases the rippling instability threshold ($\widetilde{\omega}=0$).

2.5 CURVED SWCNT

Continuum mechanics and an elastic beam model were employed in the nonlinear force vibrational analysis of an embedded, curved SWCNT. The analysis considered the effects of the curvature or waviness and midplane stretching of the nanotube on the nonlinear frequency. By utilizing HEBM, the relationships of the nonlinear amplitude and frequency were expressed for a curved, SWCNT. The amplitude frequency response curves of the nonlinear free vibration were obtained for a curved, SWCNT embedded in a Pasternak elastic foundation. Finally, the influence of the amplitude of the waviness, midplane stretching nonlinearity, shear foundation modulus, surrounding elastic medium, radius, and length of the curved CNT on the amplitude frequency response characteristics are discussed (*this section has been worked by I. Mehdipour, A. Barari, A. Kimiaeifar, and G. Domairry in nonlinear dynamics team in Mechanical Engineering Department, 2012-2013*).

2.5.1 INTRODUCTION

It is important to have accurate theoretical models for the vibrational behavior of CNTs. The natural frequencies of CNTs play an important role in nanomechanical resonators. According to the literature, controlling experiments at the nanoscale is difficult; MD simulation is limited to systems with a small number of atoms (e.g., $<10^{16}$) and remains time consuming and expensive. Recent literature shows an increased utilization of modeling methods based on elastic continuum mechanics theories for studying the vibration of CNTs and nanodevices (see Gibson et al., 2007; Ranjbartoreh et al., 2007; Wu and Wang, 2008). The continuum modeling approach requires much less computational effort and is much cheaper than MD simulations and experimental verification, respectively.

Beams and cylindrical shells resting on elastic and Pasternak foundations have wide application in modern engineering and pose great technical problems in structural design. As a result, numerous researches have been presented for the calculation and analysis of beams on elastic foundations (see Gibson et al., 2007; Kim et al., 2007; Ranjbartoreh et al., 2007; Wu and Wang, 2008).

Free axisymmetric vibrations of polar orthotropic annular plates of variable thickness resting on a Pasternak-type elastic foundation have been studied on the basis of classical plate theory (Gupta et al., 2008). Gupta et al. (2008) utilized Hamilton's energy principle to derive the governing differential equations of motion. They obtained frequency equations for an annular plate with two different combinations of edge conditions, employing the Chebyshev collocation approach. Kerr (1964) provided an excellent discussion of intensive studies on plates and beams resting on different soil models, which were regarded as soil-structure interaction problems. The Winker model has been studied as a 1-parameter model, in which the soil layer is represented by unconnected, closely spaced elastic springs. In contrast, the Pasternak model has been studied as a 2-parameter model, in which a membrane is assumed that resists only transverse shear deformations and is attached to the layer of the spring (Guller, 2004). In addition, Fwa et al. (1996) developed theoretical solutions for the analysis of load-induced concrete pavement deflections and bending stresses, as well as for thermal stresses caused by the warping of slabs.

In the present study, a nonlinear, elastic beam model with midplane stretching nonlinearity was developed for the transverse vibration of a clamped-clamped (C-C) and curved SWCNT on a Pasternak elastic foundation with a transverse harmonic excitation. The important aims of this study are summarized as follows:

(1) to express the governing equations of motion for a harmonic transverse load of a curved, SWCNT on a Pasternak elastic foundation;
(2) to use a single-mode Galerkin approximation to derive a second-order governing differential equation;
(3) to obtain the amplitude frequency response curves of the nonlinear force vibration for the curved, SWCNT on a Pasternak elastic foundation by a very effective and convenient method, the HEBM (He, 2002); and
(4) to discuss the effects of the amplitude of CNT's waviness, surrounding elastic medium, midplane stretching nonlinearity, shear foundation modulus, diameter, and length of CNT on the amplitude frequency response characteristics.

2.5.2 VIBRATIONAL MODEL

Figure 2.22 shows a schematic diagram of a CNT embedded in a Pasternak elastic foundation. The CNT is considered as a hollow, curved, cylindrical tube that has a sinusoidal curvature with a small rise function described by $Z = e.\sin(\pi x/L)$ (see Mayoof and Hawwa, 2009), where e is the amplitude of its waviness. Assume that the transverse displacement is $w(x, t)$, in terms of the spatial coordinate x and the time variable t.

The governing equation of motion for a harmonic transverse load of a curved, SWCNT on a Pasternak elastic foundation can be expressed as:

$$EI\frac{\partial^4 w}{\partial x^4} - k_p\frac{\partial^2 w}{\partial x^2} + \rho A\frac{\partial^2 w}{\partial t^2} + kw = F\cos(\tilde{\omega}t) + \frac{EA}{L}\int_0^l\left[\frac{\partial Z}{\partial x}\frac{\partial w}{\partial x} + \frac{1}{2}\left(\frac{\partial w}{\partial x}\right)^2\right]\mathrm{d}x\left(\frac{\partial^2 w}{\partial x^2} + \frac{\partial^2 Z}{\partial x^2}\right) \tag{2.67}$$

where L is the length of the CNT, A is the cross-sectional area, I is the cross-sectional inertial moment, E is the Young's modulus, ρ is the density, F is the spatial distribution of the transverse load, $\tilde{\omega}$ is the

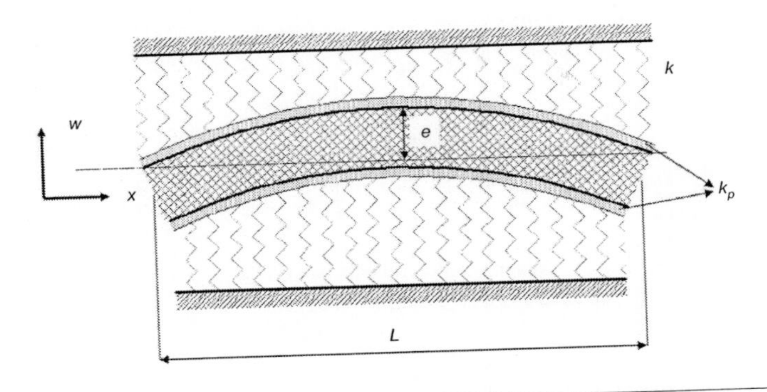

FIGURE 2.22

Model of an embedded curved carbon nanotube in an elastic Pasternak-type foundation.

nonlinear resonance frequency, k_p is the shear foundation modulus, and k is the elastic foundation modulus.

The governing equation of motion (2.67) is subject to the following boundary conditions:

$$w(0,t)=\frac{\partial w(0,t)}{\partial x}=0,\quad w(l,t)=\frac{\partial w(l,t)}{\partial x}=0 \tag{2.68}$$

Assume that the CNT is C-C at the two ends. To perform a separation of variables analysis, the transverse displacement can be written as:

$$w(x,t)=T(t)W(x) \tag{2.69}$$

The basis function $W(x)$ is assumed to take the fundamental mode shape of the linear vibration version of the problem defined by Equations (2.67) and (2.68). So, according to the boundary conditions, the unknown function $W(x)$ is given as:

$$W(x)=\sqrt{\frac{2}{3}}\left[1-\cos\left(\frac{2\pi x}{L}\right)\right] \tag{2.70}$$

It is clear that this assumption satisfies the boundary conditions. Substituting this shape function $W(x)$ and the curvature equation $Z(x)$ into Equation (2.67), multiplying the result by $W(x)$, and then integrating over the domain $[0-L]$ leads to the following equation:

$$\frac{d^2T(t)}{dt^2}+\alpha_1 T(t)+\alpha_2 T(t)^2+\alpha_3 T(t)^3=\sqrt{\frac{2}{3}}\frac{F}{\rho A}\cos(\tilde{\omega}t) \tag{2.71}$$

where

$$\alpha_1=\frac{4}{3}\frac{\pi^2 k_p}{\rho A L^2}+\frac{128}{27}\frac{\pi^2 E e^2}{\rho L^4}+\frac{k}{\rho A}+\frac{16}{3}\frac{\pi^4 EI}{\rho A L^4} \tag{2.72}$$

$$\alpha_2 = \frac{16\sqrt{6}\pi^3 Ee}{9 \quad \rho L^4} \tag{2.73}$$

and

$$\alpha_3 = \frac{8\pi^4 E}{9\rho L^4} \tag{2.74}$$

In Equation (2.71), the parameter α_2 shows the curved shape (geometry) of the CNT, and α_3 considers the influence of midplane stretching on the vibrational equation of motion for a curved CNT embedded in elastic, Pasternak-type foundation.

2.5.3 SOLUTION METHODOLOGY

In the present analysis, a very effective and convenient method (HEBM) is employed to seek the analytical solution of Equation (2.71). HEBM can be applied to nonlinear vibrations and oscillations. This method does not require linearization or a small parameter like the normal perturbation technique. Finally, the results will reveal the simplicity of the method.

According to Equation (2.71), its variational principle can be easily obtained as follows (see Mehdipour et al., 2010):

$$J(T(t)) = \int_0^t \left(-\frac{1}{2}\dot{T}(t)^2 + \alpha_1 \frac{T(t)^2}{2} + \frac{\alpha_2}{3}T(t)^3 + \frac{\alpha_3}{4}T(t)^4 - \sqrt{\frac{2}{3\rho A}}F\cos(\tilde{\omega}t)T(t) \right) dt \tag{2.75}$$

Its Hamiltonian, therefore, can be written in the form:

$$H = \frac{1}{2}\dot{T}(t)^2 + \alpha_1 \frac{T(t)^2}{2} + \frac{\alpha_2}{3}T(t)^3 + \frac{\alpha_3}{4}T(t)^4 - \sqrt{\frac{2}{3\rho A}}F\cos(\tilde{\omega}t)T(t) = \alpha_1\frac{a^2}{2} + \frac{\alpha_2}{3}a^3 + \frac{\alpha_3}{4}a^4 - \sqrt{\frac{2}{3\rho A}}F\cos(\tilde{\omega}t)a \tag{2.76}$$

or

$$R(t) = \frac{1}{2}\dot{T}(t)^2 + \alpha_1 \frac{T(t)^2}{2} + \frac{\alpha_2}{3}T(t)^3 + \frac{\alpha_3}{4}T(t)^4 - \sqrt{\frac{2}{3\rho A}}F\cos(\tilde{\omega}t)T(t) - \alpha_1\frac{a^2}{2} - \frac{\alpha_2}{3}a^3 - \frac{\alpha_3}{4}a^4 + \sqrt{\frac{2}{3\rho A}}F\cos(\tilde{\omega}t)a = 0 \tag{2.77}$$

Oscillatory systems contain two important physical parameters, i.e., the frequency $\tilde{\omega}$ and the amplitude of oscillation, a. Therefore, the following initial conditions are considered:

$$T(0) = a, \quad \dot{T}(0) = 0 \tag{2.78}$$

It is assumed that the initial approximate guess can be expressed as:

$$T(t) = a\cos\tilde{\omega}t \tag{2.79}$$

Substituting Equation (2.79) into Equation (2.77) yields

$$R(t)=\frac{1}{2}(a\sin\tilde{\omega}t)^2\tilde{\omega}^2+\alpha_1\frac{(a\cos\tilde{\omega}t)^2}{2}+\frac{\alpha_2}{3}(a\cos\tilde{\omega}t)^3+\frac{\alpha_3}{4}(a\cos\tilde{\omega}t)^4$$
$$-\sqrt{\frac{2}{3}}\frac{F}{\rho A}\cos(\tilde{\omega}t)(a\cos\tilde{\omega}t)-\alpha_1\frac{a^2}{2}-\frac{\alpha_2}{3}a^3-\frac{\alpha_3}{4}a^4+\sqrt{\frac{2}{3}}\frac{F}{\rho A}\cos(\tilde{\omega}t)a=0 \tag{2.80}$$

By collecting at $\tilde{\omega}t=(\pi/4)$, the following expression is obtained:

$$\tilde{\omega}=\sqrt{\alpha_1+\left(\frac{4}{3}-\frac{\sqrt{2}}{3}\right)\alpha_2 a+\frac{3}{4}\alpha_3 a^2+\left(\frac{2\sqrt{6}}{3}-\frac{4\sqrt{3}}{3}\right)\frac{F}{\rho A a}} \tag{2.81}$$

Substituting Equations (2.72)–(2.74) into Equation (2.81) results in:

$$\tilde{\omega}=\sqrt{\left(\left(\frac{4}{3}\frac{\pi^2 k_p}{\rho A L^2}+\frac{128}{27}\frac{\pi^2 E e^2}{\rho L^4}+\frac{k}{\rho A}+\frac{16}{3}\frac{\pi^4 EI}{\rho A L^4}\right)+\left(\frac{4}{3}-\frac{\sqrt{2}}{3}\right)\left(\frac{16\sqrt{6}\pi^3 E e}{9\rho L^4}\right)a+\frac{3}{4}\left(\frac{8\pi^4 E}{9\rho L^4}\right)a^2+\left(\frac{2\sqrt{6}}{3}-\frac{4\sqrt{3}}{3}\right)\frac{F}{\rho A a}\right)} \tag{2.82}$$

2.5.4 NUMERICAL RESULTS AND DISCUSSION

The linear free vibrational frequency is assumed to be $\overline{\omega}$ in Equation (2.71), and $\overline{\omega}^2=\alpha_1$. The parameters of the material and geometry are taken as $E=3.3$ TPa, $\rho=2.27\times10^3$ kg/m^3, $L=60$ nm, outside radius (r_{out})$=0.8$ nm, and inside radius (r_{in})$=0.7$ nm (effective thickness [h] of SWCNTs is: $h=r_{out}-r_{in}=0.1$ nm) (Batra and Gupta, 2008).

Figure 2.23 shows the results that were obtained with the present theoretical approach for the influence of the midplane stretching nonlinearity and the amplitude of CNT's waviness on the amplitude frequency response curves for the force vibration of a CNT embedded in an elastic Pasternak-type medium. The effects of the combination of waviness and stretching nonlinearity on the nonlinear frequency of the force vibration of the CNT were larger than the effects of waviness or stretching nonlinearity alone. Increasing the amplitude increased the effect of nonlinearity on the amplitude frequency response curves of the SWCNT.

Figure 2.24 compares the amplitude frequency response curves between straight and curved CNTs. For $e=0$, the vibrational behavior of the CNT was similar to that of a straight beam. The nonlinear frequency increased relative to the linear frequency, due to an increase in the amplitude of curvature (e). In other words, when e was increased, the effects of the curvature or waviness of the CNT on nonlinear frequency were increased.

Figure 2.25 depicts the effects of the amplitude of the CNT's waviness on the nonlinear frequency relative to the linear frequency of a C-C SWCNT, with different Winkler constants (k) and amplitude frequencies. When the amplitude frequency was increased, increased effects of the amplitude of CNT's waviness and midplane stretching were seen on the vibration of the SWCNT. However, when the amplitude frequency remained constant and the value of k was increased, decreased effects of the amplitude of CNT's waviness and midplane stretching were seen on the nonlinear frequency. In addition, a

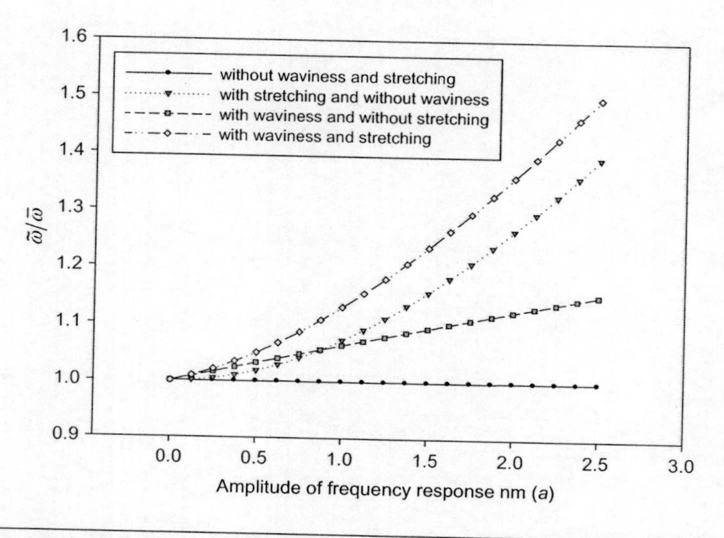

FIGURE 2.23

Effects of waviness and midplane stretching on nonlinear frequency when $k=10^7$, $k_p=10^{-8}$, and $F=1$ pN.

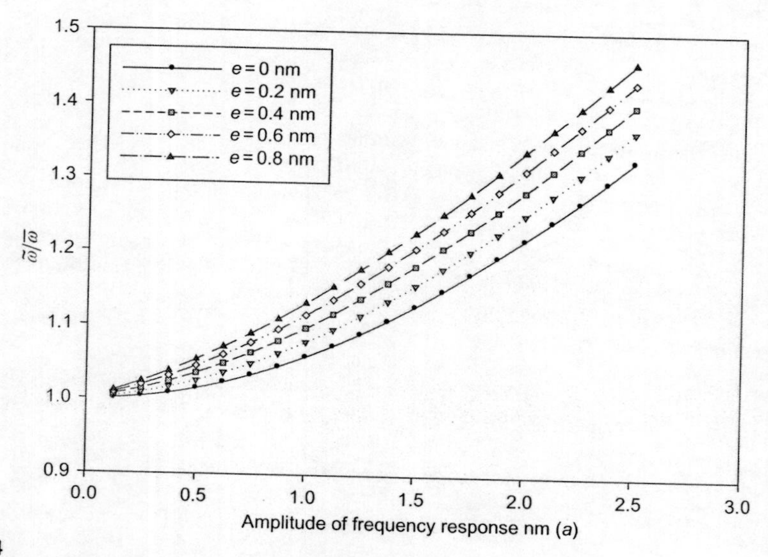

FIGURE 2.24

Effects of amplitude of CNT's waviness on nonlinear frequency when $k=10^7$, $k_p=10^{-8}$, and $F=1$ pN.

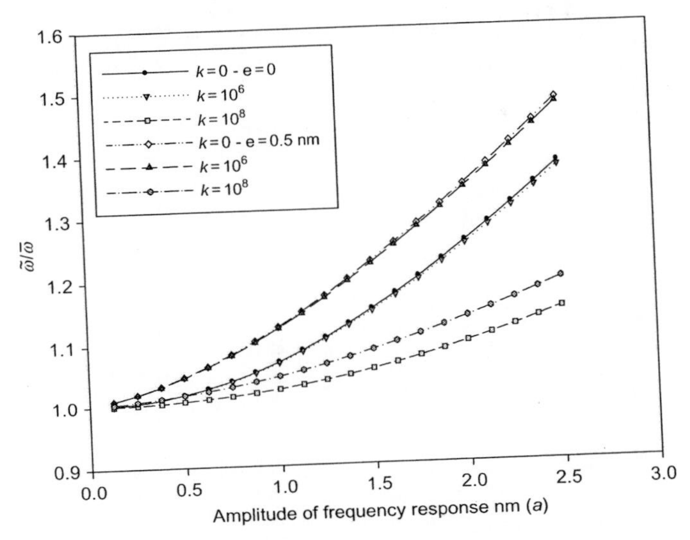

FIGURE 2.25

Effects of amplitude of CNT's waviness and elastic medium on nonlinear frequency when $k_p = 10^{-8}$ and $F = 1$ pN.

comparison between straight ($e = 0$) and curved ($e \neq 0$) CNTs for large values of k showed that the effects of the surrounding elastic medium on the nonlinear frequency of the straight CNT were larger than those of the curved CNT. Therefore, for the same value of k, the graph shows a higher curve for curved CNTs relative to straight CNTs. It can be concluded that the cause of this difference is the amplitude of CNT's waviness.

Figure 2.26 shows the influence of the amplitude of CNT's waviness and midplane stretching on the nonlinear frequency relative to the linear frequency of the C-C SWCNT, with different values of the shear foundation modulus (k_p) and amplitude frequency. When the amplitude frequency was increased, increased effects of the amplitude of CNT's waviness and midplane stretching were seen on the vibration of the SWCNT. However, when the amplitude frequency remained constant and the value of k_p was increased, decreased effects of the amplitude of CNT's waviness and midplane stretching were seen on the nonlinear frequency. Additionally, for small values of k_p, the effects of the shear (Pasternak) foundation on the nonlinear frequency of the straight CNT were larger than those of the curved CNT. Thus, for the same value of k_p, the graph shows a higher curve for the curved CNT relative to the straight CNT. The cause of this difference is the amplitude of CNT's waviness. According to the graph, the difference between the curves for curved and straight CNTs decreases as the value of k_p increases.

Figure 2.27 shows the influence of the outer radius on the effects of the amplitude of CNT's waviness and midplane stretching on the nonlinear frequency of the C-C SWCNT. When the radius of CNT was increased, the effects of the amplitude of CNT's waviness and midplane stretching were decreased, because increasing the radius makes the nanotube stiffer. When the outer radius of CNT was small, the

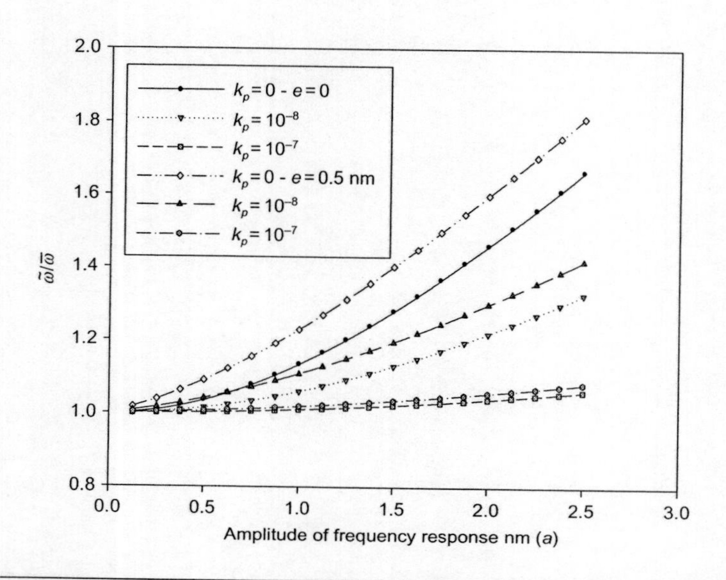

FIGURE 2.26

Effects of amplitude of CNT's waviness and Pasternak medium on nonlinear frequency when $k=10^7$ and $F=1$ pN.

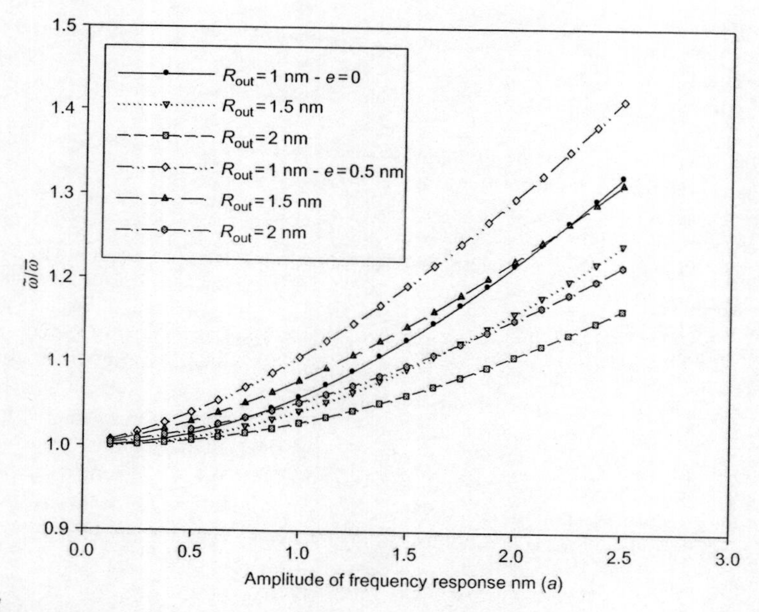

FIGURE 2.27

Effects of amplitude of CNT's waviness and radius on nonlinear frequency when $k=10^7$, $k_p=10^{-8}$, and $F=1$ pN.

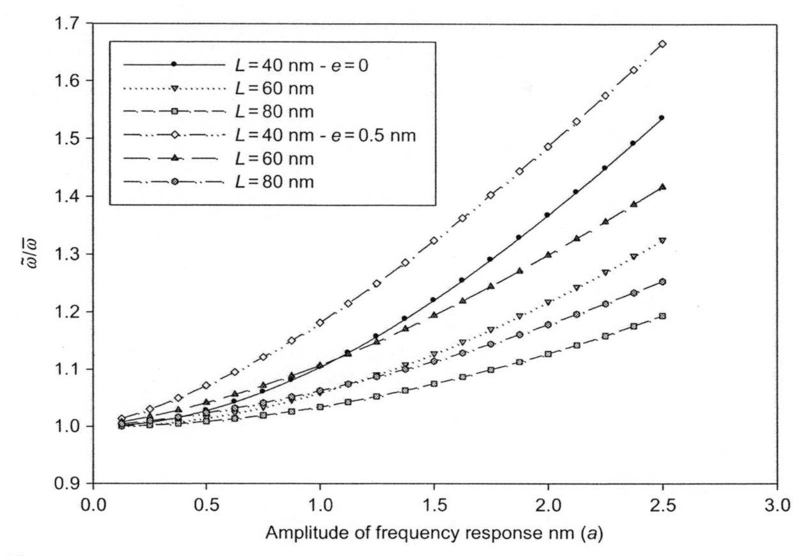

FIGURE 2.28

Effects of amplitude of CNT's waviness and length of CNT on nonlinear frequency when $k=10^7$, $k_p=10^{-8}$, and $F=1$ pN.

nonlinear frequency of the curved CNT was larger than that of a straight CNT when the amplitude frequency (a) was constant, because a CNT with a small radius is relatively soft.

The length of the curved CNT also affects the nonlinear frequency of the C-C curved CNT with midplane stretching nonlinearity on an elastic Pasternak-type foundation (Figure 2.28). When the length of the curved CNT was increased, the ratio of the nonlinear frequency to the linear frequency for a curved CNT was greater than that obtained for a straight CNT. In other words, the comparison between the curved and straight CNTs shows that, for the same length of CNT, the effect of midplane stretching nonlinearity on the nonlinear frequency of the curved CNT is more than that of the straight CNT. According to this result, it was not unexpected that a curved CNT with a large length had a vibrational behavior that was close to that of a straight CNT. In other words, for the same amplitude frequency response (a) and a large length, the vibrational behavior of a curved CNT is close to that of a straight CNT.

2.5.5 CONCLUSION

Based on the midplane stretching nonlinearity, a nonlinear elastic beam model was presented for the transverse vibration of a curved SWCNT embedded in a Pasternak elastic foundation. The analysis indicated that the midplane stretching nonlinearity, elastic constant, shear foundation modulus, amplitude of CNT's waviness, amplitude frequency, and CNT length and radius significantly influence the nonlinear frequency of the C-C SWCNT with nonlinear curvature. The results indicate that:

(1) The combination effects of waviness and stretching nonlinearity on the nonlinear frequency of the SWCNT were larger than the effects of waviness or stretching nonlinearity alone.
(2) Increasing the amplitude of curvature increased the effects of the curvature and waviness of the CNT on the nonlinear frequency.
(3) Increasing the Winkler constant (k) decreased the effects of the midplane stretching nonlinearity on the nonlinear frequency of the curved SWCNT.
(4) Increasing the shear foundation modulus (k_p) decreased the effects of midplane stretching nonlinearity on the nonlinear frequency of the curved SWCNT.
(5) For a CNT with a small outer radius, the nonlinear frequency of a curved CNT was larger than that of a straight CNT.
(6) A curved CNT with a large length had vibrational behavior that was close to that of a straight CNT.

2.6 CNT WITH RIPPLING DEFORMATIONS

Based on the rippling deformation, a nonlinear beam model is developed for transverse vibration of SWCNTs on elastic foundation. The nonlinear natural frequency has been calculated for typical boundary conditions using the perturbation method of multiscales. It is shown that the nonlinear resonant frequency due to rippling is the function of the stiffness of the foundation, the boundary conditions, the excitation load-to-damping ratio and the diameter-to-length ratio. Moreover, the rippling instability of CNTs and the parameters influence on it are briefly discussed (*this chapter has been worked by P. Soltani, D.D. Ganji, I. Mehdipour, A. Farshidianfar in nonlinear dynamics team in Mechanical Engineering Department, 2012-2013*).

2.6.1 INTRODUCTION

As mentioned in the previous section, the effective Young modulus of a nanotube may be determined indirectly from its measured natural frequencies or mode shapes if a sufficiently precise theoretical model is used. The high elastic modulus of CNT (higher than 1 TPa) and remarkable bending flexibility (up to 20%) without breaking, exhibit new phenomena in bending vibration of nanotubes, called rippling. The rippling affects directly on the resonant frequencies and elastic modulus of CNTs. Liu et al. (2001) determined the Young's modulus of CNTs by measuring resonance frequency and using the modulus-frequency relation resulting from the linear vibration theory. They showed that the Young's modulus of CNT decreases sharply, from about 1 to 0.1 TPa with the diameter D increasing from 8 to 40 nanometers, corresponding to unusual bending mode called rippling mode. Furthermore, the investigators confirmed that the result from linear bending theory might be invalid in such measurements. Nonlinear generalized local quasi-continuum method has been used to investigate the effective bending, torsional stiffness of MWCNTs with rippling deformations by Arroyo and Belytschko (2003).

In recent years, a lot of researches have been devoted to the application of CNTs as a reinforced phase in nanocomposites, therefore, considerable attention is focused on mechanical and vibrational behaviors of CNTs embedded in an elastic matrix such as polymer and metal (see Ranjbartoreh et al., 2007; Thostenson et al., 2001). In the present work, a nonlinear elastic beam model is developed for transverse vibration of a SWCNT on elastic foundation on the basis of rippling deformation. Following this model, the nonlinear fundamental frequency is derived using the perturbation method of multiscales for the case of typical boundary conditions and the influences of the stiffness of the foundation, boundary conditions, the excitation load-to-damping ratio, and the diameter-to-length ratio on it are discussed. In addition, the instability caused by rippling deformation is introduced and the conditions that cause happening this kind of instability are determined.

2.6.2 VIBRATION MODEL

The equation of vibration of a SWCNT on an elastic foundation with a harmonic excitation force $F(x, t)$ can be expressed as:

$$M''(x,t) + k_e w(x,t) + 2\mu \dot{w}(x,t) + \rho A \ddot{w}(x,t) = F(x,t) \tag{2.83a}$$

$$F(x,t) = G(x)\cos(\tilde{\omega} t) \tag{2.83b}$$

where $M(x, t)$ denotes the bending moment, $w(x, t)$ is the beam deflection function, k_e is the Winkler constant determined by the material constants of the surrounding elastic medium, μ is the damping coefficient, $F(x, t)$ *is* the excitation load measured per unit length, and $w'(x,t)$ and $\dot{w}(x,t)$ are the partial derivatives $\partial w(x,t)/\partial x$ and $\partial w(x,t)/\partial t$, respectively.

2.6.2.1 Boundary conditions

In this research typical boundary conditions such as C-C, cantilever or clamped-free (C-F), hinged-hinged (H-H), and clamped-hinged (C-H) boundary conditions have been noticed as follows:

C-C boundary conditions:

$$w(0,t) = w'(0,t) = 0, \quad w(L,t) = w'(L,t) = 0 \tag{2.84}$$

C-F boundary conditions:

$$w(0,t) = w'(0,t) = 0, \quad w''(L,t) = w'''(L,t) = 0 \tag{2.85}$$

H-H boundary conditions:

$$w(0,t) = w''(0,t) = 0, \quad w(L,t) = w''(L,t) = 0 \tag{2.86}$$

C-H boundary conditions:

$$w(0,t) = w'(0,t) = 0, \quad w(L,t) = w''(L,t) = 0 \tag{2.87}$$

2.6.2.2 Nonlinear vibration model

When CNT bends, the rippling formation occurs specially for the relatively and locally large deformation. In these cases, the linear relationship between bending moment and curvature of the CNT does not match anymore. The nonlinear effect of rippling mode for a bent CNT has been calculated using FEM simulation (see Wang et al., 2005). Figure 2.29 shows the bending moment M versus the bending curvature κ with the length-to-diameter ratios $L/D = 10$, and 20 from FEM simulating result.

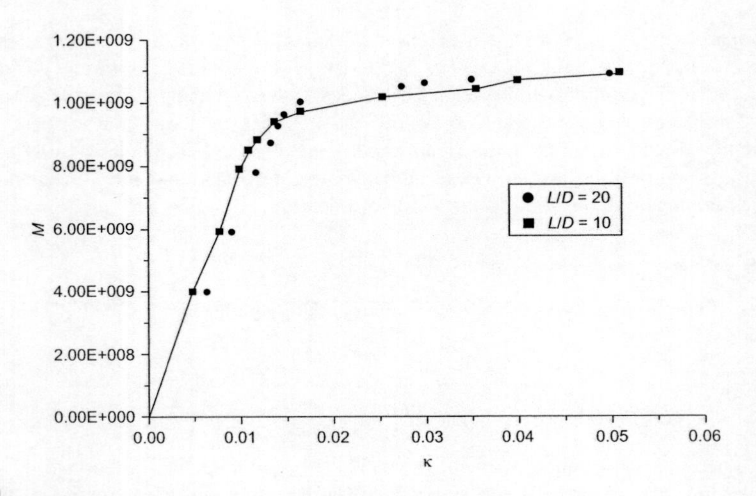

FIGURE 2.29

The bending moment of CNT against the bending curvature for two different *L/D*.

Using beam bending theory and curve fitting technique, a ninth-order polynomial equation can be introduced the behavior of bending moment M against the bending curvature κ.

$$M(x,t) = EI\kappa\left(1 - a_3D^2\kappa^2 + a_5D^4\kappa^4 - a_7D^6\kappa^6 + a_9D^8\kappa^8\right) \tag{2.88}$$

where $\kappa = \kappa(x,t)$, $a_3 = 1.755 \times 10^3$, $a_5 = 2.0122 \times 10^6$, $a_7 = 1.115 \times 10^9$, and $a_9 = 2.266 \times 10^{11}$.

The two orders of partial derivatives to x variable in Equations (2.83a) and (2.83b) gives

$$\begin{aligned} M''(x,t) = EI\big[&\kappa''\left(1 - 3a_3D^2\kappa^2 + 5a_5D^4\kappa^4 - 7a_7D^6\kappa^6 + 9a_9D^8\kappa^8\right) \\ &+ (\kappa')^2\left(-6a_3D^2\kappa + 20a_5D^4\kappa^3 - 42a_7D^6\kappa^6 - 72a_9D^8\kappa^7\right)\big] \end{aligned} \tag{2.89}$$

Based on the large deflection deformation, the relation between the bending curvature $\kappa(x,t)$ and the beam deflection $w(x,t)$ can be expressed as:

$$\begin{aligned} \kappa(x,t) &= \frac{w''(x,t)}{\left[1 + w'(x,t)^2\right]^{3/2}} \\ &= w''\left[1 - \frac{3}{2}(w')^2 + \frac{15}{8}(w')^4 - \cdots\right] \\ &\cong w''(x,t)\left[1 - r(w')^2\right], \quad r = 1.5 \end{aligned} \tag{2.90}$$

where substituting Equation (2.90) into Equation (2.89) yields

$$M''(x,t) = EI\left\{w'''' - 3a_3D^2\left[2w''(w''')^2 + (w'')^2 w''''\right] - r\left[2(w'')^3 + 6w'w''w''' + (w')^2 w''''\right]\right\} \tag{2.91}$$

The nonlinear vibration equation will be obtained by substituting Equation (2.91) into Equation (2.83):

$$EIw'''' + k_e w + 2\mu\dot{w} + \rho A\ddot{w} = EIN(w) + F(x,t) \tag{2.92a}$$

where $N(w)$ denotes for the nonlinear part of equation as follows:

$$N(w) = 3a_3D^2\left[2w''(w''')^2 + (w'')^2 w''''\right] + r\left[2(w'')^3 + 6w'w''w''' + (w')^2 w''''\right] \tag{2.92b}$$

Making all the variables in Equations (2.92a) and (2.92b) dimensionless by using the characteristic length L, time $L^2\sqrt{\rho A/EI}$, and force EI/L^3 gives

$$\overline{w}'''' + \overline{k}_e\overline{w} + 2\overline{\mu}\dot{\overline{w}} + \ddot{\overline{w}} = \overline{N} + \overline{F}(x^*, t^*) \tag{2.93a}$$

$$\overline{F}(x^*, t^*) = \overline{G}(x^*)\cos(\widetilde{\omega}^* t^*) \tag{2.93b}$$

where

$$x^* = x/L,\ \ t^* = t\sqrt{EI/\rho A}/L^2,\ \ \overline{w} = w/L,\ \ \overline{k}_e = k_e L^4/EI, \overline{\mu} = \mu L^2/\sqrt{EI\rho A},\ \ \overline{N} = N/L^3,\ \ \overline{G} = GL^3/EI \tag{2.94}$$

The associated dimensionless boundary conditions handled in this paper are given as follows:

C-C boundary conditions:

$$\overline{w}(0,t^*) = \overline{w}'(0,t^*) = 0,\ \ \overline{w}(1,t^*) = \overline{w}'(1,t^*) = 0 \tag{2.95}$$

C-F boundary conditions:

$$\overline{w}(0,t^*) = \overline{w}'(0,t^*) = 0,\ \ \overline{w}''(1,t^*) = \overline{w}'''(1,t^*) = 0 \tag{2.96}$$

H-H boundary conditions:

$$\overline{w}(0,t^*) = \overline{w}''(0,t^*) = 0,\ \ \overline{w}(1,t^*) = \overline{w}''(1,t^*) = 0 \tag{2.97}$$

C-H boundary conditions:

$$\overline{w}(0,t^*) = \overline{w}'(0,t^*) = 0,\ \ \overline{w}(1,t^*) = \overline{w}''(1,t^*) = 0 \tag{2.98}$$

2.6.2.3 Nonlinear analysis

The fundamental linear frequency of a vibrating CNT corresponding to each previous boundary condition is introduced as ω^*. Pursuant to nonlinearity, the nonlinear frequency $\widetilde{\omega}^*$ deviates slightly from ω^*. The perturbation method of multiscales has been applied to calculate the resonance frequency $\widetilde{\omega}^*$ for a CNT on elastic foundation with rippling deformation. The beam deflection w can be expanded, using small perturbation parameter ε, into $u = u_0 + \varepsilon u$ where u_0 should be zero. To make all terms in

Equations (2.92a) and (2.92b) be of the same order in w, the parameters $\overline{\mu}=\varepsilon^2 \upsilon$ and $\overline{G}(x^*)=\varepsilon^3 g(x^*)$ are determined and the nonlinear Equations (2.92a) and (2.92b) becomes (see Wang et al., 2005):

$$u'''' + \overline{k}_e u + 2\varepsilon^2 \upsilon \dot{u} + \ddot{u} = \varepsilon^2 N_u + \varepsilon^2 g\cos(\widetilde{\omega}^* t^*) \tag{2.99}$$

where

$$N_u = 3a_3 (D/L)^2 \left[2u''(u''')^2 + (u'')^2 u''''\right] + r\left[2(u'')^3 + 6u'u''u''' + (u')^2 u''''\right] \tag{2.100}$$

In the above formula, $u(x^*, t^*)$ and $g(x^*)$ are, respectively, expanded as:

$$u(x^*, t^*) = \sum_{n=1}^{\infty} q_n(t^*)\varphi_n(x^*), \quad g(x^*) = \sum_{n=1}^{\infty} g_n \varphi_n(x^*) \tag{2.101}$$

where φ_n for $n=1, 2, 3\ldots$ represent the normalized mode functions of the beam from the linear vibration analysis due to the specified boundary condition. Meanwhile the mode function φ_n satisfies the following formula.

$$\int_0^1 \varphi_i(x^*)\varphi_j(x^*)\mathrm{d}x^* = \delta_{ij} \quad i,j = 1,2,\ldots \tag{2.102}$$

and δ_{ij} represents Kronecker delta. Substituting Equation (2.101) into Equation (2.100) and utilizing Equation (2.102), we have

$$\ddot{q}_1 + 2\varepsilon^2 \upsilon \dot{q}_1 + (\omega_1^{*2} + \overline{k_e})q_1 = \varepsilon^2 \alpha q_1^3 + \varepsilon^2 g_1 \cos(\widetilde{\omega}^* t^*) \tag{2.103}$$

According to Equation (2.103), it is clear that fundamental linear frequency of CNT is $\omega^{*2} = \omega_1^* 2 + \overline{k}_e$

Based on linear analysis of the beams, the first order of inherence frequency corresponding to the fundamental mode of vibration ω_1^*, have been given according to Table 2.5.

Equation (2.103) represents a third-order nonlinear oscillation equation (called damping forced Duffing's equation) (see Nyfeh and Mook, 1997), which is yielded by the parameter α. α is the parameter depends on normalized mode functions of the beam, the associated boundary condition and diameter-to-length of CNT. This parameter is calculated for each case as follows:

For C-C condition

$$\beta_1 = \int_0^1 \varphi_1 \left[2\varphi_1''(\varphi_1''')^2 + (\varphi'')^2 \varphi''''_1\right] \mathrm{d}x^* = 1.546729013 \times 10^5 \tag{2.104a}$$

Table 2.5 The First Order of Inherence Frequency Due To Boundary Conditions

Boundary Condition	ω_1^*
C-F	3.516
C-C	22.37328786
C-H	15.41820327
H-H	9.8696044

$$\beta_2 = \int_0^1 \varphi_1 \left[2(\varphi'')^2 + 6\varphi_1'\varphi_1''\varphi_1''' + (\varphi_1')^2\varphi_1''''\right] dx^* = 2846.49998 \tag{2.104b}$$

$$\begin{aligned} \alpha &= 3a_3(D/L)^2\beta_1 + r\beta_2 \\ &\approx 3a_3(D/L)^2 \times 1.546729013 \times 10^5 + r \times 2846.499985 \\ &\approx (28536.86783D/L)^2 + 4269.749978 \end{aligned} \tag{2.105}$$

For C-F condition

$$\beta_1 = \int_0^1 \varphi_1 \left[2\varphi_1''(\varphi_1''')^2 + (\varphi_1'')^2\varphi_1''''\right] dx^* = 119.6 \tag{2.106a}$$

$$\beta_2 = \int_0^1 \varphi_1 \left[2(\varphi_1'')^2 + 6\varphi_1'\varphi_1''\varphi_1''' + (\varphi_1')^2\varphi_1''''\right] dx^* = 20.1939 \tag{2.106b}$$

$$\begin{aligned} \alpha &= 3a_3(D/L)^2\beta_1 + r\beta_2 \\ &\approx 3a_3(D/L)^2 \times 119.6 + r \times 20.1939 \\ &\approx (793.533D/L)^2 + 30.3555 \end{aligned} \tag{2.107}$$

For H-H condition

$$\beta_1 = \int_0^1 \varphi_1 \left[2\varphi_1''(\varphi_1''')^2 + (\varphi_1'')^2\varphi_1''''\right] dx^* = \frac{1}{2}\pi^8 \tag{2.108a}$$

$$\beta_2 = \int_0^1 \varphi_1 \left[2(\varphi_1'')^2 + 6\varphi_1'\varphi_1''\varphi_1''' + (\varphi_1')^2\varphi_1''''\right] dx^* = \frac{1}{2}\pi^6 \tag{2.108b}$$

$$\begin{aligned} \alpha &= 3a_3(D/L)^2\beta_1 + r\beta_2 \\ &\approx 3a_3(D/L)^2 \times \frac{1}{2}\pi^8 + r \times \frac{1}{2}\pi^6 \\ &\approx \left(51.30789413\pi^4 D/L\right)^2 + 0.75\pi^6 \end{aligned} \tag{2.109}$$

For C-H condition

$$\beta_1 = \int_0^1 \varphi_1 \left[2\varphi_1''(\varphi_1''')^2 + (\varphi_1'')^2\varphi_1''''\right] dx^* = 31818.01028 \tag{2.110a}$$

$$\beta_2 = \int_0^1 \varphi_1 \left[2(\varphi_1'')^2 + 6\varphi_1'\varphi_1''\varphi_1''' + (\varphi_1')^2\varphi_1''''\right] dx^* = 1321.677879 \tag{2.110b}$$

$$\begin{aligned} \alpha &= 3a_3(D/L)^2\beta_1 + r\beta_2 \\ &\approx 3a_3(D/L)^2 \times 31818.01028 + r \times 1321.677879 \\ &\approx (12943.02222D/L)^2 + 1982.516818 \end{aligned} \tag{2.111}$$

Moreover, the nonlinear frequency $\widetilde{\omega}^*$ can be expressed by perturbation parameter ε_1 as

$$\widetilde{\omega}^* = \sqrt{\omega_1^{*2} + \bar{k}_e} - \sigma\varepsilon_1 \tag{2.112}$$

The solution for Equation (2.103) and dimensionless excitation force can be, respectively, stated as:

$$q_1(t^*, \varepsilon) = q_{10}(T_0, T_1) + \varepsilon_1 q_{11}(T_0, T_1) + \cdots \tag{2.113a}$$

$$\varepsilon^2 g_1 \cos(\widetilde{\omega}^* t^*) = \varepsilon_1 g_1 \cos\left(\sqrt{\omega_1^{*2} + \bar{k}_e T_0} - \sigma T_1\right) \tag{2.113b}$$

Substituting Equations (2.113a) and (2.113b) into Equation (2.103), and comparing coefficients of the identical power of ε_1, we have

$$(\varepsilon_1^0): D_0^2 q_{10} + \left((\omega_1^*)^2 + \bar{k}_e\right) q_{10} = 0 \tag{2.114a}$$

$$(\varepsilon_1^1): D_0^2 q_{11} + \left((\omega_1^*)^2 + \bar{k}_e\right) q_{11} = -2D_0 D_1 q_{10} - 2\upsilon D_0 q_{10} + \alpha q_{10}^3 + g_1 \cos\pi\left(\sqrt{\omega_1^{*2} + \bar{k}_e T_0} - \sigma T_1\right) \tag{2.114b}$$

where $D_n = \dfrac{\partial}{\partial T_n} (n = 0, 1)$.

The generating solution can be obtained from Equation (2.114a)

$$q_{10} = A(T_1) e^{i\sqrt{\omega_1^{*2} + \bar{k}_e T_0}} + \bar{A}(T_1) e^{-i\sqrt{\omega_1^{*2} + \bar{k}_e T_0}} \tag{2.115}$$

Substituting for q_{10} into Equation 2.114b gives

$$\begin{aligned} D_0^2 q_{11} + \left((\omega_1^*)^2 + \bar{k}_e\right) q_{11} &= \left(-2\left(\left(\frac{d}{dT_1} A(T_1)\right) + \upsilon A(T_1)\right) i\sqrt{\omega_1^{*2} + \bar{k}_e} + 3\alpha A(T_1)^2 \bar{A}(T_1)\right) e^{i\sqrt{\omega_1^{*2} + \bar{k}_e T_0}} \\ &\quad + \alpha A(T_1)^3 e^{3i\sqrt{\omega_1^{*2} + \bar{k}_e T_0}} + \frac{1}{2} g_1 e^{i\left(\sqrt{\omega_1^{*2} + \bar{k}_e T_0 - \sigma T_1}\right)} + cc \end{aligned} \tag{2.116}$$

where cc stands for the complex conjugate of preceding terms. Secular terms will be eliminated from the particular solution of Equation (2.116) if we choose A to be a solution of

$$2\left(\left(\frac{d}{dT_1} A(T_1)\right) + \upsilon A(T_1)\right) i\sqrt{\omega_1^{*2} + \bar{k}_e} - 3\alpha A(T_1)^2 \bar{A}(T_1) - \frac{1}{2}\frac{g_1}{e^{i\sigma T_1}} = 0 \tag{2.117}$$

To solve Equation (2.117), we write A in the polar form

$$A(T_1)=\frac{1}{2}a(T_1)\exp(i\beta(T_1)) \tag{2.118}$$

where a and β are real function of slow time scale T_1. Then by separating, the result into its real and imaginary parts:

$$\begin{aligned} a' &= -\upsilon a-\frac{1}{2}\frac{g_1}{\sqrt{\omega_1^{*2}+\bar{k}_e}}\sin(\gamma(T_1)), \\ a\beta' &= -\frac{3}{8}\frac{\alpha}{\sqrt{\omega_1^{*2}+\bar{k}_e}}a^3+\frac{1}{2}\frac{g_1}{\sqrt{\omega_1^{*2}+\bar{k}_e}}\cos(\gamma(T_1)) \end{aligned} \tag{2.119}$$

where $\gamma(T_1)=\beta(T_1)+\sigma T_1$.

For steady-state response in the neighborhoods of singular points, every small perturbation motion has to decay and this occurs when $a'=\gamma'=0$ therefore

$$\begin{aligned} \upsilon a &= \frac{1}{2}\frac{g_1}{\sqrt{\omega_1^{*2}+\bar{k}_e}}\sin(\gamma), \\ \frac{3}{8}\frac{\alpha}{\sqrt{\omega_1^{*2}+\bar{k}_e}}a^3-a\sigma &= \frac{1}{2}\frac{g_1}{\sqrt{\omega_1^{*2}+\bar{k}_e}}\cos(\gamma) \end{aligned} \tag{2.120}$$

By omitting γ form Equations (2.120), it can be written as:

$$\upsilon^2\left(\omega_1^{*2}+\bar{k}_e\right)a^2+\left(\sigma\sqrt{\omega_1^{*2}+\bar{k}_e}-\frac{3}{8}\alpha a^2\right)a^2=\frac{1}{4}g_1^2 \tag{2.121}$$

In previous equation, the maximum vibration amplitude a and phase angle β will be calculated simply by removing secular term as:

$$a=\frac{g_1}{2\sqrt{\omega_1^{*2}+\bar{k}_e\upsilon}} \tag{2.122}$$

and

$$\sigma\sqrt{\omega_1^{*2}+\bar{k}_e}=\frac{3}{8}\alpha a^2 \tag{2.123}$$

In Equation (2.123), σ represents a variable of maximum vibration amplitude a and dimensionless parameter of Winkler constant $\bar{k}_e$, and is written as:

$$\sigma=\frac{3\alpha}{8\sqrt{\omega_1^{*2}+\bar{k}_e}}a^2=\frac{3\alpha}{8\sqrt{\omega_1^{*2}+\bar{k}_e}}\left(\frac{g_1}{2\sqrt{\omega_1^{*2}+\bar{k}_e\upsilon}}\right)^2=\frac{3\alpha g_1^2}{32\left(\omega_1^{*2}+\bar{k}_e\right)^{3/2}\upsilon^2} \tag{2.124}$$

At last, the nonlinear resonance frequency of the CNT on Winkler-like foundation with rippling effect $\widetilde{\omega}^*$ can be determined as:

$$\widetilde{\omega}^* = \sqrt{\omega_1^{*2} + \overline{k}_e} - \sigma\varepsilon_1 = \sqrt{\omega_1^{*2} + \overline{k}_e} - \frac{3\alpha g_1^2}{32\left(\omega_1^{*2} + \overline{k}_e\right)^{3/2} \upsilon^2} \varepsilon_1 \tag{2.125}$$

where

$$\varepsilon_1 = \varepsilon^2, \quad g_1 = \frac{\overline{G}_1}{\varepsilon^3}, \quad \text{and} \quad \upsilon = \frac{\overline{\mu}}{\varepsilon^2} \tag{2.126}$$

Substituting Equation (2.126) into Equation (2.125).

$$\widetilde{\omega}^* = \sqrt{\omega_1^{*2} + \overline{k}_e} - \frac{3\alpha \overline{G}_1^2}{32\left(\omega_1^{*}2 + \overline{k}_e\right)^{3/2} \overline{\mu}^2} \tag{2.127}$$

In addition, the nonlinear resonant frequency ratio r can be introduced as:

$$r = \frac{\widetilde{\omega}^*}{\omega^*} = 1 - \frac{3\alpha \overline{G}_1^2}{32\left(\omega_1^{*}2 + \overline{k}_e\right)^{2} \overline{\mu}^2} \tag{2.128}$$

2.6.3 RESULTS AND DISCUSSION

Equation (2.128) shows the ratio of nonlinear resonant frequency relative to linear frequency caused by rippling configuration of a SWCNT. Actually, the second term of this equation stands for the nonlinearity of the model so this nonlinearity caused by rippling formation is a function of linear natural frequency $\omega_{1_1}^{*}{}^{*}$, the Winkler constant $\overline{k}_e$, exciting load-to-damping ratio D/L, boundary conditions and diameter-to-length ratio. According to this model and for verification, the resultant nonlinear frequency has been adopted with results from Fu et al. (2006) and Wang et al. (2004) which shows good agreement in the case of long slender CNTs without elastic medium.

In Figure 2.30 the nonlinear frequency ratio r is presented as a function of Winkler constant $\overline{k}_e$ for different values of D/L with C-C boundary condition.

It is seen $\overline{G}_1/\overline{\mu}$ when the excitation load-to-damping ratio remains constant the nonlinear frequency can decrease as the diameter-to-length ratio D/L increases. Moreover, for a CNT with a given value of D/L, as the stiffness of medium increases, the nonlinear effects of rippling formation decreases and the nonlinear frequency ratio r tend to 1. Meanwhile for long slender CNTs with D/L smaller than 0.02, the nonlinear frequency ratio is approximately equal one ($r \cong 1$) for all values of Winkler constants $\overline{k}_e$. In these cases, The nonlinear effects of rippling deformation can be neglected and the geometric nonlinearity of the structure can be out of account.

The effect of the excitation load-to-damping ratio on the nonlinear frequency has been plotted in Figure 2.31 for $\overline{k}_e = 5$. From this figure, it can be seen that the nonlinear frequency ratio r decreases sharply for large values of D/L especially when the excitation load-to-damping ratio $\overline{G}_1/\overline{\mu}$ is almost high. In other word, the nonlinear effect of rippling deformation exaggerates with raising the excitation load amplitude or with the lessening of the damping of the model.

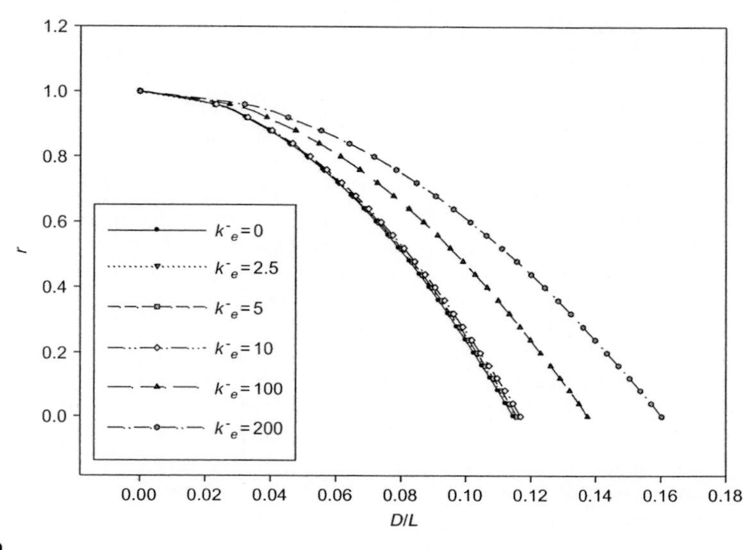

FIGURE 2.30

The effects of Winkler stiffness on the nonlinear frequency ratio in C-C condition when $\overline{G}_1/\overline{\mu}=0.5$.

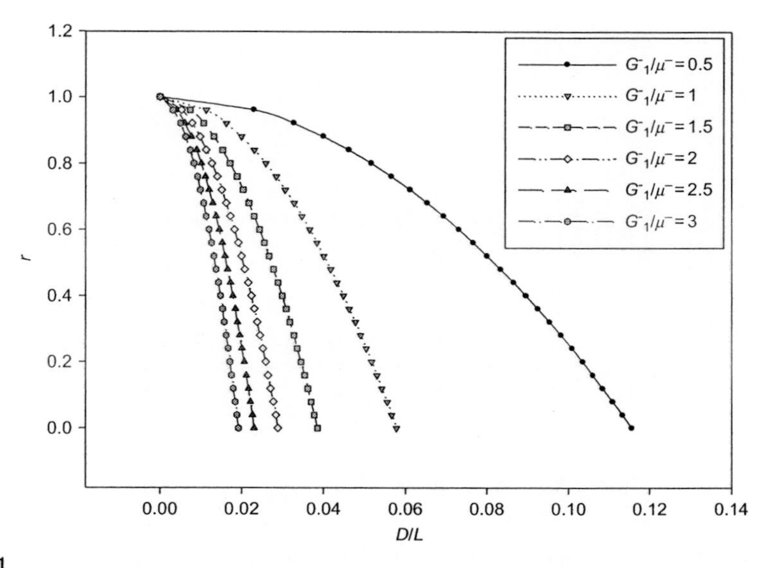

FIGURE 2.31

The effects of excitation load-to-damping ratio on the nonlinear frequency ratio in C-C condition when $\overline{k}_e=5$.

Another important factor affects on the nonlinear resonant frequency of CNT, is boundary conditions. α is the parameter which controls this effect in Equations (2.108a) and (2.108b) and causes changing in the nonlinear frequency ratio r. Figures 2.32–2.34 present the variation of r respect to D/L for four typical boundary conditions mentioned in previous section. In all of these figures, the excitation load-to-damping ratio$\overline{G}_1/\overline{\mu}$ remains equal to 1 while Winkler constant $\overline{k}_e$ is equal to 5, 20, and 30, respectively.

According to these figures, for a CNT with a given diameter to length ratio D/L, by reducing the stiffness of the system caused by the boundary conditions, the nonlinearity effect due to rippling deformation decreases and the ratio r becomes closer to 1. Furthermore, for long slender CNTs with small values of D/L (say), the nonlinear frequency ratio is very close to 1 especially when the stiffness of medium is relatively small (see Figure 2.32). Hence, for long slender CNTs embedded in a soft elastic medium, the rippling nonlinearity can be ignored.

Rippling formation may cause instability on CNT vibration under certain conditions, especially in large deformations. Indeed, in this situation, locally wave-like configurations appear in internal radius of a bent CNT, should diminish bending stiffness to zero and lead the vibrational amplitude to infinity. When nonlinear frequency ratio r takes nonpositive values, the model shows rippling instability thus the instability threshold will be $r=0$. Consequently, the effect of variables influences on rippling the instability phenomenon, can be calculated from Equation (2.128). In the instability threshold ($r=0$), the variation of diameter-to-length D/L against Winkler stiffness and boundary conditions is shown in Figures 2.35–2.37 while load-to-damping ratio $\overline{G}_1/\overline{\mu}$ remains constant in each figure. It is seen that

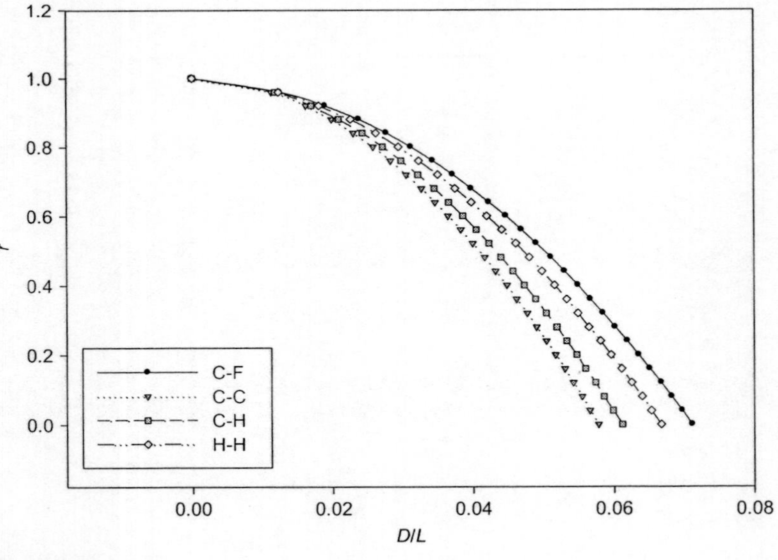

FIGURE 2.32

The effects of boundary conditions on the nonlinear frequency ratio when $\overline{k}_e=5, \overline{G}_1/\overline{\mu}=1$.

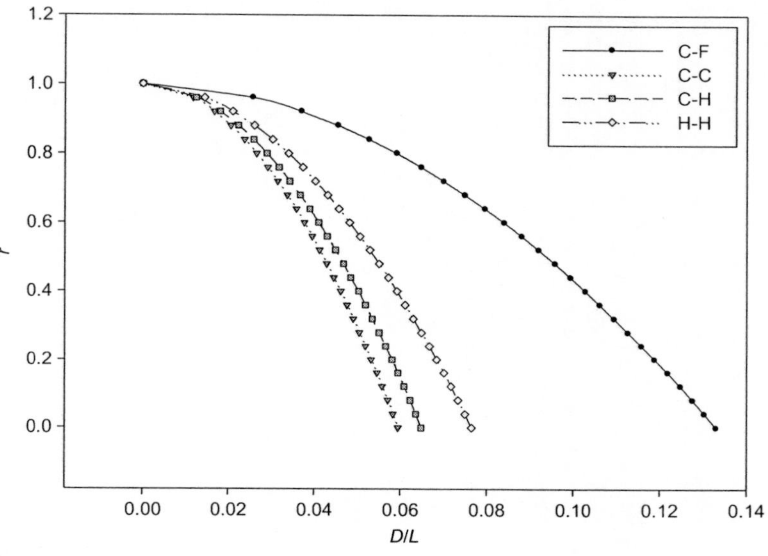

FIGURE 2.33

The effects of boundary conditions on the nonlinear frequency ratio when $\overline{k}_e = 20, \overline{G}_1/\overline{\mu} = 1$.

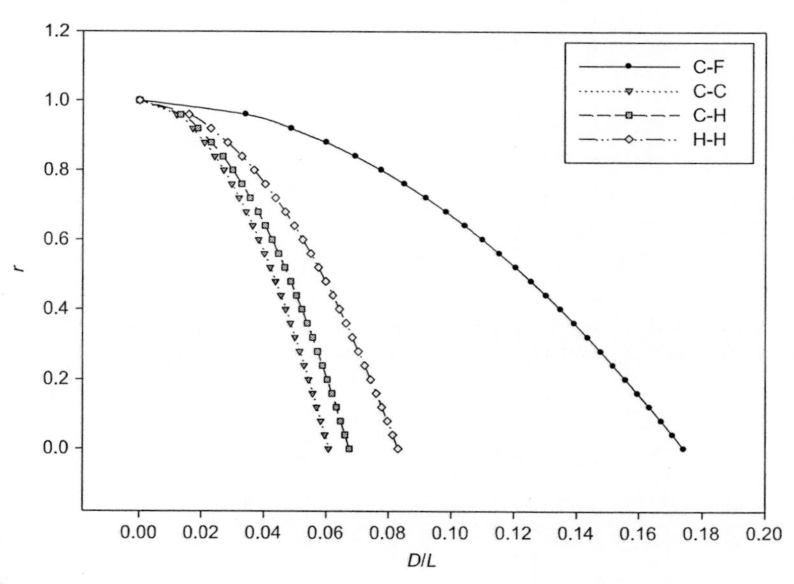

FIGURE 2.34

The effects of boundary conditions on the nonlinear frequency ratio when $\overline{k}_e = 30, \overline{G}_1/\overline{\mu} = 1$.

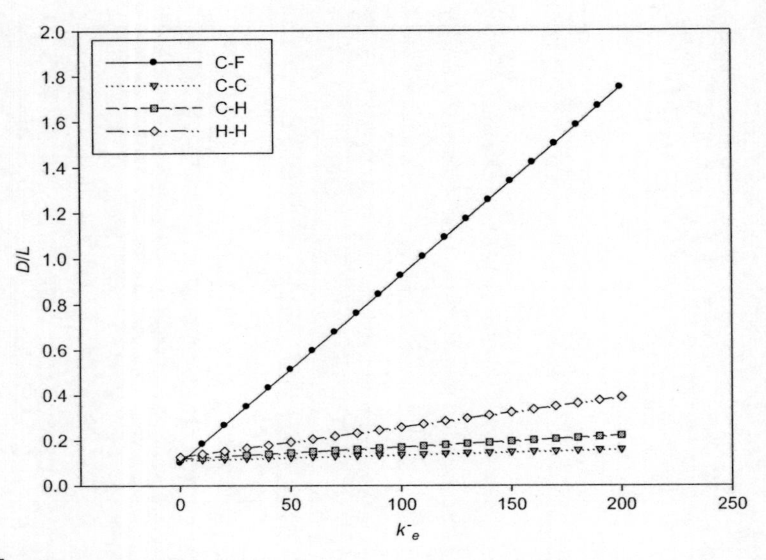

FIGURE 2.35

The effects of Winkler stiffness and boundary condition on the stability of CNT due to rippling when $\overline{G}_1/\overline{\mu}=0.5$.

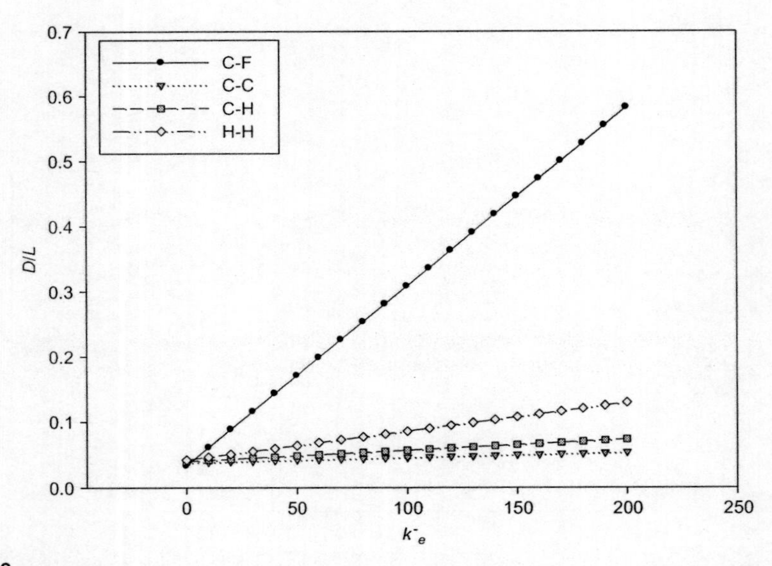

FIGURE 2.36

The effects of Winkler stiffness and boundary condition on the stability of CNT due to rippling when $\overline{G}_1/\overline{\mu}=1.5$.

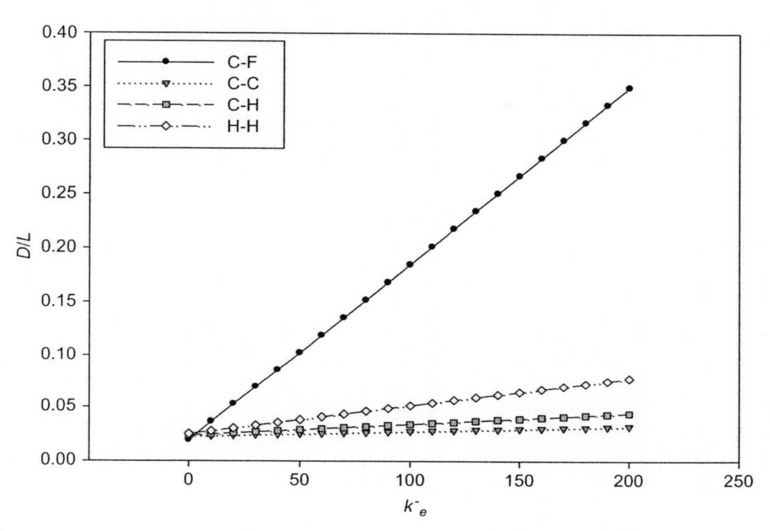

FIGURE 2.37

The effects of Winkler stiffness and boundary condition on the stability of CNT due to rippling when $\overline{G}_1/\overline{\mu}=2.5$.

rippling instability occurs in larger values of D/L as the surrounding stiffness rises. In other words, as the foundation of vibrating CNT becomes stiffer, the instability due to rippling happens in CNTs with larger diameter in a constant length.

Another important factor, which influences on the rippling instability phenomenon, is boundary condition.

As the stiffness due to the boundary conditions decreases from C-C to C-F, rippling instability occurs in larger values of D/L in the same Winkler constant. Mean while, when the stiffness of medium is relatively small, the different between curves decreases and rippling instability happens almost in the same D/L for all kinds of boundary conditions.

For very compliant and soft mediums, instability should occur in the same values of D/L. In addition, as it is seen in these figures, the slopes of C-C curves are near to zero, and in consequence, the effect of Winkler's stiffness on the rippling instability can be ignored while for C-F, the Winkler constant plays an important role on the rippling instability.

2.6.4 CONCLUSION

Based on rippling deformation, a nonlinear elastic beam model was presented for transverse vibration of a SWCNT. The stiffness of medium was incorporated in the formulation to determine the vibration behavior of CNT more precisely. The nonlinear frequency ratio has been solved for C-C, C-H, H-H, and C-F boundary conditions. The stiffness of medium (Winkler's coefficient), boundary conditions, diameter-to-length ratio and the excitation load-to-damping ratio effect on nonlinear frequency directly. Results predicted by present model show that as the Winkler coefficient decreases or the excitation load-to-damping ratio increases the nonlinearity effect due to rippling lifts up.

In addition, the rippling instability has been introduced as a geometrical instability and the results show that as the stiffness of the medium and the stiffness due to boundary conditions decrease the rippling instability happens in smaller values of the diameter-to-length ratio.

REFERENCES

Allen, B.L., Kichambare, P.D., Star, A., 2007. Review: carbon nanotube field-effect-transistor-based biosensors. Adv. Mater. 19 (11), 1439.

Arroyo, M., Belytschko, T., 2003. Nonlinear mechanical response and rippling of thick multiwalled carbon nanotubes. Phys. Rev. Lett. 91 (21), 215505.

Batra, R.C., Gupta, S.S., 2008. Wall thickness and radial breathing modes of single-walled carbon nanotubes. J. Appl. Mech. 75, 0610101–0610106.

Baughman, R.H., Cui, C., Zakhidov, A.A., Iqbal, Z., Barisci, J.N., Spinks, G.M., Wallace, G.G., Mazzoldi, A., De Rossi, D., Rinzler, A.G., Jaschinski, O., Roth, S., Kertesz, M., 1999. Carbon nanotube actuators. Science 284 (5418), 1340.

Bethune, D.S., Kiang, C.H., de Vries, M.S., Gorman, G., Savoy, R., Vazquez, J., Beyers, R., 1993. Cobalt-catalysed growth of carbon nanotubes with single-atomic-layer walls. Nature 363, 605–607.

Boresi, A.P., Schmidt, R.J., 2002. Advanced Mechanics of Materials, sixth ed. John Wiley & Sons Inc., New York.

Chowdhury, R., Adhikari, S., Mitchell, J., 2009. Vibrating carbon nanotube based bio-sensors. Physica E 42, 104–109.

Coffin, D.W., Carlsson, L.A., Pipes, R.B., 2006. On the separation of carbon nanotubes. Compos. Sci. Technol. 66, 1132–1140.

Fereidoon, A., Kordani, N., Ahangari, M.G., Ashoory, M., 2011. Damping augmentation of epoxy using carbon nanotubes. Int. J. Polym. Mater. 60, 11–26.

Fu, Y.M., Hong, J.W., Wang, X.Q., 2006. Analysis of nonlinear vibration for embedded carbon nanotubes. J. Sound Vib. 296, 746–756.

Fwa, T.F., Shi, X.P., Tan, S.A., 1996. Use of Pasternak foundation model in concrete pavement analysis. J. Transport. Eng. 122, 323–328.

Gibson, R.F., Ayorinde, E.O., Wen, Y.-F., 2007. Vibrations of carbon nanotubes and their composites: a review. Compos. Sci. Technol. 67, 1–28.

Gu, L., Elkin, T., Jiang, X., Li, H., Lin, Y., Qu, L., Tzeng, T.-R.J., Joseph, R., Sun, Y.-P., 2005. Single-walled carbon nanotubes displaying multivalent ligands for capturing pathogens. Chem. Commun. 7, 874.

Guller, K., 2004. Circular elastic plate resting on tensionless Pasternak foundation. J. Eng. Mech-ASCE 130, 1251–1254.

Gupta, U.S., Lai, R., Sharma, S., 2008. Effect of Pasternak foundation on axisymmetric vibration of non-uniform polar orthotropic annular plate. Int. J. Appl. Math. Mech. 4, 9–25.

He, J.H., 2002. Preliminary report on the energy balance for nonlinear oscillations. Mech. Res. Commun. 29 (2-3), 107–111.

He, J.H., 2004. Variational principles for some nonlinear partial differential equations with variable coefficients. Chaos, Solitons Fractals 19 (4), 847–851.

He, J.-H., 2007. Variational approach for nonlinear oscillators. Chaos, Solitons Fractals 34, 1430–1439.

Iijima, S., 1991. Helical microtubules of graphitic carbon. Nature 354, 56–58.http://www.nanocyl.com/en/CNT-Expertise-Centre/Carbon-Nanotubes.

Iijima, S., Ichihashi, T., 1993. Single-shell carbon nanotubes of 1-nm diameter. Nature 356, 776–778.

Joshi, A.Y., Harsha, S.P., Sharma, S.C., 2010. Vibration signature analysis of single walled carbon nanotube based nanomechanical sensors. Physica E 42, 2115–2123.

Kerr, D., 1964. Elastic and viscoelastic foundation models. J. Appl. Mech. 31, 491–498.
Kim, N.-I., Fu, C.C., Kim, M.-Y., 2007. Dynamic stiffness matrix of non-symmetric thin-walled curved beam on Winkler and Pasternak type foundations. Adv. Eng. Softw. 38, 158–171.
Kudin, K.N., Scuseria, G.E., Yakobson, B.I., 2001. C2F, BN, and C nanoshell elasticity from ab initio computations. Phys. Rev. B 64, 235406–235410.
Lau, K.T., Chipara, M., Ling, H.Y., Hui, D., 2004. On the effective elastic moduli of carbon nanotubes for nanocomposite structures. Compos. Part B-Eng. 35, 35–101.
Li, Z., Dharap, P., Nagarajaiah, S., Nordgren, R.P., Yakobson, B., 2004. Nonlinear analysis of a SWCNT over a bundle of nanotubes. Int. J. Solids Struct. 41, 6925–6936.
Liu, J.Z., Zheng, Q.S., Jiang, Q., 2001. Effect of a rippling mode on resonances of carbon nanotubes. Phys. Rev. Lett. 86 (21), 4843–4846.
Liu, J.Z., Zheng, Q., Jiang, Q., 2003. Effect of bending instabilities on the measurements of mechanical properties of multiwall carbon nanotubes. Phys. Rev. B 67, 075414.
Lu, F., Gu, L., Meziani, M.J., Wang, X., Luo, P.G., Veca, L.M., Cao, Li, Sun, Y.-P., 2009. Advances in bioapplications of carbon nanotubes. Adv. Mater. 21 (2), 139–152.
Mattson, M.P., Haddon, R.C., Rao, A.M., 2000. Molecular functionalization of carbon nanotubes and use as substrates for neuronal growth. J. Mol. Neurosci. 14 (3), 175.
Mayoof, F.N., Hawwa, M.A., 2009. Chaotic behavior of a curved carbon nanotube under harmonic excitation. Chaos, Solitons Fractals 42, 1860–1867.
Mehdipour, I., Ganj, D.D., Mozaffari, M., 2010. Application of the energy balance method to nonlinear vibrating equations. Curr. Appl. Phys. 10, 104–112.
Mohammad, R.H., Mohammad, J.A., Malekzadeh, P., 2007. A differential quadrature analysis of unsteady open channel flow. Appl. Math. Model. 31, 1594–1608.
Nyfeh, A.H., Mook, D.T., 1997. Perturbation Methods. Wiley, New York.
Overney, G., Zhong, W., Tománek, D., 1993. Structural rigidity and low frequency vibrational modes of long carbon tubules. Z. Phys. D: At. Mol. Clusters 27, 93–96.
Pradhan, S.C., Murmu, T., 2009. Thermo-mechanical vibration of FGM sandwich beam under variable elastic foundation using differential quadrature method. J. Sound Vib. 321, 342–362.
Qian, D., Liu, W.K., Ruoff, R.S., 2003. Load transfer mechanism in carbon nanotube ropes. Compos. Sci. Technol. 63, 1561–1569.
Ranjbartoreh, A.R., Ghorbanpour, A., Soltani, B., 2007. Double-walled carbon nanotube with surrounding elastic medium under axial pressure. Physica E 39, 230–239.
Rossit, C.A., Laura, P.A.A., 2001. Technical note. Free vibrations of a cantilever beam with a spring–mass system attached to the free end. Ocean Eng. 28, 933–939.
Rossit, C.A., Bambill, D.V., Laura, P.A.A., 1998. Letters to the Editor. Nonharmonic Fourier expansions in the case of a cantilever beam with a mass–spring combination at the free end. J. Sound Vib. 191 (5), 973–975.
Saether, E., Frankland, S.J.V., Pipes, R.B., 2003. Transverse mechanical properties of single-walled carbon nanotube crystals. Pt. I: Determination of elastic moduli. Compos. Sci. Technol. 63, 1543–1550.
Salvetat, J.P., Kulik, A.J., Bonard, J.M., Briggs, G.A.D., Stockli, T., Metenier, K., Bonnamy, S., Beguin, F., Burnham, N.A., Forro, L., 1999. Elastic modulus of ordered and disordered multiwalled carbon nanotubes. Adv. Mater. 11, 161–165.
Sammalkorpi, M., Krasheninnikov, A.V., Kuronen, A., Nordlund, K., 2005. Irradiation-induced stiffening of carbon nanotube bundles. Nucl. Inst. Methods Phys. Res. B 228, 142–145.
Shigley, J.E., 1988. Mechanical Engineering Design. McGraw-Hill, USA.
Spitalskya, Z., Tasisb, D., Papagelisb, K., Galiotis, C., 2010. Carbon nanotube–polymer composites: chemistry, processing, mechanical and electrical properties. Prog. Polym. Sci. 35, 357–401.
Thostenson, E.T., Ren, Z.F., Chou, T.W., 2001. Advances in the science and technology of carbon nanotubes and their composites a review. Compos. Sci. Technol. 61, 1899–1912.
Timoshenko, S.P., Gere, J.M., 1961. Theory of Elastic Stability. McGraw-Hill, New York.

Wang, J., Musameh, M., 2003. Carbon nanotube/teflon composite electrochemical sensors and biosensors. Anal. Chem. 75 (9), 2075.

Wang, X., Zhang, Y.C., Xia, X.H., Huang, C.H., 2004. Effective bending modulus of carbon nanotubes with rippling deformation. Int. J. Solids Struct. 41, 6429–6439.

Wang, X., Wang, Y., Xiao, J., 2005. A non-linear analysis of the bending modulus of carbon nanotubes with rippling deformations. Compos. Struct. 69, 315–321.

Wong, E.W., Sheehan, P.E., Lieber, C.M., 1997. Nanobeam mechanics: elasticity, strength, and toughness of nanorods and nanotubes. Science 277, 1971–1975.

Wu, T.Y., Liu, G.R., 1999. The differential quadrature as a numerical method to solve the differential equation. Comput. Mech. 24, 197–205.

Wu, H.A., Wang, X.X., 2008. An atomistic-continuum inhomogeneous material model for the elastic bending of metal nanocantilevers. Adv. Eng. Softw. 39, 764–769.

Wu, D.H., Chien, W.T., Chen, C.S., Chen, H.H., 2006. Resonant frequency analysis of fixed-free single-walled carbon nanotube-based mass sensor. Sensors Actuators A 126, 117–121.

Yaghmaei, K., Rafii-Tabar, H., 2009. Observation of fluid layering and reverse motion in double-walled carbon nanotubes. Curr. Appl. Phys. 9, 1411–1422.

Yogeswaran, U., Chen, S.M., 2008. A review on the electrochemical sensors and biosensors composed of nanowires as sensing material. Sensors 8 (1), 290.

Yogeswaran, U., Thiagarajan, S., Chen, S.M., 2008. Recent updates of DNA incorporated in carbon nanotubes and nanoparticles for electrochemical sensors and biosensors. Sensors 8 (11), 7191.

Zhang, H.W., Zhang, Z.Q., Wang, L., 2009. Molecular dynamics simulations of electrowetting in double-walled carbon nanotubes. Curr. Appl. Phys. 9, 750–754.

CHAPTER

PHYSICAL RELATIONSHIPS BETWEEN NANOPARTICLE AND NANOFLUID FLOW

3

CHAPTER CONTENTS

3.1 Turbulent Natural Convection Using Cu/Water Nanofluid 71
3.1.1 Introduction 72
3.1.2 Numerical Method 73
3.1.2.1 Problem Statement 73
3.1.2.2 LBM 73
3.1.2.3 LES Method 75
3.1.2.4 LBM Based on LES Model 75
3.1.2.5 LBM for Nanofluid 76
3.1.2.6 Boundary Conditions 77
3.1.3 Code Validation and Mesh Results 77
3.1.4 Result and Discussion 78
3.1.5 Conclusions 81
3.2 Heat Transfer of Cu-Water Nanofluid Flow Between Parallel Plates 83
3.2.1 Introduction 83
3.2.2 Governing Equations 84
3.2.3 Analysis of the HPM 86
3.2.4 Implementation of the Method 87
3.2.5 Results and Discussion 88
3.2.6 Conclusion 91
3.3 Slip Effects on Unsteady Stagnation Point Flow of a Nanofluid over a Stretching Sheet 92
3.3.1 Introduction 94
3.3.2 Governing Equations 94
3.3.3 Result and Discussion 97
3.3.4 Conclusion 106
References 106

3.1 TURBULENT NATURAL CONVECTION USING Cu/WATER NANOFLUID

In this section, Lattice Boltzmann simulation of turbulent natural convection with large-eddy simulations (LESs) in a square cavity which is filled by water/copper nanofluid has been investigated. The present results are validated by finds of an experimental research at $Ra = 1.58 \times 10^9$. Calculations

Application of Nonlinear Systems in Nanomechanics and Nanofluids. http://dx.doi.org/10.1016/B978-0-323-35237-6.00003-0

were performed for high Rayleigh numbers ($Ra=107$-109) and volume fractions of nanoparticles change from 0 to 0.05 ($0 \leq \varphi \leq 0.05$). This investigation is tried to present Large-eddy turbulence nanofluid flow model by Lattice Boltzmann method (LBM) with a clear and simple statement. Effects of nanoparticles are displayed on streamlines, isotherm counters, local Nusselt number, and average Nusselt number. The average Nusselt number enhances with augmentation of nanoparticle volume fraction in the base fluid while this manner has an erratic trend toward different Rayleigh numbers (*this section has been worked by H. Sajjadi, M. Gorji, G.H.R. Kefayati, D.D. Ganji in nonlinear dynamics team in Mechanical Engineering Department, 2011-2012*).

3.1.1 INTRODUCTION

Turbulence in fluids is ubiquitous in nature and technological systems and represents one of the most challenging aspects in fluid mechanics. The difficulty stems from the inherent presence of many scales that are generally inseparable among many other factors. Nevertheless, considerable progress has been made over the years toward more fundamental physical understanding of turbulence phenomena through measurements, statistical phenomenological theories, modeling, and computation (see Barakos et al., 1994; Frisch, 1995). Also, a lot of experimental investigations have been done on it; for instance, Ampofo and Karayiannis (2003) studied low-level turbulence natural convection in an air-filled vertical square cavity while the hot and cold walls of the cavity were isothermal at 50 and 10 °C, respectively giving a Rayleigh number of $Ra=1.58\times 10^9$.

Applying a fluid with high heat transfer in systems of diverse industries such as cooling systems for electronic devices, chemical vapor deposition instruments, furnace engineering, solar energy collectors, phase change material, and so forth was a permanent parameter, whereas the fluids which were utilized in these industries had low thermal conductivity and heat transfer; so, millimeter- and micrometer-sized solid particles with high thermal conductivity were used and improved this problem. But these particles cause many troublesome problems such as poor suspension stability and clogging in different systems. At first, Choi (1995) solved the uncomplimentary phenomenon by producing nanoparticles. So, fluids with nanoparticles suspended in them are called nanofluids. Many numerical, experimental, and theoretical investigations were performed about natural convection flow of nanofluid in different shapes (see Ho et al., 2008; Jahanshahi et al., 2010; Kefayati et al., 2011; Khanafer et al., 2003; Xuan and Li, 2003).

Turbulent flows are modeled by various methods and the most tradition of method is large-eddy (see Dixit and Babu (2006)). This model is applied at various applications such as analysis of geophysical phenomena in the atmosphere, oceans, and magnetosphere and provides a starting point for modeling these phenomena; the confinement of thermonuclear plasmas and in superfluid and superconductive behavior of thin films. Chen (2009) proposed a novel and simple large-eddy-based Lattice Boltzmann model to simulate two-dimensional turbulence. He showed that the model is efficient, stable, and simple for two-dimensional turbulence simulation. Recently, Sajjadi et al. (2011) studied numerical analysis of turbulent natural convection in square cavity using LES in LBM. They exhibited this method in acceptable agreement with other verifications of such a flow.

The aim of this section is to study effects of turbulence on flow field and temperature distribution in nanofluid-filled enclosure. Also, it is to present the ability of LBM for solving problems of nanofluid and various models of turbulence, while no previous study on effects of turbulence with a simple large-eddy-based Lattice Boltzmann model on natural convection in nanofluid-filled enclosure has been studied so far.

3.1.2 NUMERICAL METHOD

3.1.2.1 Problem statement

In this section, the proposed model is applied to simulate natural convection in a square cavity with side walls maintained at different temperatures (Figure 3.1).

The left vertical wall is maintained at a high temperature T_H while the right vertical wall is kept at a low temperature T_C. The horizontal walls are assumed to be insulated, nonconducting, and impermeable to mass transfer. The cavity is filled with a mixture of water and solid copper (Cu). The nanofluid in the cavity is Newtonian, incompressible, and laminar. Thermophysical properties of the nanofluid are assumed to be constant (Table 3.1).The density variation in the nanofluid is approximated by the standard Boussinesq model.

3.1.2.2 LBM

For the incompressible flow, if the transport coefficients are independent of the temperature, the energy equation can be decoupled from the mass and momentum equations. For the incompressible thermal problem, f and g are two functions called as flow distribution function and temperature distribution function, respectively. These functions are utilized to obtain macroscopic characteristics of the flow like velocity, pressure, temperature, etc.

In this section, a square grid and D2Q9 model is used for both flow and temperature functions (see Hou et al., 1996). By detachment of Navier-Stokes equations, governing equations for flow and temperature functions are as follows:

For the flow field:

$$f_i(x+c_i\Delta t, t+\Delta t) - f_i(x,t) = -\frac{1}{\tau_v}\left[f_i(x,t) - f_i^{eq}(x,t)\right] + \Delta t F \tag{3.1}$$

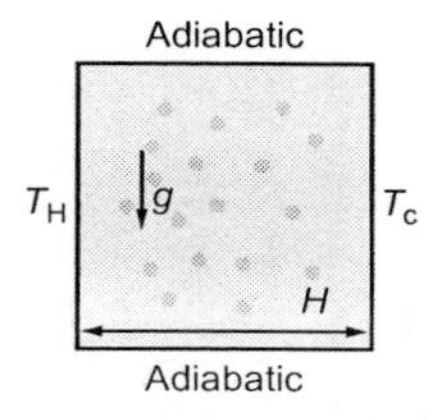

FIGURE 3.1

Geometry of the present study.

Table 3.1 Thermophysical Properties of Water and Copper

Property	Water	Copper
μ (kg/ms)	8.9×10^{-4}	–
C_p (J/kg K)	4179	383
ρ (kg/m^3)	997.1	8954
β (k^{-1})	2.1×10^{-4}	1.67×10^{-5}
k (W/m K)	0.6	400

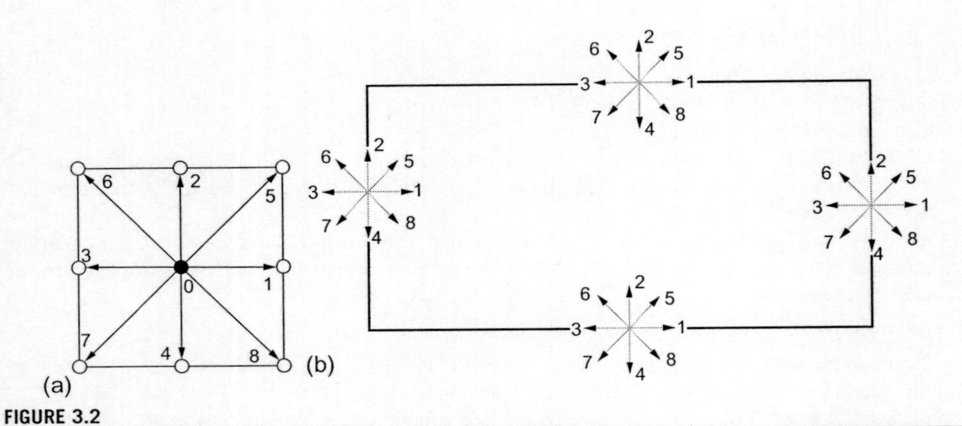

FIGURE 3.2

square grid and D2Q9 model (a) The discrete velocity vectors for D2Q9 (b) Domain boundaries.

For the temperature field:

$$g_i(x+c_i\Delta t, t+\Delta t) - g_i(x,t) = -\frac{1}{\tau_c}\left[g_i(x,t) - g_i^{eq}(x,t)\right] \tag{3.2}$$

where the discrete particle velocity vectors are defined c_i (Figure 3.2a); Δt denotes lattice time step which is set to unity; τ_v and τ_c are the relaxation time for the flow and temperature fields, respectively; and f_i^{eq} and g_i^{eq} are the local equilibrium distribution functions that have an appropriately prescribed functional dependence on the local hydrodynamic properties which are calculated with Equations (3.3) and (3.4) for flow and temperature fields, respectively. Also, F_i is an external force term.

$$f_i^{eq}(x,t) = \omega_i \rho\left[1 + \frac{c_i \cdot u}{c_s^2} + \frac{1}{2}\frac{(c_i \cdot u)^2}{c_s^4} - \frac{1}{2}\frac{u \cdot u}{c_s^2}\right] \tag{3.3}$$

$$g_i^{eq} = \omega_i T\left[1 + \frac{c_i \cdot u}{c_s^2}\right] \tag{3.4}$$

For the 2-D case, applying third-order Gauss-Hermite quadrature leads to the D2Q9 model with the following discrete velocities c_i:

$$C_i = \begin{cases} 0 & i=0 \\ c\left(\cos\left[(i-1)\frac{\pi}{2}\right], \sin\left[(i-1)\frac{\pi}{2}\right]\right) & i=1-4 \\ c\sqrt{2}\left(\cos\left[(i-5)\frac{\pi}{2}+\frac{\pi}{4}\right], \sin\left[(i-5)\frac{\pi}{2}+\frac{\pi}{4}\right]\right) & i=5-8 \end{cases} \tag{3.5}$$

where $\omega_0 = 4/9$, $\omega_{1\text{-}4} = 1/9$, $\omega_{5\text{-}9} = 1/36$, and $c = \sqrt{3RT_m}$ (to improve numerical stability, T_m is the mean value of temperature for the calculation of c).

Using a Chapman-Enskog expansion, the Navier-Stokes equations can be recovered with the described model. The kinematic viscosity ϑ and the thermal diffusivity α are then related to the relaxation times by:

$$\vartheta = \left[\tau_v - \frac{1}{2}\right] c_s^2 \Delta t \text{ and } \alpha = \left[\tau_c - \frac{1}{2}\right] c_s^2 \Delta t \tag{3.6}$$

where c_s is the speed of sound and is equal to $c/\sqrt{3}$.

In the simulation, the Boussinesq approximation is applied to the buoyancy force term. In that case, the external force F_i appearing in Equation (3.1) is given by:

$$F_i = 3\omega_i g_y \beta \Delta T \tag{3.7}$$

where g_y, β, and ΔT are gravitational acceleration, thermal expansion coefficient and temperature difference, respectively.

Finally, the macroscopic variables ρ, u, and T can be calculated as follows:

$$\text{Flow density}: \rho = \sum_i f_i \tag{3.8}$$

$$\text{Momentum}: \rho u_j = \sum_i f_i c_i \tag{3.9}$$

$$\text{Temperature}: T = \sum_i g_i \tag{3.10}$$

3.1.2.3 LES method

In this model, the main aim is obtaining ν_t and $\alpha_t = (\nu_t/Pr_t)$, where Pr_t is turbulent Prandtl number which is assumed to be 0.5 for air and 4 for water. In order to evaluate ν_t, we perform as follows:

$$\nu_t = (C\Delta)^2 \left(|\overline{S}|^2 + \frac{Pr}{Pr_t} \nabla T \cdot \frac{\vec{g}}{|\vec{g}\,|} \right)^{1/2} \tag{3.11}$$

C is considered as Smagorinsky constant and in this chapter it is assumed as 0.1. It is gained from $\Delta = \sqrt{(\Delta x)^2 + (\Delta y)^2}$; Δx and Δy are grid extents in X and Y directions.

For $|\overline{S}|$, we have:

$$|\overline{S}| = \sqrt{2\overline{S}_{\alpha\beta}\overline{S}_{\alpha\beta}} \tag{3.12}$$

$$\overline{S}_{\alpha\beta} = \frac{(\partial_\alpha \overline{u}_\beta + \partial_\beta \overline{u}_\alpha)}{2} \tag{3.13}$$

3.1.2.4 LBM based on LES model

Large-eddy model is easily applied in LBM the way ν_t affects relaxation time.

$$\nu_{\text{total}} = c_s{}^2(\tau_m - 0.5) = \nu_0 + \nu_t \tag{3.14}$$

where ν_{total} and ν_0 are total viscosity and initial viscosity, respectively.

$$\tau_m = \frac{(\nu_0 + \nu_t)}{c_s^2} + 0.5 = \frac{\nu_0}{c_s^2} + 0.5 + \frac{\nu_t}{c_s^2} = \tau_0 + \frac{\nu_t}{c_s^2} \tag{3.15}$$

To obtain ν_t in LBM, we have:

$$|\overline{S}| = \frac{3}{2\tau_m}|Q| \tag{3.16}$$

$$Q = \sum_{i=0}^{8} e_{i\alpha} e_{i\beta} \left(f_i - f_i^{eq}\right) \tag{3.17}$$

If we put $|\overline{S}|$ in Equation (3.12):

$$\nu_t = (C\Delta)^2 \left(\frac{9}{4\tau_m^2}|Q|^2 + \frac{Pr}{Pr_t} \nabla T \cdot \frac{\vec{g}}{|\vec{g}|} \right)^{1/2} \tag{3.18}$$

And if we substitute the above equation in Equation (3.14):

$$\tau_{total} = \tau_0 + \frac{(C\Delta)^2 \left(\frac{9}{4\tau_m^2}|Q|^2 + \frac{Pr}{Pr_t} \nabla T \cdot \frac{\vec{g}}{|\vec{g}|} \right)^{1/2}}{c_s^2} \tag{3.19}$$

To obtain relaxation time in temperature function equation, we have:

$$\tau_h = \tau_{D0} + \frac{\alpha_t}{c_s{}^2} = \tau_{D0} + \frac{\nu_t / Pr_t}{c_s{}^2} \tag{3.20}$$

where $\tau_{D0} = (\alpha_0 / c_s{}^2) + 0.5$.

Substituting new relaxation time in Equations (3.1) and (3.2) yields Lattice Boltzmann equations based on large-eddy model.

3.1.2.5 LBM for nanofluid

The dynamical similarity depends on two dimensionless parameters: the Prandtl number Pr and the Rayleigh number Ra, whereas it is assumed that nanofluid is similar to a pure fluid and then nanofluid qualities are gotten and they are applied for the two parameters.

The thermophysical properties of the nanofluid are assumed to be constant (Table 3.1) except for the density variation, which is approximated by the Boussinesq model.

The effective properties of the nanofluid are defined as follows:

Density:

$$\rho_{nf} = (1-\varphi)\rho_f + \varphi\rho_s \tag{3.21}$$

Heat capacitance:

$$C_{nf} = \frac{\varphi\rho_s C_{nf} + (1-\varphi)\rho_f C_f}{\rho_{nf}} \tag{3.22}$$

Effective thermal conductivity:

$$\frac{k_{nf}}{k_f} = \frac{k_s + 2k_f + 2\varphi(k_f - k_s)}{k_s + 2k_f - \varphi(k_f - k_s)} \tag{3.23}$$

Viscosity:

$$\mu_{nf} = \frac{\mu_f}{(1-\varphi)^{2.5}} \tag{3.23}$$

whereas Equations (3.25a), (3.25b), (3.25c) and (3.26) are appropriate for spherical and equal nanoparticles.

3.1.2.6 Boundary conditions

3.1.2.6.1 Flow

Implementation of boundary conditions is very important for the simulation. The unknown distribution functions pointing to the fluid zone at the boundary nodes must be specified (Figure 3.2). Concerning the no-slip boundary condition, bounce-back boundary condition is used on the solid boundaries. For instance, the unknown density distribution functions at the boundary east can be determined by the following conditions:

$$f_{6,n} = f_{8,n},\quad f_{7,n} = f_{5,n},\quad f_{3,n} = f_{1,n} \tag{3.24}$$

where n is the lattice on the boundary.

3.1.2.6.2 Temperature

If the north and the south of the boundaries are adiabatic, then the bounce-back boundary condition is used on them. Temperature at the west and the east wall are known, in the west wall $T_H = 1.0$. Since we are using D2Q9, the unknowns are g_1, g_5, and g_8 at west wall which are evaluated as follows:

$$g_1 = T_H(\omega_1 + \omega_3) - g_3 \tag{3.25a}$$

$$g_5 = T_H(\omega_5 + \omega_7) - g_7 \tag{3.25b}$$

$$g_8 = T_H(\omega_8 + \omega_6) - g_6 \tag{3.25c}$$

Nusselt number (Nu) is one of the most important dimensionless parameters in the description of the convective heat transport. The local Nusselt number and the average value at the hot wall are calculated as:

$$Nu_y = -\frac{H}{\Delta T}\frac{\partial T}{\partial x} \tag{3.26}$$

$$Nu_{avg} = \frac{1}{H}\int_0^H Nu_y \, dy \tag{3.27}$$

Because of the convenience, a normalized average Nusselt number is defined as the ratio of Nusselt number at any volume fraction of nanoparticles to that of pure water that is as follows:

$$Nu_{avg\,(db)}(\varphi) = \frac{Nu_{avg}(\varphi)}{Nu_{avg}(\varphi = 0)} \tag{3.28}$$

3.1.3 CODE VALIDATION AND MESH RESULTS

The nanofluid in the cavity with different heated vertical sides is chosen as Cu-water mixture. Calculations were made for various values of volume fraction of nanoparticle ($0 \le \varphi \le 0.06$) and Rayleigh numbers ($10^7 < Ra < 10^9$). The present numerical method was validated at two topics of this problem

Table 3.2 Comparison of Mean *Nu* with Previous Works

Ra Number	Mesh	Mean *Nu* (This Work)	Mean *Nu* [18]	Mean *Nu* [19]
10^7	256 × 256	17.2	–	16.8
10^8	512 × 512	31.2	32.3	30.5
10^9	1024 × 1024	58.1	60.1	57.4

former. The two topics are effect of nanoparticles and turbulence. Table 3.2 shows the comparison of average Nusselt numbers for different Rayleigh numbers between present results and finds of Barakos et al. (1994) and Dixit and Babu (2006) as cavity was filled by air with $Pr=0.71$.

A comparison with velocity at the middle section of the cavity and local Nusselt number on the hot wall were considered with experimental results of Ampofo and Karayiannis (2003) in Figure 3.3. Clearly, it is seen that the results match previous work. Furthermore, this table demonstrates needful various meshes to utilize for different Rayleigh numbers. For second part, the method of solution for nanofluid by LBM was validated against results of Khanafer et al. (2003) and Jahanshahi et al. (2010).

As Figure 3.4 shows a comparison with temperature at mid section of the cavity for Cu-water nanofluid whereas volume fraction is $\varphi=0.1$.

3.1.4 RESULT AND DISCUSSION

Figure 3.5 shows a comparison between pure fluid ($\varphi=0$) and nanofluid ($\varphi=0.05$) for various Rayleigh numbers in term of the isotherms and the streamlines. When Rayleigh number increases, the symmetry state wastes and the centralization of the streamlines in the core of the cavity tend the hot wall. The heat transfer process increases when Rayleigh number enhances. This process is

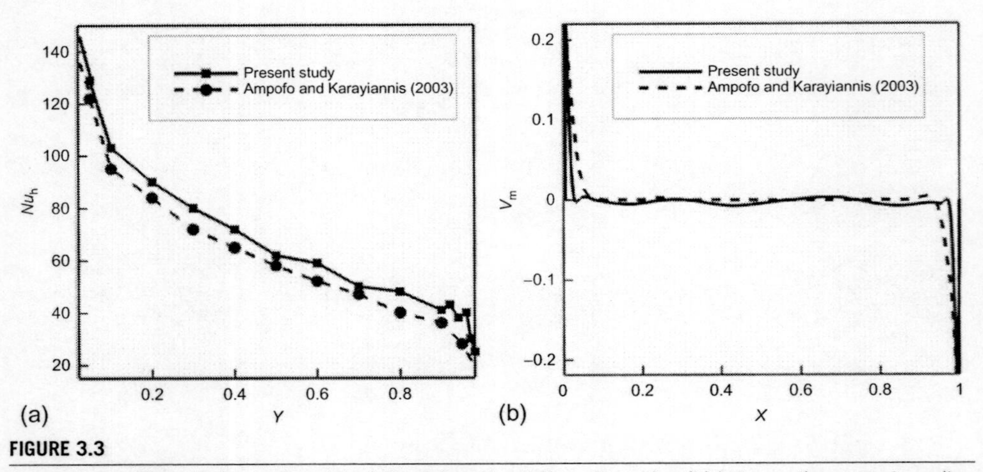

FIGURE 3.3

Comparison of the velocity on the axial midline (a) and local Nusselt number (b) between the present results and numerical results by Ampofo and Karayiannis (2003) ($Ra=1.58\times10^9$).

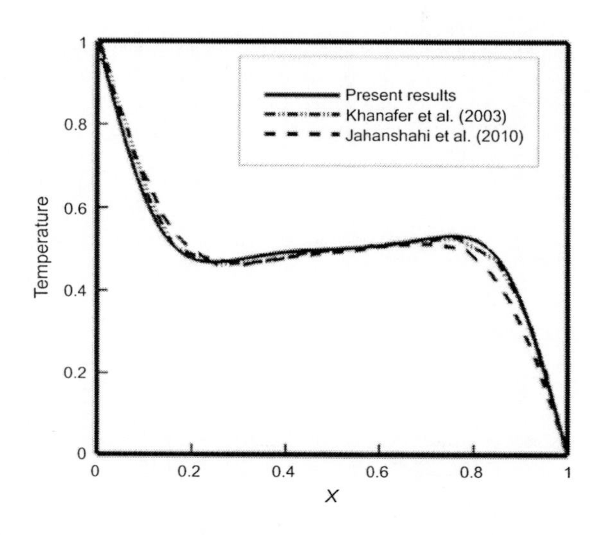

FIGURE 3.4

Comparison of the temperature on axial midline between the present results and numerical results by Khanafer et al. (2003) and Jahanshahi et al. (2010) ($Pr=6.2$, $\varphi=0.1$, $Gr=10^4$).

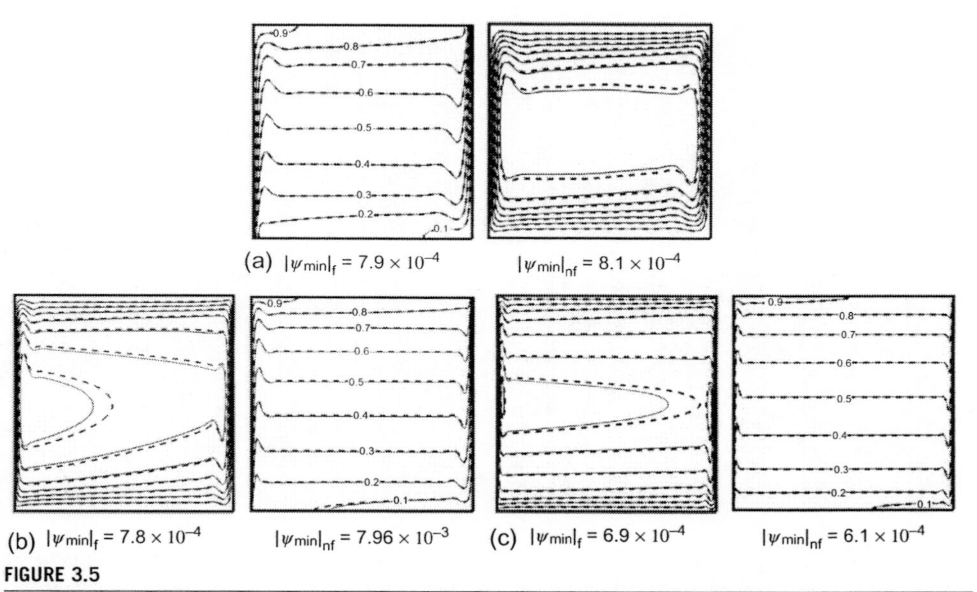

FIGURE 3.5

Comparison of the isotherms and streamline between nanofluids (- - -) ($\varphi=0.05$) and base fluid (—) ($\varphi=0$) at various Rayleigh numbers (a) $Ra=10^7$, (b) $Ra=10^8$, and (c) $Ra=10^9$.

obvious where two isotherms of $T=0.1$ and 0.9 move to the cold wall and the hot wall, respectively when Rayleigh number increases. The effect of nanoparticles on the streamlines is clear, whereas the streamlines for nanofluid expand more than pure fluid.

This phenomenon shows the increment of heat transfer in the presence of nanoparticles. On the other hand, the augmentation of the maximum streamline values in the presence of nanoparticle declines with the increment of Rayleigh numbers whereas at $Ra=10^9$, this value for nanofluid is less than pure fluid.

Figure 3.6 examines the local Nusselt number on the hot wall for different volume fractions and Rayleigh numbers. Generally, the local Nusselt numbers of nanofluids have an incremental rate with

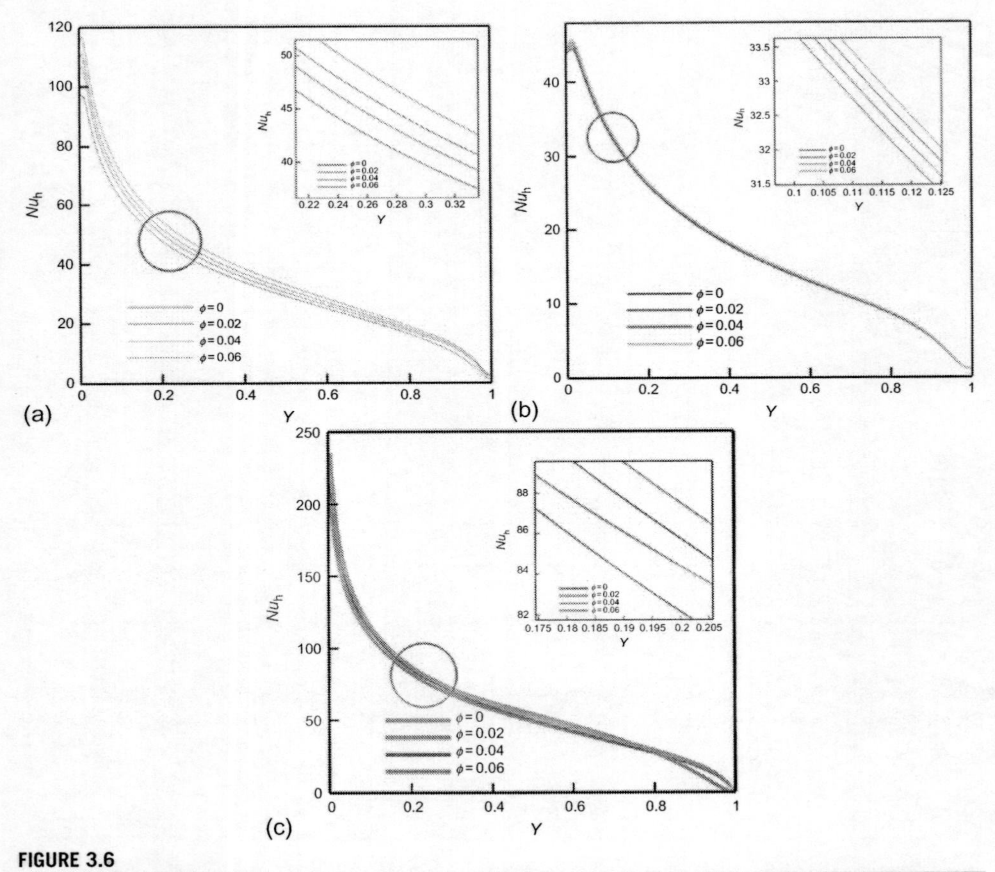

FIGURE 3.6

Nusselt number distributions on the hot wall at different volume fractions and Rayleigh numbers (a) $Ra=10^7$, (b) $Ra=10^8$, and (c) $Ra=10^9$.

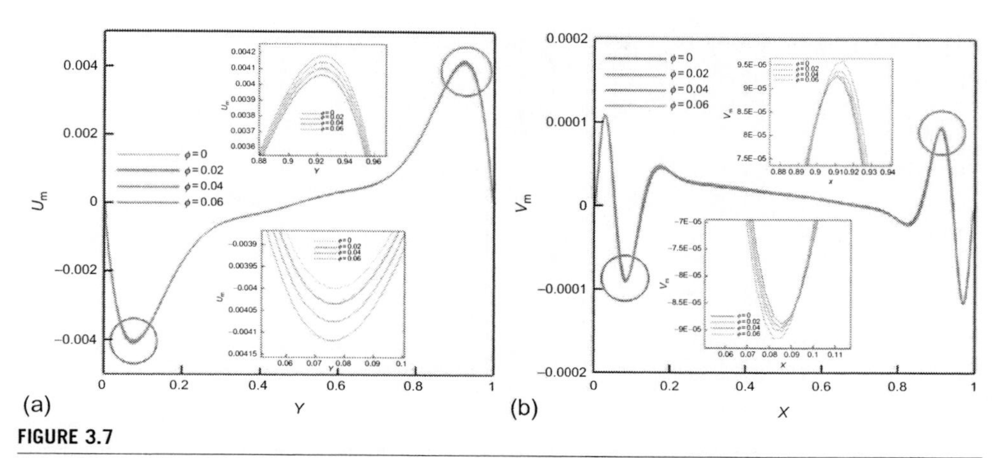

FIGURE 3.7

Values of the horizontal (a) and vertical (b) velocity on the axial midline for $Ra = 10^7$.

the augmentation of volume fraction. It demonstrates that the effect of nanoparticles decreases as they move upward the hot wall where the local Nusselt number value for the nanofluid with $\varphi = 0.04$ is lower than that for pure fluid. Moreover, the most impact of nanoparticles on the local Nusselt number is observed in the middle of the plot that is exhibited by the circles.

Figure 3.7 displays the values of the vertical and horizontal velocity on the axial midline for $Ra = 10^7$. The plot has erratic manners but at the minimum and the maximum values, obtains regular behavior toward the growth of the volume fractions. As the pictures are zoomed, the gradual trend in the plot can be observed. This plot proves that the changes expanses for the both velocities enhances. When the velocities of the fluid are increased by the nanoparticles, the heat transfer process improves.

Figure 3.8 illustrates the influence of the nanoparticle volume fraction φ on the average Nusselt number (Nu_{avg}) and the normalized average Nusselt number ($Nu_{avg(db)}$) along the heated surface for different volume fractions and Rayleigh numbers.

It shows the values of the average Nusselt number and normalized average Nusselt number for different volume fractions and Raleigh numbers. The average Nusselt number increases with the augmentation of nanoparticles. The increment is different for various Rayleigh numbers and the volume fractions. The average Nusselt number indicates a linearly manner toward the increase in the volume fractions at $Ra = 10^7$, but there is a jump at $\varphi = 0.01$ for $Ra = 10^8$ then the plot has a like trend with $Ra = 10^7$. It is obvious that the effects of nanoparticles at $Ra = 10^9$ are very weak on the average Nusselt number than the lower Rayleigh numbers.

3.1.5 CONCLUSIONS

Natural convection in a cavity which is filled with a water/Cu nanofluid has been conducted numerically by LBM. This study has been carried out for the pertinent parameters in the following ranges: the Rayleigh number of base fluid, $Ra = 10^7$-10^9, the volume fractions 0-5% and some conclusions were summarized as follows:

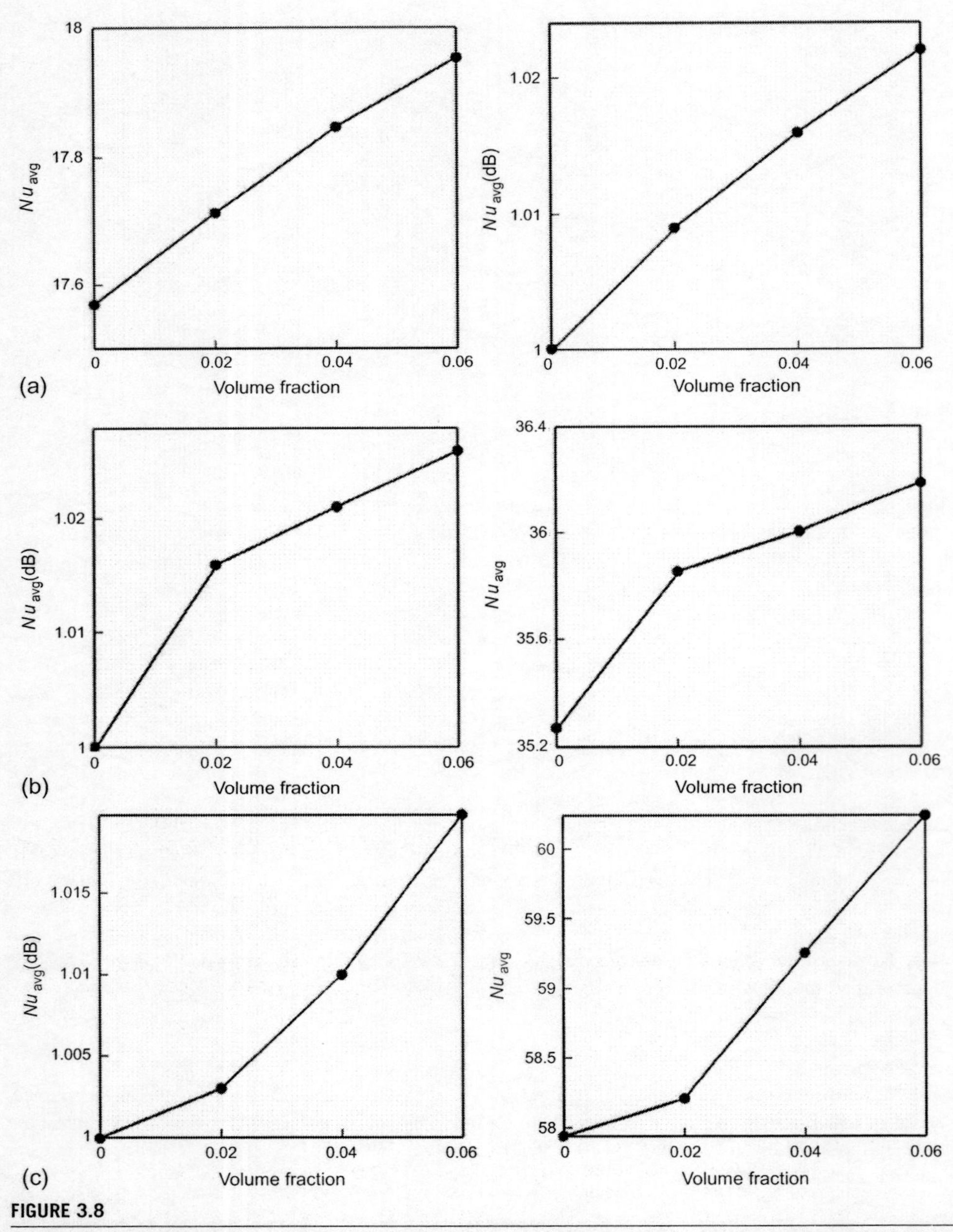

FIGURE 3.8

Values of the average Nusselt number (the left side) and normalized average Nusselt number (the right side) for different volume fractions and Raleigh numbers (a) $Ra=10^7$, (b) $Ra=10^8$, and (c) $Ra=10^9$.

(a) A proper validation with previous numerical investigations demonstrates that LBM is an appropriate method for turbulent and multiphase flow problems.
(b) Generally, the increase in the volume fractions and the Rayleigh numbers results in the augmentation of heat transfer.
(c) The effect of nanoparticles on the average Nusselt number at $Ra = 10^9$ is less than other considered Rayleigh numbers.
(d) The nanoparticles increase the velocity of the base fluid with a regular manner toward the increment of the volume fractions for different Rayleigh numbers.

3.2 HEAT TRANSFER OF Cu-WATER NANOFLUID FLOW BETWEEN PARALLEL PLATES

Heat transfer of a nanofluid flow which is squeezed between parallel plates is investigated analytically using homotopy perturbation method (HPM). Copper as nanoparticle with water as its base fluid has been considered. The effective thermal conductivity and viscosity of nanofluid are calculated by the Maxwell-Garnett (MG) and Brinkman models, respectively. This investigation compared other numerical methods and found to be in excellent agreement. The effects of the squeeze number, the nanofluid volume fraction, Eckert number, and δ on Nusselt number are investigated. The results show that Nusselt number has direct relationship with nanoparticle volume fraction, δ, the squeeze number, and Eckert number when two plates are separated, but it has reverse relationship with the squeeze number when two plates are squeezed (*this chapter has been worked by M. Sheikholeslami, D.D. Ganji in nonlinear dynamics team in Mechanical Engineering Department, 2012-2013*).

3.2.1 INTRODUCTION

Fluid heating and cooling are important in many industrial fields such as power, manufacturing, and transportation. Effective cooling techniques are absolutely needed for cooling any sort of high energy device. Common heat transfer fluids such as water, ethylene glycol, and engine oil have limited heat transfer capabilities due to their low heat transfer properties. In contrast, in metals, thermal conductivities are up to three times higher than the fluids, so it is naturally desirable to combine the two substances to produce a heat transfer medium that behaves like a fluid but has the thermal conductivity of a metal.

The study of heat transfer for unsteady squeezing viscous flow between two parallel plates has been regarded as one of the most important research topics due to its wide range of scientific and engineering applications such as hydrodynamical machines, polymer processing, lubrication system, chemical processing equipment, formation and dispersion of fog, damage of crops due to freezing, food processing, and cooling towers. The first research on the squeezing flow in lubrication system was reported by Stefan (1874). Mahmood et al. (2007) investigated the heat transfer characteristics in the squeezed flow over a porous surface. Mustafa et al. (2012) studied heat and mass transfer characteristics in a viscous fluid which is squeezed between parallel plates.

Most of the engineering problems, especially some heat transfer equations, are nonlinear; therefore, some of them are solved using numerical solution and some are solved using the different analytic method, such as perturbation method (PM), HPM, and variational iteration method (VIM). Therefore, many different methods have recently introduced some ways to eliminate the small parameter. One of the semi-exact methods which do not need small parameters is the HPM. The HPM is proposed and improved by He (2004). This method yields a very rapid convergence of the solution series in most cases.

The main purpose of this section is to apply HPM to find approximate solutions of nonlinear differential equations governing the problem of heat transfer in the unsteady squeezing nanofluid flow between parallel plates.

3.2.2 GOVERNING EQUATIONS

We considered the heat transfer analysis in the unsteady two-dimensional squeezing nanofluid flow between the infinite parallel plates (Figure 3.9). The two plates are placed at $z=\pm\ell(1-\alpha t)^{1/2}=\pm h(t)$. For $\alpha>0$, the two plates are squeezed until they touch $t=1/\alpha$ and for $\alpha<0$ the two plates are separated. The viscous dissipation effect, the generation of heat due to friction caused by shear in the flow, is retained. This effect is quite important in the case when the fluid is largely viscous or flowing at a high speed. This behavior occurs at high Eckert number ($\gg 1$). Further, the symmetric nature of the flow is adopted. The fluid is a water-based nanofluid containing Cu (copper). The nanofluid is a two-component mixture with the following assumptions: incompressible; no-chemical reaction; negligible viscous dissipation; negligible radiative heat transfer; nanosolid particles and the base fluid are in thermal equilibrium and no slip occurs between them. The thermophysical properties of the nanofluid are given in Table 3.3 (Sheikholeslami et al., 2012).

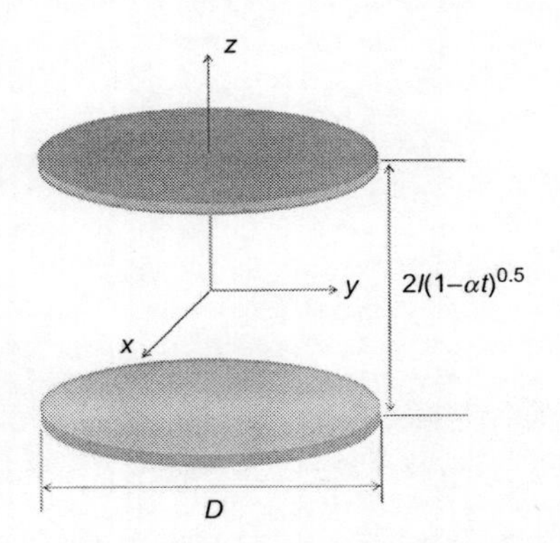

FIGURE 3.9

Geometry of problem.

Table 3.3 Thermo Physical Properties of Water and Nanoparticles

	ρ (kg/m^3)	C_p (J/kg K)	k (W/m K)
Pure water	997.1	4179	0.613
Copper (Cu)	8933	385	401

The governing equations for momentum and energy in unsteady two-dimensional flow of a nanofluid are:

$$\frac{\partial u}{\partial x}+\frac{\partial v}{\partial y}=0 \tag{3.29}$$

$$\rho_{\mathrm{nf}}\left(\frac{\partial u}{\partial t}+u\frac{\partial u}{\partial v}+v\frac{\partial u}{\partial y}\right)=-\frac{\partial p}{\partial x}+\mu_{\mathrm{nf}}\left(\frac{\partial^2 u}{\partial x^2}+\frac{\partial^2 u}{\partial y^2}\right) \tag{3.30}$$

$$\rho_{\mathrm{nf}}\left(\frac{\partial v}{\partial t}+u\frac{\partial v}{\partial v}+v\frac{\partial v}{\partial y}\right)=-\frac{\partial p}{\partial y}+\mu_{\mathrm{nf}}\left(\frac{\partial^2 v}{\partial x^2}+\frac{\partial^2 v}{\partial y^2}\right) \tag{3.31}$$

$$\frac{\partial T}{\partial t}+u\frac{\partial T}{\partial x}+v\frac{\partial T}{\partial y}=\frac{k_{\mathrm{nf}}}{(\rho C_p)_{\mathrm{nf}}}\left(\frac{\partial^2 T}{\partial x^2}+\frac{\partial^2 T}{\partial y^2}\right)+\frac{\mu_{\mathrm{nf}}}{(\rho C_p)_{\mathrm{nf}}}\left(4\left(\frac{\partial u}{\partial x}\right)^2+\left(\frac{\partial u}{\partial x}+\frac{\partial u}{\partial y}\right)^2\right) \tag{3.32}$$

Here, u and v are the velocities in the x and y directions, respectively; T is the temperature; p is the pressure; effective density (ρ_{nf}), the effective dynamic viscosity (μ_{nf}), the effective heat capacity $(\rho C_p)_{\mathrm{nf}}$ and the effective thermal conductivity k_{nf} of the nanofluid are defined as:

$$\rho_{\mathrm{nf}}=(1-\varphi)\rho_{\mathrm{f}}+\varphi\rho_{\mathrm{s}},\quad \mu_{\mathrm{nf}}=\frac{\mu_{\mathrm{f}}}{(1-\varphi)^{2.5}},\quad (\rho C_p)_{\mathrm{nf}}=(1-\varphi)(\rho C_p)_{\mathrm{f}}+\varphi(\rho C_{\mathrm{p}})_{\mathrm{s}},$$
$$\frac{k_{\mathrm{nf}}}{k_{\mathrm{f}}}=\frac{k_{\mathrm{s}}+2k_{\mathrm{f}}-2\varphi(k_{\mathrm{f}}-k_{\mathrm{s}})}{k_{\mathrm{s}}+2k_{\mathrm{f}}+2\varphi(k_{\mathrm{f}}-k_{\mathrm{s}})} \tag{3.33}$$

The relevant boundary conditions are:

$$\begin{aligned} &v=v_{\mathrm{w}}=\frac{\mathrm{d}h}{\mathrm{d}t},\quad T=T_{\mathrm{H}} && \text{at } y=h(t)\\ &v=\frac{\partial u}{\partial y}=\frac{\partial T}{\partial y}=0 && \text{at } y=0 \end{aligned} \tag{3.34}$$

We introduce these parameters:

$$\eta=\frac{y}{\left[l(1-\alpha t)^{1/2}\right]},\quad u=\frac{\alpha x}{[2(1-\alpha t)]}f'(\eta),\quad v=-\frac{\alpha l}{\left[2(1-\alpha t)^{1/2}\right]}f(\eta),\quad \theta=\frac{T}{T_{\mathrm{H}}},\quad A_1=(1-\phi)+\phi\frac{\rho_{\mathrm{s}}}{\rho_{\mathrm{f}}} \tag{3.35}$$

Substituting the above variables into Equations (3.30) and (3.31) and then eliminating the pressure gradient from the resulting equations give:

$$f^{iv}-SA_1(1-\phi)^{2.5}(\eta f'''+3f''+f'f''-ff''')=0 \tag{3.36}$$

Using Equation (3.35), Equations (3.31) and (3.32) reduce to the following differential equations:

$$\theta''+PrS\left(\frac{A_2}{A_3}\right)(f\theta'-\eta\theta')+\frac{PrEc}{A_3(1-\phi)^{2.5}}\left(f''^2+4\delta^2 f'^2\right)=0 \tag{3.37}$$

Here A_2 and A_3 are constants given by:

$$A_2=(1-\varphi)+\varphi\frac{(\rho C_p)_{\mathrm{s}}}{(\rho C_p)_{\mathrm{f}}},\quad A_3=\frac{k_{\mathrm{nf}}}{k_{\mathrm{f}}}=\frac{k_{\mathrm{s}}+2k_{\mathrm{f}}-2\varphi(k_{\mathrm{f}}-k_{\mathrm{s}})}{k_{\mathrm{s}}+2k_{\mathrm{f}}+2\varphi(k_{\mathrm{f}}-k_{\mathrm{s}})} \tag{3.38}$$

With these boundary conditions:

$$\begin{aligned} &f(0)=0 \quad f''(0)=0 \\ &f(1)=1 \quad f'(1)=0 \\ &\theta'(0)=0 \quad \theta(1)=1 \end{aligned} \tag{3.39}$$

where S is the squeeze number, Pr is the Prandtl number, and Ec is the Eckert number, which are defined as:

$$S=\frac{\alpha l^2}{2\upsilon_f},\quad Pr=\frac{\mu_f(\rho C_p)_f}{\rho_f k_f},\quad Ec=\frac{\rho_f}{(\rho C_p)_f}\left(\frac{\alpha x}{2(1-\alpha t)}\right)^2,\quad \delta=\frac{l}{x} \tag{3.40}$$

Physical quantities of interest are the skin fraction coefficient and Nusselt number which are defined as:

$$Cf=\frac{\mu_{nf}\left(\frac{\partial u}{\partial y}\right)_{y=h(t)}}{\rho_{nf}v_w^2},\quad Nu=\frac{-lk_{nf}\left(\frac{\partial T}{\partial y}\right)_{y=h(t)}}{kT_H} \tag{3.41}$$

In terms of Equation (3.35), we obtain

$$\begin{aligned} &C_f^*=l^2/x^2(1-\alpha t)Re_x C_f=A_1(1-\phi)^{2.5}f''(1) \\ &Nu^*=\sqrt{1-\alpha t}Nu=-A_3\theta'(1) \end{aligned} \tag{3.42}$$

3.2.3 ANALYSIS OF THE HPM

To illustrate the basic ideas of this method, we consider the following equation:

$$A(u)-f(r)=0,\quad r\in\Omega \tag{3.43}$$

with the boundary condition of:

$$B\left(u,\frac{\partial u}{\partial n}\right)=0,\quad r\in\Gamma \tag{3.44}$$

where A is the general differential operator, B is the boundary operator, $f(r)$ is the known analytical function, and Γ is the boundary of the domain Ω.

A can be divided into two parts which are L and N, where L is linear and N is nonlinear. Equation (3.43) can therefore be rewritten as follows:

$$L(u)+N(u)-f(r)=0,\quad r\in\Omega \tag{3.45}$$

Homotopy perturbation structure is shown as follows:

$$H(\nu,p)=(1-p)[L(\nu)-L(u_0)]+p[A(\nu)-f(r)]=0 \tag{3.46}$$

where

$$\nu(r,p):\Omega\times[0,1]\rightarrow R \tag{3.47}$$

In Equation (3.46), $p \in [0, 1]$ is an embedding parameter and u_0 is the first approximation that satisfies the boundary condition. We can assume that the solution of Equation (3.46) can be written as a power series in p as following:

$$\nu = \nu_0 + p\nu_1 + p^2\nu_2 + \cdots \tag{3.48}$$

and the best approximation for solution is:

$$u = \lim_{p\to 1}\nu = \nu_0 + \nu_1 + \nu_2 + \cdots \tag{3.49}$$

3.2.4 IMPLEMENTATION OF THE METHOD

According to the so-called HPM, we construct a homotopy suppose the solution of Equation (3.46) has the form:

$$H(f,p) = (1-p)\left(f^{iv} - f_0^{iv}\right) + p\left(f^{iv} - SA_1(1-\phi)^{2.5}\left(\eta f''' + 3f'' + f'f'' - ff'''\right)\right) = 0 \tag{3.50}$$

$$H(\theta,p) = (1-p)(\theta'' - \theta_0'') + p\left(\theta'' + PrS\left(\frac{A_2}{A_3}\right)(f\theta' - \eta\theta') + \frac{PrEc}{A_3()(1-\phi)^{2.5}}\left(f''^2 + 4\delta^2 f'^2\right)\right) = 0 \tag{3.51}$$

We consider f and θ as follows:

$$f(\eta) = f_0(\eta) + f_1(\eta) + \cdots = \sum_{i=0}^{n} f_i(\eta) \tag{3.52}$$

$$\theta(\eta) = \theta_0(\eta) + \theta_1(\eta) + \cdots = \sum_{i=0}^{n} \theta_i(\eta) \tag{3.53}$$

with substituting f, θ from Equations (3.52) and (3.53) and Equations (3.50) and (3.51) into and some simplification and rearranging based on powers of p-terms, we have:

$$p^0:$$

$$\begin{aligned} &f_0^{iv} = 0 \\ &\theta_0'' = 0 \end{aligned} \tag{3.54}$$

And boundary conditions are:

$$\begin{aligned} &f_0(0) = 0,\ f_0''(0) = 0,\ \theta_0'(0) = 0 \\ &f_0(1) = 1,\ f_0'(1) = 0,\ \theta_0(1) = 1 \end{aligned} \tag{3.55}$$

$$p^1:$$

$$\begin{aligned} &f_1^{iv} + SA_1(1-\phi)^{2.5}\left(f_0'f_0'' - 3f_0'' - \eta f_0''' + f_0'' - f_0f_0'''\right) = 0 \\ &\theta_1'' + PrEc\left(f_0''\right)^2 - PrS\left(\frac{A_2}{A_3}\right)\eta\theta_0' + 4\frac{PrEc}{A_3(1-\phi)^{2.5}}\delta^2\left(f_0'\right)^2 + PrS\left(\frac{A_2}{A_3}\right)f_0\theta_0' = 0 \end{aligned} \tag{3.56}$$

And boundary conditions are:

$$f_1(0)=0,\ f_1''(0)=0,\ \theta_1'(0)=0$$
$$f_1(1)=1,\ f_1'(1)=0,\ \theta_1(1)=1 \tag{3.57}$$

Solving Equations (3.54) and (3.56) with boundary conditions, we have:

$$f_0(\eta)=-0.5\eta^3+1.5\eta$$
$$\theta_0(\eta)=1 \tag{3.58}$$

$$f_1(\eta)=SA_1(1-\phi)^{2.5}(0.003571428571\eta^7-0.1\eta^5+0.1892857143\eta^3-0.09285714286\eta)$$
$$\theta_1(\eta)=\frac{PrEc}{A_3(1-\phi)^{2.5}}(-0.75\eta^4-0.3\delta^2\eta^6+1.5\delta^2\eta^4-4.5\delta^2\eta^2+0.75+3.3\delta^2) \tag{3.59}$$

The terms $f_i(\eta)$ and $\theta_i(\eta)$ when $i \geq 2$ are too large that is mentioned graphically.

The solution of this equation, when $p \to 1$, will be as follows:

$$f(\eta)=f_0(\eta)+f_1(\eta)+\cdots$$
$$\theta(\eta)=\theta_0(\eta)+\theta_1(\eta)+\cdots \tag{3.60}$$

3.2.5 RESULTS AND DISCUSSION

Heat transfer in the unsteady squeezing nanofluid flow between parallel plates is studied using HPM. Figure 3.10a shows the average of error for different steps of HPM; as seen, HPM is converged in step 10 and error has been minimized. The error of f, θ in different location is shown in Figure 3.10b. It can

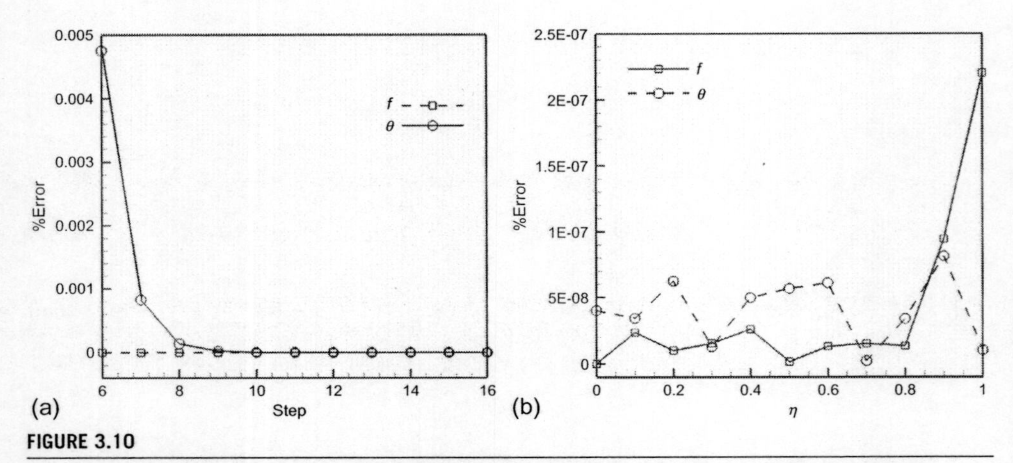

FIGURE 3.10

%Error for $\theta(\eta)$ (a) for different steps of HPM; (b) versus η when $S=0.5$, $Ec=0.1$, $\delta=1$, $\phi=0.06$ and $Pr=6.2$; and

continued

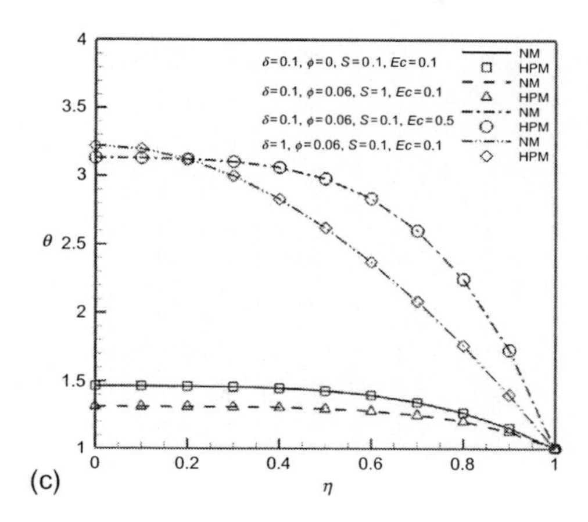

FIGURE 3.10, cont'd

(c) comparison between numerical and HPM solution results for different values of effective parameters.

Table 3.4 Comparison Between the Numerical Results and HPM Solution for $f(\eta)$ and $\theta(\eta)$ when $S=0.5$, $Ec=0.1$, $\delta=1$, $\phi=0.06$ and $Pr=6.2$

	$f(\eta)$			$\theta(\eta)$		
η	NM	HPM	%Error	NM	HPM	%Error
0	0	0	0	2.955692	2.955692393	7.80501E−08
0.1	0.144419	0.144419	6.17833E−08	2.931896	2.931896113	4.85167E−08
0.2	0.286441	0.286441	3.5474E−08	2.861462	2.861461527	6.93453E−08
0.3	0.423607	0.423607	3.58913E−08	2.747105	2.747105304	6.4213E−08
0.4	0.553338	0.553338	2.26784E−08	2.592878	2.592878059	4.04975E−08
0.5	0.672872	0.672872	3.58606E−10	2.403478	2.403477941	1.528E−08
0.6	0.7792	0.7792	4.19237E−08	2.18333	2.18333034	4.55199E−08
0.7	0.869002	0.869002	5.25162E−08	1.935434	1.935433836	3.477E−08
0.8	0.938582	0.938582	4.67179E−08	1.659916	1.659915701	9.80953E−09
0.9	0.983799	0.983799	3.67931E−09	1.352136	1.352136285	7.38832E−10
1	1	1	1.1E−07	1	1.000000001	1E−07

be seen that both functions errors have fluctuating behavior. Validity of HPM is shown in Tables 3.4 and 3.5 and Figure 3.10c.

Excellent agreement between numerical solution obtained by fourth-order Runge-Kutta method and analytical solution is obtained. In these tables and figures, the $\%\text{Error} = \left|\frac{f(\eta)_{\text{NM}} - f(\eta)_{\text{HPM}}}{f(\eta)_{\text{NM}}}\right|$.

Table 3.5 Comparison Between the Numerical Results and HPM Solution for $f(\eta)$ and $\theta(\eta)$ when $S=0.5$, $Ec=0.4$, $\delta=1.5$, $\phi=0.06$ and $Pr=6.2$

	$f(\eta)$			$\theta(\eta)$		
η	NM	HPM	%Error	NM	HPM	%Error
0	0	0	0	14.92923	14.92920334	0.000173028
0.1	0.140402	0.140402	5.48842E−07	14.72736	14.72733466	0.000175225
0.2	0.278858	0.278858	5.12836E−07	14.13412	14.13409813	0.000182378
0.3	0.413324	0.413324	4.74251E−07	13.18484	13.18481286	0.000194316
0.4	0.541556	0.541556	4.3055E−07	11.93268	11.93265657	0.000211946
0.5	0.661003	0.661003	3.72779E−07	10.44102	10.44099373	0.000234478
0.6	0.768691	0.768691	3.26071E−07	8.773987	8.773963841	0.000259448
0.7	0.861096	0.861096	2.80714E−07	6.985854	6.985834397	0.000280704
0.8	0.934001	0.934001	2.31716E−07	5.108728	5.108713075	0.000285736
0.9	0.982336	0.982336	2.35631E−07	3.136528	3.136520514	0.000249339
1	1	1	3E−07	1	1.000000029	2.9E−06

This study is completed by depicting the effects of the squeeze number, the nanofluid volume fraction, and Eckert number on heat transfer characteristics. Effect of squeeze number, volume fraction of nanofluid, Eckert number, and the squeeze number on the temperature profile is shown in Figures 3.11 and 3.12. Also, Figure 3.11c shows the effect of the squeeze number and nanoparticle volume fraction on Nusselt number.

The sensitivity of thermal boundary layer thickness to volume fraction of nanoparticles is related to the increased thermal conductivity of the nanofluid. In fact, higher values of thermal conductivity are accompanied by higher values of thermal diffusivity. The high values of thermal diffusivity cause a drop in the temperature gradients and accordingly increase the boundary thickness. This increase in thermal boundary layer thickness reduces the Nusselt number; however, the Nusselt number is a multiplication of temperature gradient and the thermal conductivity ratio (conductivity of the nanofluid to the conductivity of the base fluid). Since the reduction in temperature gradient due to the presence of nanoparticles is much smaller than thermal conductivity ratio, an enhancement in Nusselt number is taking taken place by increasing the volume fraction of nanoparticles. An increase in the squeeze number can be related with the decrease in the kinematic viscosity, an increase in the distance between the plates and an increase in the speed at which the plates move. Thermal boundary layer thickness increases as the absolute magnitude of the squeeze number decreases when $S<0$ but it increases with increase in $|S|$ when $S>0$. It is obvious that the temperature boundary layer thickness is relatively high when the plates are moving toward each other. Also, it can be seen that increasing Eckert number and δ lead to increase in temperature.

Figure 3.13 shows the effect of squeeze number, Eckert number, and δ on Nusselt number. The presence of viscous dissipation effects significantly increases the temperature. Thus Nusselt number increases with increase of Eckert number. Nusselt number is an increasing function of Eckert number and δ.

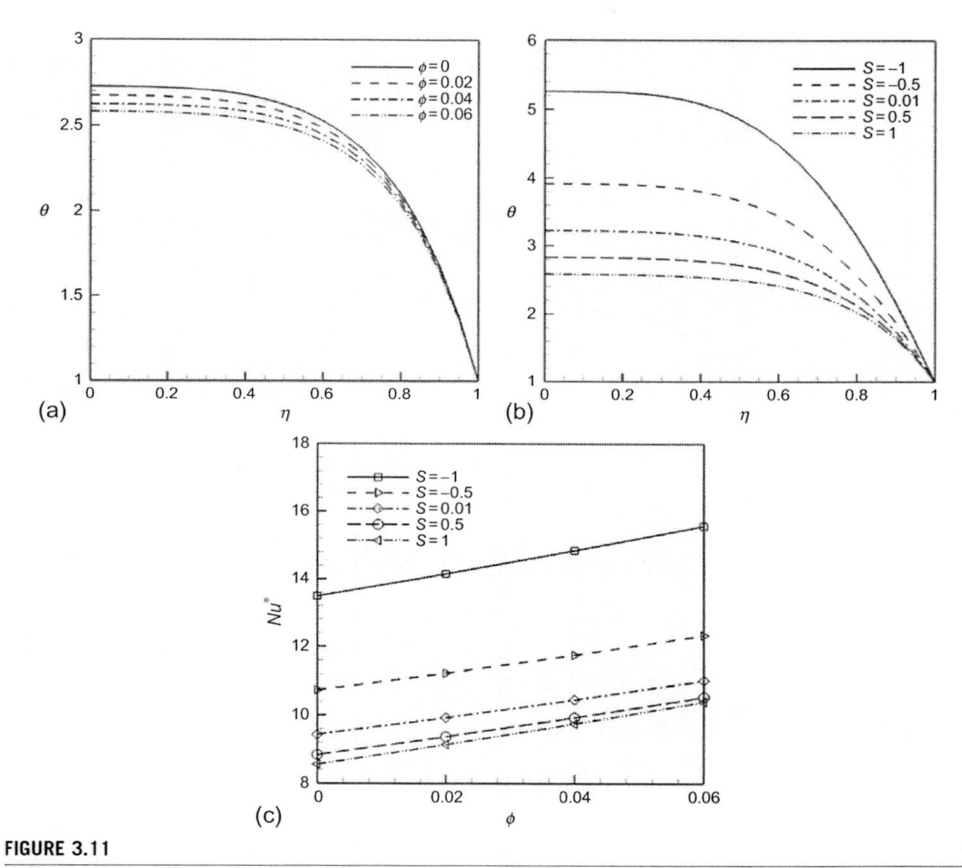

FIGURE 3.11

Effects of the squeeze number and nanoparticle volume friction on the temperature profile and Nusselt number when (a) $Ec = 0.5$, $Pr = 6.2$, $\delta = 0.1$ and $\phi = 0.06$; (b) $Ec = 0.5$, $Pr = 6.2$, $\delta = 0.1$, $S = 1$.

3.2.6 CONCLUSION

In this section, the heat transfer in the unsteady squeezing nanofluid flow between parallel plates was investigated using HPM. The effects of the squeeze number, the nanofluid volume fraction, and Eckert number on temperature profile and Nusselt number have been investigated. The results obtained from numerical solutions via fourth grade order Runge-Kutta are in excellent agreement with those obtained from HPM. The results show Nusselt number increases with increase of nanoparticle volume fraction, δ, the squeeze number when $S < 0$, while it decreases with increase of the squeeze number when $S > 0$.

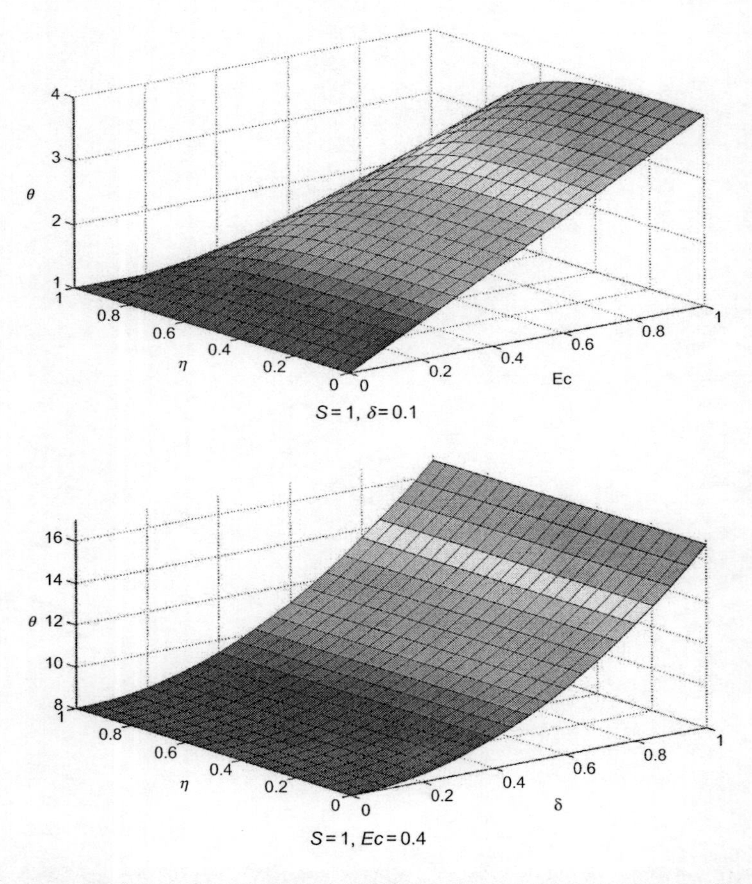

FIGURE 3.12

Effect of the Eckert number and δ on the temperature profile at $\phi = 0.06$ and $Pr = 6.2$.

3.3 SLIP EFFECTS ON UNSTEADY STAGNATION POINT FLOW OF A NANOFLUID OVER A STRETCHING SHEET

Unsteady two-dimensional stagnation point flow of a nanofluid over a stretching sheet is investigated numerically. In contrast to the conventional no-slip condition at the surface, Navier's slip condition has been applied. Nanofluid's behavior was investigated for three different nanoparticles in the water-base fluid, namely, copper, alumina, and titania. Employing similarity variables, the governing partial differential equations including continuity, momentum, and energy have been reduced to ordinary ones and solved via Runge-Kutta-Fehlberg scheme. It was shown that dual solution exists for negative values of A and as it

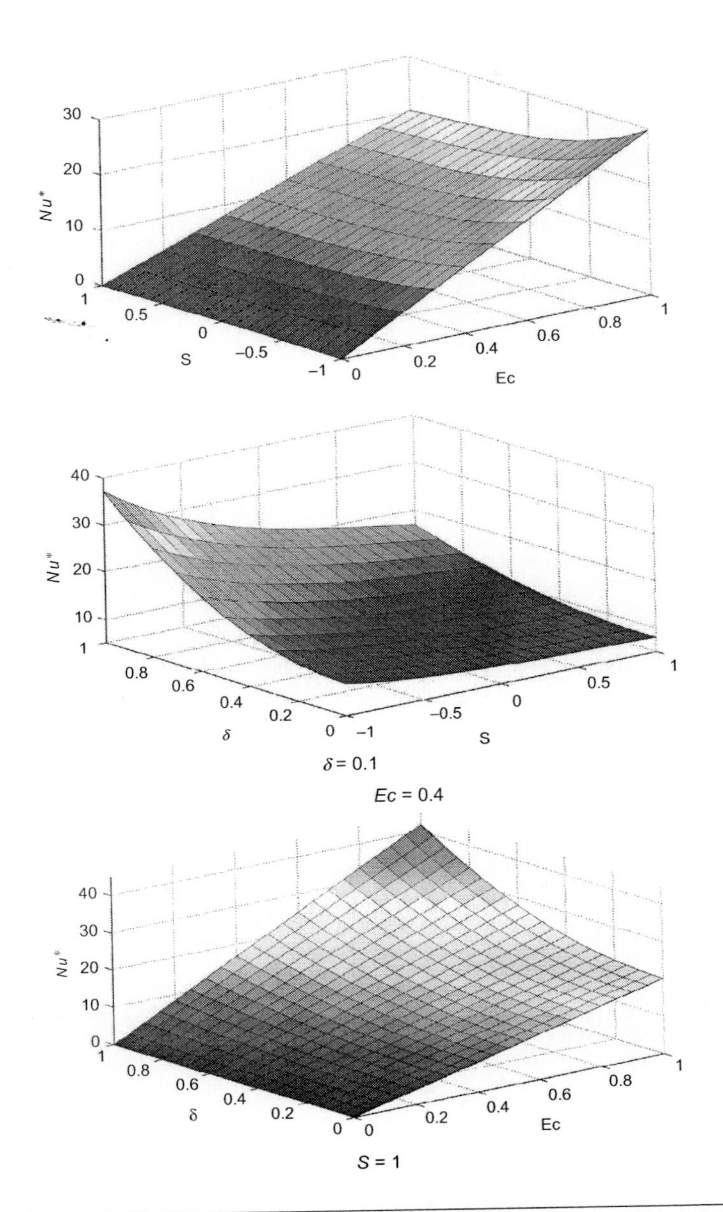

FIGURE 3.13

Effects of the squeeze number, Eckert number, and δ on Nusselt number when $Pr = 6.2$ and $\phi = 0.06$.

increases, skin friction C_{fr} increases but heat transfer rate Nu_r takes a decreasing trend. The results also indicate that unlike the stretching parameter ε, increasing in the values of slip parameter λ may widen the ranges of unsteadiness parameter A for which the solution exist. Furthermore, it was found that an increase in both ε and λ intensifies heat transfer rate (*this chapter has been worked by A. Malvandi, F. Hedayati, D. D. Ganji in nonlinear dynamics team in Mechanical Engineering Department, 2013-2014*).

3.3.1 INTRODUCTION

For years, many researchers have paid a lot attention to viscous fluid motion near the stagnation region of a solid body, where "body" corresponds to either fixed or moving surfaces in a fluid. This multi-disciplinary concept has frequent applications in high speed flows, thrust bearings, and thermal oil recovery (see Bhattacharyya and Vajravelu, 2012; Hiemenz, 1911; Homann and Angew, 1936; Ziabakhsh et al., 2010).

Beside stagnation point flow, stretching surfaces have a wide range of applications in engineering and several technical purposes particularly in metallurgy and polymer industry. For instance, gradual cooling of continuous stretching metal or plastic strips which have multiple applications in mass production. Needles to say, the final quality of the product strongly depends on the rate of heat transfer from the stretching surface.

Improving the technology, it was realized that the energy consumption of the industrial devices and their volumes have to be optimized. So, the idea of adding particles to a conventional fluid in order to enhance its heat transfer characteristics was emerged. Among all dimensions of particles such as macro, micro, and nano, due to some obstacles in pressure drop of the system (Ko et al., 2007) or keeping the mixture homogeneous (Jalaal et al., 2010), nano-scaled particles have attracted more attention. These tiny particles have high thermal conductivity; so, the mixed fluids have better thermal properties. In a different study, stagnation point flow over a stretching/shrinking sheet in a nanofluid has been considered by Bachok et al. (2011a). They have concluded that skin friction and heat transfer coefficients increase as nanoparticle volume climbs up.

This section deals with the effects of slip velocity and stretching parameter on unsteady stagnation point flow of a nanofluid over a stretching sheet where velocity of the sheet and free stream vary continuously with time. The employed model for nanofluid incorporates the effects of unsteadiness parameter, slip parameter, stretching parameter, and solid volume fraction simultaneously. The basic boundary layer equations have been reduced into a two-point boundary value problem via similarity variables, and solved numerically.

3.3.2 GOVERNING EQUATIONS

Consider an incompressible unsteady viscous flow of nanofluids being confined to $y > 0$ toward a stretching sheet coinciding with the plane $y = 0$ with a fixed stagnation point at $x = 0$ as shown in Figure 3.14.

We have assumed that the free stream and sheet's velocity vary with time from a fixed stagnation point in the form of $U_e(x,t) = ax(1-ct)^{-1}$ and $U_w(x,t) = bx(1-ct)^{-1}$, respectively where a, b, c are positive constants. It is also assumed that the temperature at the surface has constant value of T_w, while the ambient temperature beyond boundary layer has constant value of T_∞. The basic unsteady conservation equations of mass, momentum, and thermal energy can be expressed as (see Bachok et al., 2011b)

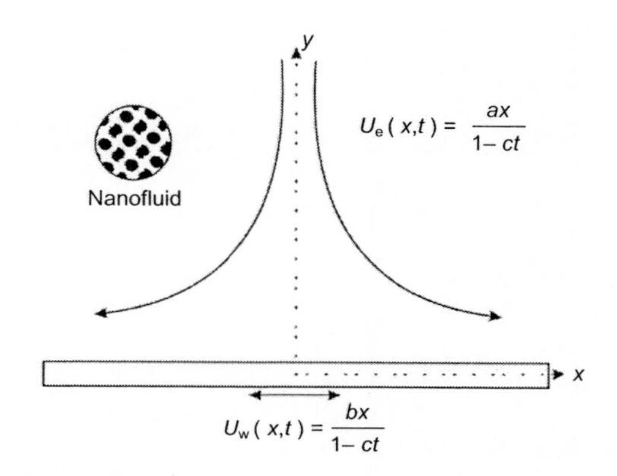

FIGURE 3.14

Physical model and coordinate system.

$$\frac{\partial u}{\partial x}+\frac{\partial v}{\partial y}=0 \tag{3.61}$$

$$\frac{\partial u}{\partial t}+\left(u\frac{\partial u}{\partial x}+v\frac{\partial u}{\partial y}\right)=\frac{\partial U_e}{\partial t}+U_e\frac{dU_e}{dx}+\frac{\mu_{nf}}{\rho_{nf}}\left(\frac{\partial^2 u}{\partial x^2}+\frac{\partial^2 u}{\partial y^2}\right) \tag{3.62}$$

$$\frac{\partial T}{\partial t}+\left(u\frac{\partial T}{\partial x}+v\frac{\partial T}{\partial y}\right)=\alpha_{nf}\left(\frac{\partial^2 T}{\partial x^2}+\frac{\partial^2 T}{\partial y^2}\right) \tag{3.63}$$

Subject to the boundary conditions

$$\begin{aligned} &v=0, \quad u=U_w(x,t)+U_{Slip}(x,t), \quad T=T_w \quad \text{at } y=0 \\ &u=U_e(x,t)=ax(1-ct)^{-1}, \qquad T=T_\infty \quad \text{as } y\to\infty \end{aligned} \tag{3.64}$$

Here, u and v are the velocity components along the x and y directions, respectively, $U_{Slip}(x,t)=N_1\nu\frac{\partial u}{\partial y}$ where $N_1=N\sqrt{t}$ is slip velocity factor and t is the time. T is the temperature, μ_{nf} is the viscosity of nanofluid, ρ_{nf} is the density of nanofluid, α_{nf} is the thermal diffusivity which for spherical nanoparticles are (see Oztop and Abu-Nada, 2008):

$$\begin{aligned} &\alpha_{nf}=\frac{k_{nf}}{(\rho C_p)_{nf}}, \quad \rho_{nf}=(1-\phi)\rho_f+\phi\rho_s, \quad \mu_{nf}=\frac{\mu_f}{(1-\phi)^{2.5}} \\ &(\rho C_p)_{nf}=(1-\phi)(\rho C_p)_f+\phi(\rho C_p)_s, \quad \frac{k_{nf}}{k_f}=\frac{k_s+2k_f-2\phi(k_f-k_s)}{k_s+2k_f+\phi(k_f-k_s)} \end{aligned}$$

Seeking for similarity solution of Equations (3.61)–(3.63) with the following parameters

$$\eta=\left(\frac{a}{\nu_f(1-ct)}\right)^{1/2}y, \quad \psi=\left(\frac{\nu_f a}{(1-ct)}\right)^{1/2}xf(\eta), \quad \theta(\eta)=\frac{T-T_\infty}{T_w-T_\infty} \tag{3.65}$$

The governing equations (3.61)–(3.63) collapse into

$$f''' + (1-\phi)^{2.5}\left(1-\phi+\phi\frac{\rho_s}{\rho_f}\right)\left(f''\left(f-A\frac{\eta}{2}\right) - f'(A+f') + A + 1\right) = 0 \tag{3.66}$$

$$\frac{k_{nf}}{k_f}\theta'' + Pr\left(1-\phi+\phi\frac{(\rho C_p)_s}{(\rho C_p)_f}\right)\left(f-\frac{A}{2}\eta\right)\theta' = 0 \tag{3.67}$$

Subjected to the following dimensionless forms of boundary condition of Equation (3.64)

$$\begin{array}{l} \text{At } \eta=0: f=0, f'=\varepsilon+\lambda f'', \ \theta=1 \\ \text{At } \eta=\infty: f'=1, \ \theta=0 \end{array} \tag{3.68}$$

In the above equations, η is the similarity variable, λ is the slip parameter, $\varepsilon = b/a$ is the stretching parameter, ψ is the usual stream function, i.e., $u = \partial\psi/\partial y$ and $v = -\partial\psi/\partial x$ and prime $'$ denotes differentiation with respect to η, ϕ is the nanoparticle volume fraction and Pr stands for Prandtl number. As a consequential parameter in this study, the local skin-friction coefficient Cf_x can be defined as:

$$Cf_x = \frac{\tau_w}{\rho_f u_w^2} \tag{3.69}$$

where τ_w is the surface shear stress which is given by

$$\tau_w = \mu_{nf}\left(\frac{\partial u}{\partial y}\right)_{y=0} \tag{3.70}$$

Implementing similarity variables of Equation (3.65) into Equation (3.61) we obtain

$$C_{fr} = Re_x{}^{1/2} Cf_x = \frac{1}{(1-\phi)^{2.5}} f''(0) \tag{3.71}$$

where $Re = (u_w x)/\nu_f$ is the local Reynolds number. The physical thermal quantity of interest is local Nusselt number Nu_x, defined as:

$$Nu_x = \frac{x\, q_w}{k_f\,(T_w - T_\infty)} \tag{3.72}$$

where q_w is the surface heat flux which is

$$q_w = -k_{nf}\left(\frac{\partial T}{\partial y}\right)_{y=0} \tag{3.73}$$

Using Equations (3.72) and (3.73) we obtain

$$Nu_r = Re_x{}^{-1/2} Nu_x = -\frac{k_{nf}}{k_f}\theta'(0) \tag{3.74}$$

Like Bachok et al. (2010), in the present context C_{fr} and Nu_r are referred as the reduced skin-friction coefficient and reduced Nusselt number which are represented by Equations (3.71) and (3.74), respectively.

3.3.3 RESULT AND DISCUSSION

The system of Equations (3.66) and (3.67) with boundary conditions of Equation (3.68) have been solved numerically via shooting method based on fourth-order Runge-Kutta. The variations of reduced Nusselt number and skin-friction coefficient are investigated for three different nanofluids, namely, TiO_2, Cu, and Al_2O_3 suspended in water as the base fluid for which $Pr = 6.2$ and $0 < \phi < 0.2$ (see Khanafer et al., 2003; Oztop and Abu-Nada, 2008). Thermophysical properties of the base fluid and the nanoparticles are shown in Table 3.6.

Following branch solutions were obtained by different initial guesses for the missing values of $f'(0)$ and $\theta'(0)$. The best accuracy of the present results has been shown in Table 3.7 where $f''(0)$, $-\theta'(0)$ are compared with reported data of Bachok et al. (2012) for stationary sheets with no-slip condition.

It should be stated at the outset that one error is inevitable, because the physical domain is unbounded whereas the computational domain has to be finite. Here, for our bulk computations, the far field boundary conditions denoted by η_{max} set to $\eta_{max} = 15$ which was sufficient to achieve the far field boundary conditions asymptotically.

Table 3.6 Thermophysical Properties of the Base Fluid and the Nanoparticles

Physical Properties	Fluid Phase (water)	Cu	Al_2O_3	TiO_2
C_p (J/kg K)	4179	385	765	686.2
ρ (kg/m^3)	997.1	8933	3970	4250
k (W/m K)	0.613	400	40	8.9538
$\alpha \times 10^{-7}$ (m^2/s)	1.47	1163.1	131.7	30.7

Table 3.7 Comparison of the Values of $f''(0)$ and $-\theta'(0)$ when $\varepsilon=\lambda=0$

			$f''(0)$		$-\theta'(0)$	
A	Material	ϕ	Bachok et al. (2012)	Present Result	Bachok et al. (2012)	Present Result
1	Cu	0.1	1.7604	1.76039	0.4681	0.46870
		0.2	–	1.82528	–	*0.46779*
	Al_2O_3	0.1	1.4967	1.49669	0.4032	0.40320
		0.2	–	1.43248	–	*0.38477*
	TiO_2	0.1	1.5128	1.51280	0.4076	0.40757
		0.2	–	1.45746	–	*0.39199*
−1	TiO_2	0.1	0.932	0.93199	1.491	1.50207
			−0.9945	*−0.99451*	*0.4491*	*0.45298*
		0.2	–	0.89790	–	1.35558
			–	*−0.95813*	–	*0.41189*

"Italic" *second solution.*

Effects of different nanoparticles, solid volume fraction, stretching, and slip parameters on velocity and temperature gradients for the range of unsteadiness parameter $A < 1$ have been shown in Figures 3.15–3.22. The figures reveal that for $A_c < A < 0$ the solutions are nonunique, i.e., dual solution exists, where A_c is the critical value of A beyond no solution exists. Based on stability analysis, only the first solution (upper branch) is stable and physically accurate. The impacts of slip and stretching parameters on the value of A_c are cited in Table 3.8; clearly, unlike the slip parameter λ, increasing in the values of stretching parameter ε causes a decrease in the absolute values of A_c; in contrast, types of nanoparticles and different solid volume fractions have negligible effects on A_c.

Figures 3.15, 3.17, and 3.19 illustrate that for upper solution, as positive values of A pass to negative ones, the values of $f''(0)$ keep decreasing while for lower solution, it starts with a declining trend to special values of A named A_m beyond climbs up to reach the upper solution value at A_c.

Considering the temperature gradients (Figures 3.16, 3.18, 3.20, and 3.22), lower solution's behavior is similar to $f''(0)$ one, however, the upper solution keeps increasing when A passes from its positive values to negative ones which is supported by Bachok et al. (2012).

Figures 3.23 and 3.24 show the effects of unsteadiness parameter, types of nanoparticle and solid volume fraction for accelerating flow ($A > 0$) on reduced skin-friction coefficient C_{fr}, and reduced Nusselt number Nu_r, respectively. As it is obvious, increase in the value of A leads to a rise in C_{fr} and fall in Nu_r.

In addition, inclusion of nanoparticles enhances C_{fr} and Nu_r. It is not surprising that due to the highest thermal conductivity of Cu among all other particles (Table 3.6), the values of reduced Nusselt number for Cu/water mixture is the greatest.

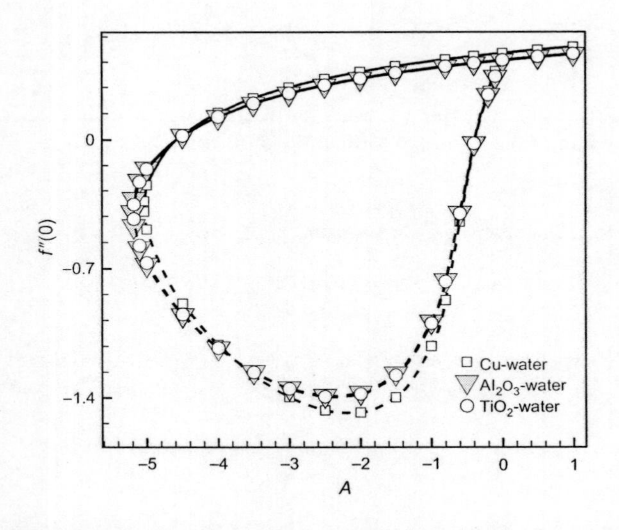

FIGURE 3.15

Variations of $f''(0)$ for different nanoparticles with unsteadiness parameter A when $\varepsilon = \lambda = 0.5, \phi = 0.1$ (dash lines represent the second solution).

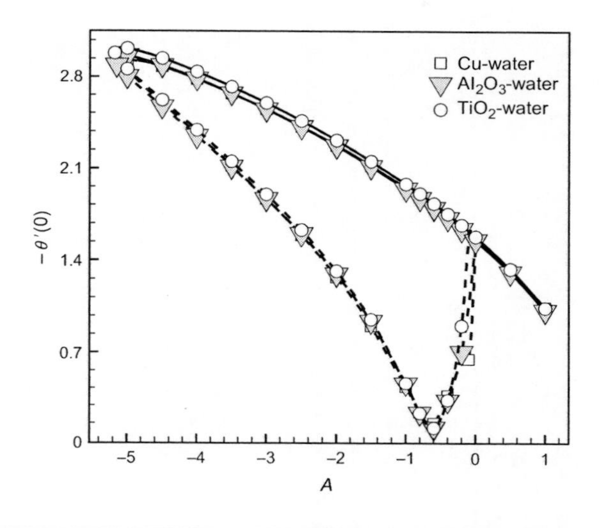

FIGURE 3.16

Variations of $-\theta'(0)$ for different nanoparticles with unsteadiness parameter A when $\varepsilon = \lambda = 0.5, \phi = 0.1$ (dash lines represent the second solution).

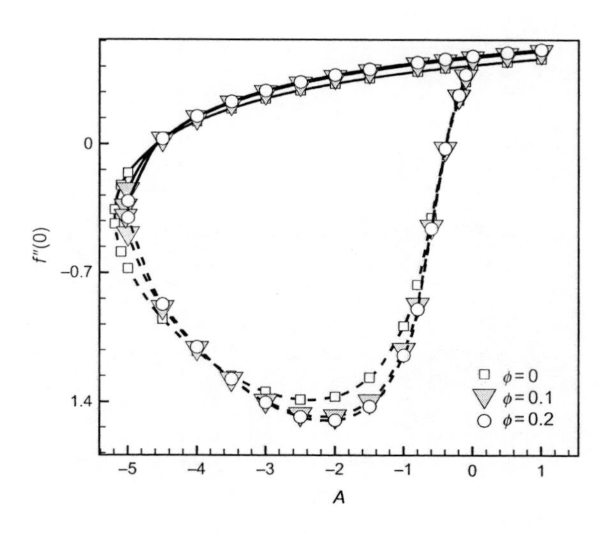

FIGURE 3.17

Variations of $f''(0)$ for Cu-water working fluid with unsteadiness parameter A when $\varepsilon = \lambda = 0.5$ (dash lines represent the second solution).

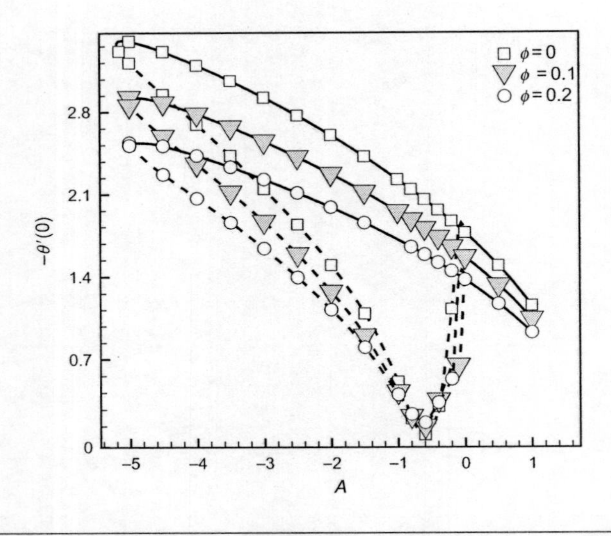

FIGURE 3.18

Variations of $-\theta'(0)$ for Cu-water working fluid with unsteadiness parameter A when $\varepsilon=\lambda=0.5$ (dash lines represent the second solution).

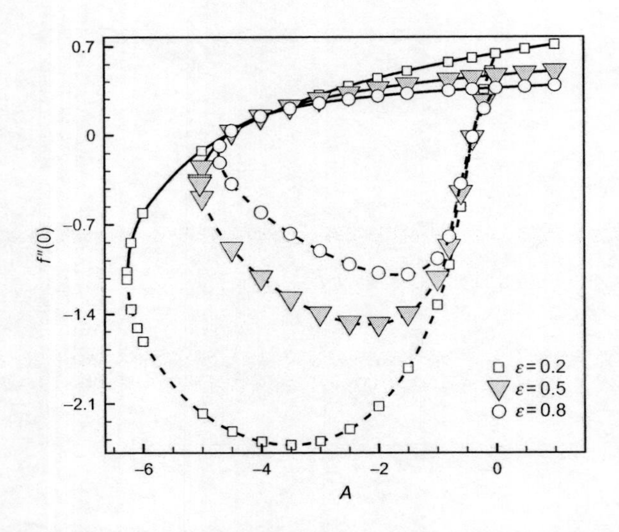

FIGURE 3.19

Variations of $f''(0)$ for Cu-water working fluid with unsteadiness parameter A when $\lambda=0.5, \phi=0.1$ (dash lines represent the second solution).

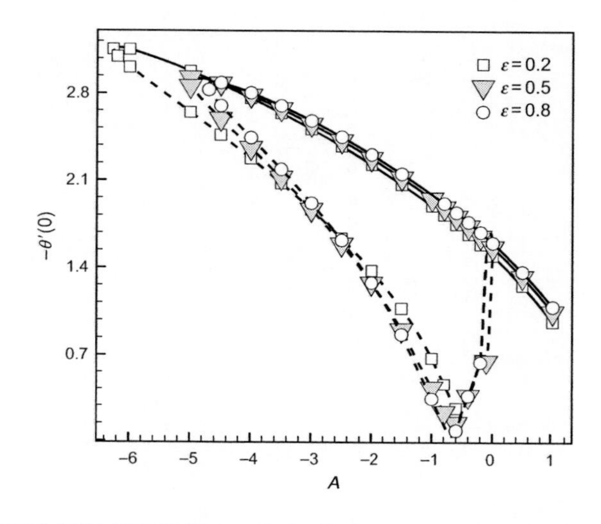

FIGURE 3.20

Variations of $-\theta'(0)$ for Cu-water working fluid with unsteadiness parameter A when $\lambda=0.5, \phi=0.1$ (dash lines represent the second solution).

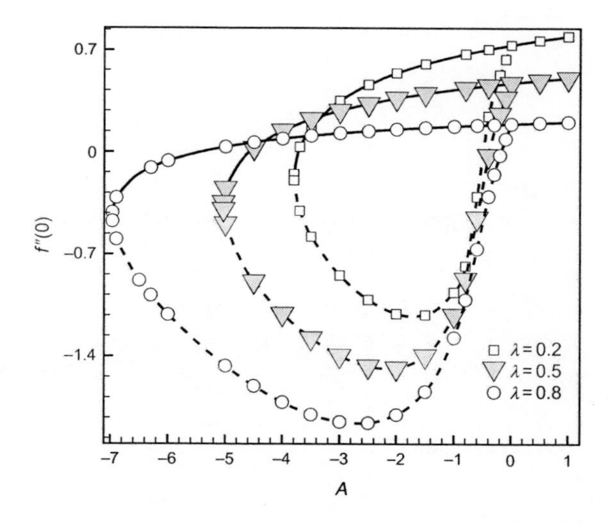

FIGURE 3.21

Variations of $f''(0)$ for Cu-water working fluid with unsteadiness parameter A when $\varepsilon=0.5, \phi=0.1$ (dash lines represent the second solution).

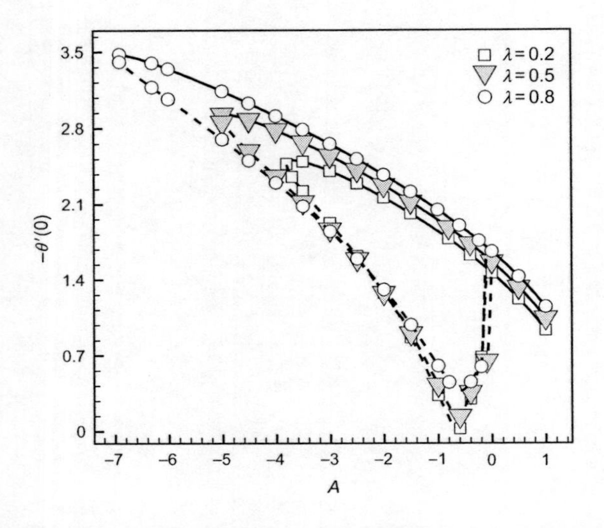

FIGURE 3.22

Variations of $-\theta'(0)$ for Cu-water working fluid with unsteadiness parameter A when $\varepsilon=0.5, \phi=0.1$ (dash lines represent the second solution).

Table 3.8 Critical Value of A_c for Various Values of ε and λ for Cu-Water Nanofluid and $\phi=0.1$

ε	λ	A_c
0.5	0.2	−3.801
0.5	0.5	−5.037
0.5	0.8	−6.973
0.2	0.5	−6.271
0.8	0.5	−4.722

In micro/nanoscale problems, depending on the interfacial roughness and fluid properties, the empirical no-slip boundary condition may no longer exist. Therefore, the slip velocity should be taken into account and there is a certain need to introduce a slip boundary condition λ. Also, the stretching parameter ε, the ratio between the sheet and free stream velocity, may affect the hydrodynamic and thermal boundary layer thickness and change the C_{fr} and Nu_r, so, they should all be considered. From Equation (3.68), it can be mathematically predicted that ε and λ must have similar effects on C_{fr} and Nu_r. Figures 3.25 and 3.26 depict the variations of C_{fr} and Nu_r with stretching parameter ε and slip parameter λ for different nanoparticles. As the eyes see, ε and λ have same effects on C_{fr} and Nu_r; an increase in ε and λ reduces C_{fr} and intensifies Nu_r.

The samples of velocity and temperature profiles for selective values of unsteadiness parameter are presented in Figures 3.27 and 3.28.

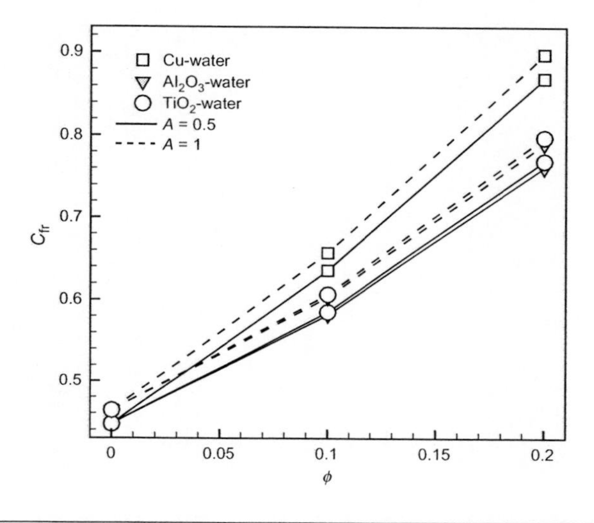

FIGURE 3.23

Effects of unsteadiness parameter for different nanoparticles on reduced skin-friction coefficient C_{fr} when $\varepsilon = \lambda = 0.5, \phi = 0.1$.

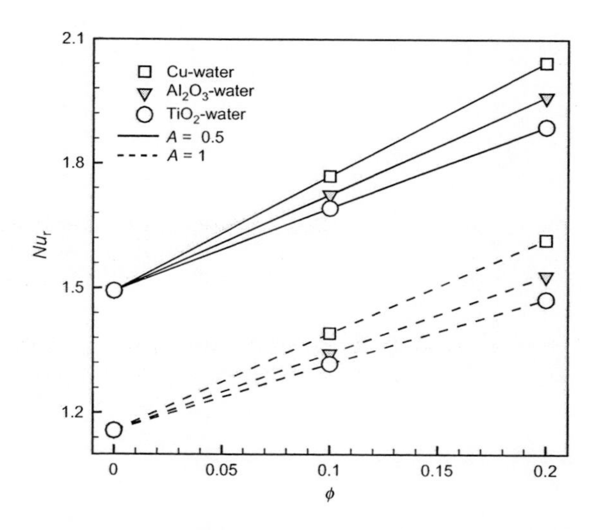

FIGURE 3.24

Effects of unsteadiness parameter for different nanoparticles on reduced Nusselt number Nu_r when $\varepsilon = \lambda = 0.5, \phi = 0.1$.

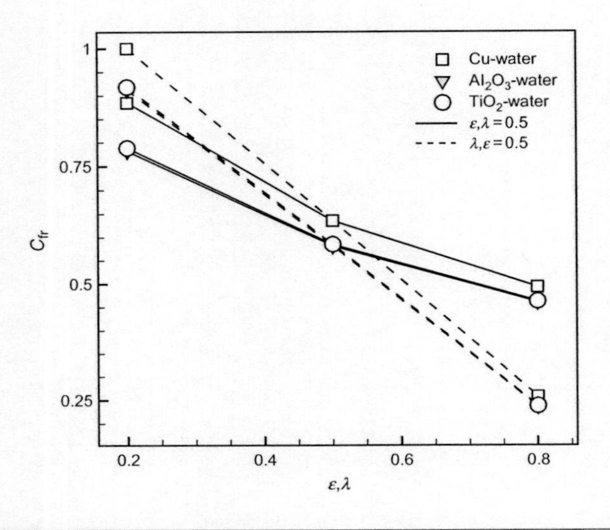

FIGURE 3.25

Effects of stretching parameter ε and slip parameter λ for different nanoparticles on reduced skin-friction coefficient C_{fr} when $\phi = 0.1$.

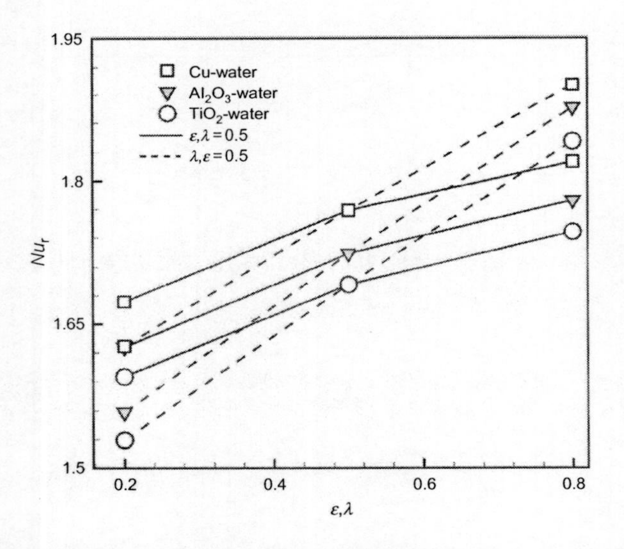

FIGURE 3.26

Effects of stretching parameter ε and slip parameter λ for different nanoparticles on reduced Nusselt number Nu_r when $\phi = 0.1$.

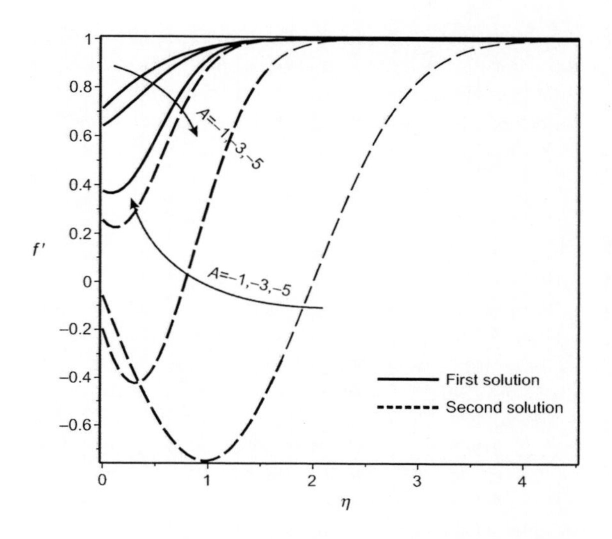

FIGURE 3.27

Effects of unsteadiness parameter A on hydrodynamic boundary layer for Cu-water working fluid when $\varepsilon=\lambda=0.5, \phi=0.1$.

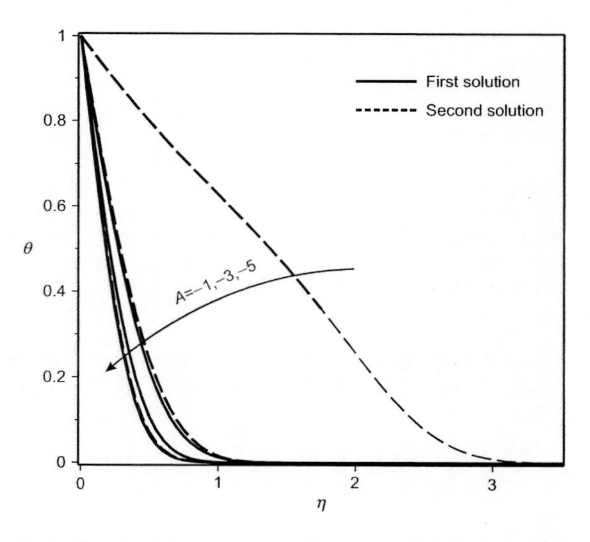

FIGURE 3.28

Effects of unsteadiness parameter A on thermal boundary layer for Cu-water working fluid when $\varepsilon=\lambda=0.5, \phi=0.1$.

These profiles have essentially the same form as that of regular fluids. The figures show that the first and second solutions satisfy far field boundary conditions and can be used for validity of our results as well as supporting the existence of the dual solution. Further, it can be observed that as A tends to A_c, first and second solution profiles get closer together which is evident that at $A = A_c$, the profiles overlap, i.e., one solution exists in A_c.

3.3.4 CONCLUSION

Unsteady two-dimensional stagnation point flow of a nanofluid over a stretching sheet is investigated. In contrast to the conventional no-slip condition at the surface, Navier's slip condition has been applied. Reducing the governing partial differential equations to ordinary ones via similarity variables, the obtained equations have been solved numerically. Assuming water as the base fluid $Pr = 6.2$ and unsteadiness parameter $A < 1$, results have been obtained for three different type of nanoparticles, namely, copper (Cu), alumina (Al_2O_3), and titania (TiO_2) with solid volume fraction $\phi < 0.2$. The outcomes reveal that unlike stretching parameter ε, increase in the values of slip parameter λ may widen the ranges of unsteadiness parameter A for which the solution exists. In addition, increase in the values of slip parameter λ and stretching parameter ε leads to drop in the values of C_{fr} and rise in Nu_r. Moreover, as the acceleration parameter ($A > 0$) increases, reduced Nusselt number Nu_r decreases while reduced skin-friction coefficient C_{fr} grows.

REFERENCES

Ampofo, F., Karayiannis, T.G., 2003. Experimental benchmark data for turbulent natural convection in an air filled square cavity. Int. J. Heat Mass Transfer 46, 3551–3572.

Bachok, N., Ishak, A., Pop, I., 2010. Boundary-layer flow of nanofluids over a moving surface in a flowing fluid. Int. J. Therm. Sci. 49 (9), 1663–1668.

Bachok, N., Ishak, A., Pop, I., 2011a. On the stagnation-point flow towards a stretching sheet with homogeneous-heterogeneous reactions effects. Commun. Nonlinear Sci. Numer. Simul. 16 (11), 4296–4302.

Bachok, N., Ishak, A., Pop, I., 2011b. Stagnation-point flow over a stretching/shrinking sheet in a nanofluid. Nanoscale Res. Lett. 6.

Bachok, N., Ishak, A., Pop, I., 2012. The boundary layers of an unsteady stagnation-point flow in a nanofluid. Int. J. Heat Mass Transfer 55 (23), 6499–6505.

Barakos, G., Mitsoulis, E., Assimacopoulos, D., 1994. Natural convection flow in a square cavity revisited: laminar and turbulent models with wall function. Int. J. Numer. Method Fluids 18, 695–719.

Bhattacharyya, K., Vajravelu, K., 2012. Stagnation-point flow and heat transfer over an exponentially shrinking sheet. Commun. Nonlinear Sci. Numer. Simul. 17 (7), 2728–2734.

Chen, S., 2009. A large-eddy-based Lattice Boltzmann model for turbulent flow simulation. Appl. Math. Comput. 215, 591–598.

Choi, S.U.S., 1995. Enhancing thermal conductivity of fluids with nanoparticles. In: Siginer, D.A., Wang, H.P. (Eds.), Developments and Applications of Non-Newtonian Flows. FED-V.231/MD-V.66. ASME, New York, NY, USA, pp. 99–105.

Dixit, H.N., Babu, V., 2006. Simulation of high Rayleigh number natural convection in a square cavity using the Lattice Boltzmann method. Int. J. Heat Mass Transfer 46, 727–739.

Frisch, U., 1995. Turbulence: The Legacy of A.N. Kolmogorov. Cambridge University Press, New York.

He, J.H., 2004. Comparison of homotopy perturbation method and homotopy analysis method. Appl. Math. Comput. 156, 527–539.

Hiemenz, K., 1911. Die Grenzschicht an einem in den gleichförmigen Flüssigkeitsstrom eingetauchten geraden Kreiszylinder. Dingl. Polytech. J. 326, 321–410.

Ho, C.J., Chen, M.W., Li, Z.W., 2008. Numerical simulation of natural convection of nanofluid in a square enclosure: effects due to uncertainties of viscosity and thermal conductivity. Int. J. Heat Mass Transfer 51, 4506–4516.

Homann, F., Angew, Z., 1936. Der Einfluss Grosser Zahighkeit bei der Stromung um den Zylinder und um die KugelVol. Math. Mech. 16, 11.

Hou, S., Sterling, J., Chen, S., Doolen, G.D., 1996. A Lattice Boltzmann subgrid model for high Reynolds number flows. Fields Inst. Commun. 6, 151–166.

Jahanshahi, M., Hosseinizadeh, S.F., Alipanah, M., Dehghani, A., Vakilinejad, G.R., 2010. Numerical simulation of free convection based on experimental measured conductivity in a square cavity using water/SiO_2 nanofluid. Int. Commun. Heat Mass Transfer 37, 687–694.

Jalaal, M., Ganji, D.D., Ahmadi, G., 2010. Analytical investigation on acceleration motion of a vertically falling spherical particle in incompressible Newtonian media. Adv. Powder Technol. 21 (3), 298–304.

Kefayati, G.H.R., Hosseinizadeh, S.F., Gorji, M., Sajjadi, H., 2011. Lattice Boltzmann simulation of natural convection in tall enclosures using water/SiO_2 nanofluid. Int. Commun. Heat Mass Transfer 38, 798–805.

Khanafer, K., Vafai, K., Lightstone, M., 2003. Buoyancy-driven heat transfer enhancement in a two-dimensional enclosure utilizing nanofluids. Int. J. Heat Mass Transfer 46 (19), 3639–3653.

Ko, G.H., Heo, K., Lee, K., Kim, D.S., Kim, C., Sohn, Y., Choi, M., 2007. An experimental study on the pressure drop of nanofluids containing carbon nanotubes in a horizontal tube. Int. J. Heat Mass Transfer 50 (23-24), 4749–4753.

Mahmood, M., Asghar, S., Hossain, M.A., 2007. Squeezed flow and heat transfer over a porous surface for viscous fluid. Heat Mass Transfer 44, 165–173.

Mustafa, M., Hayat, T., Obaidat, S., 2012. On heat and mass transfer in the unsteady squeezing flow between parallel plates. Meccanica, 1581–1589.

Oztop, H.F., Abu-Nada, E., 2008. Numerical study of natural convection in partially heated rectangular enclosures filled with nanofluids. Int. J. Heat Fluid Flow 29 (5), 1326–1336.

Sajjadi, H., Gorji, M., Hosseinizadeh, S.F., Kefayati, G.H.R., Ganji, D.D., 2011. Numerical analysis of turbulent natural convection in square cavity using large-eddy simulation in Lattice Boltzmann method. Iran. J. Sci. Technol. 35, 133–142.

Sheikholeslami, M., Gorji-Bandpay, M., Ganji, D.D., 2012. Magnetic field effects on natural convection around a horizontal circular cylinder inside a square enclosure filled with nanofluid. Int. Commun. Heat Mass Transfer 39, 978–986.

Stefan, M.J., 1874. Versuch Über die scheinbare adhesion. Akad. Wissensch. Wien Math. Natur. 69, 713–721.

Xuan, Y., Li, Q., 2003. Investigation on convective heat transfer and flow features of nanofluids. ASME J. Heat Transfer 125 (1), 151–155.

Ziabakhsh, Z., Domairry, G., Mozaffari, M., Mahbobifar, M., 2010. Analytical solution of heat transfer over an unsteady stretching permeable surface with prescribed wall temperature. J. Taiwan Inst. Chem. Eng. 41 (2), 169–177.

This page intentionally left blank

CHAPTER

HEAT TRANSFER IN NANOFLUID

4

CHAPTER CONTENTS

4.1 Boundary-Layer Flow of Nanofluids Over a Moving Surface in a Flowing Fluid 110
4.1.1 Introduction 110
4.1.2 Mathematical Model 111
4.1.3 Analytical Solution by Homotopy Analysis Method 113
4.1.3.1 Zeroth-Order Deformation Problems 113
4.1.3.2 mth-Order Deformation Problems 114
4.1.4 Convergence of the HAM Solution 115
4.1.5 Results and Discussion 115
4.1.6 Conclusions 127
4.2 Heat Transfer in a Liquid Film of Nanofluid on an Unsteady Stretching Sheet 127
4.2.1 Introduction 127
4.2.2 Problem Formulation and Governing Equation 128
4.2.3 Numerical Procedure and Validation 131
4.2.4 Results and Discussion 132
4.2.5 Conclusions 137
4.3 Investigation of Squeezing Unsteady Nanofluid Flow Using ADM 138
4.3.1 Introduction 138
4.3.2 Governing Equations 138
4.3.3 Fundamentals of Adomian Decomposition Method (ADM) 141
4.3.4 Solution with Adomian Decomposition Method 142
4.3.5 Results and Discussion 143
4.3.6 Conclusion 149
4.4 Investigation on Entropy Generation of Nanofluid Over a Flat Plate 149
4.4.1 Introduction 150
4.4.2 Governing Equation 150
4.4.3 Entropy Generation 153
4.4.4 Results and Discussion 154
4.4.5 Conclusion 161
4.5 Viscous Flow and Heat Transfer of Nanofluid Over Nonlinearly Stretching Sheet 162
4.5.1 Introduction 162
4.5.2 Basic Concepts of HPM 163
4.5.3 Formulation of Problem 164

Application of Nonlinear Systems in Nanomechanics and Nanofluids. http://dx.doi.org/10.1016/B978-0-323-35237-6.00004-2

4.5.4 Homotopy Perturbation Solution 166
4.5.5 Padé Approximation 169
4.5.6 Results and Discussion 169
4.5.7 Conclusions 176
References 178

4.1 BOUNDARY-LAYER FLOW OF NANOFLUIDS OVER A MOVING SURFACE IN A FLOWING FLUID

The steady boundary-layer flow of a nanofluid over a moving surface in a flowing fluid is investigated analytically by using the Homotopy Analysis Method (HAM). The plate is assumed to move in the same direction to the free stream. Analytical results are obtained for the skin-friction coefficient, namely, the local Nusselt number and the local Sherwood number. The results show that the reduced Nusselt number is a decreasing function of each dimensionless number, while the reduced Sherwood number is an increasing function of higher Pr and a decreasing function of lower Pr number for each Le, Nb, and Nt numbers like the results presented by Bachok. Contrary to the results presented by Bachok, it is found that the reduced Nusselt number increases with the increase in Pr for a special Nb number. Also, whenever the velocity of surface is greater than the velocity of the free stream, this conclusion would be correct. As the results obtained, the reduced Sherwood number has reverse behavior whenever the velocity of surface is greater than the velocity of the free stream (*this chapter has been worked by M. Hassani, M. Mohammad Tabar, A. Mohammad Tabar, G. Domairry in non-linear dynamics team in Mechanical Engineering Department, 2012-2013*).

4.1.1 INTRODUCTION

The problem of viscous boundary-layer flow on a moving or fixed flat plate is a classical problem, and it has been considered by many researchers, for example, see Afzal (1993), Weidman et al. (2006), and Cortell (2007a). The flow and heat transfer of a viscous fluid over a moving surface has many important applications in the modern industry such as polymer industry, glass fiber drawing, crystal growing, plastic extrusion, continuous casting, etc. (see Magyari and Keller, 2000).

Much literature on this problem has been cited in the recent paper by Ishak et al. (2009). Khan and Pob (2010) analyzed the development of the steady boundary layer flow, heat transfer, and nanoparticle fraction over a stretching surface in a nanofluid. Their solution depended on a *Pr*andtl number Pr, a Lewis number Le, a Brownian motion number Nb, and a thermophoresis number Nt. The dependency of the local Nusselt and local Sherwood numbers on these four parameters was numerically investigated. Bachok et al. (2010) investigated the steady boundary-layer flow of a nanofluid past a moving semi-infinite flat plate in a uniform free stream. Numerical results were obtained for the skin-friction coefficient, the local Nusselt number, and the local Sherwood number as well as the velocity, temperature, and the nanoparticle volume fraction profiles for some values of the governing parameters, namely, the plate velocity parameter, Prandtl number, Lewis number, the Brownian motion parameter, and the thermophoresis parameter.

In this section, the development of boundary-layer flow on a fixed or moving surface in a nanofluid is studied analytically by using the HAM (see Liao, 2003, 2004). The results are discussed more in detail and compared with the results presented by Bachok et al. (2010) in some cases. This solution depends on a Prandtl number Pr, a Lewis number Le, a Brownian motion number Nb, and a thermophoresis number Nt. The dependency of the local Nusselt and local Sherwood numbers on these four parameters is analytically investigated. The dependency of the results to parameters mentioned above has clearly been shown by more figures than the results presented by Bachok et al. (2010). Also the discrepancy of our results and Bachok's results are clearly discussed by some figures.

4.1.2 MATHEMATICAL MODEL

The steady boundary-layer flow of a nanofluid past a moving semi-infinite flat plate is considered in a uniform free stream as shown in Figure 4.1.

It is assumed that the velocity of the uniform free stream is U and that of the flat plate is $U_w = \lambda U$, where λ, is the plate velocity parameter (see Bachok et al., 2010). The flow takes place at $y \geq 0$, where y is the coordinate measured normal to the moving surface. Following the conservation of mass, momentum, thermal energy, and nanoparticle fraction, the basic equations are:

$$\nabla.\mathbf{v} = 0 \tag{4.1}$$

$$\rho_f\left(\frac{\partial \mathbf{v}}{\partial t} + \mathbf{v}.\nabla\mathbf{v}\right) = -\nabla p + \mu\nabla^2\mathbf{v} \tag{4.2}$$

$$(\rho c)_f\left(\frac{\partial T}{\partial t} + \mathbf{v}.\nabla T\right) = k\nabla^2 T + (\rho c)_p[D_B\nabla C.\nabla T + (D_T/T_\infty)\nabla T.\nabla T] \tag{4.3}$$

$$\frac{\partial C}{\partial t} + \mathbf{v}.\nabla C = D_B\nabla^2 C + (D_T/T_\infty)\nabla^2 T \tag{4.4}$$

We consider a steady-state flow and make the standard boundary layer approximations, based on a scale analysis, and write the governing Equations (4.1)-(4.4) as:

$$\frac{\partial u}{\partial x} + \frac{\partial v}{\partial y} = 0 \tag{4.5}$$

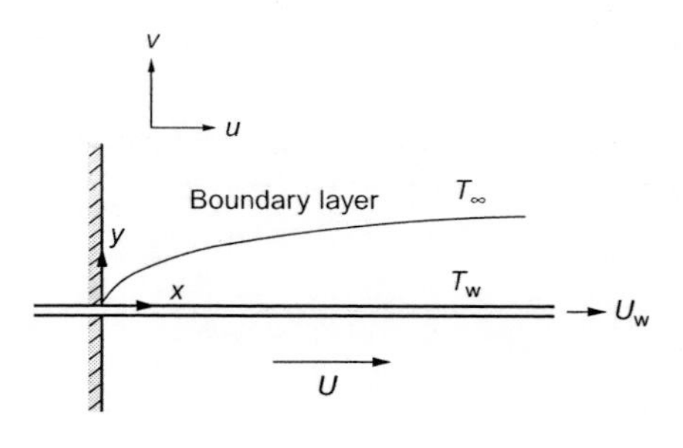

FIGURE 4.1

Physical model and coordinate system.

$$u\frac{\partial u}{\partial x}+v\frac{\partial u}{\partial y}=\upsilon\left(\frac{\partial^2 u}{\partial y^2}\right) \tag{4.6}$$

$$u\frac{\partial T}{\partial x}+v\frac{\partial T}{\partial y}=\alpha\nabla^2 T+\tau\left\{D_{\mathrm{B}}\frac{\partial C}{\partial y}\frac{\partial T}{\partial y}+\left(\frac{D_{\mathrm{T}}}{T_\infty}\right)\left(\frac{\partial T}{\partial y}\right)^2\right\} \tag{4.7}$$

$$u\frac{\partial C}{\partial x}+v\frac{\partial C}{\partial y}=D_{\mathrm{B}}\frac{\partial^2 C}{\partial y^2}+\left(\frac{D_{\mathrm{T}}}{T_\infty}\right)\frac{\partial^2 T}{\partial y^2} \tag{4.8}$$

The boundary conditions of Equations (4.5)-(4.8) are taken to be

$$\begin{aligned} &v=0,\ u=U_{\mathrm{w}}=\lambda U,\ T=T_{\mathrm{w}},\ C=C_{\mathrm{w}} && \text{at } y=0\\ &u\to U,\ T\to T_\infty,\ C\to C_\infty && \text{at } y\to\infty \end{aligned} \tag{4.9}$$

Where the stream function ψ, is defined in the usual way as $u=\partial\psi/\partial y$ and $v=-\partial\psi/\partial x$, which identically satisfies Equation (4.5). Equations (4.6)-(4.8) can then be written as:

$$\frac{\partial\psi}{\partial y}\frac{\partial^2\psi}{\partial x\partial y}-\frac{\partial\psi}{\partial x}\frac{\partial^2\psi}{\partial y^2}=\upsilon\frac{\partial^3\psi}{\partial y^3} \tag{4.10}$$

$$\frac{\partial\psi}{\partial y}\frac{\partial T}{\partial x}-\frac{\partial\psi}{\partial x}\frac{\partial T}{\partial y}=\alpha\frac{\partial^2 T}{\partial y^2}+\tau\left\{D_{\mathrm{B}}\frac{\partial C}{\partial y}\frac{\partial T}{\partial y}+\left(\frac{D_{\mathrm{T}}}{T_\infty}\right)\left(\frac{\partial T}{\partial y}\right)^2\right\} \tag{4.11}$$

$$\frac{\partial\psi}{\partial y}\frac{\partial C}{\partial x}-\frac{\partial\psi}{\partial x}\frac{\partial C}{\partial y}=D_{\mathrm{B}}\frac{\partial^2 C}{\partial y^2}+\left(\frac{D_{\mathrm{T}}}{T_\infty}\right)\frac{\partial^2 T}{\partial y^2} \tag{4.12}$$

Further, we look for a similarity solution of Equations (4.10)-(4.12) by introducing the following similarity transformation:

$$\begin{aligned} &\psi=(2U\upsilon x)^{1/2}f(\eta) && \theta(\eta)=\frac{T-T_\infty}{T_{\mathrm{w}}-T_\infty}\\ &\varphi(\eta)=\frac{C-C_\infty}{C_{\mathrm{w}}-C_\infty} && \eta=(u/2\upsilon x)^{1/2}y \end{aligned} \tag{4.13}$$

On substituting Equation (4.13) into Equations (4.10)-(4.12), we obtain the following nonlinear ordinary differential equations (ODEs):

$$f'''+ff''=0 \tag{4.14}$$

$$\frac{1}{Pr}\theta''+f\theta'+Nb\varphi'\theta'+Nt\theta'^2=0 \tag{4.15}$$

$$\varphi''+Le\,f\varphi'+\frac{Nt}{Nb}\theta''=0 \tag{4.16}$$

Subject to the boundary conditions

$$f(0)=0,\ f'(0)=\lambda,\ \theta(0)=1,\ \varphi(0)=1,\ f'(\infty)\to 1,\ \theta(\infty)\to 0,\ \varphi(\infty)\to 0 \tag{4.17}$$

where primes denote differentiation with respect to η and the four parameters are defined by

$$Pr=\frac{\upsilon}{\alpha},\quad Le=\frac{\upsilon}{D_B},\quad Nb=\frac{(\rho C)_p D_B(\varphi_w-\varphi_\infty)}{(\rho C)_f \upsilon},\quad Nt=\frac{(\rho C)_p D_T(T_w-T_\infty)}{(\rho C)_f T_\infty \upsilon} \tag{4.18}$$

Here Pr, Le, Nb, and Nt denote the Prandtl number, Lewis number, the Brownian motion parameter, and the thermophoresis parameter, respectively. We notice that when Nb and Nt are zero, Equations (4.14) and (4.15) involve just two dependent variables, namely, $f(\eta)$ and $\theta(\eta)$, and the boundary-value problem for these two variables reduces to the classical problem of Weidman et al. (2006) for an impermeable moving surface in a Newtonian fluid. (The boundary-value problem for $\varphi(\eta)$ then becomes ill-posed and is of no physical significance). Quantities of practical interest are the skin-friction coefficient C_f, the local Nusselt number Nu_x and the local Sherwood number Sh_x which are defined as:

$$C_f=\frac{\tau_w}{\rho U^2},\quad Nu_x=\frac{x q_w}{k(T_w-T_\infty)},\quad Sh_x=\frac{x q_m}{D_B(C_w-C_\infty)} \tag{4.19}$$

where τ_w, q_w and q_m are the shear stress, heat flux, and mass flux at the surface. Using variables Equation (4.13), we obtain

$$(2Re_x)^{1/2}C_f=f''(0),\quad (Re_x/2)^{-1/2}\ Nu_x=-\theta'(0),\quad (Re_x/2)^{-1/2}\ Sh=-\varphi'(0) \tag{4.20}$$

where $Re_x=Ux/\upsilon$ is the local Reynolds number. In the present context $Re_x^{-1/2}Nu_x$ and $Re_x^{-1/2}Sh_x$ are referred to as the reduced Nusselt number and reduced Sherwood number, which are represented by $-\theta'(0)$ and $-\varphi'(0)$, respectively.

4.1.3 ANALYTICAL SOLUTION BY HOMOTOPY ANALYSIS METHOD

For HAM solutions, we choose the initial guesses and auxiliary linear operator in the following form:

$$f_0(\eta)=\eta+(\lambda-1)(1-e^{-\eta}),\quad \phi_0(\eta)=e^{-\eta},\quad \theta_0(\eta)=e^{-\eta} \tag{4.21}$$

$$L(f)=f'''-f',\quad L(\phi)=\phi''-\phi,\quad L(\theta)=\theta''-\theta \tag{4.22}$$

$$L(c_1+c_2e^{\eta}+c_3e^{-\eta})=L(c_1e^{\eta}+c_2e^{-\eta})=L(c_1e^{\eta}+c_2e^{-\eta})=0 \tag{4.23}$$

And c_i ($i=1$-3) are constants and $p\in[0,1]$ denotes the embedding parameter and $\hbar$ indicate the none-zero auxiliary parameters. We then construct the following problems:

4.1.3.1 Zeroth-order deformation problems

$$\begin{aligned}(1-p)L_1[f(\eta,p)-f_0(\eta)]&=p\hbar_1 N_1[f(\eta,p),\phi(\eta,p),\theta(\eta,p)]\\(1-p)L_2[\phi(\eta,p)-\phi_0(\eta)]&=p\hbar_2 N_2[f(\eta,p),\phi(\eta,p),\theta(\eta,p)]\\(1-p)L_3[\theta(\eta,p)-\theta_0(\eta)]&=p\hbar_3 N_3[f(\eta,p),\phi(\eta,p),\theta(\eta,p)]\end{aligned} \tag{4.24}$$

$$\begin{aligned}&f(0,p)=0\ \ f'(0,p)=\lambda\ \ f'(\infty,p)=0\\&\phi(0,p)=0\ \ \phi(\infty,p)=0\\&\theta(0,p)=1\ \ \theta(\infty,p)=0\end{aligned} \tag{4.25}$$

$$
\begin{aligned}
N_1[f(\eta,p),\phi(\eta,p),\theta(\eta,p)] &= \frac{\partial^3 f(\eta,p)}{\partial\eta^3} + f(\eta,p)\frac{\partial^2 f(\eta,p)}{\partial\eta^2} \\
N_2[f(\eta,p),\phi(\eta,p),\theta(\eta,p)] &= \frac{\partial^2\phi(\eta,p)}{\partial\eta^2} + Le\left(\frac{\partial\phi(\eta,p)}{\partial\eta}\right)f(\eta,p) + \frac{Nt}{Nb}\left(\frac{\partial^2\theta(\eta,p)}{\partial\eta^2}\right) \\
N_3[f(\eta,p),\phi(\eta,p),\theta(\eta,p)] &= \frac{1}{Pr}\frac{\partial^2\theta(\eta,p)}{\partial\eta^2} + f(\eta;p)\frac{\partial\theta(\eta,p)}{\partial\eta} \\
&\quad + Nb\left(\frac{\partial\phi(\eta,p)}{\partial\eta}\right)\left(\frac{\partial\theta(\eta,p)}{\partial\eta}\right) + Nt\left(\frac{\partial\theta(\eta,p)}{\partial\eta}\right)^2,
\end{aligned} \tag{4.26}
$$

For $p=0$ and $p=1$ we have

$$
\begin{aligned}
f(\eta,0) &= f_0(\eta) \quad f(\eta,1) = f(\eta) \\
\phi(\eta,0) &= \phi_0(\eta) \quad \phi(\eta,1) = \phi(\eta) \\
\theta(\eta,0) &= \theta_0(\eta) \quad \theta(\eta,1) = \theta(\eta)
\end{aligned} \tag{4.27}
$$

Due to tailor's series with respect to p, we have:

$$
\begin{aligned}
f(\eta,p) &= f_0(\eta) + \sum_{m=1}^{\infty} f_m(\eta)p^m \\
\phi(\eta,p) &= \phi_0(\eta) + \sum_{m=1}^{\infty} \phi_m(\eta)p^m \\
\theta(\eta,p) &= \theta_0(\eta) + \sum_{m=1}^{\infty} \theta_m(\eta)p^m
\end{aligned} \tag{4.28}
$$

$$
\begin{aligned}
f_m(\eta) &= \frac{1}{m!}\frac{\partial^m(f(\eta,p))}{\partial p^m} \\
\phi_m(\eta) &= \frac{1}{m!}\frac{\partial^m(\phi(\eta,p))}{\partial p^m} \\
\theta_m(\eta) &= \frac{1}{m!}\frac{\partial^m(\theta(\eta,p))}{\partial p^m}
\end{aligned} \tag{4.29}
$$

4.1.3.2 mth-order deformation problems

$$
\begin{aligned}
L_1[f_m(\eta) - \chi_m f_{m-1}(\eta)] &= \hbar_1 R_m^f(\eta) \\
L_2[\phi_m(\eta) - \chi_m \phi_{m-1}(\eta)] &= \hbar_2 R_m^\phi(\eta) \\
L_3[\theta_m(\eta) - \chi_m \theta_{m-1}(\eta)] &= \hbar_3 R_m^\theta(\eta)
\end{aligned} \tag{4.30}
$$

$$
\begin{aligned}
f_m(0) = f'_m(0) &= f'_m(\infty) = 0 \\
\phi_m(0) &= \phi_m(\infty) = 0 \\
\theta_m(0) &= \theta_m(\infty) = 0
\end{aligned} \tag{4.31}
$$

$$
\begin{aligned}
R_m^f &= f'''_{m-1} + \sum_{n=0}^{m-1} f_{m-1-n} f''_n \\
R_m^\phi &= \phi''_{m-1} + Le\sum_{n=0}^{m-1} f_{m-1-n}\phi'_n + \frac{Nt}{Nb}\theta''_{m-1} \\
R_m^\theta &= \frac{1}{Pr}\theta''_{m-1} + \sum_{n=0}^{m-1} f_{m-1-n}\theta'_n + Nb\sum_{n=0}^{m-1}\phi'_{m-1-n}\theta'_n + Nt\sum_{n=0}^{m-1}\theta'_{m-1-n}\theta'_n
\end{aligned} \tag{4.32}
$$

$$\chi_m \begin{cases} 0 & m \leq 1 \\ 1 & m > 1 \end{cases} \tag{4.33}$$

which $\hbar$ is chosen in such a way that these three series are convergent at $p = 1$, therefore we have through Equations (4.28) and (4.29) that

$$\begin{aligned} f(\eta) &= f_0(\eta) + \sum_{m=1}^{\infty} f_m(\eta) \\ \phi(\eta) &= \phi_0(\eta) + \sum_{m=1}^{\infty} \phi_m(\eta) \\ \theta(\eta) &= \theta_0(\eta) + \sum_{m=1}^{\infty} \theta_m(\eta) \end{aligned} \tag{4.34}$$

4.1.4 CONVERGENCE OF THE HAM SOLUTION

As Liao pointed out, the convergence rate of approximation for the HAM solution strongly depend on the value of auxiliary parameter $\hbar$. Figure 4.2 clearly depict that the range, for admissible values of $\hbar$ is $(-1 < \hbar < -0.2)$. Our calculations for this case clearly indicate that for whole region of η when $\hbar = -0.4$, respectively.

4.1.5 RESULTS AND DISCUSSION

The system of Equations (4.14)-(4.17) has been solved analytically by using the HAM. The results depend on a Prandtl number Pr, a Lewis number Le, a Brownian motion number Nb, and a thermophoresis number Nt. Most nanofluids have large values of the Lewis number Le ($Le > 1$). The physical

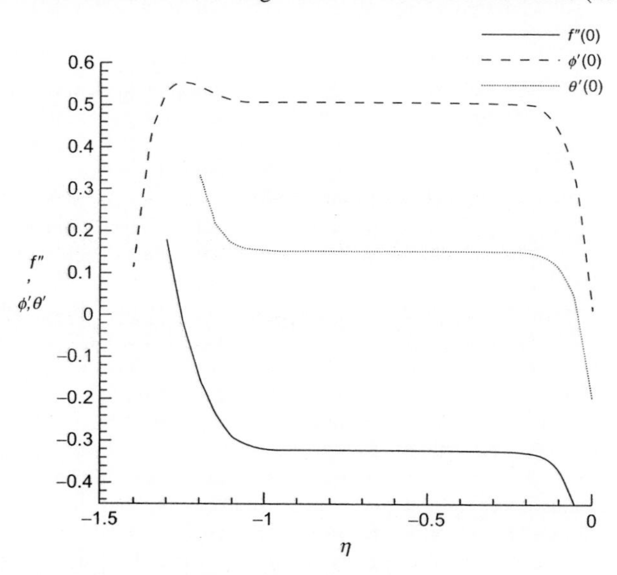

FIGURE 4.2

$\theta'(0), \varphi'(0)$ and $f''(0)$ plots for determining of optimum of $\hbar$ coefficient.

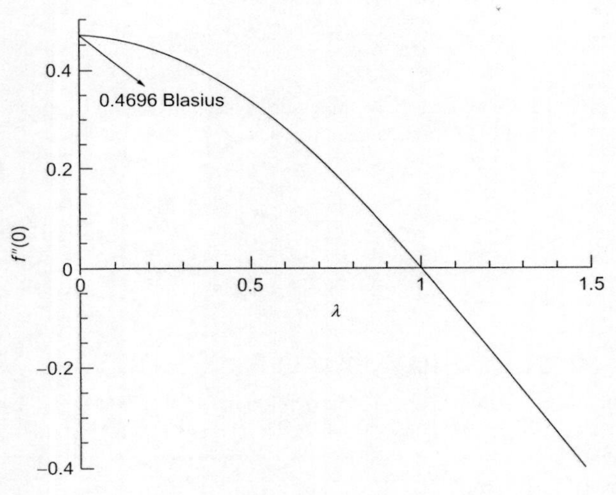

FIGURE 4.3

Variation of the reduced skin-friction coefficient $f''(0)$ with λ.

domain in this problem is unbounded; therefore we apply the far field boundary conditions for the similarity variable η at a finite value denoted here by η_{max}. The maximum value of the similarity variable is selected equal to 12. The variation with λ of the reduced skin-friction coefficient $f''(0)$ is shown in Figure 4.3.

The values of $f''(0)$ are positive when $\lambda < 1$, while they are negative $(\lambda > 1)$. Physically, a positive sign for $f''(0)$ implies that the fluid exerts a drag force on the plate and a negative sign implies the opposite. The value $f''(0) = 0.4696$ for a fixed plate ($\lambda = 0$, Blasius problem, (Blasius, 1908a) as given in Figure 4.3 is in excellent agreement with that reported by White (2006a). Figure 4.4 shows that the results of analytical method have good agreements with the numerical results presented by Bachok et al. (2010).

The variations of the reduced Nusselt number (heat transfer rate) $-\theta'(0)$ and the reduced Sherwood number (mass flux rate) $-\varphi'(0)$ with λ for $Nt = 0.5$, $Pr = 2$, $Le = 2$ and different values of Nb are presented in Figures 4.5 and 4.6.

It can be seen that the heat transfer rate $-\theta'(0)$ increases with the increase of λ. However whenever Nb number increases, the reduced Nusselt number decreases. The results show that the mass flux rate $-\varphi'(0)$ increases as both λ and Nb increase. The variations of the reduced Nusselt number $-\theta'(0)$ with number Nt for different values of Pr are depicted in Figures 4.7-4.9.

Figure 4.7 shows that the reduced Nusselt number decreases with the increase of both Nt and Pr numbers. However Figure 4.8 presents that the reduced Nusselt number decreases with the increase of Nt for lower Pr number but for higher Pr number decreases at first and then increase. It is clear that the reduced Nusselt number increases with the increase of Pr number. The only difference between Figures 4.7 and 4.8 is related to λ (Figure 4.8: $\lambda = 1.5$, Figure 4.7: $\lambda = 0.5$). This results show whenever the velocity of surface is greater than the velocity of the free stream, the reduced number increases with

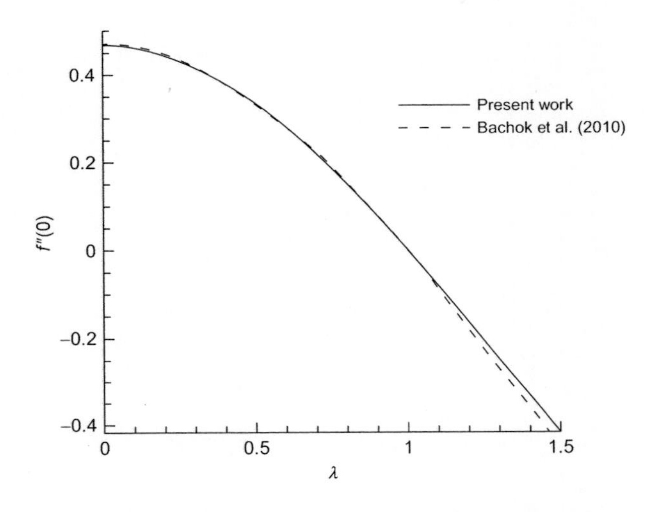

FIGURE 4.4

Comparison of analytical and numerical results of present work and Bachok et al. (2010).

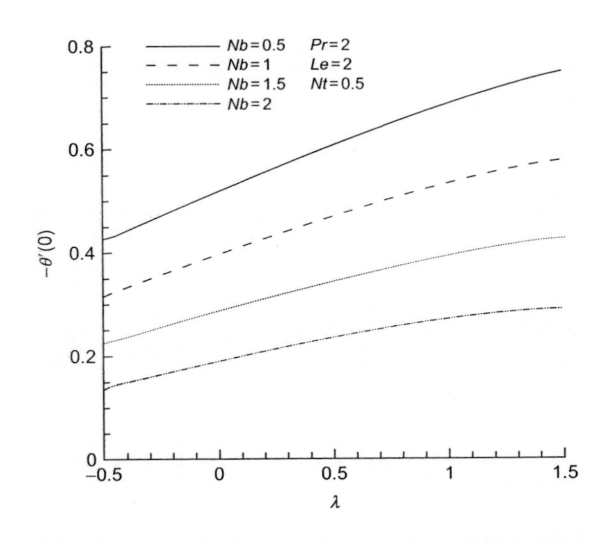

FIGURE 4.5

Variation of the reduced Nusselt number $-\theta'(0)$ with λ for various values of Nb numbers.

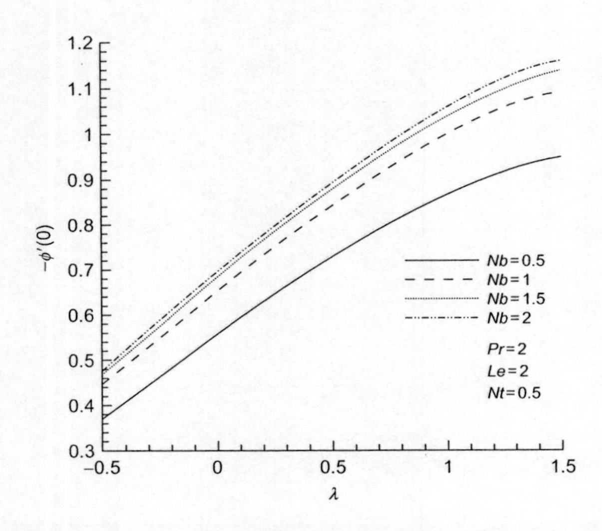

FIGURE 4.6

Variation of the reduced Sherwood number $-\varphi'(0)$ with λ for various values of *Nb* numbers.

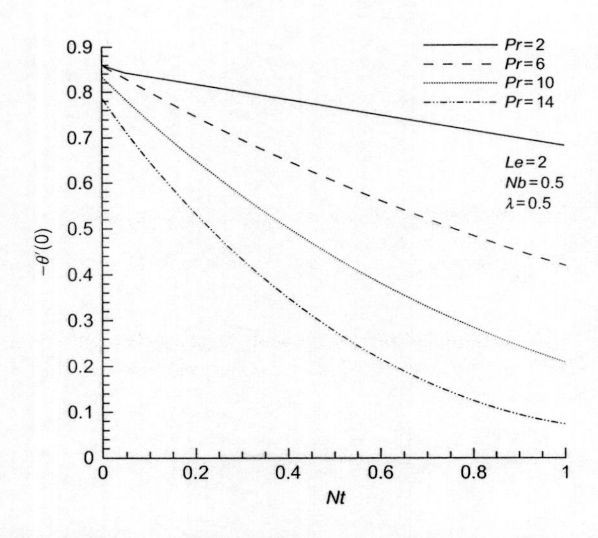

FIGURE 4.7

Variation of the reduced Nusselt number $-\theta'(0)$ with *Nt* for various values of *Pr* numbers ($\lambda=0.5$, $Nb=0.5$).

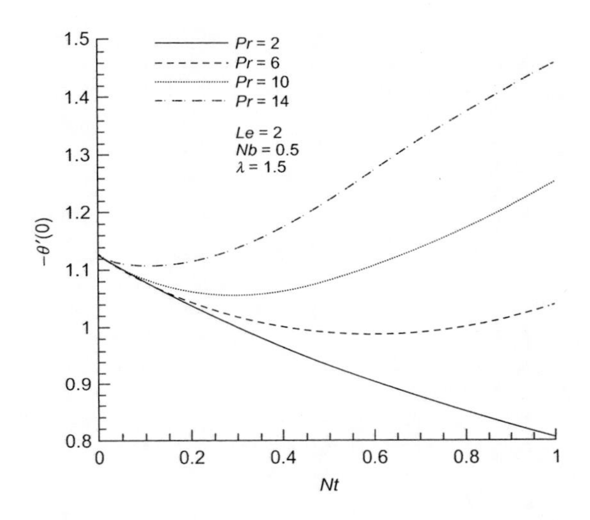

FIGURE 4.8

Variation of the reduced Nusselt number $-\theta'(0)$ with Nt for various values of Pr numbers ($\lambda = 1.5$, $Nb = 0.5$).

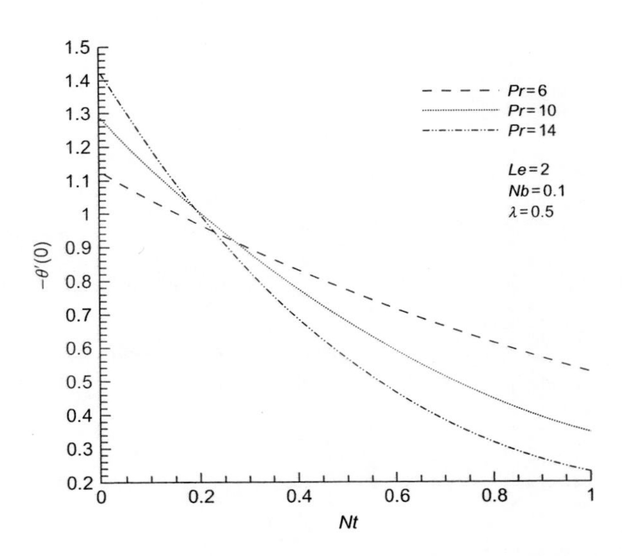

FIGURE 4.9

Variation of the reduced Nusselt number $-\theta'(0)$ with Nt for various values of Pr numbers ($\lambda = 0.5$, $Nb = 0.1$).

the increase of Pr number. Figure 4.9 is depicted with the same conditions with Figure 4.7 exception $Nb=0.5$. Nb in Figure 4.9 is equal to 0.1. Figure 4.9 shows that for Nt less than 0.2 the reduced number increases with the increase of Pr number but for Nt greater than 2 the reduced number decreases as Pr number increases. The results obtained for Nb greater than 0.5 have the same treatment like $Nb=0.5$.

The variations of the reduced Sherwood number $-\varphi'(0)$ with Nt number for different values of Pr are depicted in Figures 4.10 and 4.11. Figure 4.10 shows that the reduced Sherwood number is an increasing function of higher Pr and a decreasing function of lower Pr number. Also the Sherwood number increases with the increase of Pr number. However Figure 4.11 shows that for Nt less than 0.1 the Sherwood number increases with the increase of Pr number but for Nt greater than 0.1 the Sherwood number decreases as Pr number increases. In Figure 4.11 λ is equal to 1.5 but in Figure 4.10 λ is equal to 0.5. The results for $Nb=0.1$ are the same to all of other Nb numbers. Figures 4.12 and 4.13 show that the reduced Nusselt number and the reduced Sherwood number decreases and increases with the increase of Le number, respectively. The results obtained for $\lambda=1.5$ and $Nb=0.1$ are the same to the results presented in Figures 4.12 and 4.13.

Plots of the dependent similarity variables $f(\eta)$, $\theta(\eta)$, and $\varphi(\eta)$ for a typical case, chosen as that for $Pr=10$; $Le=10$ and $Nb=Nt=0.5$ are shown in Figure 4.14.

Figure 4.14 shows that the results of analytical method have good agreements with the numerical results presented by Bachok et al. (2010). The variations of the velocity profile, the reduced Nusselt number and the reduced Sherwood number are depicted in Figures 4.15-4.17.

Figure 4.15 clearly shows that the velocity profile increases with the increase of parameter λ. It is clear that the velocity profile for $\lambda<1$ increases as parameter η increases. The velocity profile has the same velocity to the velocity of surface in $\eta=0$ and then increases to reach to the velocity of free stream

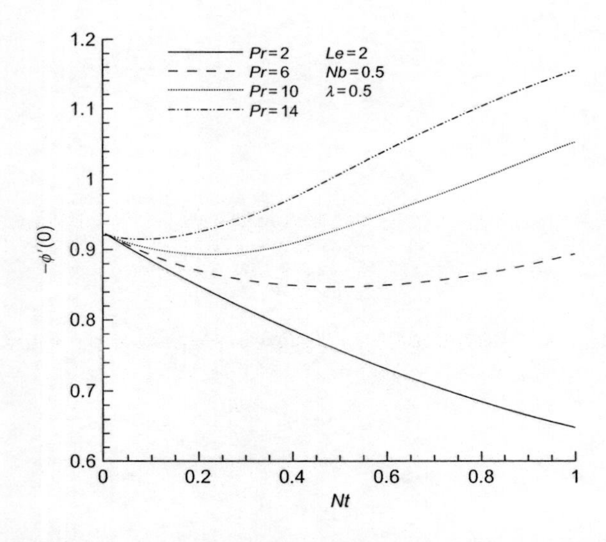

FIGURE 4.10

Variation of the reduced Sherwood number $-\varphi'(0)$ with Nt for various values of Pr numbers ($\lambda=0.5$, $Nb=0.5$).

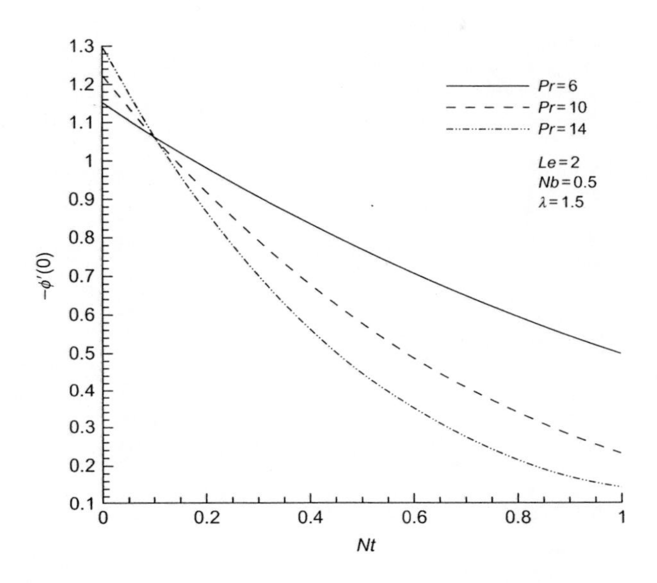

FIGURE 4.11

Variation of the reduced Sherwood number $-\varphi'(0)$ with Nt for various values of Pr numbers ($\lambda = 1.5$, $Nb = 0.5$).

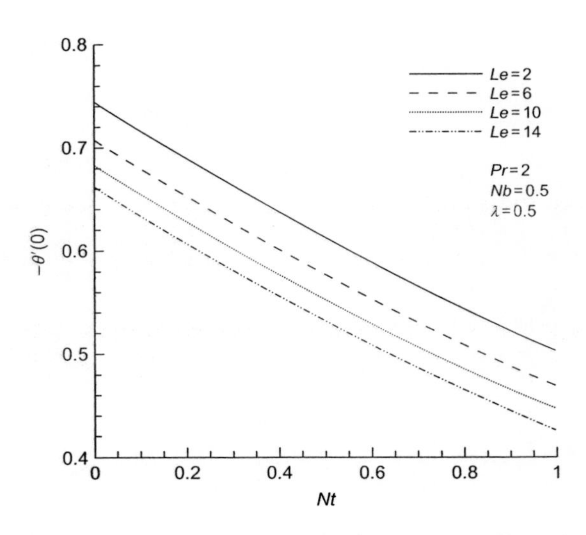

FIGURE 4.12

Variation of the reduced Nusselt number $-\theta'(0)$ with Nt for various values of Le numbers.

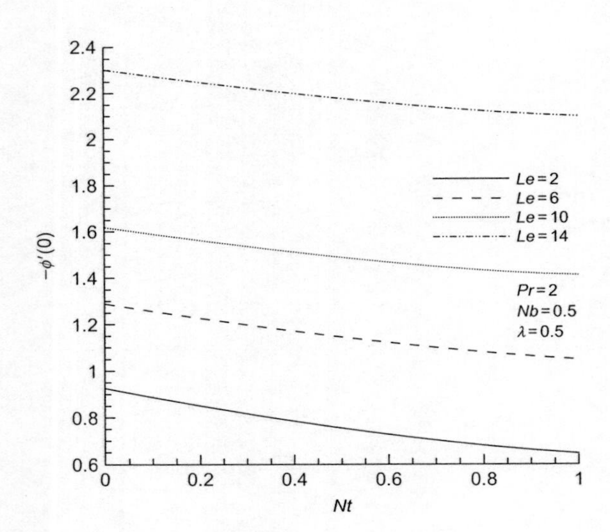

FIGURE 4.13

Variation of the reduced Sherwood number $-\varphi'(0)$ with Nt for various values of Le numbers.

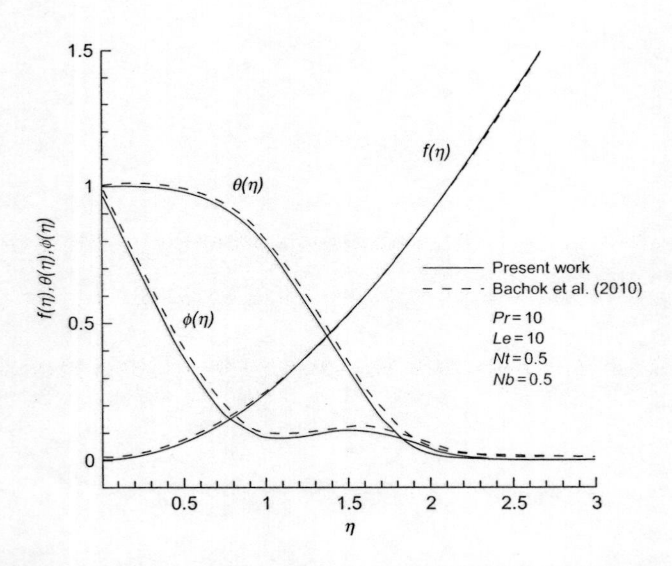

FIGURE 4.14

Profiles of $f(\eta), \theta(\eta)$ and $\varphi(\eta)$ for the case $Pr=10$; $Le=10$; $Nb=0.5$ and $Nt=0.5$ when $\lambda=0$.

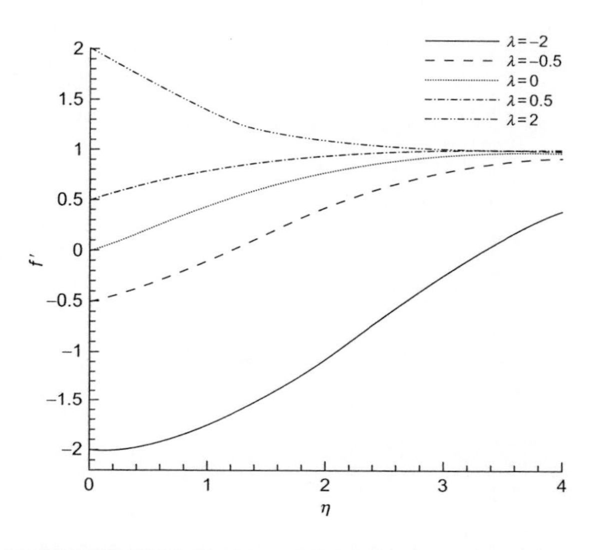

FIGURE 4.15

Variation of the velocity Profiles with η for various values of λ parameters.

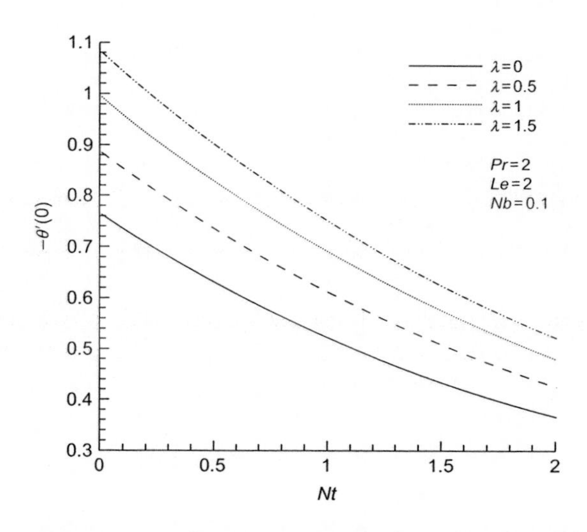

FIGURE 4.16

Variation of the reduced Nusselt number $-\theta'(0)$ with Nt for various values of λ parameters.

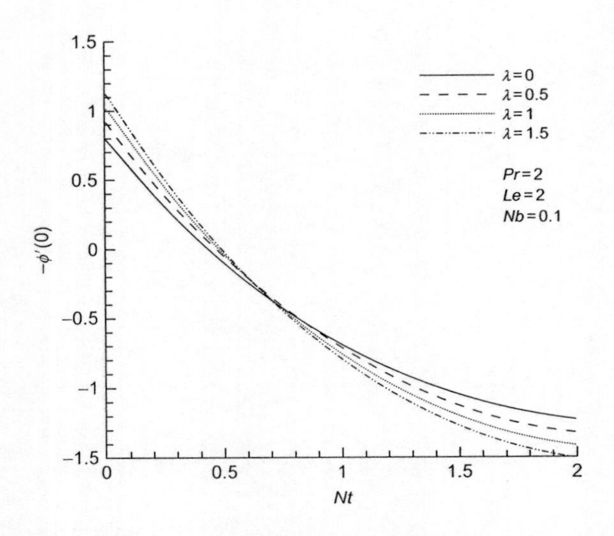

FIGURE 4.17

Variation of the reduced Sherwood number $-\varphi'(0)$ with Nt for various values of λ parameters.

velocity. For $\lambda > 1$, the velocity profile decreases with the increase of parameter η. The boundary-layer profiles for the stream function $f(\eta)$ and the temperature function $\theta(\eta)$ have essentially the same form as in the case of a regular (Newtonian) fluid. As one would expect, the thermal boundary-layer thickness is less than the momentum boundary-layer thickness when $Pr > 1$. The thickness of the boundary layer for the mass fraction function $\varphi(\eta)$ is smaller than the thermal boundary-layer thickness when $Le > 1$. The plots of the reduced Nusselt number and the reduced Sherwood number with Nt for different values of λ are depicted in Figures 4.16 and 4.17. It should be noted that the results are for $Nb = 0.1$. Figure 4.16 clearly shows that the reduced Nusselt number increases with the increase of parameter λ. Figure 4.17 shows that the reduced Sherwood number increases with the increase of parameter λ for Nt less than 0.7, but for Nt greater than 0.7 the reduced Sherwood number decreases as parameter λ increases. We know that the reduced Sherwood number increases with the increase of Nb but decreases with the increase of Nt. Therefore, for Nt less than 0.7 Nb number has dominant effect but for Nt greater than 0.7 Nt number has dominant effect.

The samples of velocity, temperature and the nanoparticle volume fraction profiles are given in Figures 4.18-4.21, respectively.

Figure 4.18 shows that the effects of Le numbers on the temperature distribution for the selected values of Pr, λ, Nb, and Nt parameters. It is observed that the temperature decreases with the increase of Le numbers.

The effects of Le numbers on the concentration profiles for the selected parameters are shown in Figure 4.19. It is clear that the concentration decreases as the Le numbers increase.

Figure 4.20 shows the effects of Nb numbers on the temperature distribution for the selected values of Pr, λ, Le, and Nt parameters. It is observed that the temperature increases with the increase in Nb numbers.

The effects of Nb numbers on the concentration profiles for the selected parameters are shown in Figure 4.21. It is clear that the concentration decreases as the Nb numbers increase.

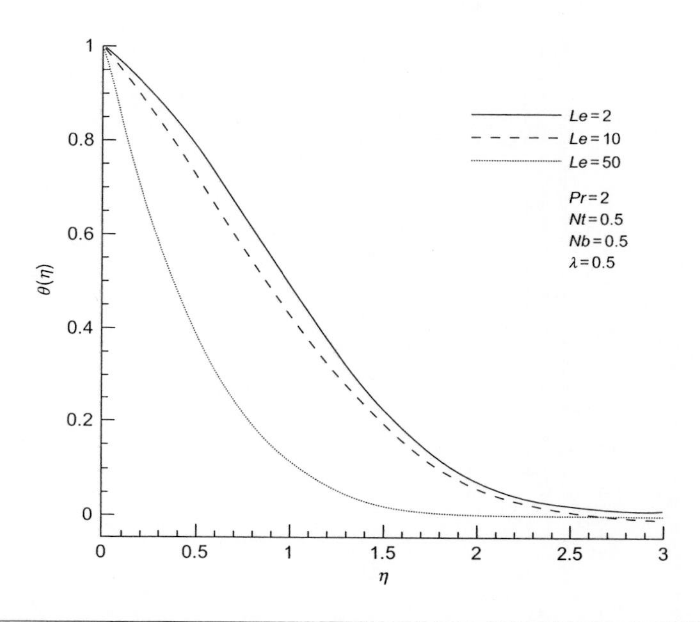

FIGURE 4.18

Temperature profiles $\theta(\eta)$ for various values of Le when $Pr=2$; $Nb=0.5$; $Nt=0.5$ and $\lambda=0.5$.

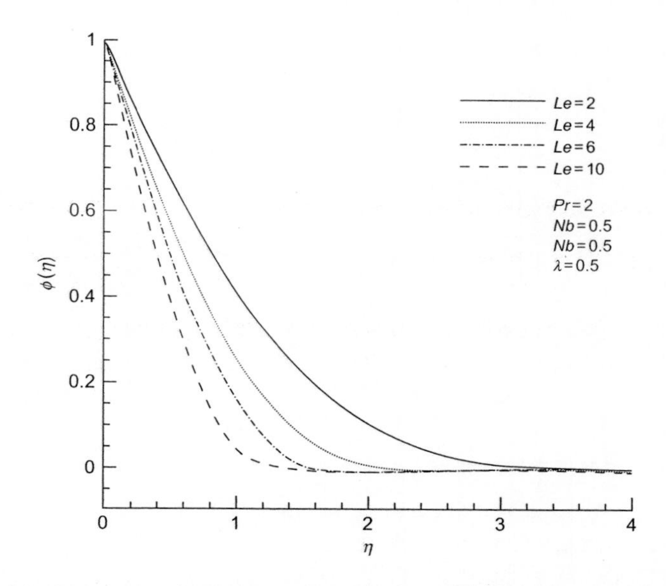

FIGURE 4.19

Nanoparticle fraction profiles $\varphi(\eta)$ for various values of Le when $Pr=2$; $Nb=0.5$; $Nt=0.5$ and $\lambda=0.5$.

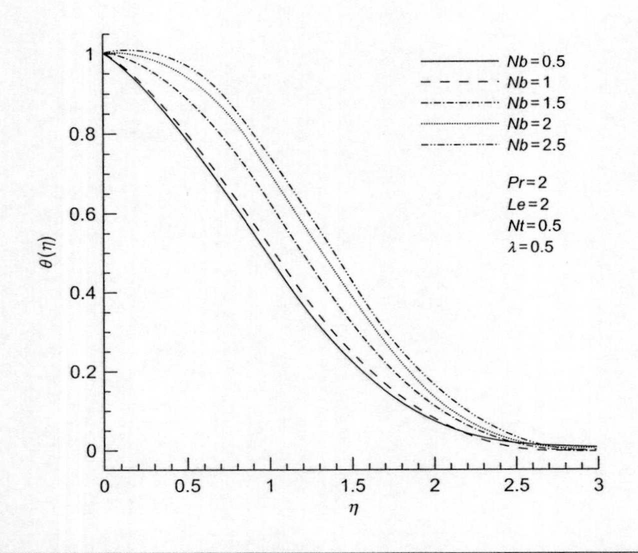

FIGURE 4.20

Temperature profiles $\theta(\eta)$ for various values of Nb when $Pr=2$; $Le=2$; $Nt=0.5$ and $\lambda=0.5$.

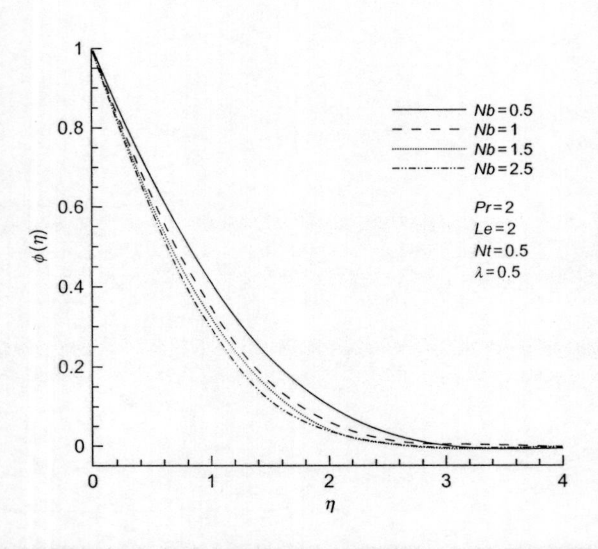

FIGURE 4.21

Nanoparticle fraction profiles $\varphi(\eta)$ for various values of Nb when $Pr=2$; $Le=2$; $Nt=0.5$ and $\lambda=0.5$.

4.1.6 CONCLUSIONS

In the present section, the problem of steady boundary-layer flow of a nanofluid past a moving semi-infinite flat plate in a uniform free stream is studied analytically by using the HAM. The dependency of the local Nusselt and local Sherwood numbers depends on a Prandtl number Pr, a Lewis number Le, a Brownian motion number Nb, and a thermophoresis number Nt. The reduced Nusselt number decreases with the increase of both Nt and Pr numbers for $\lambda = 0.5$ and Nb greater than 0.1. However, the reduced Nusselt number increases with the increase in Pr for $Nb = 0.1$. Also whenever the velocity of surface is greater than the velocity of the free stream, this conclusion would be correct ($\lambda = 1.5$). The reduced Sherwood increases as Pr number increases for $\lambda = 0.5$.Whenever the velocity of surface is greater than the velocity of the free stream, the Sherwood number increases with the increase of Pr number for Nt less than 0.1 but for Nt greater than 0.1 the Sherwood number decreases as Pr number increases. The reduced Nusselt number increases with the increase of parameter λ. The reduced Sherwood number increases with the increase of parameter λ for Nt less than 0.7, but for Nt greater than 0.7 the reduced Sherwood number decreases as parameter λ increases.

4.2 HEAT TRANSFER IN A LIQUID FILM OF NANOFLUID ON AN UNSTEADY STRETCHING SHEET

In this section, the thermal behavior of a liquid film of nanofluid driven by an unsteady stretching surface has been studied. The governing nonlinear ODEs are solved using fourth-order Runge-Kutta method. Comparisons with previously published works show the accuracy of the obtained results. Effects of dispersion of different nanoparticles along with changes in nanoparticles volume fraction on flow and heat characteristics are presented through temperature and velocity profiles and Nusselt number at the surface. It was found that increasing the volume fraction of nanoparticles angle enhanced the heat transfer rate and reduces the skin-friction coefficient and at the surface. Moreover, Cu-water has the highest heat transfer rate and the lowest skin-friction coefficient at the surface compared with others (*this chapter has been worked by M. Toomaj, D.D. Ganji, M. Abdollahzadeh, M. Esmaeilpour in nonlinear dynamics team in Mechanical Engineering Department, 2012-2013*).

4.2.1 INTRODUCTION

Fluid and heat flow induced by continuous stretching heated surfaces is often encountered in many engineering and industrial disciplines including extrusion process, wire and fiber coating, polymer processing, design of various heat exchangers, and chemical processing equipment, etc. In order to gain some fundamental understanding of such processes, the analysis of momentum and thermal transports of a viscous fluid over a stretching surface have been extensively investigated.

The flow field and heat transfer can be unsteady due to a sudden stretching of the flat sheet or by a step change of the temperature of the sheet. Several authors studied various aspects of this problem. The hydrodynamics of a flow in a thin liquid film driven by an unsteady stretching surface was first considered by Wang (1990). This problem was then extended by several researchers (see Andersson et al., 2000, Dandapata et al., 2008, Liu and Andersson, 2008, and Wang, 2006).

Since nanofluid behaves like a fluid than a mixture, several researchers have tried to explain the physics behind the increase in thermal conductivity. To this end, numerous models and methods have been proposed by different authors to study transport properties of nanofluid considering different approaches (see Das et al., 2003, Hamilton and Crosser, 1962, Wasp, 1977). Yu and Choi (2003), Kumar et al. (2004), Prasher et al. (2006), and Patel et al. (2005) has improved the model given in previous researches. Recently, Buongiorno (2006) considered seven slip mechanisms that can produce a relative velocity between the nanoparticles and the base fluid, namely, inertia, Brownian diffusion, thermophoresis, diffusiophoresis, Magnus effect, fluid drainage, and gravity. Recently, several authors (see Kandasamya et al., 2011 and Makinde and Aziz, 2011) used the nanofluid model proposed by Buongiorno (2006).

The present section investigates the thermal behavior of a liquid film of nanofluid driven by an unsteady stretching surface. The resulting governing nonlinear ODEs are integrated numerically using a Rung-Kutta shooting method. Effects of dispersion of different nanoparticles along with changes in nanoparticles volume fraction on flow and heat characteristics are presented through temperature and velocity profiles and Nusselt number at the surface. Results are provided for representative values of the unsteadiness parameter (S) and nanoparticle volume fraction ($\phi = 0, 0.1, 0.2$). Furthermore, the variation of the velocity gradient due to nanoparticles dispersion is presented in graphical forms.

4.2.2 PROBLEM FORMULATION AND GOVERNING EQUATION

Let us consider the thin elastic sheet that emerges from a narrow slit at the origin of the Cartesian coordinate system shown in Figure 4.22.

The continuous sheet aligned with the x-axis at $y=0$ moves in its own plane with a velocity $U_s(x,t)$ and the temperature distribution $T_s(x,t)$ varies both along the sheet and with time. A thin liquid film of nanofluid with uniform thickness $h(t)$ rests on the horizontal sheet. It was assumed that the flow is steady, and that the base fluid and nanoparticles are in thermal equilibrium and no slip occurs between

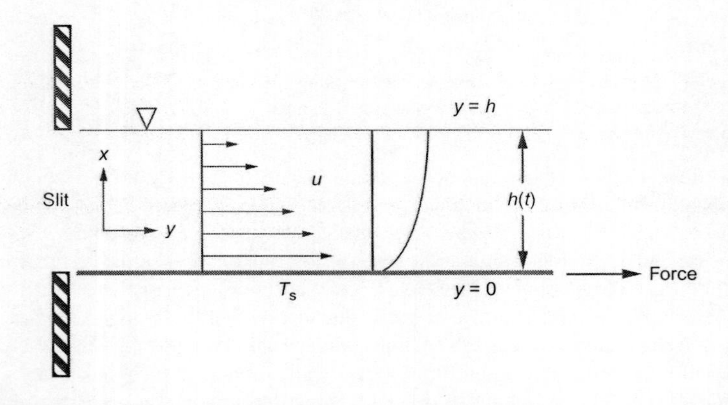

FIGURE 4.22

Schematic representation of problem.

Table 4.1 Thermophysical Properties of Nanoparticles and Base Fluid (Water)

Prosperities	Base Fluid (Water)	Cu	Al_2O_3	TiO_2
ρ	997.1	8933	3970	4250
C_p	4179	385	765	686.2
k	0.613	400	40	8.9538

them. Thermo-physical properties of the nanofluid are assumed to be constant. The thermo-physical properties of the base fluid and nanoparticle which were used for simulation are given in Table 4.1.

The governing time-dependent equations for mass, momentum, and energy conservation are given by

$$\frac{\partial u}{\partial x}+\frac{\partial v}{\partial y}=0 \tag{4.35}$$

$$\frac{\partial u}{\partial t}+u\frac{\partial u}{\partial x}+v\frac{\partial u}{\partial y}=\frac{\mu_{\text{nf}}}{\rho_{\text{nf}}}\left(\frac{\partial^2 u}{\partial y^2}+\frac{\partial^2 u}{\partial x^2}\right) \tag{4.36}$$

$$\frac{\partial T}{\partial t}+u\frac{\partial T}{\partial x}+v\frac{\partial T}{\partial y}=\frac{k_{\text{nf}}}{(\rho C_p)_{\text{nf}}}\left(\frac{\partial^2 T}{\partial y^2}+\frac{\partial^2 T}{\partial x^2}\right) \tag{4.37}$$

The appropriate boundary conditions for the above boundary layer equations are

$$u=U_s,\ v=0,\ T=T_s,\ \text{at}\ y=0 \tag{4.38}$$

$$\frac{\partial u}{\partial y}=\frac{\partial T}{\partial y}=U_s,\ v=\frac{dh}{dt},\ \text{at}\ y=h(t) \tag{4.39}$$

where $h(t)$ is the free surface elevation of the liquid film, i.e., the film thickness. The first part of Equation (4.39) reflects the absence of viscous shear stress and heat flux at the free surface, while the last part is a Kinematic free-surface condition.

The sheet moves in its own plane, causes the fluid motion within the liquid film, with the stretching velocity $U_s(x,t)$ which is defined as:

$$U_s=\frac{bx}{1-ct} \tag{4.40}$$

where both b and c are positive constants with dimension per time. The surface temperature of the stretching sheet is assumed to vary with the horizontal coordinate x and time t as

$$T_s=T_0-T_{\text{ref}}\frac{dx^r}{\nu_f}(1-ct)^{-m} \tag{4.41}$$

where T_0 is the temperature at the origin of slit and T_{ref} is the constant reference temperature such that $0\le T_{\text{ref}}\le T_0$. It should be noticed that expressions (4.40) and (4.41) are valid only for time $0<t<c^{-1}$. The particular form of the expressions for $T_s(x,t)$ were introduced to consider a variety of different situations, both with respect to spatial and temporary temperature variations through the definition of power indices r and m. $r>0$ and $m>0$ represents a situation in which the sheet temperature decreases from T_0 at the slit with the rate of x^{r-2} and such that the amount of temperature reduction along the sheet increases with time in proportion to $(1-ct)^{-m}$

In the above equations, ρ_{nf}, the density of the nanofluid is given by:

$$\rho_{nf} = (1-\varphi)\rho_f + \varphi\rho_s \tag{4.42}$$

whereas the heat capacitance of the nanofluid and thermal expansion coefficient of the nanofluid can be determined by:

$$(\rho C_p)_{nf} = (1-\varphi)(\rho C_p)_f + \varphi(\rho C_p)_s \tag{4.43}$$

$$(\rho\beta)_{nf} = (1-\varphi)(\rho\beta)_f + \varphi(\rho\beta)_s \tag{4.44}$$

with φ being the volume fraction of the solid particles and subscripts f, nf, and s stand for base fluid, nanofluid, and solid, respectively. The effective dynamic viscosity of the nanofluid is:

$$\mu_{nf} = \frac{\mu_f}{(1-\phi)^{2.5}} \tag{4.45}$$

Abu-Nada and Chamkha (2010) compared different thermal conductivity and viscosity models for natural convection of CuO-EG-water nanofluid. Also Santra et al. (2008) studied the non-Newtonian effects of copper-water nanofluid on heat transfer in a differentially heat square cavity.

The thermal conductivity of the stagnant (subscript 0) nanofluid and the effective thermal conductivity of the nanofluid for spherical nanoparticles, according to Maxwell (1904), are:

$$\frac{k_{nf\ 0}}{k_f} = \frac{k_s + 2k_f - 2\phi(k_f - k_s)}{k_s + 2k_f + \phi(k_f - k_s)} \tag{4.46}$$

Equations (4.36)-(4.39) can be converted to nondimensional forms, using the following nondimensional parameters and the similarity variable η

$$\begin{gathered}
\psi = x\left[\frac{\nu_f b}{1-\alpha t}\right]^{1/2} f(\eta) \\
u = \frac{\partial\psi}{\partial x} = \frac{bx}{1-ct} f'(\eta),\quad v = -\frac{\partial\psi}{\partial y} = -\left(\frac{b}{1-ct}\right)^{1/2} f(\eta) \\
T = T_0 - T_{ref}\frac{dx^r}{\nu_f}(1-ct)^{-m}\theta(\eta) \\
Pr = \frac{\nu_f}{\alpha_f} \\
\eta = \left(\frac{b}{\nu_f}\right)^{1/2}(1-ct)^{-1/2} y
\end{gathered} \tag{4.47}$$

The velocity components u and v in Equation (4.42) automatically satisfy the continuity Equation (4.35). By using the transformation Equation (4.42), the governing Equations (4.36) and (4.37) and the boundary conditions Equations (4.38) and (4.39) are written as follows

$$sf' + \eta\frac{s}{2}f'' + f'f' - ff'' = \frac{1}{(1-\varphi)^{2.5}\left((1-\varphi) + \frac{\rho_s}{\rho_f}\varphi\right)} f''' \tag{4.48}$$

$$\frac{\left((1-\varphi) + \frac{(\rho C_p)_s}{(\rho C_p)_f}\varphi\right)Pr}{\frac{k_s + 2k_f - 2\varphi(k_f - k_s)}{k_s + 2k_f + 2\varphi(k_f - k_s)}}\left(ms\theta + \theta'\frac{1}{2}s\eta + rf'\theta - f\theta'\right) = \theta'' \tag{4.49}$$

And the dimensionless boundary conditions, used to solve the Equations (4.47)-(4.50) are as follows:

$$f(0)=0,\ f'(0)=1,\ \theta(0)=1 \tag{4.50}$$

$$f''(\beta)=0,\ \theta'(\beta)=0,\ f(\beta)=\frac{S}{2}\beta \tag{4.51}$$

Here $S\equiv\frac{c}{b}$ is the unsteadiness parameter and the prime indicates differentiation with respect to η. Further, β denotes the value of the similarity variable η at the free surface so that Equation (4.46) gives

$$\beta=\left(\frac{b}{\nu_{\mathrm{f}}}\right)^{1/2}(1-ct)^{-1/2}h \tag{4.52}$$

where β is a yet unknown constant which should be determined as an integral part of the boundary-value problem.

The local skin-friction coefficient, which is of practical importance, is given by

$$C_{\mathrm{f}}=-\frac{2\mu_{\mathrm{nf}}}{\rho_{\mathrm{f}}U_{\mathrm{s}}^{2}}\frac{\partial u}{\partial y}\bigg|_{y=0}=-2Re_{x}^{-1/2}\frac{f''(0)}{(1-\varphi)^{2.5}} \tag{4.53}$$

and local Nusselt number (the heat transfer between the surface and the fluid) is given by

$$Nu_{x}=-\frac{k_{\mathrm{nf}}}{k_{\mathrm{f}}}\frac{x}{T_{\mathrm{ref}}}\frac{\partial T}{\partial y}\bigg|_{y=0}=-\frac{d}{b}x^{r-2}Re_{x}^{3/2}(1-ct)^{1-m}\frac{k_{\mathrm{nf}}}{k_{\mathrm{f}}}\theta'(0) \tag{4.54}$$

Here, $Re_{x}=\frac{U_{s}x}{\nu_{\mathrm{f}}}$ is the local Reynolds number.

4.2.3 NUMERICAL PROCEDURE AND VALIDATION

The numerical solution of nonlinear differential Equations (4.48) and (4.49) with boundary conditions given in Equations (4.50) and (4.51) is accomplished by shooting technique with fourth-order Runge-Kutta algorithm (Gerald and Weatly, 1989). The nonlinear differential Equations (4.49) and (4.50) are first decomposed to a system of first-order differential equations in the form;

$$\begin{aligned}
&f_0=f\frac{\mathrm{d}f_0}{\mathrm{d}\eta}=f_1\\
&\frac{\mathrm{d}f_1}{\mathrm{d}\eta}=f_2\\
&sf_1+\eta\frac{s}{2}f_2+f_1^2-f_0f_2=\frac{1}{(1-\varphi)^{2.5}\left((1-\varphi)+\frac{\rho_{\mathrm{s}}}{\rho_{\mathrm{f}}}\varphi\right)}\frac{\mathrm{d}f_2}{\mathrm{d}\eta}\\
&\theta_0=\theta\ \ \frac{\mathrm{d}\theta_0}{\mathrm{d}\eta}=\theta_1\\
&\frac{\left((1-\varphi)+\frac{(\rho C_p)_{\mathrm{s}}}{(\rho C_p)_{\mathrm{f}}}\varphi\right)Pr}{\frac{k_{\mathrm{s}}+2k_{\mathrm{f}}-2\varphi(k_{\mathrm{f}}-k_{\mathrm{s}})}{k_{\mathrm{s}}+2k_{\mathrm{f}}+2\varphi(k_{\mathrm{f}}-k_{\mathrm{s}})}}\left(ms\theta_0+\theta_1\frac{1}{2}s\eta+rf_1\theta_0-f\theta_1\right)=\frac{\mathrm{d}\theta_1}{\mathrm{d}\eta}
\end{aligned} \tag{4.55}$$

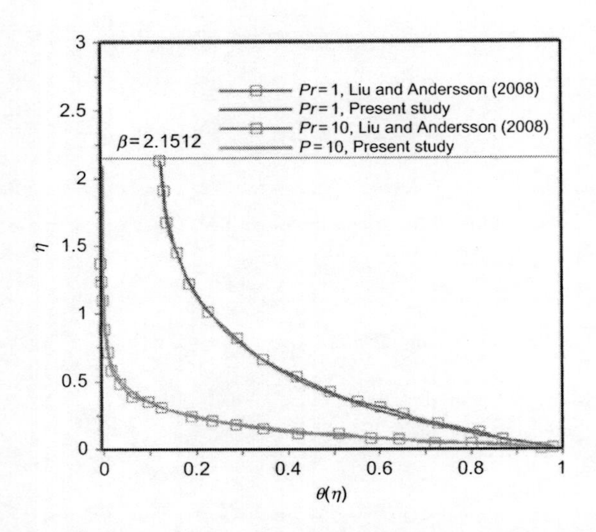

FIGURE 4.23

Comparison of the results of the present work and numerical results of Liu and Andersson (2008) with $S=0.8$ and $r=5$, $m=0$.

Corresponding boundary conditions take the form,

$$\begin{aligned} &f_0(0)=0,\ \ f_1(0)=1,\ \ \theta_0(0)=1 \\ &f_1'(\beta)=0,\ \ \theta_1(\beta)=0,\ \ f_0(\beta)=\frac{S}{2}\beta \end{aligned} \tag{4.56}$$

Appropriate guessing of the missing slopes $f_2(0)$ and $\theta_1(0)$ and the film thickness β, the boundary value problem is first converted into an initial value problem which is solved by a fourth-order Runge-Kutta Method. Then the values of $f_2(0)$ and $\theta_1(0)$ are updated using shooting method and the value of β is adjusted by a simple iteration method so that the condition $f(\beta)=S\beta/2$ holds. The convergence criterion largely depends on fairly good guesses of the initial conditions in the shooting technique. The convergence criteria were to reduce the maximum relative error in the values of relative difference between two calculated $f(\beta)$ and $S\beta/2$ in successive iterations below 10^{-7}. Present results are compared with some of the earlier published results (see Liu and Andersson, 2008) which are depicted in Figure 4.23.

4.2.4 RESULTS AND DISCUSSION

Heat transfer in a liquid film of nanofluid on an unsteady stretching sheet is studied numerically for different nanoparticle over a range of particle volume fraction. Adopting appropriate similarity transformation, the governing partial differential equations of flow and heat transfer are transferred into a system of nonlinear ODEs and then are solved using shooting method. The numerical results are obtained for $0 \le S \le 2$ as the solution exists only for a small range of values of unsteadiness parameter. Further, the effects sheet-temperature characteristics are not affected by the presence of nanoparticles. Hence, we omit the discussion on the results of r and m as they are extensively studied by Liu and Andersson (2008).

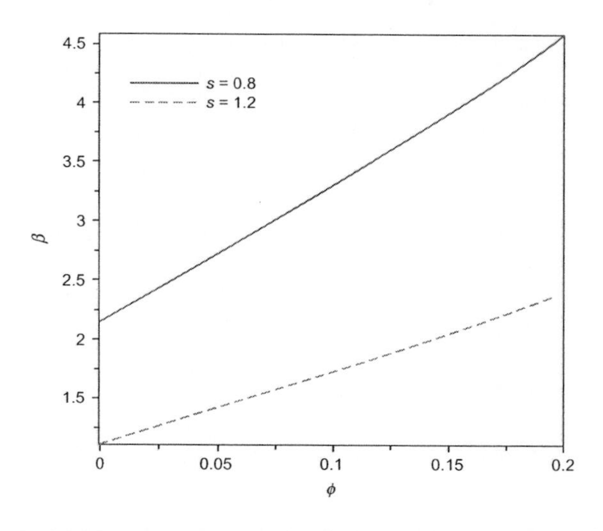

FIGURE 4.24

Variation of film thickness β for different values of nanoparticle volume fraction with $r=0.1$ and $m=1$.

The effects of nanoparticle volume fraction on various fluid dynamic quantities are shown in Figures 4.24-4.31 for different parameter. In all cases, the default value is Cu nanoparticle, unless otherwise stated.

The variation of film thickness β with respect to the nanoparticle volume fraction φ is projected in Figure 4.24 for different values of unsteadiness parameter. It is clear from this plot that the increasing value of volume fraction increases the film thickness. And the result is true for different values of unsteadiness parameter S.

The variation of free surface velocity $f'(\beta)$ with respect to φ is shown in Figure 4.24. The free surface velocity behaves almost as a constant function of φ as can be seen from Figure 4.25. This observation is ascribed to the increase in film thickness which prevents changes in free surface velocity. The effect of φ reduced skin-friction coefficient $-f''(0)/(1-\varphi)^{2.5}$ is illustrated in Figure 4.26. Clearly, increasing values of φ results in decreasing the skin-friction coefficient.

Figure 4.27 demonstrates the effect of φ on the free surface temperature $\theta(\beta)$. From this plot it is evident that the free surface temperature increases almost linearly with φ. Figure 4.28 highlights the effect of φ on the dimensionless wall heat flux $-\frac{k_{nf}}{k_f}\theta'(0)$. It is found from this plot that the dimensionless wall heat flux $-\frac{k_{nf}}{k_f}\theta'(0)$ decreases with the increasing values of φ.

The effect of φ on the axial velocity is depicted in Figures 4.29a and b for two different values of S. From these plots it is clear that the increasing values of nanoparticle volume fraction decrease the axial velocity as compared to pure water. This is due to the fact that the presence of solid nanoparticles leads to further thinning of the boundary layer. The drop in horizontal velocity as a consequence of increase in the nanoparticle volume fraction is observed for both the values of $S=0.8$ and $S=1.2$.

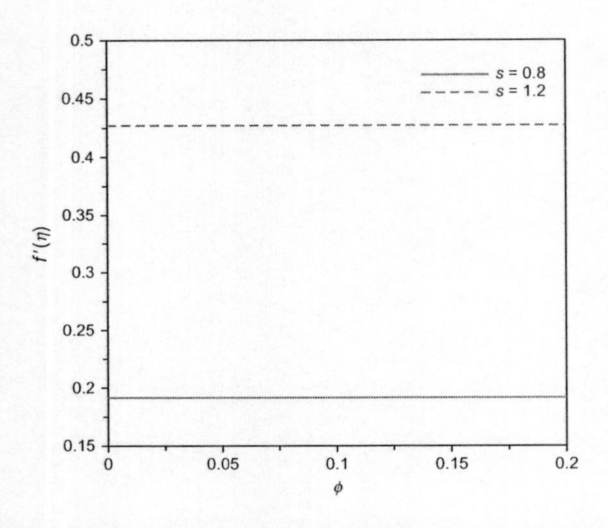

FIGURE 4.25

Variation of free surface velocity $f'(\beta)$ for different values of nanoparticle volume fraction with $r = 0.1$ and $m = 1$.

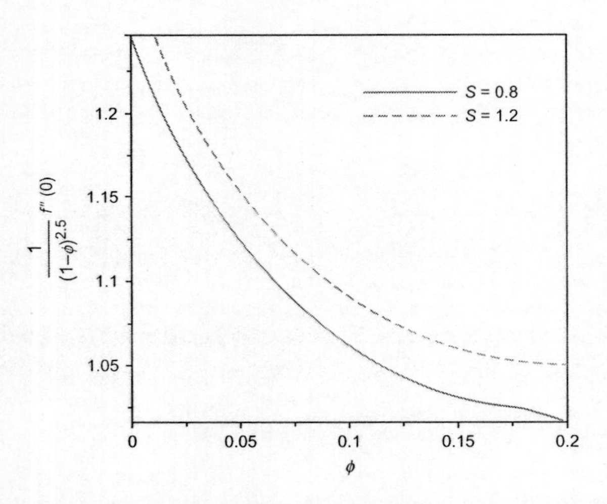

FIGURE 4.26

Variation of reduced skin-friction coefficient $-f''(0)/(1-\varphi)^{2.5}$ for different values of nanoparticle volume fraction for $r = 0.1$ and $m = 1$.

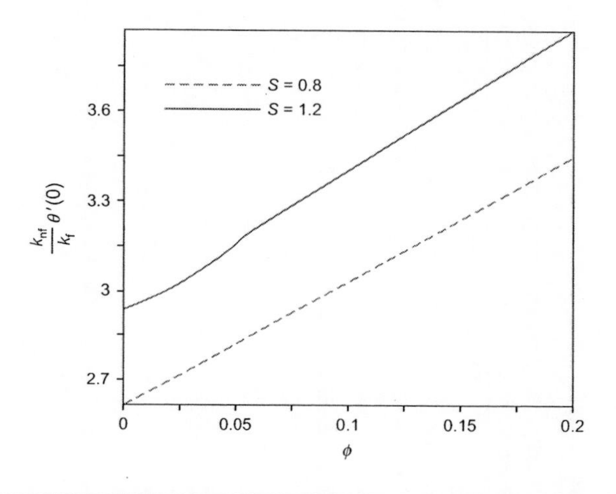

FIGURE 4.27

Variation of free surface temperature $\theta(\beta)$ for different values of nanoparticle volume fraction with $r=0.1$ and $m=1$.

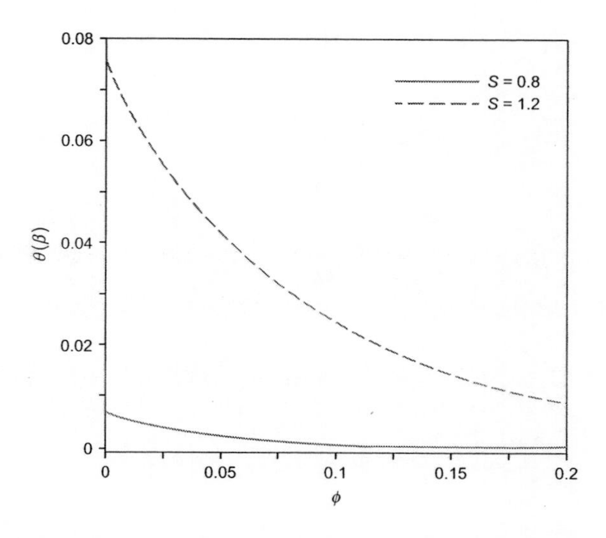

FIGURE 4.28

Variation of reduced Nusselt number $-\frac{k_{nf}}{k_f}\theta'(0)$ for different values of nanoparticle volume fraction with $r=0.1$ and $m=1$.

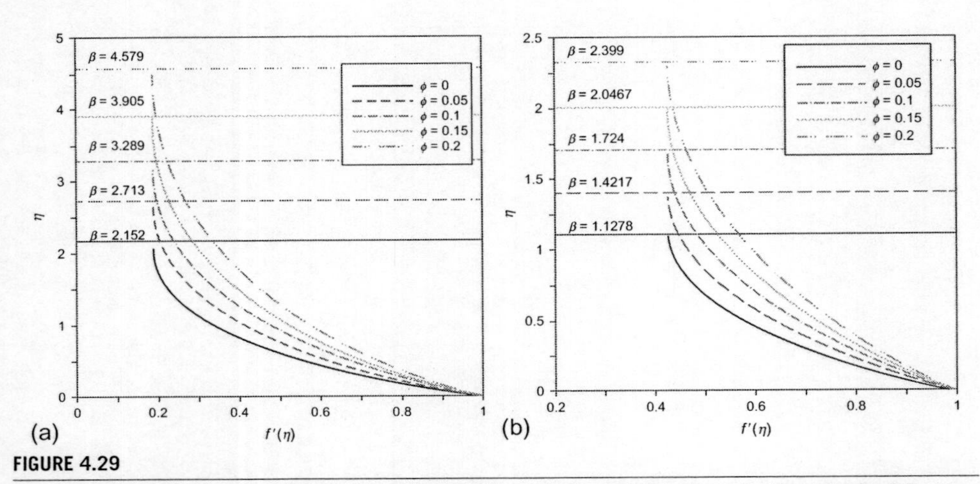

FIGURE 4.29

(a) Variation in the velocity profiles $f'(\eta)$ for different values of nanoparticle volume fraction with $S=0.8$ with $r=0.1$ and $m=1$. (b) Variation in the velocity profiles $f'(\eta)$ for different values of nanoparticle volume fraction with $S=1.2$ with $r=0.1$ and $m=1$.

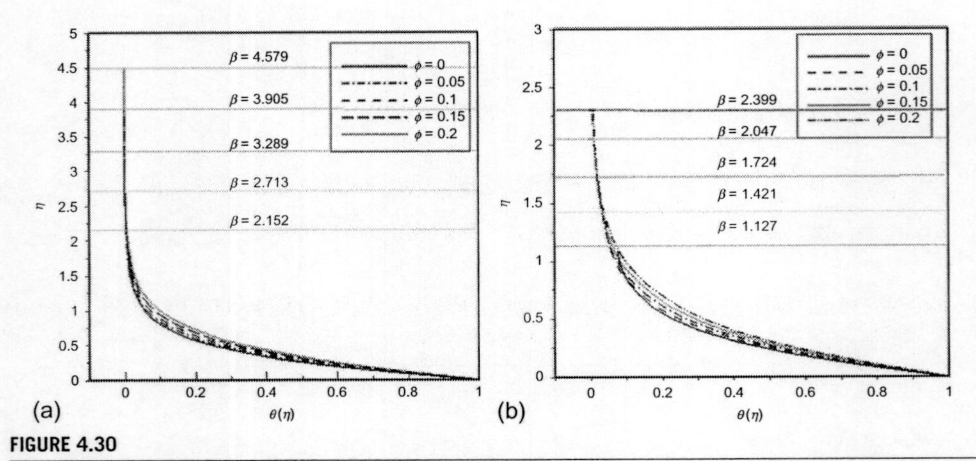

FIGURE 4.30

(a) Variation of dimensionless temperature $\theta(\eta)$ for different values of nanoparticle volume fraction with $S=0.8$ With $r=0.1$ and $m=1$. (b) Variation of dimensionless temperature $\theta(\eta)$ for different values of nanoparticle volume fraction with $S=1.2$ with $r=0.1$ and $m=1$.

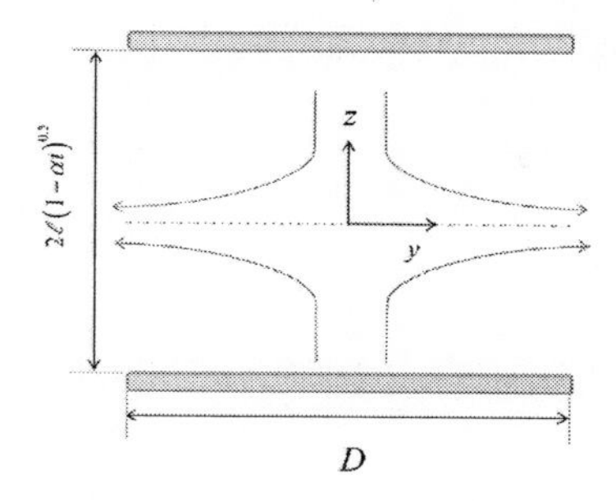

FIGURE 4.31

Geometry of problem.

Table 4.2 Show the Variation of the Skin-Friction Coefficient and the Local Nusselt Number for Different Types of Nanofluids with $m=1$, $r=0.1$

		$\frac{1}{(1-\varphi)^{2.5}}f''(0)$			$\frac{k_{nf}}{k_f}\theta'(0)$		
S	φ	Cu	Al_2O_3	TiO_2	Cu	Al_2O_3	TiO_2
0	0	0.99932	0.99932	0.99932	1.9003	1.9003	1.9003
	0.05	0.9	0.993367	0.98732	2.070026	2.03778	1.971417
	0.1	0.8478	0.99900	0.99000	52.2349545	2.2027684	2.032466
0.5	0	1.1672	1.1672	1.01700	2.2349545	2.23495450	2.349545
	0.05	1.05256	1.1609695	1.154	2.542587	2.5275929	2.431778
	0.1	0.99358	1.1686458	1.15620	2.734781	2.7099622	2.502238
1	0	1.32052	1.320522	1.03407	2.774624	2.774624	2.774624
	0.05	1.15226	1.31346	1.30551	2.9903097	2.9801157	2.867712
	0.1	1.0877	1.32214	1.30807	3.0189484	2.39563	2.947445

Figures 4.30a and b depicts the effect of φ on temperature profiles for two different values of S. The results show that the thermal boundary layer thickness increases with the increasing values of nanoparticle volume fraction φ. This is because solid particles have high thermal conductivity, so the thickness of the thermal boundary layer increases.

Table 4.2 show the variation of the skin-friction coefficient and the local Nusselt number for different types of nanofluids. It is observed that the skin-friction coefficient and the local Nusselt number are respectively lowest and highest for Cu compared to TiO_2 and Al_2O_3. This is because solid particles have high thermal conductivity.

4.2.5 CONCLUSIONS

We have numerically studied the boundary layer behavior in a liquid film of nanofluid over an unsteady stretching sheet. The governing partial differential equations were transformed into a system of nonlinear ODEs using a similarity transformation, before being solved numerically by Rung-Kutta shooting method. Three different types of nanoparticles, namely, copper Cu, alumina Al_2O_3, and titania TiO_2 with water as the base fluid were considered. It was found that increasing the volume fraction of nanoparticles angle enhanced the heat transfer rate and reduces the skin-friction coefficient and at the surface. Moreover, Cu-water has the highest heat transfer rate and the lowest skin-friction coefficient at the surface compared with others.

4.3 INVESTIGATION OF SQUEEZING UNSTEADY NANOFLUID FLOW USING ADM

In this section, unsteady squeezing nanofluid flow between parallel plates is investigated. Adomian Decomposition Method (ADM) is used to solve this problem. The base fluid in the enclosure is water containing different types of nanoparticles: copper, silver, alumina, titanium oxide. The effective thermal conductivity and viscosity of nanofluid are calculated by the MG and Brinkman models, respectively. The analytical investigation is carried out for different governing parameters, namely, the squeeze number, nanoparticle volume fraction, and Eckert number. The results show that for that case in which two plates moving together, Nusselt number increases with increase of nanoparticle volume fraction and Eckert number but it decreases with increase of the squeeze number (*this chapter has been worked by M. Sheikholeslami, D.D. Ganji, H. R. Ashorynejad in nonlinear dynamics team in Mechanical Engineering Department, 2012-2013*).

4.3.1 INTRODUCTION

One of the most important research topics due to its wide range of scientific and engineering applications, such as hydrodynamical machines, polymer processing, lubrication system, chemical processing equipment, formation and dispersion of fog, damage of crops due to freezing, food processing and cooling towers, is investigation of heat and mass transfer for unsteady squeezing viscous flow between two parallel plates. The first research on the squeezing flow in lubrication system was reported by Stefan (1874). Mahmood et al. (2007) investigated the heat transfer characteristics in the squeezed flow over a porous surface. Mustafa et al. (2012) studied heat and mass transfer characteristics in a viscous fluid which is squeezed between parallel plates. They found that the magnitude of local Nusselt number is an increasing function of Pr and Ec. Magnetohydrodynamic squeezing flow of a viscous fluid between parallel disks was analyzed by Domairry and Aziz (2009).

In this section, ADM is applied to find approximate solutions of nonlinear differential equations governing the problem of unsteady squeezing nanofluid flow and heat transfer (Adomian, 1988). The effects of the squeeze number, the nanofluid volume fraction and Eckert number on Nusselt number and skin fraction coefficient are investigated.

4.3.2 GOVERNING EQUATIONS

The flow and heat transfer analysis in the unsteady two-dimensional squeezing nanofluid flow between the infinite parallel plates is considered. The two plates are placed at $z = \pm\ell(1-\alpha t)^{1/2} = \pm h(t)$. For $\alpha > 0$, the two plates are squeezed until they touch $t = 1/\alpha$ and for $\alpha < 0$ the two plates are separated. The viscous dissipation effect, the generation of heat due to friction caused by shear in the flow, is retained. This effect is quite important in the case when the fluid is largely viscous or flowing at a high speed. This behavior occurs at high Eckert number ($>> 1$). Eckert number expresses the relationship

Table 4.3 Thermo Physical Properties of Water and Nanoparticles

	ρ(kg/m³)	C_p(J/kg K)	k(W/m K)
Pure Water	997.1	4179	0.613
Copper(Cu)	8933	385	401
Silver(Ag)	10,500	235	429
Alumina(Al_2O_3)	3970	765	40
Titanium Oxide(TiO_2)	4250	686.2	8.9538

between a flow's kinetic energy and enthalpy. The fluid is a water-based nanofluid containing copper, silver, alumina, and titanium oxide. The nanofluid is a two-component mixture with the following assumptions: incompressible, no-chemical reaction, negligible viscous dissipation, negligible radiative heat transfer, nanosolid particles and the base fluid are in thermal equilibrium and no slip occurs between them. The thermo physical properties of the nanofluid are given in Table 4.3.

The governing equations for momentum and energy in unsteady two-dimensional flow of a nanofluid are:

$$\frac{\partial u}{\partial x}+\frac{\partial v}{\partial y}=0, \tag{4.57}$$

$$\rho_{\text{nf}}\left(\frac{\partial u}{\partial t}+u\frac{\partial u}{\partial v}+v\frac{\partial u}{\partial y}\right)=-\frac{\partial p}{\partial x}+\mu_{\text{nf}}\left(\frac{\partial^2 u}{\partial x^2}+\frac{\partial^2 u}{\partial y^2}\right), \tag{4.58}$$

$$\rho_{\text{nf}}\left(\frac{\partial v}{\partial t}+u\frac{\partial v}{\partial v}+v\frac{\partial v}{\partial y}\right)=-\frac{\partial p}{\partial y}+\mu_{\text{nf}}\left(\frac{\partial^2 v}{\partial x^2}+\frac{\partial^2 v}{\partial y^2}\right), \tag{4.59}$$

$$\frac{\partial T}{\partial t}+u\frac{\partial T}{\partial x}+v\frac{\partial T}{\partial y}=\frac{k_{\text{nf}}}{(\rho C_p)_{\text{nf}}}\left(\frac{\partial^2 T}{\partial x^2}+\frac{\partial^2 T}{\partial y^2}\right)+\frac{\mu_{\text{nf}}}{(\rho C_p)_{\text{nf}}}\left(4\left(\frac{\partial u}{\partial x}\right)^2+\left(\frac{\partial u}{\partial x}+\frac{\partial u}{\partial y}\right)^2\right), \tag{4.60}$$

Here u and v are the velocities in the x and y directions, respectively, T is the temperature, P is the pressure, effective density ρ_{nf}, the effective dynamic viscosity μ_{nf}, the effective heat capacity $(\rho C_p)_{\text{nf}}$ and the effective thermal conductivity k_{nf} of the nanofluid are defined as (see Domairry and Aziz, 2009):

$$\begin{aligned}
\rho_{\text{nf}}&=(1-\phi)\rho_{\text{f}}+\phi\rho_{\text{s}},\\
(\rho C_p)_{\text{nf}}&=(1-\phi)(\rho C_p)_{\text{f}}+\phi(\rho C_p)_{\text{s}}\\
\mu_{\text{nf}}&=\frac{\mu_{\text{f}}}{(1-\phi)^{2.5}},\quad \text{(Brinkman)}\\
\frac{k_{\text{nf}}}{k_{\text{f}}}&=\frac{k_{\text{s}}+2k_{\text{f}}-2\phi(k_{\text{f}}-k_{\text{s}})}{k_{\text{s}}+2k_{\text{f}}+2\phi(k_{\text{f}}-k_{\text{s}})}\quad \text{(Maxwell – Garnetts)}
\end{aligned} \tag{4.61}$$

The relevant boundary conditions are:

$$\begin{aligned}
v&=v_{\text{w}}=\mathrm{d}h/\mathrm{d}t,\quad T=T_{\text{H}} \text{ at } y=h(t)\\
v&=\partial u/\partial y=\partial T/\partial y=0 \text{ at } y=0.
\end{aligned} \tag{4.62}$$

We introduce these parameters:

$$\eta = \frac{y}{\left[l(1-\alpha t)^{1/2}\right]}, \quad u = \frac{\alpha x}{[2(1-\alpha t)]} f'(\eta)$$

$$v = -\frac{\alpha l}{\left[2(1-\alpha t)^{1/2}\right]} f(\eta), \quad \theta = \frac{T}{T_H} \tag{4.63}$$

$$A_1 = (1-\phi) + \phi \frac{\rho_s}{\rho_f}.$$

Substituting the above variables into Equations (4.58) and (4.59) and then eliminating the pressure gradient from the resulting equations give:

$$f^{iv} - SA_1(1-\phi)^{2.5}(\eta f''' + 3f'' + f'f'' - ff''') = 0, \tag{4.64}$$

Using Equation (4.63), Equations (4.59) and (4.60) reduce to the following differential equations:

$$\theta'' + Pr\, S\left(\frac{A_2}{A_3}\right)(f\theta' - \eta\theta') + \frac{PrEc}{A_3(1-\phi)^{2.5}}\left(f''^2 + 4\delta^2 f'^2\right) = 0, \tag{4.65}$$

Here A_2 and A_3 are dimensionless constants given by:

$$A_2 = (1-\varphi) + \varphi \frac{(\rho C_p)_s}{(\rho C_p)_f}, \quad A_3 = \frac{k_{nf}}{k_f} = \frac{k_s + 2k_f - 2\varphi(k_f - k_s)}{k_s + 2k_f + 2\varphi(k_f - k_s)} \tag{4.66}$$

With these boundary conditions:

$$\begin{gathered} f(0) = 0, \ f''(0) = 0 \\ f(1) = 1, \ f'(1) = 0 \\ \theta'(0) = 0, \ \theta(1) = 1. \end{gathered} \tag{4.67}$$

where S is the squeeze number, Pr is the Prandtl number, and Ec is the Eckert number, which are defined as:

$$S = \frac{\alpha l^2}{2\upsilon_f}, \quad Pr = \frac{\mu_f (\rho C_p)_f}{\rho_f k_f}, \quad Ec = \frac{\rho_f}{(\rho C_p)_f}\left(\frac{\alpha x}{2(1-\alpha t)}\right)^2, \quad \delta = \frac{l}{x}, \tag{4.68}$$

Physical quantities of interest are the skin-friction coefficient and Nusselt number which are defined as:

$$C_f = \frac{\mu_{nf}\left(\dfrac{\partial u}{\partial y}\right)_{y=h(t)}}{\rho_{nf} v_w^2}, \quad Nu = \frac{-lk_{nf}\left(\dfrac{\partial T}{\partial y}\right)_{y=h(t)}}{kT_H}. \tag{4.69}$$

In terms of Equation (4.63), we obtain

$$\begin{gathered} C_f^* = l^2/x^2(1-\alpha t)Re_x C_f = A_1(1-\phi)^{2.5} f''(1) \\ Nu^* = \sqrt{1-\alpha t}\, Nu = -A_3\theta'(1). \end{gathered} \tag{4.70}$$

4.3.3 FUNDAMENTALS OF ADOMIAN DECOMPOSITION METHOD (ADM)

Consider equation $Fu(t)=g(t)$, where F represents a general nonlinear ordinary or partial differential operator including both linear and nonlinear terms. The linear terms are decomposed into $L+R$, where L is easily invertible (usually the highest order derivative) and R is the remained of the linear operator. Thus, the equation can be written as (Adomian, 1988):

$$Lu+Nu+Ru=g \tag{4.71}$$

where Nu indicates the nonlinear terms. By solving this equation for Lu, since L is invertible, we can write:

$$L^{-1}Lu=L^{-1}g-L^{-1}Ru-L^{-1}Nu \tag{4.72}$$

If L is a second-order operator, L^{-1} is a twofold indefinite integral. By solving Equation (4.72), we have:

$$u=A+Bt+L^{-1}g-L^{-1}Ru-L^{-1}Nu \tag{4.73}$$

where A and B are constants of integration and can be found from the boundary or initial conditions. Adomian method assumes the solution u can be expanded into infinite series as:

$$u=\sum_{n=0}^{\infty}u_n \tag{4.74}$$

Also, the nonlinear term Nu will be written as:

$$Nu=\sum_{n=0}^{\infty}A_n \tag{4.75}$$

where A_n are the special Adomian polynomials. By specified A_n, next component of u can be determined:

$$u_{n+1}=L^{-1}\sum_{n=0}^{n}A_n \tag{4.76}$$

Finally, after some iteration and getting sufficient accuracy, the solution can be expressed by Equation (4.73).

In Equation (4.73), the Adomian polynomials can be generated by several means. Here, we used the following recursive formulation:

$$A_n=\frac{1}{n!}\left[\frac{d^n}{d\lambda^n}\left[N\left(\sum_{i=0}^{n}\lambda^i u_i\right)\right]\right]_{\lambda=0},\quad n=0,1,2,3,\cdots \tag{4.77}$$

Since the method does not resort to linearization or assumption of weak nonlinearity, the solution generated is in general more realistic than those achieved by simplifying the model of the physical problem.

4.3.4 SOLUTION WITH ADOMIAN DECOMPOSITION METHOD

According to Equation (4.71), Equations (4.64)-(4.66) must be written as following:

$$L_1 f = A_1(1-\phi)^{2.5} S(\eta f''' + 3f'' + f'f'' - ff''')$$
$$L_2\theta = -Pr\, S\left(\frac{A_2}{A_3}\right)(f\theta' - \eta\theta') - \frac{PrEc}{A_3(1-\phi)^{2.5}}\left(f''^2 + 4\delta^2 f'^2\right). \tag{4.78}$$

where the differential operator L_1, L_2, and L_3 are given by $L_1 = \frac{d^4}{d\eta^4}$ and $L_2 = \frac{d^2}{d\eta^2}$, respectively. Assume the inverse of the operator $L_i (i = 1,2)$ exists and it can be integrated from 0 to η, i.e.,

$$L_1^{-1} = \int_0^\eta \int_0^\eta \int_0^\eta \int_0^\eta (\bullet) d\eta d\eta d\eta d\eta, \quad L_2^{-1} = \int_0^\eta \int_0^\eta (\bullet) d\eta d\eta. \tag{4.79}$$

Operating with L_i^{-1} on Equation (4.77) and after exerting boundary condition on it, we have:

$$f(\eta) = f(0) + f'(0)\eta + f''(0)\frac{\eta^2}{2} + f'''(0)\frac{\eta^3}{6} + L^{-1}(N_1 u),$$
$$\theta(\eta) = \theta(0) + \theta'(0)\eta + L^{-1}(N_2 u). \tag{4.80}$$

where $N_i u$ are introduced as:

$$N_1 u = A_1(1-\phi)^{2.5} S\left(\eta f_m''' + 3f_m'' + \sum_{n=0}^{m} f_n' f_{m-n}'' - \sum_{n=0}^{m} f_n''' f_{m-n}\right)$$
$$N_2 u = -Pr\, S\left(\frac{A_2}{A_3}\right)\left(\sum_{n=0}^{m} \theta_n' f_{m-n} - \eta\theta_m'\right) - \frac{PrEc}{A_3(1-\phi)^{2.5}}\left(\sum_{n=0}^{m} f_n'' f_{m-n}'' + 4\delta^2 \sum_{n=0}^{m} f_n' f_{m-n}'\right). \tag{4.81}$$

ADM introduced the following expression:

$$f(\eta) = \sum_{m=0}^{\infty} f_m(\eta), \quad f(\eta) = \sum_{m=0}^{\infty} f_m = f_0 + L^{-1}(N_1 u)$$
$$\theta(\eta) = \sum_{m=0}^{\infty} \theta_m(\eta), \quad \theta(\eta) = \sum_{m=0}^{\infty} \theta_m = \theta_0 + L^{-1}(N_2 u) \tag{4.82}$$

To determine the components of $f_m(\eta)$ and $\theta_m(\eta)$ the $f_0(\eta)$ and $\theta_0(\eta)$ are defined by applying the boundary condition of Equation (4.67):

$$f_0(\eta) = a_1 \frac{\eta^6}{6} + a_2 \eta$$
$$\theta_0(\eta) = a_3. \tag{4.83}$$

$$f_1(\eta) = \frac{1}{30}SA_1(1-\phi)^{2.5}\eta^5 a_1 + \frac{1}{2520}SA_1(1-\phi)^{2.5}a_1^2\eta^7$$
$$\theta_1(\eta) = -\frac{1}{30}PrEc\left(\frac{A_2}{A_3}\right)\delta^2\eta^6 a_1^6 + \left(\frac{A_2}{A_3}\right)\left(-\frac{1}{12}PrEca_1^2 - \frac{1}{3}PrEc\delta^2 a_1 a_2\right)\eta^4 - 2PrEc\delta^2 a_2^2\eta^2. \quad (4.84)$$

$f_m(\eta)$ and $\theta_m(\eta)$ for $m \geq 2$ be determined in similar way from Equation (4.81). Then using $f(\eta) = \sum_{m=0}^{\infty} f_m(\eta)$ and $\theta(\eta) = \sum_{m=0}^{\infty}\theta_m(\eta)$ lead to following equations:

$$f(\eta) = \sum_{m=0}^{\infty} f_m(\eta) = a_1\frac{\eta^6}{6} + a_2\eta + \frac{1}{30}S\eta^5 a_1 + \frac{1}{2520}Sa_1^2\eta^7 + \ldots$$
$$\theta(\eta) = \sum_{m=0}^{\infty}\theta_m(\eta) = a_3{}^4 - \frac{1}{30}PrEc\left(\frac{A_2}{A_3}\right)\delta^2\eta^6 a_1^6 + \left(\frac{A_2}{A_3}\right)\left(-\frac{1}{12}PrEca_1^2 - \frac{1}{3}PrEc\delta^2 a_1 a_2\right)\eta^4 - 2PrEc\left(\frac{A_2}{A_3}\right)\delta^2 a_2^2\eta^2 + \ldots. \quad (4.85)$$

According to Equation (4.82), the accuracy of ADM solution increases by increasing the number of solution terms (m). For the complete solution of Equation (4.85), $a_i, (i = 1,2,3)$ should be determined, with boundary condition at $\eta = 1$.

For example when $S = 1$, $Pr = 6.2$, $Ec = 0.01$, $\phi = 0.02$(Cu − water), and $\delta = 0.01$ constant values are obtained as follow: $a_1 = -2.028835932$, $a_2 = 1.416975001$, and $a_3 = 1.031977530$.

4.3.5 RESULTS AND DISCUSSION

In this study, ADM is used to solve the problem of unsteady squeezing nanofluid flow (see Figure 4.31).

The effects of active parameters such as the squeeze number, the nanofluid volume fraction, and Eckert number on flow and heat transfer characteristics are investigated. The present code is validated by comparing the obtained results with other works reported in literature, Mustafa et al., 2012 (see Table 4.4).

Comparison between results obtained by the numerical method (fourth-order Runge-Kutta) and ADM for different value of active parameters is shown in Figure 4.32 and Table 4.3. All these comparisons illustrate that ADM offers highly accurate solution (Table 4.5).

Figure 4.33 shows the effect of the squeeze number on the velocity profile and skin-friction coefficient.

It is important to note that the squeeze number (S) describes the movement of the plates ($S > 0$ corresponds to the plates moving apart, while $S < 0$ corresponds to the plates moving together (the so-called squeezing flow). The effect of positive and negative squeeze numbers has different effects on velocity profile. For the case of squeezing flow, the velocity increases due to increase in the absolute values of squeeze number when $\eta < 0.5$, while it decreases for $\eta > 0.5$. Also it can be seen that the

Table 4.4 Comparison of $-\theta'(1)$ Between the Present Results and Analytical Results Obtained by Mustafa et al. (2012) for $S=0.5$ and $\delta=0.1$

Pr	*Ec*	Mustafa et al., 2012	Present Word
0.5	1	1.522368	1.52236749518
1	1	3.026324	3.02632355855
2	1	5.98053	5.98053039715
5	1	14.43941	14.4394132325
1	0.5	1.513162	1.51316180648
1	1.2	3.631588	3.63158826816
1	2	6.052647	6.05264710721
1	5	15.13162	15.1316178324

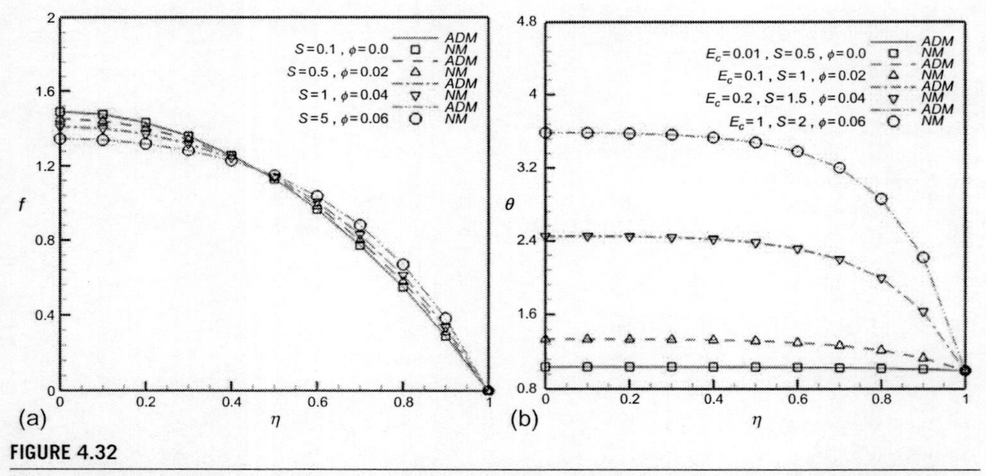

FIGURE 4.32

Comparison between results obtained via numerical solution and ADM at $Pr=6.2$, $\delta=0.1$(Cu-Water).

absolute values of squeeze number has reverse relationship with the absolute values of skin-friction coefficient for squeezing flow cases. Also Figure 4.33 shows that opposite trend is observed for the case of the plates moving apart. Besides, it can be found that velocity components of nanofluid increase as a result of an increase in the energy transport in the fluid with the increasing of volume fraction. Thus, the absolute value of skin-friction coefficient increases with increasing of volume fraction of nanofluid.

Effect of volume fraction of nanofluid and the squeeze number on the temperature profile is shown in Figures 4.34 and 4.35, respectively.

Table 4.5 Comparison Between the Results of NM and ADM Solution for $f(\eta)$ and $\theta(\eta)$ When $S=1$, $Pr=6.2$, $Ec=0.01$, $\phi=0.02$ (Cu–water) and $\delta=0.01$

	$f(\eta)$		$\theta(\eta)$	
η	NM	ADM	NM	ADM
0	0	0	1.032066589	1.031978
0.1	0.141359	0.141359	1.032064281	1.031975
0.2	0.280666	0.280666	1.032032236	1.031943
0.3	0.415781	0.415781	1.031892848	1.031804
0.4	0.544379	0.544379	1.031508324	1.031419
0.5	0.663857	0.663857	1.030664437	1.030575
0.6	0.771229	0.771229	1.029040022	1.028951
0.7	0.863016	0.863015	1.026152219	1.026064
0.8	0.93512	0.93512	1.0212593	1.021174
0.9	0.982695	0.982695	1.013186166	1.013116
1	1	1	1	1

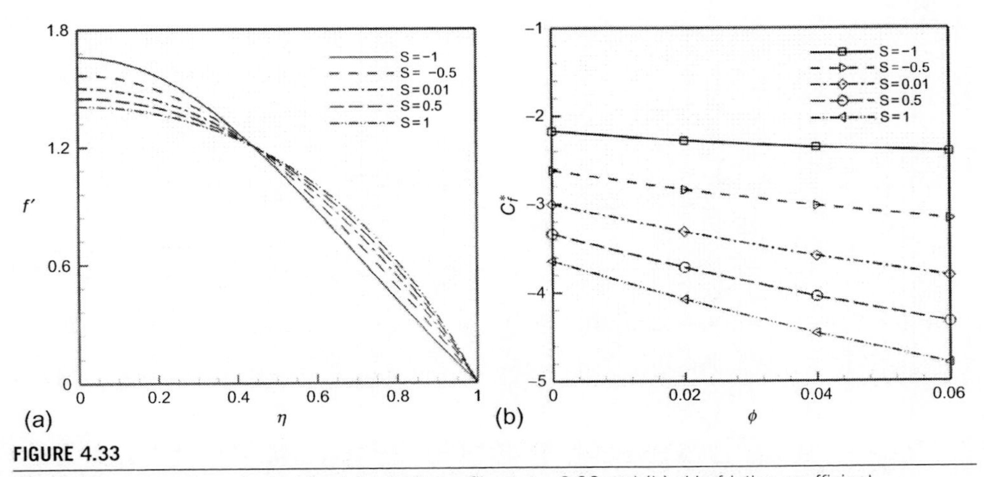

FIGURE 4.33

Effect of the squeeze number on the (a) velocity profile at $\phi=0.06$ and (b) skin-friction coefficient.

Increasing the volume fraction of nanofluid leads to decrease thermal boundary layer thickness. So Nusselt number increases as the volume fraction of nanofluid increases (Figure 4.36). An increase in the squeeze number can be related with the decrease in the kinematic viscosity, an increase in the distance between the plates and an increase in the speed at which the plates move. When two plates moving together, thermal boundary layer thickness increases as the absolute magnitude of the squeeze

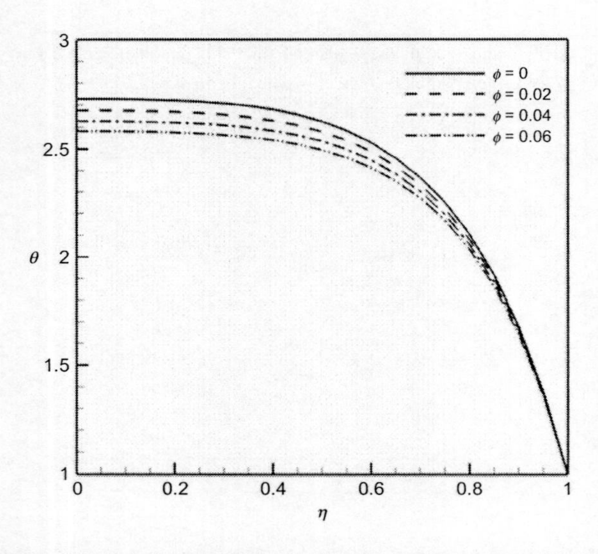

FIGURE 4.34

Effect of volume fraction of nanofluid on the temperature profile when $Ec = 0.5$, $Pr = 6.2$, $\delta = 0.1$, $S = 1$.

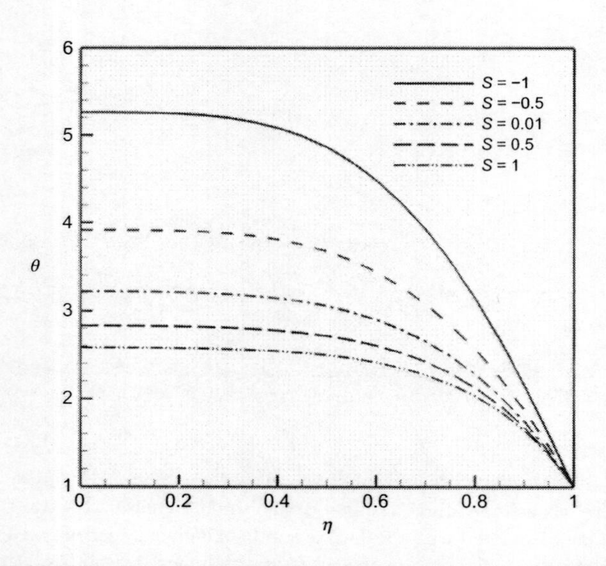

FIGURE 4.35

Effect of the squeeze number on the temperature profile when $Ec = 0.5$, $Pr = 6.2$, $\delta = 0.1$ and $\phi = 0.06$.

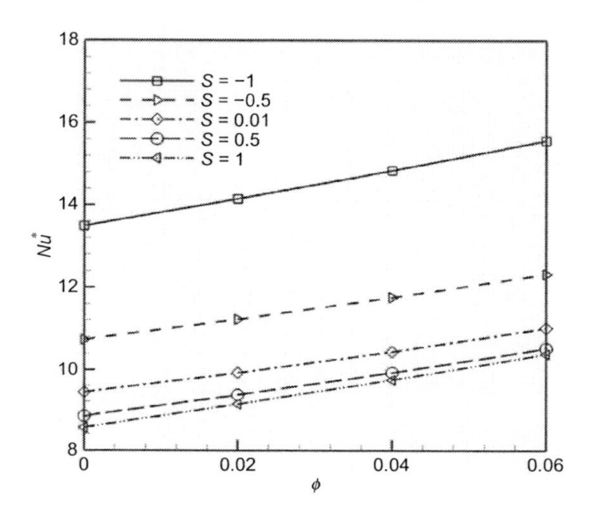

FIGURE 4.36

Effect of the squeeze number and nanoparticle volume fraction on Nusselt number.

number increases. This increase in thermal boundary layer thickness reduces the Nusselt number. Also it can be found that opposite behavior is observed when two plates moving apart. It is obvious that the temperature boundary layer thickness is relatively high when the plates are moving towards each other.

Figure 4.37 shows the effect of the Eckert number on the temperature profile and Nusselt number. The presence of viscous dissipation effects significantly increases the temperature. So, Nusselt number increase with increase of Eckert number.

The type of nanofluid is a key factor for heat transfer enhancement. So, a comparison among different types of nanoparticles is done to find selecting which of them leads to highest cooling performance for this problem. Effect of different types of nanoparticles on temperature profiles, skin-friction coefficient, and Nusselt number nanoparticles is shown in Figures 4.38 and 4.39, respectively.

The figures show that by using different types of nanofluid the values of the skin-friction coefficient and Nusselt number change. This means that the nanofluids type will be important in the cooling and heating processes. Changing in type of nanoparticles has no significant effect on velocity boundary layer thickness ($f''(1)$). But according to Equation (4.70) skin-friction coefficient is a multiplication of $(1-\phi)^{2.5}f''(1)$ and $A_1=(1-\phi)+\phi\rho_s/\rho_f$. Silver and alumina has maximum and minimum of density between these types of nanofluid, respectively. So, choosing alumina as the nanoparticle leads to the minimum amount of $|C_f^*|$, while selecting silver leads to obtain the maximum amount of it. Also according to thermal conductivity of nanoparticle, it can be seen that selecting silver as nanoparticle leads to obtain the highest Nusselt number while minimum amount of it occurs by selecting TiO_2.

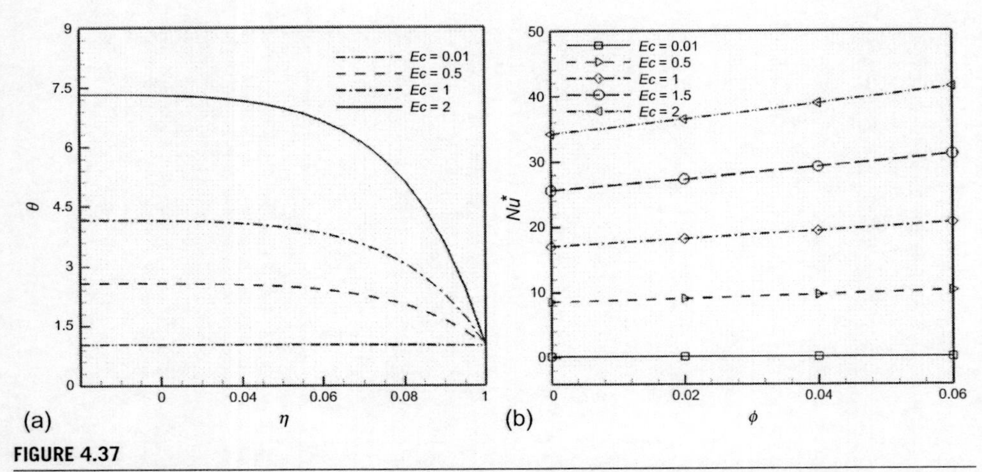

FIGURE 4.37

Effect of the Eckert number on the (a) temperature profile at $\phi = 0.06$ and (b) Nusselt number.

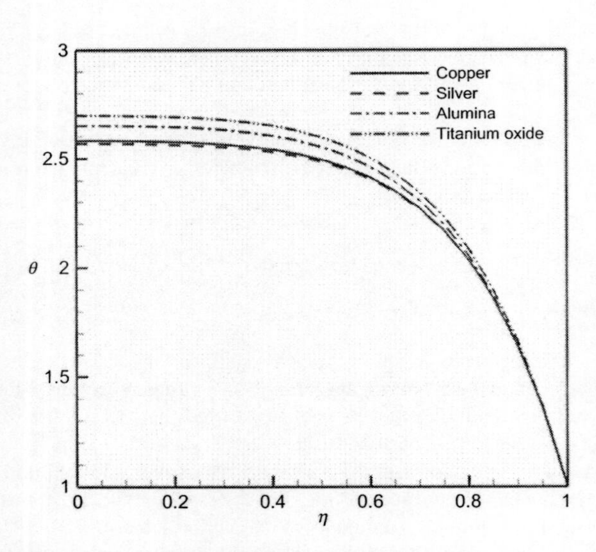

FIGURE 4.38

Temperature profiles for different types of nanoparticles when $Ec = 0.5$, $Pr = 6.2$, $\delta = 0.1$, $S = 1$ and $\phi = 0.06$.

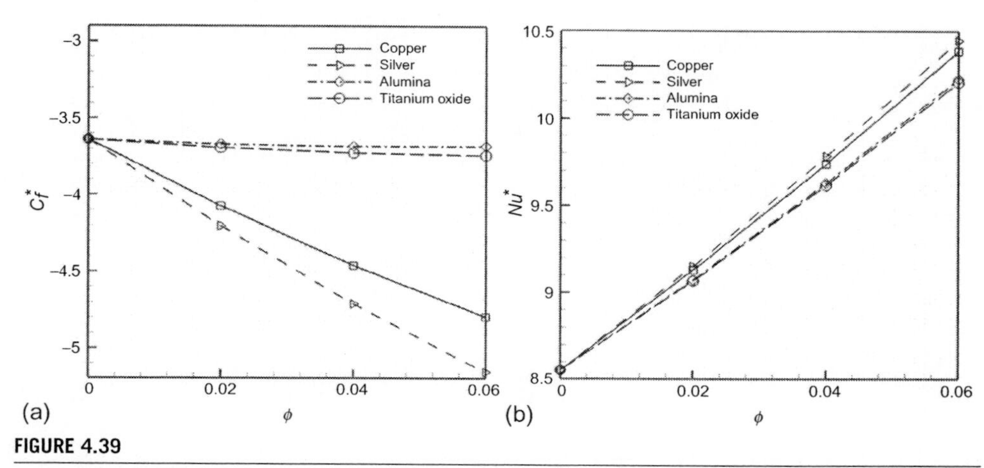

FIGURE 4.39

Effect of different types of nanoparticles on the skin-friction coefficient and Nusselt number nanoparticles when (a) $S=1$; (b) $Ec=0.5$, $Pr=6.2$, $\delta=0.1$, $S=1$.

4.3.6 CONCLUSION

In this study, unsteady squeezing flow between parallel plates which is filled with nanofluid is investigated using ADM. The effects of the squeeze number, the nanofluid volume fraction, and Eckert number on Nusselt number and skin-friction coefficient are studied. The results show that the type of nanofluid is a key factor for heat transfer enhancement. Selecting silver as nanoparticle leads to obtain the highest values of Nusselt number. Also, it can be found when two plates moving together, Nusselt number has direct relationship with nanoparticle volume fraction and Eckert number but it has reverse relationship with the squeeze number.

4.4 INVESTIGATION ON ENTROPY GENERATION OF NANOFLUID OVER A FLAT PLATE

In this section, the steady two-dimensional boundary layer flow over a flat plate is studied numerically to analyze the generated entropy inside the boundary layer at a constant wall temperature. Applying the transformation of the Basic equations of continuity, momentum, and energy to ODE ones by similarity variables, a dimensionless equation for entropy generation inside the boundary layer is presented for the first time. The novelty of this study is to combine the classical Blasius equation with a new concept in fluid mechanics and heat transfer which is nanofluids in order to investigate the entropy generation. We have found that the generated entropy strongly depends on volume fraction, type of nanoparticles, and dimensionless parameters, namely, Pr, Ec, Re. In addition, the physical interpretation of the results is explained in details and the related figures are plotted as well (*this chapter has been worked by A. Malvandi, F. Hedayati, D.D. Ganji, Y. Rostamiyan in nonlinear dynamics team in Mechanical Engineering Department, 2012-2013*).

4.4.1 INTRODUCTION

The classical concept of boundary layer corresponds to a thin region next to the wall in a flow where viscous forces are important which may affect the engineering process of producing. For example viscous forces play essential rules in glass fiber drawing, crystal growing, and plastic extrusion (see Bachok et al., 2010). The above-mentioned wall can be in various geometrical shapes. Blasius (1908a) studied the simplest boundary layer over a flat plate. He employed a similarity transformation which reduces the partial differential boundary layer equations to a nonlinear third-order ordinary differential one before solving it analytically. In contrast with the Blasius problem, Sakiadis (1961) introduced the boundary layer flow induced by a moving plate in a quiescent ambient fluid. A large amount of literatures on this problem has been cited in the books by Schlichting and Gersten (2000) and White (2006b).

Adding nanoparticles to a fluid make the base fluid inhomogeneous, as a result, thermodynamic irreversibility in the flow increases which can cause more energy and power losses in the system. Conserving useful energy depends on how to design an efficient heat transfer process from a thermodynamic point of view. Energy conversion processes are led to an irreversible increase in entropy. Thus, even though energy is conserved, the quality of the energy decreases by converting it into a different form of energy at which less work can be obtained. Reducing the generated entropy will result in more efficient designs of energy systems. In 1996, Bejan (1982, 1996) presented a method named Entropy Generation Minimization (EGM) to measure and optimize the disorder or disorganization generated during a process specifically in the fields of refrigeration and solar thermal power conversion. The method is also known as second law analysis and thermodynamic optimization. This field has been developed astoundingly during the 1990s, in both engineering and physics. Rahman (2007) has shown a method for calculating entropy generation in a human body under various environmental and physiological conditions. Later, Kolsi et al. (2010) studied the effects of an external magnetic field on the entropy generation. He concluded that in the presence of a magnetic field the generated entropy is distributed on the entire cavity.

In this section, considering the effects of nanoparticles in a fluid, we have employed the similarity variables introduced by Blasius (1908b) to obtain a relation for entropy generation through boundary layers for a nanofluid. Then considering the appropriate boundary conditions, governing equation including momentum and energy equations have been solved with the fourth-order Runge-Kutta as a reliable numerical technique. Taking into consideration the variation of different parameters such as: various volume fractions of nanoparticles and different kind of particles on the velocity and thermal gradients, the variations of Bejan number is studied to find effective parameters on the generated entropy. In addition, physical interpretations of the results are discussed in detail.

4.4.2 GOVERNING EQUATION

Consider an incompressible viscous nanofluid which flows over a flat plate, as shown in Figure 4.40. The wall temperature T_w is uniform and constant and is greater than the free stream temperature T_∞. It is assumed that the free stream velocity, U_∞, is also uniform and constant as well. Further, assuming that the flow in the laminar boundary layer is two-dimensional, and that the temperature gradients resulting from viscous dissipation are small, the continuity, momentum, and energy equations can be expressed as

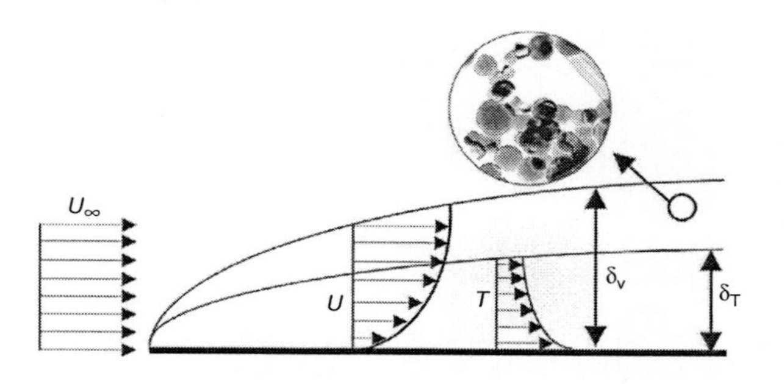

FIGURE 4.40

Velocity and thermal boundary layers for a nanofluid over a flat plate.

$$\frac{\partial u}{\partial x}+\frac{\partial u}{\partial y}=0 \tag{4.86}$$

$$u\frac{\partial u}{\partial x}+v\frac{\partial u}{\partial y}=\nu_{\mathrm{nf}}\frac{\partial^2 u}{\partial y^2} \tag{4.87}$$

$$u\frac{\partial T}{\partial x}+v\frac{\partial T}{\partial y}=\alpha_{\mathrm{nf}}\frac{\partial^2 T}{\partial y^2} \tag{4.88}$$

And the appropriate boundary conditions are

$$\begin{gathered}\text{At } y=0: u=v=0,\ \ T=T_{\mathrm{w}}\\ \lim_{y\to\infty} u=U_\infty \text{ and } \lim_{y\to\infty} T=T_\infty\end{gathered} \tag{4.89}$$

Physical properties of the nanofluids may be (Bachok et al., 2010)

$$\mu_{\mathrm{nf}}=\frac{\mu_{\mathrm{f}}}{(1-\Phi)^{2.5}},\ \ \rho_{\mathrm{nf}}=(1-\Phi)\rho_{\mathrm{f}}+\rho_{\mathrm{p}} \tag{4.90}$$

where μ_{nf} is the viscosity of the nanofluid, Φ is the solid volume fraction of the nanofluid, ρ_{f} is the density of the base fluid, ρ_{p} is the density of the solid particle, and μ_{f} is the viscosity of the base fluid and thermal properties are

$$(\rho C_p)_{\mathrm{nf}}=(1-\Phi)(\rho C_p)_{\mathrm{f}}+(\Phi\rho C_p)_{\mathrm{p}} \tag{4.91}$$

$$\frac{k_{\mathrm{nf}}}{k_{\mathrm{f}}}=\frac{(k_{\mathrm{p}}+2k_{\mathrm{f}})-2\Phi(k_{\mathrm{f}}-k_{\mathrm{p}})}{(k_{\mathrm{p}}+2k_{\mathrm{f}})+\Phi(k_{\mathrm{f}}-k_{\mathrm{p}})} \tag{4.92}$$

Here, k_{f} is the thermal conductivity of the fluid, k_{p} is the thermal conductivity of the solid, and k_{nf} is thermal conductivity of the nanofluid. $(\rho C_p)_{\mathrm{nf}}$, $(\rho C_p)_{\mathrm{f}}$ and $(\rho C_p)_{\mathrm{p}}$ are heat capacity of the nanofluid, fluid, and particle, respectively. Equation (4.90) is valid for spherical particles. These amounts for different materials are listed in Table 4.6.

Table 4.6 Thermo Physical Properties of the Base Fluid and the Nanoparticles (Bachok et al., 2010)

Physical Properties	Fluid Phase (Water)	Cu	Al_2O_3	TiO_2
C_p (J/kg K)	4179	385	765	686.2
ρ (kg/m^3)	997.1	8933	3970	4250
k (W/m K)	0.613	400	40	8.9538
$\alpha \times 10^{-7}$ (m^2/s)	1.47	1163.1	131.7	30.7

We look for a similarity solution of the Equations (4.86) and (4.87) with the boundary conditions Equation (4.89) of the following form

$$\psi = \sqrt{2\nu_f x U_\infty} f(\eta), \quad \eta = \sqrt{\frac{U_\infty}{2\nu_f x}} y \tag{4.93}$$

where ψ is the usual stream function, i.e., $u = \frac{\partial \psi}{\partial y}$ $v = -\frac{\partial \psi}{\partial x}$ and ν_f is the kinematic viscosity of the base fluid. Substituting Equations (4.92) and (4.93) into Equation (4.87), we obtained the following ODE

$$\frac{1}{(1-\Phi)^{2.5}\left(1-\Phi+\Phi\frac{\rho_p}{\rho_f}\right)} f''' + ff'' = 0 \tag{4.94}$$

With these boundary conditions

$$\text{At } \eta = 0:\ f(0) = \frac{df(\eta)}{d\eta} = 0, \quad \lim_{\eta\to\infty} \frac{df(\eta)}{d\eta} = 1 \tag{4.95}$$

The skin-friction coefficient C_f can be defined as:

$$C_f = \frac{\tau_w}{\rho_f u_\infty^2} \tag{4.96}$$

where τ_w is the surface shear stress which is given by

$$\tau_w = \mu_{nf}\left(\frac{\partial u}{\partial y}\right)_{y=0} \tag{4.97}$$

Substituting Equations (4.92) and (4.93) into Eqautions (4.96) and (4.97), we obtain

$$\sqrt{2Re_x}\, C_f = \frac{1}{(1-\Phi)^{2.5}} f''(0) \tag{4.98}$$

Looking for similarity solution for energy equation, Equation (4.88), we obtained

$$\frac{1}{Pr}\frac{\left(\frac{k_{nf}}{k_f}\right)}{\left(1-\Phi+\Phi\frac{(\rho C_p)_p}{(\rho C_p)_f}\right)} \theta'' + f\theta' = 0 \tag{4.99}$$

where

$$\theta = \frac{T - T_\infty}{T_w - T_\infty} \tag{4.100}$$

That is dimensionless temperature and $Pr = \frac{\nu_f}{\alpha_f}$. The boundary conditions are

$$\text{At } \eta = 0: \ \theta(0) = 1, \quad \lim_{\eta \to \infty} \theta(\eta) = \theta(\infty) = 0 \tag{4.101}$$

It is worth mentioning that if we substitute $\Phi = 0$ the equation will be the famous Blasius equation. The local Nusselt number Nu_x is defined as:

$$Nu_x = \frac{x q_w}{K_f (T_w - T_\infty)} \tag{4.102}$$

where q_w is the surface heat flux which is

$$q_w = -K_{nf} \left(\frac{\partial T}{\partial y} \right)_{y=0} \tag{4.103}$$

Using Equations (4.93), (4.94), (4.102), and (4.103) we obtain

$$\left(\frac{Re_x}{2} \right)^{-\frac{1}{2}} Nu_x = -\frac{k_{nf}}{k_f} \theta'(0) \tag{4.104}$$

in the present context $\left(\frac{Re_x}{2} \right)^{-\frac{1}{2}} Nu_x$ and $\sqrt{2Re_x} C_f$ are referred as the reduced Nusselt number and reduced skin-friction coefficient which are represented by $-\theta'(0)$ and $f''(0)$, respectively.

4.4.3 ENTROPY GENERATION

As Bejan (1996) suggested, entropy generation equation is:

$$S'''_{gen} = \left[\frac{K_{nf}}{T^2} (\nabla T)^2 \right] + \left[\frac{\mu_{nf}}{T} \phi \right] \tag{4.105}$$

Substituting Equations (4.92), (4.93), and (4.100) into Equation (4.106) we obtained a relation for entropy generation for a nanofluid which flows over a horizontal flat plate

$$S = \frac{k_{nf} \theta'(\eta)^2}{2Re\, PrEcK_f (\theta_\infty + \theta(\eta))^2} \left[\frac{\eta^2}{2Re} + 1 \right] + \left[\frac{1}{2} \frac{f''(\eta)^2}{(1 - \Phi)^{2.5} (\theta_\infty + \theta(\eta)) Re} \right] \tag{4.106}$$

where S is the dimensionless entropy generation and may be defined by the following relationship

$$S = \frac{\nu_f^2 \Delta T}{u_\infty^4 \mu_f} S'''_{gen} \tag{4.107}$$

Equation (4.106) denotes entropy generation in term of similarity parameters as well as similarity functions. So, it is straightforward to calculate the entropy generation with the aid of similarity solution.

It must be noted that in Equation (4.106) the first term is because of heat transfer and the latter one is due to fluid friction which are corresponded to S_h and S_f, respectively; Bejan number is defined as follows

$$Be = \frac{S_h}{S_h + S_f} \tag{4.108}$$

As it is obvious Be yields the share of S_h and S_f in total generated entropy, $Be = 1$ is the limit at which the heat transfer irreversibility dominates, $Be = 0$ is the opposite limit at which the irreversibility is dominated by fluid friction effects.

4.4.4 RESULTS AND DISCUSSION

Considering the boundary conditions of Equations (4.95) and (4.101), the system of nonlinear boundary layer Equations (4.94) and (4.99) have been solved numerically with fourth-order Runge-Kutta for different nanoparticles, namely, Cu, Al_2O_3, TiO_2 with different solid volume fractions and water as a base fluid. In order to analyze the generated entropy, variation of Bejan number is studied for different nanofluids and volume fractions. The validation of our results for $\Phi = 0$ as a conventional fluid can be seen in Figure 4.41 which is in the best agreement with previous studies (White, 2006b) and also velocity profile inside the boundary layer for Cu/water and different values of φ can be seen in Figure 4.42. It is to be said that with no loss of generality, the constant values of Pr, Ec have been chosen from Table 4.1 where $\theta_\infty = 2$ and $Re = 1000$ (see Rezaiguia et al., 2010).

Figure 4.43 shows the effects of solid volume fraction on velocity boundary layer for Cu and water mixture. Since adding the particles leads to increase in dynamic viscosity and momentum diffusion of

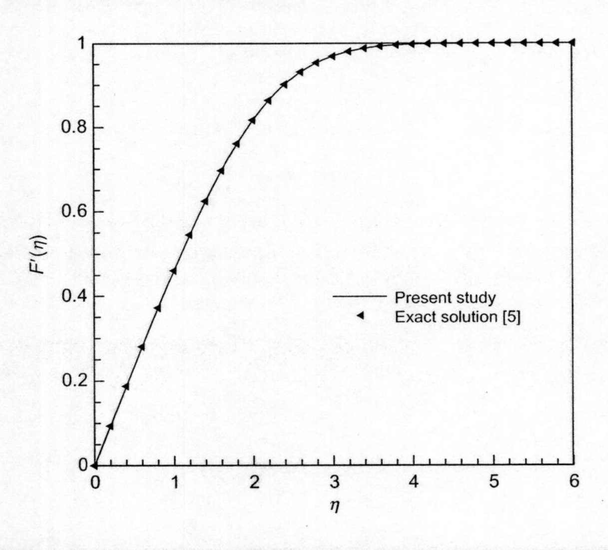

FIGURE 4.41

Validation of the developed code for velocity profile inside the boundary layer with $\phi = 0.1$.

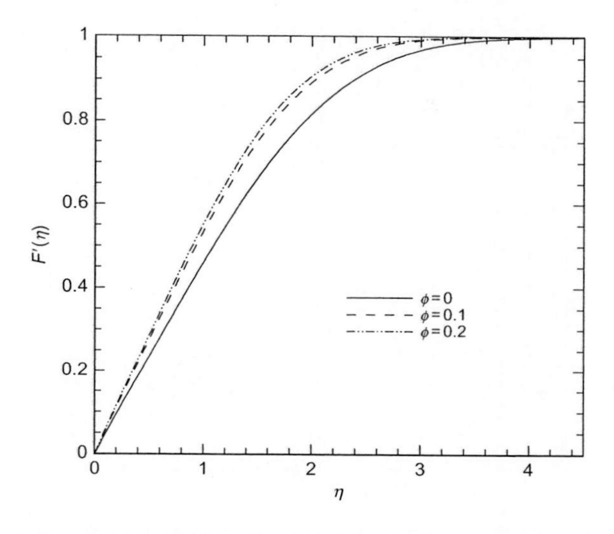

FIGURE 4.42

Velocity profile inside the boundary layer for Cu/water and different values of ϕ.

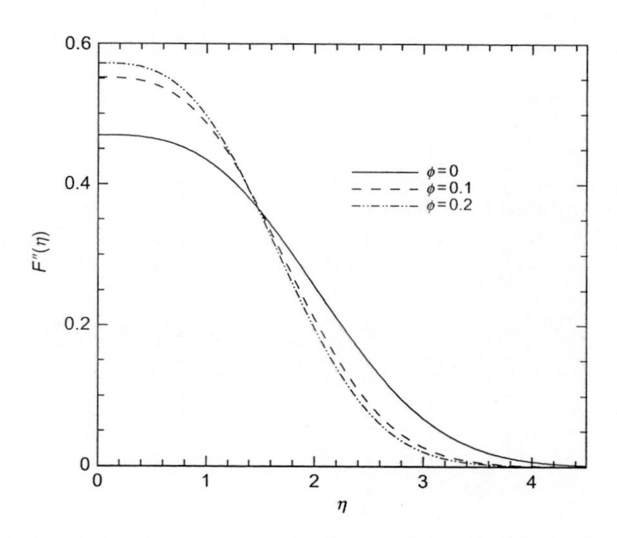

FIGURE 4.43

Velocity gradient inside the boundary layer for Cu/water and different values of ϕ.

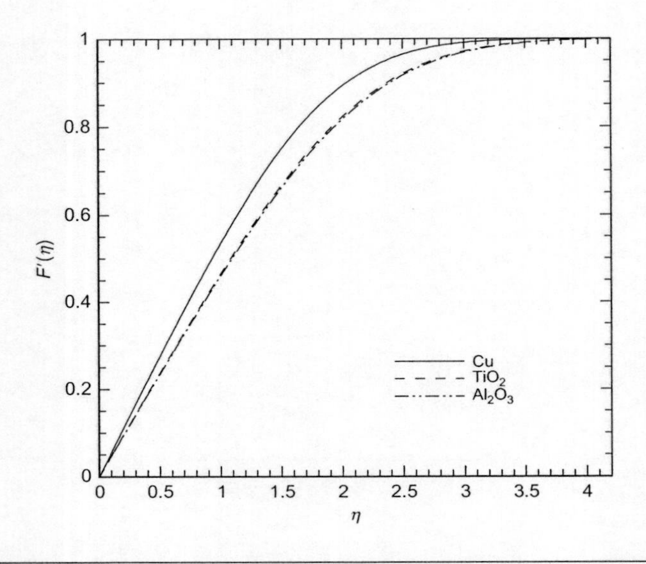

FIGURE 4.44

Velocity profile inside the boundary layer for different particles and water for $\phi = 0.1$.

the fluid, it is clear that the thickness of the boundary layer decreases with increasing in φ. Hence, the velocity gradient and skin-friction coefficient on the plate increases both which are plotted for cu/water in Figure 4.44.

Dimensionless velocity distribution for different nanofluids and constant solid volume fraction is shown in Figure 4.45. It is obvious that the velocity distribution for TiO_2/water and Al_2O_3/water are almost the same as their densities are, but due to high density of Cu, for Cu/water the dynamic viscosity increases more and leads to a thinner boundary layer than other particles. So in this case, velocity gradients on the surface are greater (see Figure 4.46).

Figures 4.47 to 4.50 show variation of thermal boundary layer thickness for different values of φ and various nanofluids, respectively. Considering Figures 4.47 and 4.48, it is obvious that increasing in Φ will lead to increase in thermal boundary layer and decrease in temperature gradient on the surface. Despite, heat transfer rate increases markedly. We have shown in Equation (4.105) that heat transfer rate is a function of θ' and K_{nf} both. When Φ increases the latter term which is more effective than the former one climbs up thus the heat transfer rate takes a similar trend. In other words when Φ increases, the thermal conduction of nanofluid rises thus we can experience the increase in heat transfer rate. Also, from Figures 4.49 and 4.50, we can conclude due to high thermal conductivity of Cu in contrast with Al_2O_3 and TiO_2, the heat transfer rate for Cu/water mixture is higher than other two nanofluids.

Bejan number is plotted for various conditions in Figures 4.51 and 4.52. It must be mentioned at the beginning of the plate Be is approximately unity ($Be \cong 0.9$), which means in the total entropy generation, heat transfer plays more important role than fluid friction but as we go along the surface, with decreasing in temperature gradient, heat transfer share in total generated entropy decreases. Another

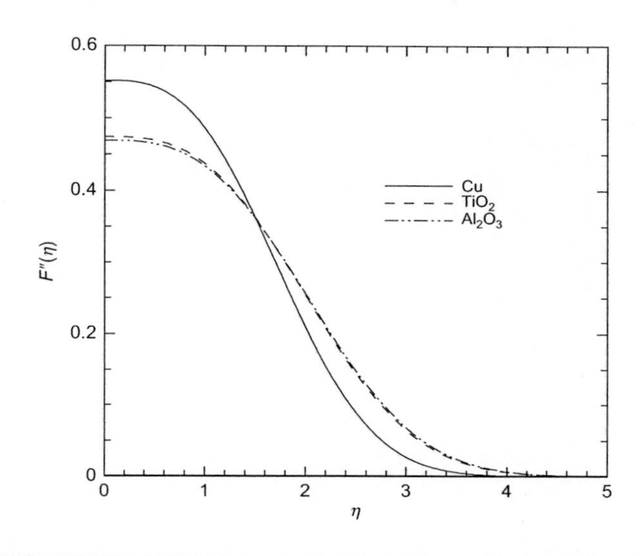

FIGURE 4.45

Velocity gradients inside the boundary layer for different particles and water for $\phi = 0.1$.

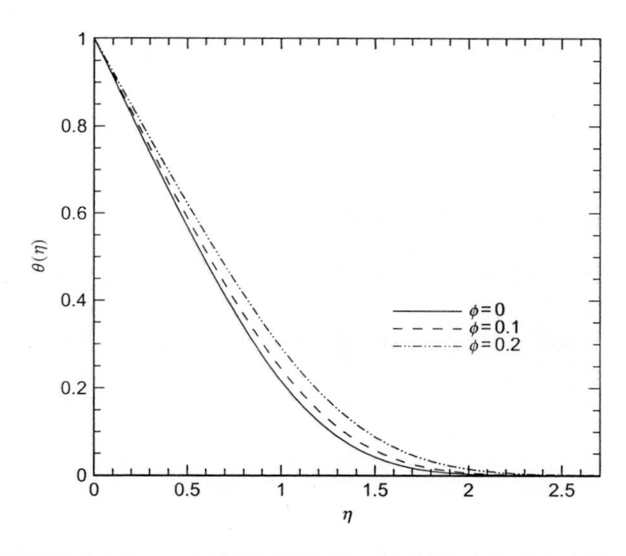

FIGURE 4.46

Temperature profile inside the boundary layer for water/Cu and different values of ϕ.

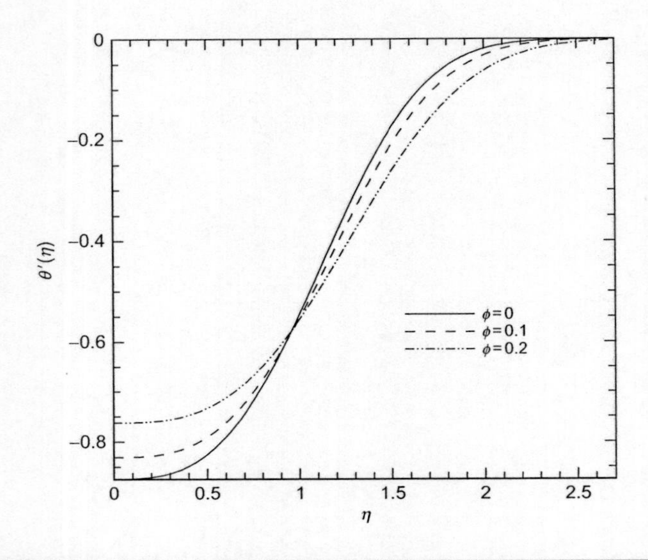

FIGURE 4.47

Temperature gradients inside the boundary layer for Cu/water and different values of ϕ.

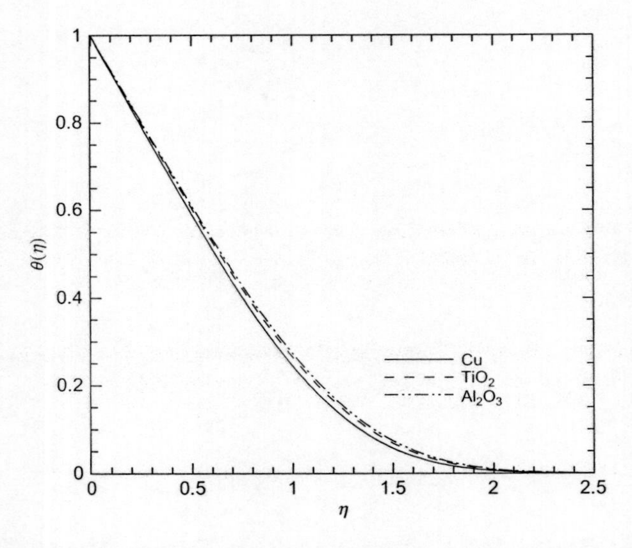

FIGURE 4.48

Temperature profile inside the boundary layer for different particles and water for $\phi = 0.1$.

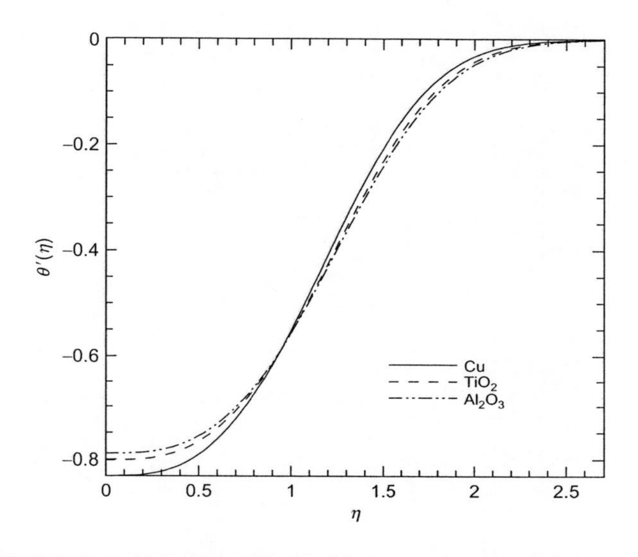

FIGURE 4.49

Temperature gradients inside the boundary layer for different particles and water for $\phi = 0.1$.

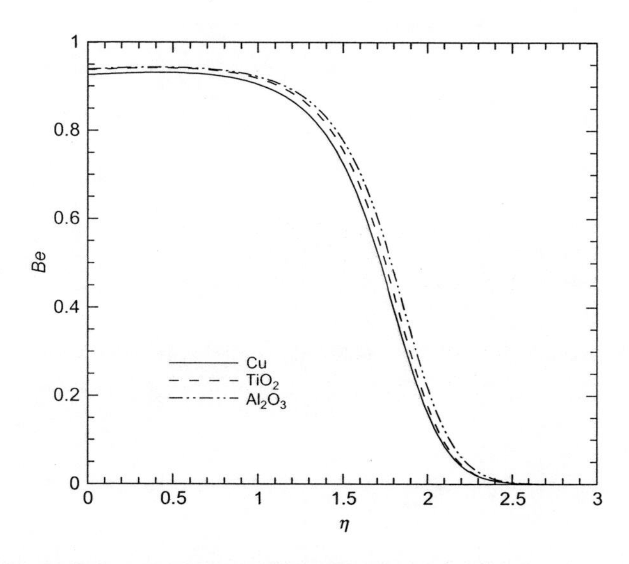

FIGURE 4.50

Be number inside the boundary layer for different particles and water for $\phi = 0.1$.

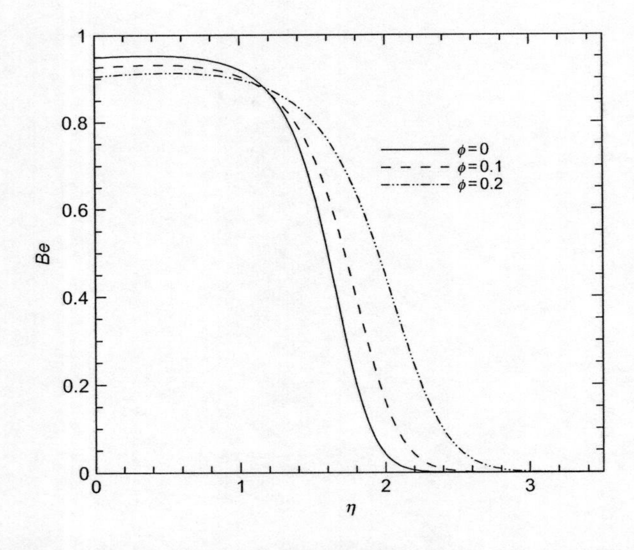

FIGURE 4.51

Be number inside the boundary layer for water/Cu and different values of ϕ.

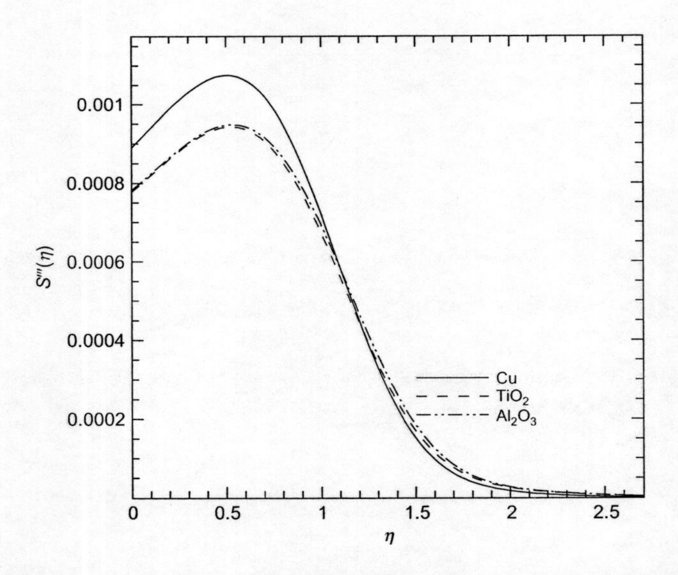

FIGURE 4.52

Volumetric entropy generation inside the boundary layer for different particles and water for $\phi = 0.1$.

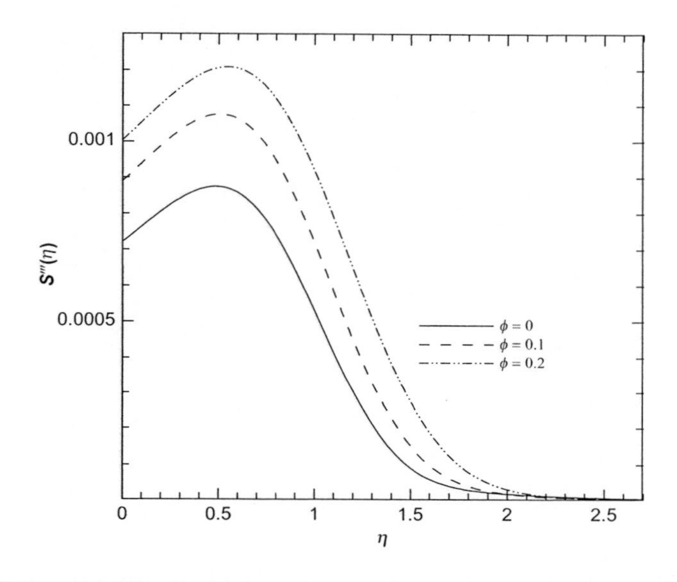

FIGURE 4.53

Volumetric entropy generation inside the boundary layer for water/Cu and different values of ϕ.

important point that can be inferred from the graph is that increasing in solid volume fraction leads to decrease in *Be* number which means the share of fluid friction in the total entropy generation increases.

Volumetric entropy generation on boundary layer is illustrated in Figure 4.53 for various types of nanofluids and solid volume fraction, respectively. Considering Figure 1.14, at the vicinity of the surface, we can see a noticeable increase in S'''_{gen} which is a result of a decrease in dimensionless temperature (Figures 4.47 and 4.49) and the constant thermal and velocity gradient in this zone (Figures 4.44, 4.46, 4.48 and 4.50) both. After $\eta \cong 0.5$, it takes a decreasing trend and finally vanishes. Considering Figure 1.13, it is obvious that *Be* number decreases along the surface; so, according to Equation (4.107), there are two possibilities: first, increasing in S_f and second, decreasing in S_h. As we showed, since the total generated entropy (Figure 4.53) inside the boundary layer decreases, the second condition is acceptable.

4.4.5 CONCLUSION

In this section, we have studied a steady two-dimensional boundary-layer flow on a flat plate for nanofluids numerically. Converting the Basic equations to ODE ones with similarity transformation, the entropy generation for a nanofluid over a horizontal plate is formulated and studied inside the boundary layer. All equations are solved numerically via shooting method as a reliable technique. The effects of different solid volume fraction with three different types of nanoparticles and water as a base fluid, on the velocity and temperature profiles and the generated entropy are discussed. *Be* Number is plotted and its variation is studied as well. It is found that generation of entropy in Boundary layer increases with increasing the solid volume fraction and due to high density of Cu, adding this nanoparticle to water causes more generated entropy in contrast with other particles. In addition, we have shown that in the total entropy generation, heat transfer term always dominates.

4.5 VISCOUS FLOW AND HEAT TRANSFER OF NANOFLUID OVER NONLINEARLY STRETCHING SHEET

In this section, the boundary-layer flow and heat transfer in a viscous fluid containing nanoparticles over a nonlinearly stretching sheet are analyzed. The stretching velocity is assumed to vary as a power function of the distance from the origin. The governing partial differential equation and auxiliary conditions are converted to couple nonlinear ODEs by a similar transformation. The obtained nonlinear ODEs are solved analytically by homotopy perturbation method (HPM) employing Padé technique. The effects of various relevant parameters, namely, the Eckert number Ec, the solid volume fraction of the nanoparticles φ and the nonlinear stretching parameter n are discussed. The solutions are compared with a numerical technique (*this chapter has been worked by D. D. Ganji, S. M. J. Hashemi in nonlinear dynamics team in Mechanical Engineering Department, 2012-2013*).

4.5.1 INTRODUCTION

The problem of viscous flow and heat transfer over a stretching sheet has important industrial applications, take for example, in metallurgical processes such as drawing of continuous filaments through quiescent fluids, annealing and tinning of copper wires, glass blowing, manufacturing of plastic and rubber sheets, crystal growing, and continuous cooling and fiber spinning, in addition to a wide range of applications in many engineering processes, such as polymer extrusion, wire drawing, continuous casting, manufacturing of foods and paper, glass fiber production, stretching of plastic films, and many others. During the manufacture of these sheets, the molten materials issues from a slit and is subsequently stretched to achieve the desired thickness. The final product with the desired characteristics strictly depends on the stretching rate, the rate of cooling in the process and the process of stretching. It is also known that nanofluids can enhance the heat transfer characteristics of the original (base) fluid. Heat transfer is an important process in physics and engineering and therefore improvements in heat transfer characteristics will improve the efficiency of many processes.

Heat, mass, and momentum transfer in the laminar boundary layer flow on a stretching sheet are important due to its applications to polymer technology and metallurgy. Gupta and Gupta (1977) stated that the stretching of the sheet may not necessarily be linear. Vajravelu (2001) studied flow and heat transfer in a viscous fluid over a nonlinear stretching sheet without viscous dissipation, but the heat transfer in this flow was analyzed only in the case when the sheet was held at a constant temperature. Raptis and Perdikis (2006) investigated the steady two-dimensional flow of an incompressible viscous and electrically conducting fluid over a nonlinearly semi-infinite stretching sheet in the presence of a chemical reaction and under the influence of a magnetic field. Bataller (2008) presented a numerical analysis in connection with the boundary layer flow induced in a quiescent fluid by a stretching sheet with velocity $u_x(x)=x^{1/3}$ along with heat transfer. Prasad and Vajravelu (2009) examined the hydromagnetic laminar boundary layer flow and heat transfer in a power law fluid over a nonisothermal stretching sheet. There have also been several recent studies, Ziabakhsh et al. (2010) employed the

HAM to compute an approximation to the solution for the problem of flow and diffusion of chemically reactive species over a nonlinearly stretching sheet immersed in a porous medium.

In the present study, the HPM (He, 2003) and Padé approximant (Adomian, 1994 and Baker, 1975) for finding approximate solution of two-dimensional flow and heat transfer of an incompressible viscous nanofluid past nonlinear stretching surface is considered. The HPM technique provides a sequence of functions which converges to the exact solution of the problem. The Padé approximant increases the convergence of the solutions obtained by the HPM. It is of interest to be noted that Padé approximants give results with no greater error bound than approximation by polynomials. The aim is to investigate the influence of various nanofluid parameters (the solid volume fraction of the nanoparticles φ) and the effect of nonlinear stretching parameter n on flow and heat-transfer characteristics.

4.5.2 BASIC CONCEPTS OF HPM

To illustrate the basic idea of this method, we consider the following nonlinear differential equation:

$$A(u)-f(r)=0, \quad r \in \Omega, \tag{4.109}$$

Considering the boundary conditions of:

$$B\left(u, \frac{\partial u}{\partial n}\right)=0, \quad r \in \Gamma, \tag{4.110}$$

where A is a general differential operator, B is a boundary operator, $f(r)$ is a known analytical function, and Γ is the boundary of the domain Ω.

The operator A can be divided into two parts of L and N, where L is the linear part, while N is a nonlinear one Equation (4.109) therefore can be rewritten as follows:

$$L(u)+N(u)-f(r)=0, \tag{4.111}$$

By the homotopy technique, we construct a homotopy as $v(r,p):\Omega \times [0,1] \rightarrow \Re$ which satisfies:

$$H(v,p)=(1-p)[L(v)-L(u_0)]+p[A(v)-f(r)]=0, \quad p \in [0,1] \quad r \in \Omega, \tag{4.112}$$

where $p \in [0,1]$ is an embedding parameter and u_0 is an initial approximation of Equation (4.111) which satisfies the boundary conditions. Obviously, considering Equation (4.112), we will have:

$$\begin{cases} H(v,0)=L(v)-L(u_0)=0 \\ H(v,1)=A(v)-f(r)=0. \end{cases} \tag{4.113}$$

The changing process of p from zero to unity is just that of $v(r,p)$ from $u_0(r)$ to $u(r)$.

In topology, this is called deformation, and $L(v)-L(u_0)$ and $A(v)-f(r)$ are called homotopy. According to HPM, we can first use the embedding parameter p as "small parameter", and assume that the solution of Equation (4.112) can be written as a power series in p:

$$v=v_0+pv_1+p^2v_2+\cdots, \tag{4.114}$$

Setting $p=1$ results in the approximate solution of Equation (4.112):

$$u=\lim_{p\rightarrow 1} v=v_0+v_1+v_2+\cdots, \tag{4.115}$$

The combination of the perturbation method and the homotopy method is called the HPM, which lacks the limitations of the traditional perturbation methods although this technique has full advantages of the traditional perturbation techniques. The series Equation (4.115) is convergent for most cases. However, the convergence rate depends on the nonlinear operator $A(v)$.

4.5.3 FORMULATION OF PROBLEM

Consider the steady laminar two-dimensional flow of an incompressible viscous fluid coinciding with the plane $y=0$, with the flow being confined to $y>0$. Two equal and opposite forces are introduced along the x-axis so that the wall is stretched whilst keeping the position of the origin fixed (see Ref. Cortell, 2007a). A schematic representation of the physical model and coordinate system is depicted in Figure 4.54.

The temperature at the stretching surface is deemed to have a constant value T_w while the ambient temperature has a constant value T_∞. It is further assumed that the regular fluid and the suspended nanoparticles are in the thermal equilibrium and no slip occurs between them. The thermo physical properties of the nanofluid are given in Table 4.7 (see Ref. Abu-Nada et al., 2008).

Under the above assumptions, the boundary layer equations governing the flow and temperature in the presence of heat source or heat sink are (using the boundary layer approximations and taking into account viscous dissipation (last term in the energy equation):

$$\frac{\partial u}{\partial x}+\frac{\partial v}{\partial y}=0, \tag{4.116}$$

$$u\frac{\partial u}{\partial x}+v\frac{\partial u}{\partial y}=\frac{\mu_{nf}}{\rho_{nf}}\frac{\partial^2 u}{\partial y^2}, \tag{4.117}$$

$$u\frac{\partial T}{\partial x}+v\frac{\partial T}{\partial y}=\alpha_{nf}\frac{\partial^2 T}{\partial y^2}+\frac{\mu_{nf}}{(\rho C_p)_{nf}}\left(\frac{\partial u}{\partial y}\right)^2 \tag{4.118}$$

where x and y are the coordinates along and perpendicular to the sheet, u and v are the velocity components in the x- and y-directions, respectively, T is the local temperature of the fluid. Further, ρ_{nf} is the effective density, μ_{nf} is the effective dynamic viscosity, $(\rho C_p)_{nf}$ is the heat capacitance, α_{nf} is the

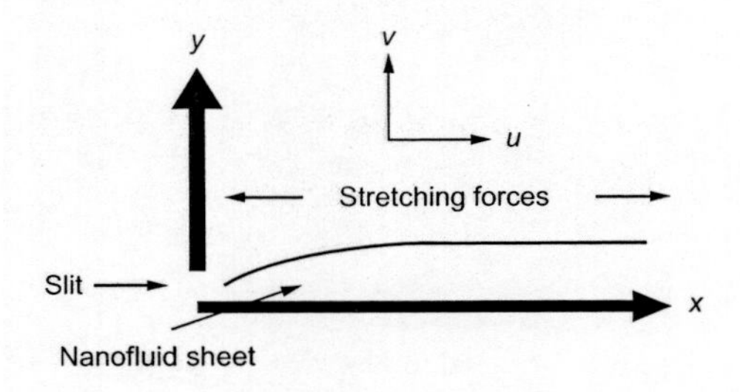

FIGURE 4.54

A schematic diagram of the physical model.

Table 4.7 Thermo-Physical Properties of Water and Nanoparticles

	ρ (kg m^{-3})	C_p (J kg^{-1} K^{-1})	k (W m^{-1} K^{-1})	$\beta \times 10^5$ (K^{-1})
Pure Water	997.1	4179	0.613	21
Copper (Cu)	8933	385	401	1.67
Silver (Ag)	10,500	235	429	1.89
Alumina (Al_2O_3)	3970	765	40	0.85
Titanium Oxide (TiO_2)	4250	686.2	8.9538	0.9

effective thermal diffusivity and k_{nf} is the effective thermal conductivity of the nanofluid which are defined as (see Refs. Aminossadati and Ghasemi, 2009 and Oztop and Abu-Nada, 2008):

$$\begin{cases} \rho_{nf} = (1-\varphi)\rho_f + \varphi\rho_s, \quad \mu_{nf} = \dfrac{\mu_f}{(1-\varphi)^{2.5}} \\ (\rho C_p)_{nf} = (1-\varphi)(\rho C_p)_f + \varphi(\rho C_p)_s, \quad \alpha_{nf} = \dfrac{\kappa_{nf}}{(\rho C_p)_{nf}} \\ k_{nf} = k_f\left(\dfrac{k_s + 2k_f - 2\varphi(k_f - k_s)}{k_s + 2k_f + 2\varphi(k_f - k_s)}\right), \end{cases} \tag{4.119}$$

where φ is the solid volume fraction of the nanoparticles. The appropriate boundary conditions for the problem are given by:

$$\begin{cases} u = u_w(x) = cx^n, \quad v = 0, \quad T = T_w(x) = T_\infty + bx^{2n} \text{ at } y = 0 \\ u \to 0, T = T_\infty \text{ as } y \to \infty, \end{cases} \tag{4.120}$$

where b and c are positive constants and n is the nonlinear stretching parameter. By introducing the following nondimensional variables (see Ref. Cortell, 2007b):

$$\begin{cases} \eta = y\sqrt{\dfrac{c(n+1)}{2\nu_f}}x^{\frac{n-1}{2}}, \quad u = cx^n F'(\eta) \\ v = -\sqrt{\dfrac{(n+1)c\nu_f}{2}}x^{\frac{n-1}{2}}\left(F(\eta) + \dfrac{n-1}{n+1}\eta F'(\eta)\right) \\ \theta(\eta) = \dfrac{T - T_\infty}{T_w - T_\infty}. \end{cases} \tag{4.121}$$

Using Equations (4.116)-(4.118) and (4.121) lead to the following nondimensional ODEs:

$$F''' + (1-\varphi)^{2.5}\left(1 - \varphi + \varphi\left(\frac{\rho_s}{\rho_f}\right)\right)\left(FF'' - \frac{2n}{n+1}F'^2\right) = 0, \tag{4.122}$$

$$\frac{1}{Pr}\left(\frac{k_{nf}}{k_f}\right)\theta'' + \frac{Ec}{(1-\varphi)^{2.5}}F''^2 + \left(1 - \varphi + \varphi\frac{(\rho C_p)_s}{(\rho C_p)_f}\right)\left(F\theta' - \frac{4n}{n+1}F'\theta\right) = 0, \tag{4.123}$$

and the corresponding boundary conditions Equation (4.120) become:

$$\begin{cases} F = 0, \quad F' = 1 \; \theta = 1 \text{ at } \eta = 0 \\ F' \to 0, \quad \theta \to 0 \text{ as } \eta \to \infty, \end{cases} \tag{4.124}$$

where $Pr=\frac{\nu}{\alpha}$ is the Prandtl number and Ec is the Eckert number which is defined as:

$$Ec=\frac{u_{\mathrm{w}}^{2}}{C_{p}(T_{\mathrm{w}}-T_{\infty})}. \tag{4.125}$$

The skin-friction coefficient C_{f} and the local Nusselt number Nu_x are defined as:

$$C_{\mathrm{f}}=\frac{\mu_{\mathrm{nf}}}{\rho_{\mathrm{f}}u_{\mathrm{w}}^{2}}\left(\frac{\partial u}{\partial y}\right)_{y=0},\quad Nu_{x}=-\frac{xk_{\mathrm{nf}}}{k_{\mathrm{f}}(T_{\mathrm{w}}-T_{\infty})}\left(\frac{\partial T}{\partial y}\right)_{y=0}. \tag{4.126}$$

Using Equations (4.121) and (4.126), we get:

$$Re_{x}^{1/2}C_{\mathrm{f}}=\frac{1}{(1-\varphi)^{2.5}}\left(\frac{n+1}{2}\right)^{\frac{1}{2}}F''(0),\quad Re_{x}^{1/2}Nu_{x}=-\frac{k_{\mathrm{nf}}}{k_{\mathrm{f}}}\left(\frac{n+1}{2}\right)^{1/2}\theta'(0). \tag{4.127}$$

where $Re_x=\frac{xu_{\mathrm{w}}}{\nu_{\mathrm{f}}}$ is the local Reynolds number based on the stretching velocity u_{w}.

4.5.4 HOMOTOPY PERTURBATION SOLUTION

In this section, we will apply HPM to nonlinear ODE Equations (4.122) and (4.123).

According to HPM, we can construct a homotopy of Equations (4.122) and (4.123) as:

$$\begin{aligned}H(F,p)&=(1-p)\left(\frac{\mathrm{d}^{3}}{\mathrm{d}\eta^{3}}F(\eta)\right)\\&+p\left(\frac{\mathrm{d}^{3}}{\mathrm{d}\eta^{3}}F(\eta)+(1-\phi)^{2.5}\left(1-\phi+\phi\frac{\rho_{\mathrm{s}}}{\rho_{\mathrm{f}}}\right)\left(F(\eta)\left(\frac{\mathrm{d}^{2}}{\mathrm{d}\eta^{2}}F(\eta)\right)-\frac{2n}{n+1}\left(\frac{\mathrm{d}}{\mathrm{d}\eta}F(\eta)\right)^{2}\right)\right),\end{aligned} \tag{4.128a}$$

$$\begin{aligned}H(\theta,p)&=(1-p)\frac{k_{\mathrm{nf}}}{Pr\,k_{\mathrm{f}}}\left(\frac{\mathrm{d}^{2}}{\mathrm{d}\eta^{2}}\theta(\eta)\right)\\&+p\left(\frac{k_{\mathrm{nf}}}{Pr\,k_{\mathrm{f}}}\left(\frac{\mathrm{d}^{2}}{\mathrm{d}\eta^{2}}\theta(\eta)\right)+\frac{Ec}{(1-\phi)^{2.5}}\left(\frac{\mathrm{d}^{2}}{\mathrm{d}\eta^{2}}F(\eta)\right)^{2}+\left(1-\phi+\phi\frac{(\rho C_{p})_{\mathrm{s}}}{(\rho C_{p})_{\mathrm{f}}}\right)\left(F(\eta)\left(\frac{\mathrm{d}}{\mathrm{d}\eta}\theta(\eta)\right)-\frac{4n}{n+1}\left(\frac{\mathrm{d}}{\mathrm{d}\eta}F(\eta)\right)\theta(\eta)\right)\right).\end{aligned} \tag{4.128b}$$

We consider:

$$F(\eta)=F_{0}(\eta)+F_{1}(\eta)p+F_{2}(\eta)p^{2}+F_{3}(\eta)p^{3}+\cdots, \tag{4.129a}$$

$$\theta(\eta)=\theta_{0}(\eta)+\theta_{1}(\eta)p+\theta_{2}(\eta)p^{2}+\theta_{3}(\eta)p^{3}+\cdots, \tag{4.129b}$$

substituting F and θ from Equations (4.129a) and (4.129b) into Equations (4.128a) and (4.128b) and some simplification and rearranging based on powers of p-terms for $\varphi=0$ and $Pr=5$, we have:

$$p^{0}:\begin{cases}\frac{\mathrm{d}^{3}}{\mathrm{d}\eta^{3}}F_{0}(\eta)=0\\ \frac{1}{5}\frac{\mathrm{d}^{2}}{\mathrm{d}\eta^{2}}\theta_{0}(\eta)=0\end{cases} \tag{4.130}$$

$$F_0(0)=0,\ F_0'(0)=1,\ F_0''(0)=\alpha_1,\ \theta_0(0)=1,\ \theta_0'(0)=\alpha_2, \tag{4.131}$$

$$p^1:\begin{cases}\frac{1}{n+1}\left(\left(\frac{d^3}{d\eta^3}F_1(\eta)\right)n+F_0(\eta)\frac{d^2}{d\eta^2}F_0(\eta)+\frac{d^3}{d\eta^3}F_1(\eta)+F_0(\eta)\left(\frac{d^2}{d\eta^2}F_0(\eta)\right)n-2n\left(\frac{d}{d\eta}F_0(\eta)\right)^2\right)=0\\ \frac{1}{n+1}\left(0.2\left(5F_0(\eta)\left(\frac{d}{d\eta}\theta_0(\eta)\right)n+\left(\frac{d^2}{d\eta^2}\theta_1(\eta)\right)n+5F_0(\eta)\frac{d}{d\eta}\theta_0(\eta)-20n\left(\frac{d}{d\eta}F_0(\eta)\right)\theta_0(\eta)+\frac{d^2}{d\eta^2}\theta_1(\eta)\right.\right.\\ \left.\left.+5Ec\left(\frac{d^2}{d\eta^2}F_0(\eta)\right)^2+5Ec\left(\frac{d^2}{d\eta^2}F_0(\eta)\right)^2 n\right)\right)=0\end{cases} \tag{4.132}$$

$$F_1(0)=0,\ F_1'(0)=0,\ F_1''(0)=0,\ \theta_1(0)=0,\ \theta_1'(0)=0, \tag{4.133}$$

$$p^2:\begin{cases}\frac{1}{n+1}\left(\left(\frac{d^3}{d\eta^3}F_2(\eta)\right)n+F_1(\eta)\frac{d^2}{d\eta^2}F_0(\eta)4n\left(\frac{d}{d\eta}F_0(\eta)\right)\frac{d}{d\eta}F_1(\eta)+F_1(\eta)\left(\frac{d^2}{d\eta^2}F_0(\eta)\right)n\right.\\ \left.+F_0(\eta)\left(\frac{d^2}{d\eta^2}F_1(\eta)\right)n+F_0(\eta)\frac{d^2}{d\eta^2}F_1(\eta)+\frac{d^3}{d\eta^3}F_2(\eta)\right)=0\\ \frac{1}{n+1}\left(0.2\left(5F_0(\eta)\frac{d}{d\eta}\theta_1(\eta)+5F_1(\eta)\frac{d}{d\eta}\theta_0(\eta)+\frac{d^2}{d\eta^2}\theta_2(\eta)-20n\left(\frac{d}{d\eta}F_0(\eta)\right)\theta_1(\eta)\right.\right.\\ +10Ec\left(\frac{d^2}{d\eta^2}F_0(\eta)\right)\frac{d^2}{d\eta^2}F_1(\eta)+10Ec\left(\frac{d^2}{d\eta^2}F_0(\eta)\right)\left(\frac{d^2}{d\eta^2}F_1(\eta)\right)n+\left(\frac{d^2}{d\eta^2}\theta_2(\eta)\right)n\\ \left.\left.+5F_0(\eta)\left(\frac{d}{d\eta}\theta_1(\eta)\right)n+5F_1(\eta)\left(\frac{d}{d\eta}\theta_0(\eta)\right)n-20n\left(\frac{d}{d\eta}F_1(\eta)\right)\theta_0(\eta)\right)\right)=0\end{cases} \tag{4.134}$$

$$F_2(0)=0,\ F_2'(0)=0,\ F_2''(0)=0,\ \theta_2(0)=0,\ \theta_2'(0)=0, \tag{4.135}$$

$$p^3:\begin{cases}\frac{1}{n+1}\left(\left(\frac{d^3}{d\eta^3}F_3(\eta)\right)n+F_2(\eta)\frac{d^2}{d\eta^2}F_0(\eta)+F_0(\eta)\left(\frac{d^2}{d\eta^2}F_2(\eta)\right)n+F_2(\eta)\left(\frac{d^2}{d\eta^2}F_0(\eta)\right)n+F_0(\eta)\frac{d^2}{d\eta^2}F_2(\eta)\right.\\ \left.+\frac{d^3}{d\eta^3}F_3(\eta)+F_1(\eta)\frac{d^2}{d\eta^2}F_1(\eta)-4n\left(\frac{d}{d\eta}F_0(\eta)\right)\frac{d}{d\eta}F_2(\eta)+F_1(\eta)\left(\frac{d^2}{d\eta^2}F_1(\eta)\right)n-2n\left(\frac{d}{d\eta}F_1(\eta)\right)^2\right)=0\\ \frac{1}{n+1}\left(0.2\left(5F_0(\eta)\frac{d}{d\eta}\theta_2(\eta)-20n\left(\frac{d}{d\eta}F_1(\eta)\right)\theta_1(\eta)+10Ec\left(\frac{d^2}{d\eta^2}F_0(\eta)\right)\left(\frac{d^2}{d\eta^2}F_2(\eta)\right)n\right.\right.\\ -20n\left(\frac{d}{d\eta}F_2(\eta)\right)\theta_0(\eta)+\frac{d^2}{d\eta^2}\theta_3(\eta)+5F_1(\eta)\frac{d}{d\eta}\theta_1(\eta)+5Ec\left(\frac{d^2}{d\eta^2}F_1(\eta)\right)^2+5Ec\left(\frac{d^2}{d\eta^2}F_1(\eta)\right)^2 n\\ +5F_2(\eta)\frac{d}{d\eta}\theta_0(\eta)+5F_2(\eta)\left(\frac{d}{d\eta}\theta_0(\eta)\right)n+\left(\frac{d^2}{d\eta^2}\theta_3(\eta)\right)n-20n\left(\frac{d}{d\eta}F_0(\eta)\right)\theta_2(\eta)\\ \left.\left.+10Ec\left(\frac{d^2}{d\eta^2}F_0(\eta)\right)\frac{d^2}{d\eta^2}F_2(\eta)+5F_0(\eta)\left(\frac{d}{d\eta}\theta_2(\eta)\right)n+5F_1(\eta)\left(\frac{d}{d\eta}\theta_1(\eta)\right)n\right)\right)=0\end{cases} \tag{4.136}$$

$$F_3(0)=0,\ F_3'(0)=0,\ F_3''(0)=0,\ \theta_3(0)=0,\ \theta_3'(0)=0, \tag{4.137}$$

Solving Equations (4.130)-(4.136) with the boundary conditions, Equations (4.131)-(4.137):

$$F_0(\eta)=\frac{1}{2}\alpha_1\eta^2+\eta, \tag{4.138}$$

$$\theta_0(\eta) = \alpha_2\eta + 1, \tag{4.139}$$

$$F_1(\eta) = \frac{1}{120}\frac{\eta^3\left(-\alpha_1^2\eta^2 - 5\alpha_1\eta + 3\eta^2\alpha_1^2 n + 15\eta\alpha_1 n + 40n\right)}{n+1}, \tag{4.140}$$

$$\theta_1(\eta) = -\frac{5}{24}\frac{\eta^2\left(-7\eta^2\alpha_2 n\alpha_1 - 12\eta\alpha_2 n + \alpha_2\eta^2\alpha_1 + 4\alpha_2\eta - 16\eta\alpha_1 n - 48n + 12Ec\alpha_1^2 + 12Ec\alpha_1^2 n\right)}{n+1}, \tag{4.141}$$

$$\begin{aligned} F_2(\eta) = {} & \frac{1}{16800000000000}\frac{1}{(n+1)^2}(\eta^5(-17500000000\alpha_1^3\eta^3 n + 11250000000n^2\eta^3\alpha_1^3 + 4583333333\alpha_1^3\eta^3 \\ & -140000000000\eta^2\alpha_1^2 n + 36666666656\,\alpha_1^2\eta^2 + 90000000080n^2\eta^2\alpha_1^2 - 420000000000\,\eta\alpha_1 n \\ & + 70000000000\alpha_1\eta + 443333333520\eta\alpha_1 n^2 - 560000000000n + 560000000000n^2)), \end{aligned} \tag{4.142}$$

$$\begin{aligned} \theta_2(\eta) = {} & -\frac{1}{42000000000}\frac{1}{(n+1)^2}(\eta^3(17583333330\eta^4\alpha_1^2\alpha_2 n - 2125000000\eta^4\alpha_1^2\alpha_2 - 16958333330n^2\eta^4\alpha_1^2\alpha_2 \\ & - 14875000000\alpha_2\eta^3\alpha_1 - 56875000000\,n^2\eta^3\alpha_2\alpha_1 + 86916666690\eta^3\alpha_2 n\alpha_1 - 61833333338n^2\eta^3\alpha_1^2 \\ & + 36166666676\eta^3\alpha_1^2 n + 89250000000n^2\eta^2\alpha_1^3\,Ec - 371000000070n^2\eta^2\alpha_1 + 59500000014\eta^2\alpha_1^3 Ecn \\ & + 216999999930\eta^2 n\alpha_1 - 26250000000\alpha_2\eta^2 - 64750000014n^2\eta^2\alpha_2 + 91000000014\eta^2\alpha_2 n \\ & -29750000007\eta^2\alpha_1^3 Ec - 105000000000Ec\alpha_1^2\eta + 350000000000\,\eta n + 140000000000n^2\eta Ec\alpha_1^2 \\ & + 35000000000\,Ec\,\alpha_1^2\,n\,\eta - 420000000000\eta n^2 + 140000000000\alpha_1 Ecn + 140000000000\alpha_1 Ecn^2)), \end{aligned} \tag{4.143}$$

$$\begin{aligned} F_3(\eta) = {} & \frac{1}{6930000000000000}\frac{1}{(n+1)^3}(\eta^7(-332812499880\alpha_1^4\eta^4 n^2 - 65104166666\alpha_1^4\eta^4 + 145312499920n^3\eta^4\alpha_1^4 \\ & + 290104166660\alpha_1^4\eta^4 n + 3191145832875\alpha_1^3\eta^3 n + 1598437500275n^3\eta^3\alpha_1^3 - 716145833095\alpha_1^3\eta^3 \\ & -3660937501375n^2\eta^3\alpha_1^3 + 10064236110500n^3\eta^2\alpha_1^2 + 13539930555250\eta^2\alpha_1^2 n - 19765625013750n^2\eta^2\alpha_1^2 \\ & -2463541667125\,\alpha_1^2\eta^2 - 2578125000000\alpha_1\eta + 27671875006875\eta\alpha_1 n - 51046875000000\eta\alpha_1 n^2 \\ & + 28703125006875n^3\eta\alpha_1 + 21999999997800n - 43999999989000n^2 + 43999999989000n^3)), \end{aligned} \tag{4.144}$$

$$\begin{aligned} \theta_3(\eta) = {} & -\frac{1}{2520000000000}\frac{1}{(n+1)^3}(\eta^4(-4213750000500\eta^4\alpha_1^2 n - 13029999996000\eta^3\alpha_1 n \\ & - 3322500000000\eta^3\alpha_2 n + 854062499700\alpha_2\eta^4\alpha_1 + 4410000000000Ec\alpha_1^2\eta^2 + 19879999994400n^2\eta^2 \\ & - 468194444550\alpha_1^3\eta^5 n + 1055833333100\,n^2\eta^5\alpha_1^3 + 9502499997000n^2\eta^4\alpha_1^2 + 26379999996000n^2\eta^3\alpha_1 \\ & - 13999999994400n\eta^2 + 25802083328\eta^6\alpha_1^3\alpha_2 + 420937499925\eta^4\alpha_1^4 Ec - 1977499999200n^3\eta^3\alpha_2 \\ & + 4200000000000Ecn^2 + 4200000000000Ecn^3 - 223621527780\eta^6\alpha_1^3\alpha_2 n - 1189687500000\eta^4\alpha_1^4 Ecn \\ & + 291072916680\eta^6\alpha_1^3 n^2\alpha_2 - 567187499700\eta^4\alpha_1^4 n^2 Ec - 619062500000n^3\eta^5\alpha_1^2\alpha_2 - 1836562500000n^3\eta^4\alpha_2\alpha_1 \\ & + 3260000001000n^3\eta^3\alpha_1^3 Ec + 9450000000000n^3\eta^2 Ec\alpha_1^2 - 112836805536n^3\eta^6\alpha_1^3\alpha_2 + 1043437500000n^3\eta^4\alpha_1^4 Ec \\ & + 6720000002100\eta\alpha_1 Ecn^3 - 4395937500000\eta^4\alpha_2 n\alpha_1 - 5530000001400Ec\alpha_1^2 n\eta^2 + 937500000000\alpha_2\eta^3 \\ & - 1768020833250\eta^5\alpha_1^2\alpha_2 n - 4880000001000\eta^3\alpha_1^3 Ecn + 1854895833000n^2\eta^5\alpha_1^2\alpha_2 + 4013437500000n^2\eta^4\alpha_2\alpha_1 \\ & - 4350000000000n^2\eta^3\alpha_1^3 Ec - 4899999999300n^2\eta^2 Ec\alpha_1^2 - 8399999997900\eta\alpha_1 Ecn - 16799999998320\eta\alpha_1 Ecn^2 \\ & + 258020833350\eta^5\alpha_1^2\alpha_2 + 2730000000000\eta^3\alpha_1^3 Ec + 3362500002000n^2\eta^3\alpha_2 - 14280000000000\eta^2 n^3 \\ & - 509305555500n^3\eta^5\alpha_1^3 - 4583750004000n^3\eta^4\alpha_1^2 - 11790000000000\eta^3\alpha_1 n^3)), \end{aligned} \tag{4.145}$$

We avoid listing the other components. However, according to Equations (4.129a) and (4.129b), we obtain $F(\eta)$ up to $O(\eta^{11})$ and $\theta(\eta)$ up to $O(\eta^{10})$ as follows:

4.5.5 PADÉ APPROXIMATION

It is well known that Padé approximations have the advantage of manipulating the polynomial approximation into a rational function of polynomials. This manipulation provides us with more information about the mathematical behavior of the solution. Besides that, power series are not useful for large values of η, say $\eta=\infty$. Boyd (1997) and others have formally shown that power series in isolation are not useful for handling boundary value problems. Therefore, the combination of the HPM with the Padé approximation provides an effective tool for handling boundary value problems on infinite or semi-infinite domains. It is a known fact that Padé approximation converges on the entire real axis if $F(\eta)$ and $\theta(\eta)$ are free of singularities on the real axis. Furthermore, it is noted that Padé approximants can be easily evaluated by using built-in functions in manipulation languages such as Maple and Mathematica. More importantly, the diagonal approximants are most accurate approximants; therefore we have to construct only diagonal approximants. Using the boundary conditions $F'(\infty)=1, \theta(\infty)=0$ the diagonal approximant [N/N] vanishes if the coefficient of η with the highest power in the numerator vanishes. By putting the coefficients of the highest power of η equal to zero, we can easily find the values of $F''(0)=\alpha_1$; $\theta'(0)=\alpha_2$ listed in Tables 4.8 and 4.9. The numerical results have been worked by Choi et al. (2001) and Bonnecaze and Brady (1990).

4.5.6 RESULTS AND DISCUSSION

In this section, the HPM-Padé is used to find approximate solutions of the transformed system of coupled nonlinear ordinary differential Equations (4.122) and (4.123) with the boundary conditions (4.124) for various values of the parameters that describe the flow characteristics and the results are illustrated graphically. In this work, the Maple Package is used to solve the differential equations. The physical quantities of interest here are the skin-friction coefficient C_f and the Nusselt number Nu_x, which are obtained and given in Equation (4.127). The distributions of the velocity $F'(\eta)$, the temperature $\theta(\eta)$, the skin friction at the surface and the Nusselt number for different types of nanofluids are shown in Figures 4.55-4.66.

We consider four different types of nanoparticles, namely, copper (Cu), silver (Ag), alumina (Al_2O_3) and titanium oxide (TiO_2) with water as the base fluid. In order to test the accuracy of our results, we compare our results with those in Ref. Cortell, 2007b, when neglecting the effects of φ. We notice that the comparison shows an excellent agreement as presented in Tables 4.8 and 4.9.

Table 4.10 depicts the skin friction at the surface $-F''(0)$ for various values of nonlinear stretching sheet n with $\varphi=0.1$ for different types of nanoparticles when the base fluid is water.

It can be seen from Table 4.10 that $|F''(0)|$ increases with an increase in the nonlinear stretching parameter n and the Ag nanoparticles are the highest skin friction, followed by Cu, TiO_2, and Al_2O_3. Figures 4.57 and 4.58 illustrate the effect of nanoparticles volume fraction φ on the nanofluid velocity and temperature profile, respectively, in the case of Cu nanoparticles and water base fluid ($Pr=6.2$) when $\varphi=0$, 0.05, 0.1, and 0.2 with $Ec=0.1$ and $n=10$. It is clear that as the nanoparticles volume fraction increases, the nanofluid velocity decreases and the temperature increases.

Table 4.8 Comparison of Results for $-F''(0)$ When $\phi = 0$

n	HPM-Padé [2/2]	HPM-Padé [3/3]	HPM-Padé [4/4]	HPM-Padé [5/5]	HPM-Padé [6/6]	HPM-Padé [7/7]	HPM-Padé [8/8]	HPM-Padé [9/9]	HPM-Padé [15/15]	Numerical
0.0	0.66667	0.65355	0.65422	0.61907	0.62914	0.62800	0.62849	0.62822	0.62835	0.62755
0.2	0.83389	0.77762	0.78307	0.76462	0.76799	0.76753	0.76759	0.76743	0.76758	0.76676
0.5	0.93121	0.88861	0.88952	0.88867	0.89033	0.89311	0.88677	0.85087	0.90946	0.88948
1.0	1.01859	0.99357	1.00076	0.99630	0.98314	0.92832	0.89486	1.26223	1.07347	1.0
3.0	1.15434	1.14627	1.15469	1.14571	1.10974	1.26967	1.08084	1.03964	2.12254	1.14859
10.0	1.23769	1.23674	1.23678	1.23253	1.18263	1.12986	0.42963	1.42892	1.00464	1.23488
20.0	1.25975	1.25905	1.25903	1.25406	1.20443	1.37618	1.12280	2.20153	1.15262	1.25742

Table 4.9 Comparison of Results for $-\theta'(0)$ When $Pr = 5$, $\phi = 0$

	$Ec = 0$					$Ec = 0.1$				
n	HPM-Padé [3/3]	HPM-Padé [4/4]	HPM-Padé [5/5]	HPM-Padé [6/6]	Numerical	HPM-Padé [3/3]	HPM-Padé [4/4]	HPM-Padé [5/5]	HPM-Padé [6/6]	Numerical
0.75	3.00646	3.15820	3.13196	3.19780	3.12498	2.90760	3.04809	3.02352	3.09207	3.01698
1.5	3.52376	3.59505	3.56553	3.56735	3.56774	3.41576	3.48169	3.45414	3.45636	3.45572
7	4.16650	4.19091	4.18115	4.16162	4.18537	4.04745	4.07019	4.06120	4.10989	4.06572
10	4.23826	4.26040	4.25198	4.28594	4.25597	4.11818	4.13881	4.13109	4.17160	4.13530

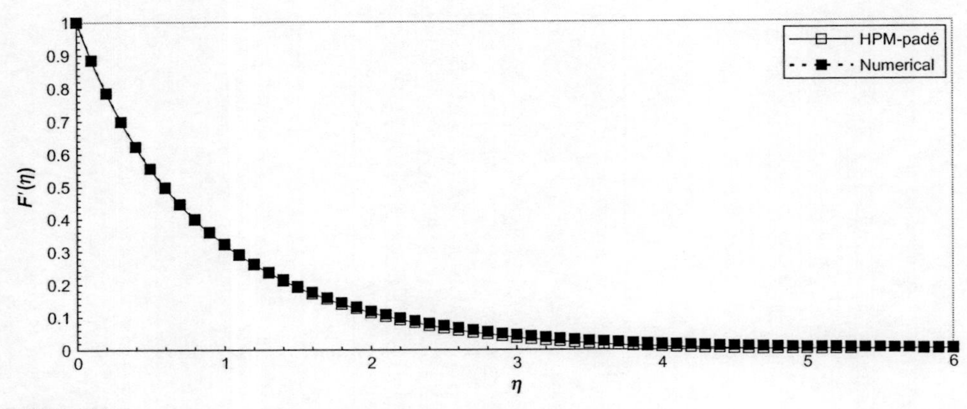

FIGURE 4.55

Comparison of HPM-Padé with numerical solution for $n=10$, $\phi=0$.

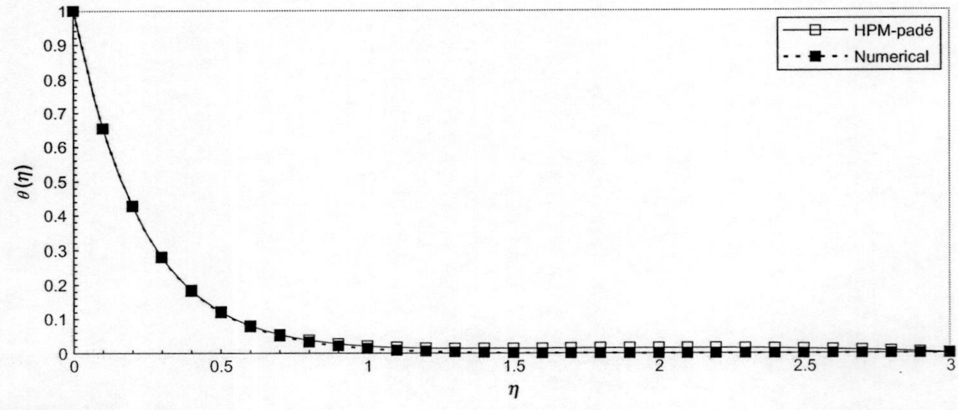

FIGURE 4.56

Comparison of HPM-Padé with numerical solution for $Pr=5$, $Ec=0$, $n=10$, $\phi=0$.

These figures illustrate this agreement with the physical behavior. When the volume of nanoparticles increases, the thermal conductivity increases, and then the thermal boundary layer thickness increases. Figures 4.59 and 4.60 depict the effect of nonlinearly stretching sheet parameter n on velocity distribution $F'(\eta)$ and temperature profile $\theta(\eta)$, respectively.

Figure 4.59 illustrates that an increase of nonlinear stretching sheet parameter n tends to decrease the nanofluid velocity in the case of Cu-water when $n=0.75$, 1.5, 3, 7, and 10 with $\varphi=0.1$. Furthermore, Figure 4.60 shows that increasing the nonlinear stretching sheet parameter n tends to decrease the temperature distribution the same values with $Ec=0.1$, thus leading to higher heat transfer rate between

Table 4.10 Values Related to the Skin Friction for Different Values of n with $\varphi = 0.1$

	$-F''(0)$			
N	Cu	Ag	Al_2O_3	TiO_2
0	0.73674	0.76809	0.62723	0.63391
1	1.17561	1.22604	0.99952	1.01026
2	1.29132	1.34226	1.09531	1.12433
3	1.34693	1.40472	1.14486	1.15720
4	1.37901	1.43811	1.17237	1.18498
5	1.40019	1.46016	1.19046	1.20327
10	1.44750	1.50946	1.23079	1.24403
20	1.47399	1.53708	1.25335	1.26682
50	1.49602	1.56020	1.27172	1.28541
100	1.50177	1.56619	1.27662	1.29036

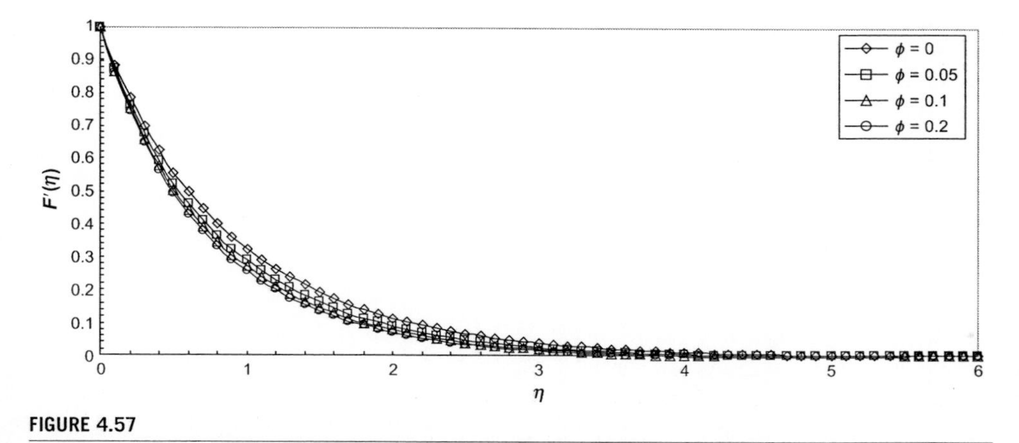

FIGURE 4.57

Effects of nanoparticle volume fraction ϕ on velocity distribution $F'(\eta)$ in the case of Cu-water for $n = 10$.

the nanofluid and the surface. The effect of the viscous dissipation parameter Ec on the temperature profile in the case of Cu-water when the Eckert number $Ec = 0$, 0.5, 1, 1.5, 2, and 2.5 with $n = 10$, and $\varphi = 0.1$ is shown in Figure 4.61.

It is clear that the temperature distribution increases with an increase in the viscous dissipation parameter Ec. The influence of Ec and n on the temperature profiles for all types of nanoparticles is shown in Figures 4.62 and 4.63, respectively.

It is found that the temperature decreases with n and increases with Ec as shown in Figures 4.60 and 4.61, respectively, and the TiO_2 nanoparticles proved to have the highest cooling performance for this problem.

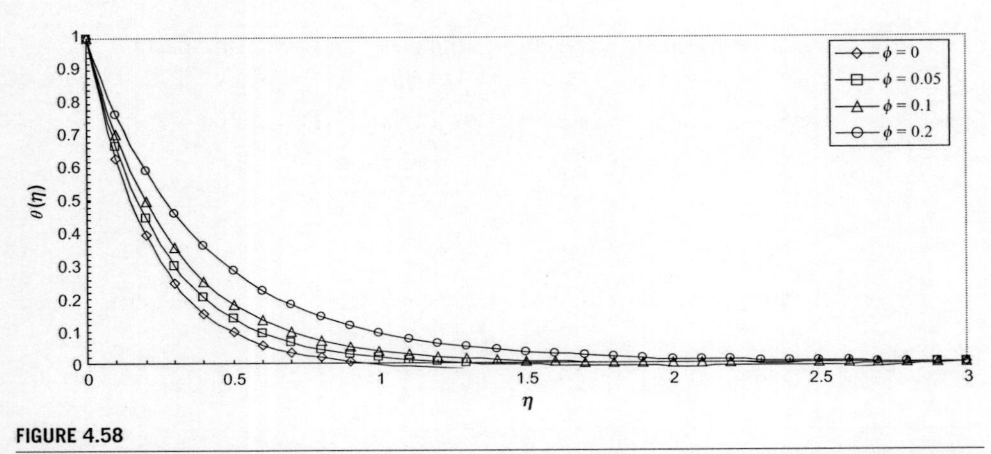

FIGURE 4.58

Effects of nanoparticle volume fraction ϕ on temperature profile $\theta(\eta)$ in the case of Cu-water for $Pr=6.2$, $Ec=0.1$, $n=10$.

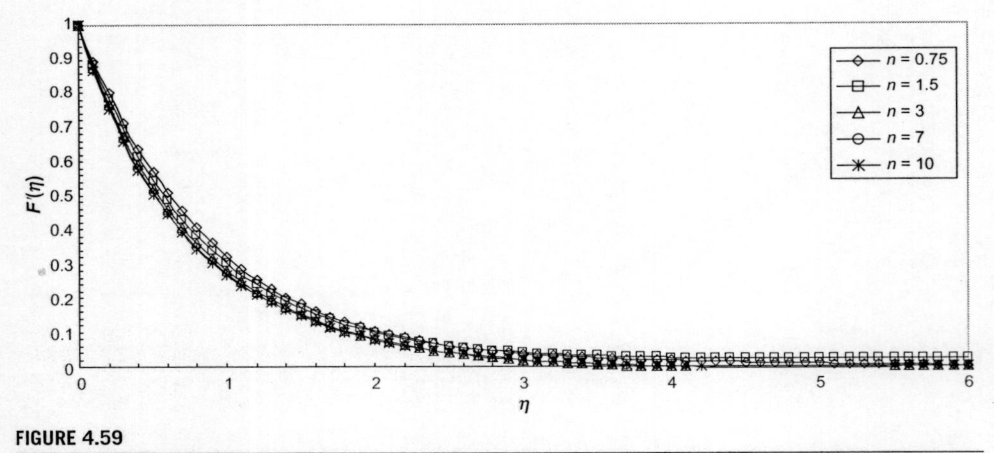

FIGURE 4.59

Effects of nonlinearly stretching sheet parameter n on velocity distribution $F'(\eta)$ in the case of Cu-water for $\phi=0.1$.

The influence of nonlinear stretching sheet n on the skin friction at the surface $-F''(0)$ with $Pr=6.2$, $\varphi=0.1$, and $Ec=0.5$ is shown in Figure 4.64.

It can be noticed that from Table 4.10 and Figure 4.64, the values of $|F''(0)|$ for different kinds of nanofluids increase with an increase in the nonlinear stretching parameter n. This implies an increment of the skin friction at the surface where Cu nanoparticles have the highest skin friction than the other nanoparticles. Figures 4.65 and 4.66 display the behavior of the heat transfer rates under the effects of Ec, and n, respectively, using different nanofluids for $Pr=6.2$ and $\varphi=0.1$.

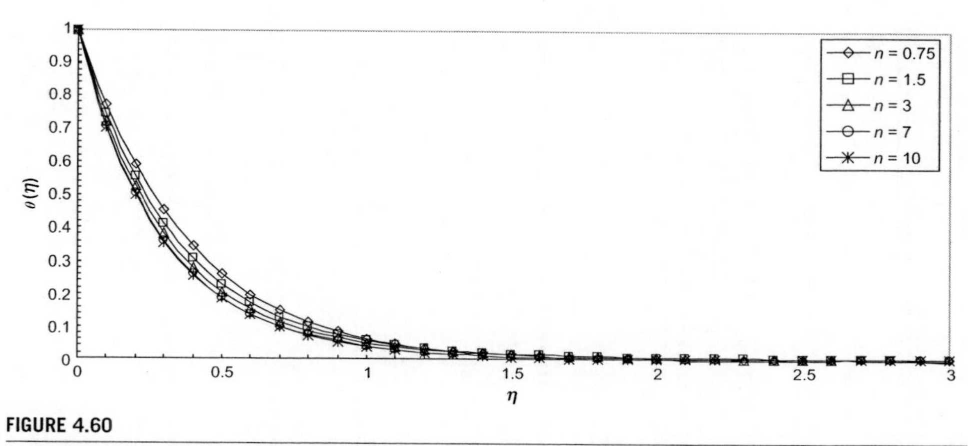

FIGURE 4.60

Effects of nonlinearly stretching sheet parameter n on temperature profile $\theta(\eta)$ in the case of Cu-water for $Pr=6.2$, $Ec=0.1$, $\phi=0.1$.

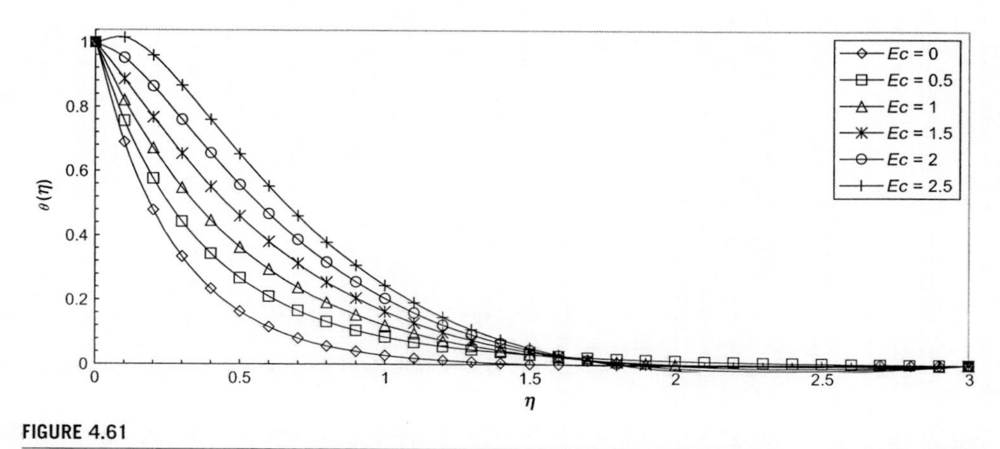

FIGURE 4.61

Effects of viscous dissipation parameter Ec on temperature profile $\theta(\eta)$ in the case of Cu-water for $Pr=6.2$, $n=10$, $\phi=0.1$.

These figures show that when using different kinds of nanofluids, the heat transfer rates change, which means that the nanofluids will be important in the cooling and heating processes. It can be noticed from the results above that, as expected, the heat transfer rate increases with an increase in the nonlinear stretching sheet parameter n, and decreases rapidly with an increase in the viscous dissipation parameter Ec.

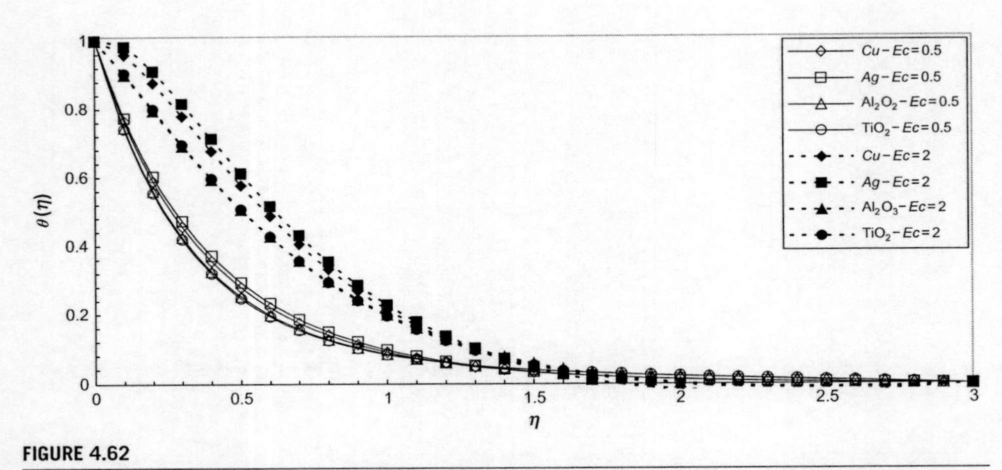

FIGURE 4.62

Effects of viscous dissipation parameter Ec on temperature profile $\theta(\eta)$ for different types of nanoparticles for $Pr=6.2$, $n=5$, $\phi=0.1$.

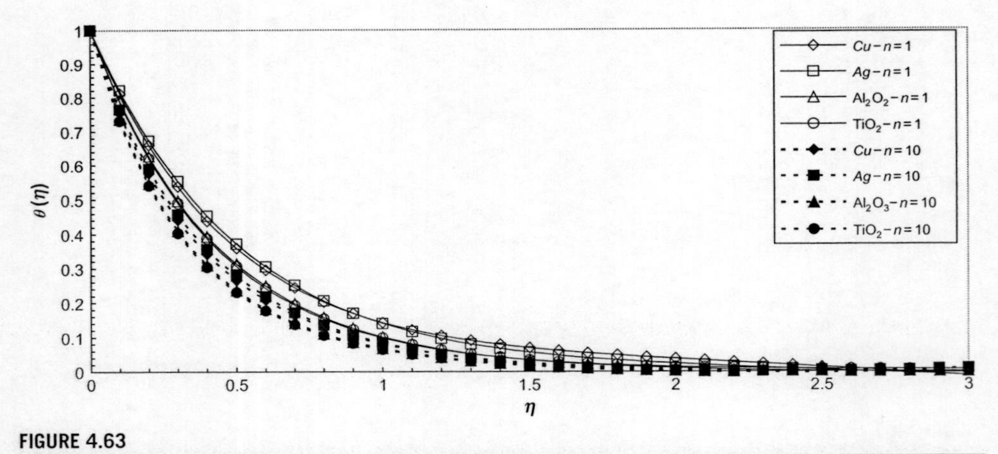

FIGURE 4.63

Effects of nonlinearly stretching sheet parameter n on temperature profile $\theta(\eta)$ for different types of nanoparticles for $Pr=6.2$, $Ec=0.5$, $\phi=0.1$.

4.5.7 CONCLUSIONS

The problem of boundary-layer flow and heat transfer in a viscous nanofluid over a nonlinearly stretched nonisothermal moving flat surface was analyzed. The governing partial differential equations were converted to ODEs by using a suitable similar transformation and then were solved

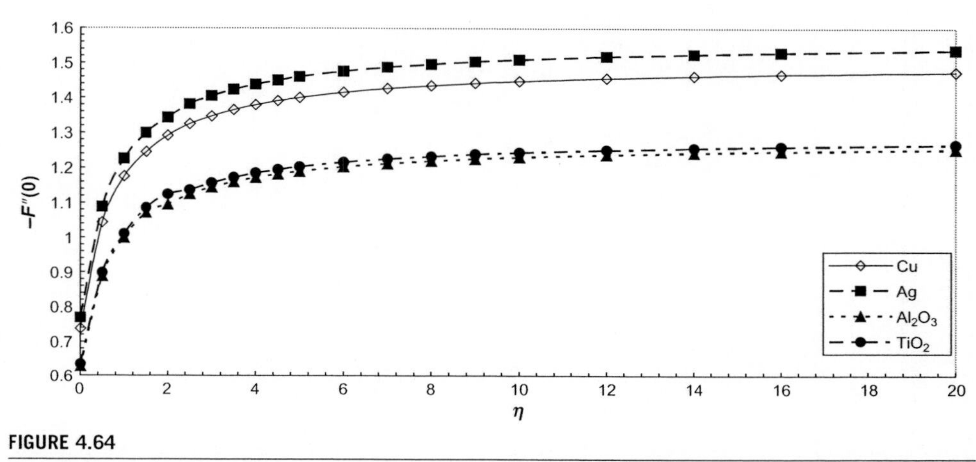

FIGURE 4.64

Effects of nonlinearly stretching sheet parameter n on skin-friction coefficient for different types of nanoparticles for $Pr=6.2$, $Ec=0.5$, $\phi=0.1$.

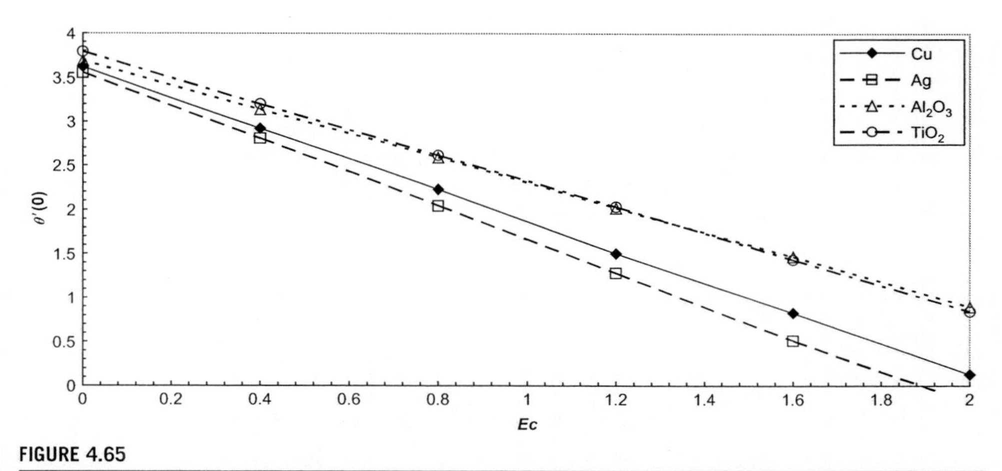

FIGURE 4.65

Effects of viscous dissipation parameter Ec on heat transfer rate for different types of nanoparticles for $Pr=6.2$, $n=5$, $\phi=0.1$.

analytically via HPM combined with Padé approximants. The effects of the solid volume fraction φ, nonlinear stretching sheet parameter n and the viscous dissipation parameter Ec on the flow and heat transfer characteristics are determined for four kinds of nanofluids: copper, silver, alumina, and titanium oxide.

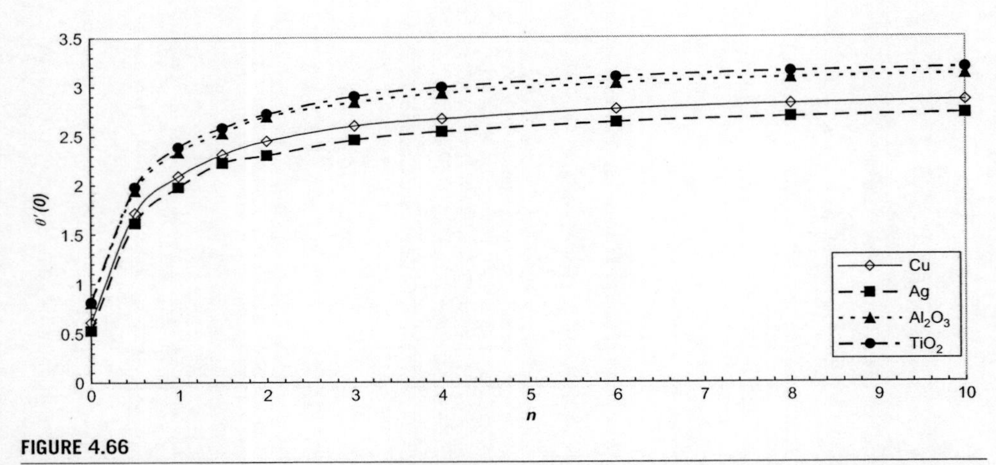

FIGURE 4.66

Effects of nonlinearly stretching sheet parameter n on heat transfer rate for different types of nanoparticles for $Pr=6.2$, $Ec=0.5$, $\phi=0.1$.

1. The increase of the solid volume fraction φ and the nonlinear stretching sheet parameter n leads to the decrease of dimensionless surface velocity; this yields to an increase in the skin friction at the surface.
2. An increment in the solid volume fraction φ and the Eckert number Ec yields to an increment in the nanofluid's temperature; this leads to a rapid reduction in the heat transfer rates.
3. An increase in the nonlinear stretching sheet parameter n yields a decrease in the nanofluid's temperature which leads to an increase in the heat transfer rates.
4. The TiO_2 nanoparticles proved to have the highest cooling performance for this problem than the other three types of nanoparticles (Cu, Ag, and Al_2O_3 nanoparticles).

REFERENCES

Abu-Nada, E., Chamkha, A.J., 2010. Effect of nanofluid variable properties on natural convection in enclosures filled with a CuO-EG-water nanofluid. Int. J. Therm. Sci. 49, 2339–2352.

Abu-Nada, E., Masoud, Z., Hijazi, A., 2008. Natural convection heat transfer enhancement in horizontal concentric annuli using nanofluids. Int. Commun. Heat Mass Transfer 35 (5), 657–665.

Adomian, G., 1988. A review of the decomposition method in applied mathematics. J. Math. Anal. Appl. 135 (2), 501–544.

Adomian, G., 1994. Solving Frontier Problems of Physics: The Decomposition Method. Kluwer, Boston, MA.

Afzal, N., 1993. Heat transfer from a stretching surface. Int. J. Heat Mass Transfer 36, 1128–1131.

Aminossadati, S.M., Ghasemi, B., 2009. Natural convection cooling of a localized heat source at the bottom of a nanofluid-filled enclosure. Eur. J. Mech. B. Fluid 28, 630–640.

Andersson, H.I., Aarseth, J.B., Dandapat, B.S., 2000. Heat transfer in a liquid film on an unsteady stretching surface. Int. J. Heat Mass Transfer 43, 69–74.

Bachok, N., Ishak, A., Pop, I., 2010. Boundary-layer flow of nanofluids over a moving surface in a flowing fluid. Int. J. Therm. Sci. 49 (9), 1663–1668.

Baker, G.A., 1975. Essentials of Padé-Approximants. Academic Press, London.

Bataller, R.C., 2008. Similarity solutions for flow and heat transfer of a quiescent fluid over a non-linearly stretching surface. J. Mater. Process. Technol. 203, 176–183.

Bejan, A., 1982. Entropy Generation Through Heat and Fluid Flow. Wiley, New York.

Bejan, A., 1996. Entropy Generation Minimization. CRC, Boca Raton, Florida.

Blasius, H., 1908a. Grenzschichten in flüssigkeiten mit kleiner reibung. Z. Angew. Math. Phys. 56, 1–37.

Blasius, H., 1908b. Grenzschichten in Flüssigkeiten mit kleiner Reibung. Z. Angew. Math. Phys. 56, 37.

Bonnecaze, R.T., Brady, J.F., 1990. A method for determining the effective conductivity of dispersions of particles. Proc. R. Soc. London, Ser. A 430, 285–313.

Boyd, J., 1997. Padé-approximant algorithm for solving nonlinear ordinary differential equation boundary value problems on an unbounded domain. Comput. Phys. 11, 299–303.

Buongiorno, J., 2006. Convective transport in nanofluids. J. Heat Transfer 128, 240–250.

Choi, S.U.S., Zhang, Z.G., Yu, W., Lockwood, F.E., Grulke, E.A., 2001. Anomalous thermal conductivity enhancement in nanotube suspensions. Appl. Phys. Lett. 79, 2252–2254.

Cortell, R., 2007a. Flow and heat transfer in a moving fluid over a moving flat surface. Theor. Comput. Fluid Dyn. 21, 435–446.

Cortell, R., 2007b. Viscous flow and heat transfer over a nonlinearly stretching sheet. Appl. Math. Comput. 184, 864–873.

Dandapata, B.S., Maity, S., Kitamura, A., 2008. Liquid film flow due to an unsteady stretching sheet. Int. J. Nonlin. Mech. 43, 880–886.

Das, S.K., Putra, N., Thiesen, P., Roetzel, W., 2003. Temperature dependence of thermal conductivity enhancement for nanofluids. J. Heat Transfer 125, 567–574.

Domairry, G., Aziz, A., 2009. Approximate analysis of MHD squeeze flow between two parallel disks with suction or injection by homotopy perturbation method. Math. Probl. Eng.. 2009, 603916.

Gerald, C.F., Weatly, P.O., 1989. Applied Numerical Analysis. Addison Wesley Publishing Company, New York.

Gupta, P.S., Gupta, A.S., 1977. Heat and mass transfer on a stretching sheet with suction or blowing. Can. J. Chem. Eng. 55, 744–746.

Hamilton, R.L., Crosser, O.K., 1962. Thermal conductivity of heterogeneous two component systems. Ind. Eng. Chem. Fundam. 1, 182–191.

He, J.H., 2003. Homotopy perturbation method: a new nonlinear analytical technique. Appl. Math. Comput. 135, 73–79.

Ishak, A., Nazar, R., Pop, I., 2009. Flow and heat transfer characteristics on a moving flat plate in a parallel stream with constant surface heat flux. Heat Mass Transfer 45, 563–567.

Kandasamya, R., Loganathan, P., Arasu, P.P., 2011. Scaling group transformation for MHD boundary-layer flow of a nanofluid past a vertical stretching surface in the presence of suction/injection. Nucl. Eng. Des. 241, 2053–2059.

Khan, W.A., Pob, I., 2010. Boundary-layer flow of a nano-fluid past a stretching sheet. Int. J. Heat Mass Transfer 53, 2477–2483.

Kolsi, L., Abidi, A., Naceur, B., Aïssia, H.B., 2010. The effect of an external magnetic field on the entropy generation in three-dimensional natural convection. Therm. Sci. 14 (2), 341–352.

Kumar, D.H., Patel, H.E., Kumar, V.R.R., Sundararajan, T., Pradeep, T., Das, S.K., 2004. Model for conduction in nanofluids. Phys. Rev. Lett. 93, 144301.1–144301.3.

Liao, S.J., 2003. Beyond perturbation: introduction to the homotopy analysis method. Chapman & Hall/CRC Press, Boca Raton.

liao, S.J., 2004. on the homotopy analysis method for nonlinear problems. Appl. Math. Comput. 47 (2), 499–513.

Liu, I.C., Andersson, H.I., 2008. Heat transfer in a liquid film on an unsteady stretching sheet. Int. J. Therm. Sci. 47, 766–772.

Magyari, E., Keller, B., 2000. Exact solutions for self-similar boundary-layer flows induced by permeable stretching walls. Eur. J. Mech. B. Fluids 19, 109–122.

Mahmood, M., Asghar, S., Hossain, M.A., 2007. Squeezed flow and heat transfer over a porous surface for viscous fluid. Heat Mass Transfer 44, 165–173.

Makinde, O.D., Aziz, A., 2011. Boundary layer flow of a nanofluid past a stretching sheet with a convective boundary condition. Int. J. Therm. Sci. 50, 1326–1332.

Maxwell, J., 1904. A Treatise on Electricity and Magnetism, second ed. Oxford University Press, Cambridge, UK.

Mustafa, M., Hayat, T., Obaidat, S., 2012. On heat and mass transfer in the unsteady squeezing flow between parallel plates. Meccanica 47 (7), 1581–1589.

Oztop, H.F., Abu-Nada, E., 2008. Numerical study of natural convection in partially heated rectangular enclosures filled with nanofluids. Int. J. Heat Fluid Flow 29, 1326–1336.

Patel, H.E., Sundarrajan, T., Pradeep, T., Dasgupta, A., Dasgupta, N., Das, S.K., 2005. Amicro-convection model for thermal conductivity of nanofluid. Pramana J. Phys. 65, 863–869.

Prasad, K.V., Vajravelu, K., 2009. Heat transfer in the MHD flow of a power law fluid over a non-isothermal stretching sheet. Int. J. Heat Mass Transfer 52, 4956–4965.

Prasher, R., Bhattacharya, P., Phelan, P.E., 2006. Brownian-motion-based convective-conductive model for the effective thermal conductivity of nanofluid. ASME J. Heat Transfer 128, 588–595.

Rahman, M.A., 2007. A novel method for estimating the entropy generation rate in a human body. Therm. Sci. 11 (1), 75–92.

Raptis, A., Perdikis, C., 2006. Viscous flow over a non-linearly stretching sheet in the presence of a chemical reaction and magnetic field. Int. J. Nonlin. Mech. 41, 527–529.

Rezaiguia, I., Mahfoud, K., Kamel, T., Belghar, N., Saouli, S., 2010. Numerical simulation of the entropy generation in a fluid in forced convection on a plane surface while using the method of Runge-Kutta. Eur. J. Sci. Res. 42 (4), 7.

Sakiadis, B.C., 1961. Boundary-layer behaviour on continuous solid surfaces. I. Boundary-layer equations for two-dimensional and axisymmetric flow. AIChE J. 7, 2.

Santra, A.K., Sen, S., Chakraborty, N., 2008. Study of heat transfer augmentation in a differentially heated square cavity using copper-water nanofluid. Int. J. Therm. Sci. 47, 1113–1122.

Schlichting, H., Gersten, K., 2000. Boundary Layer Theory. Springer, Berlin.

Stefan, M.J., 1874. Versuch Über die scheinbare adhesion. Akad. Wiss. Wien Math. Natur. 69, 713–721.

Vajravelu, K., 2001. Viscous flow over a nonlinearly stretching sheet. Appl. Math. Comput. 124, 281–288.

Wang, C.Y., 1990. Liquid film on an unsteady stretching surface. Q. Appl. Math. XLVIII, 601–610.

Wang, C., 2006. Analytic solutions for a liquid film on an unsteady stretching surface. Heat Mass Transfer 42, 759–766.

Wasp, F.J., 1977. Solid-Liquid Flow Slurry Pipeline Transportation. Trans. Tech. Publ, Berlin.

Weidman, P.D., Kubitschek, D.G., Davis, A.M.J., 2006. The effect of transpiration on self-similar boundary layer flow over moving surfaces. Int. J. Eng. Sci. 44, 730–737.

White, F.M., 2006. Viscous Fluid Flow, third ed. McGraw-Hill, New York.

Yu, W., Choi, S.U.S., 2003. The role of interfacial layer in the enhanced thermal conductivity of nanofluids: a renovated Maxwell model. J. Nanopart. Res. 5, 167–171.

Ziabakhsh, Z., Domairry, G., Bararnia, H., Babazadeh, H., 2010. Analytical solution of flow and diffusion of chemically reactive species over a nonlinearly stretching sheet immersed in a porous medium. J. Taiwan Inst. Chem. Eng. 41, 22–28.

CHAPTER

THERMAL PROPERTIES OF NANOPARTICLES

5

CHAPTER CONTENTS

5.1 Effects of Adding Nanoparticles to Water and Enhancement in Thermal Properties 181
5.1.1 Introduction 182
5.1.2 Governing Equations 182
5.1.3 Basic Idea of HAM 185
5.1.4 Application of HAM to Falkner-Skan Problem 186
5.1.4.1 Zeroth-Order Deformation Equations 187
5.1.4.2 mth-Order Deformation Equations 188
5.1.5 Convergence of HAM Solution 188
5.1.6 Results and Discussion 189
5.1.7 Conclusion 194
5.2 Temperature Variation Analysis for Nanoparticle's Combustion 194
5.2.1 Introduction 194
5.2.2 Problem Description 195
5.2.3 Applied Analytical Methods 196
5.2.3.1 Differential Transformation Method 197
5.2.3.2 Boubaker Polynomials Expansion Scheme 199
5.2.4 Results and Discussion 200
5.2.5 Conclusion 201
References 202

5.1 EFFECTS OF ADDING NANOPARTICLES TO WATER AND ENHANCEMENT IN THERMAL PROPERTIES

In the age of technology, it is vital to cool the different parts of a device to use it more beneficially. Using nanofluids is one of the most common methods which have shown very effective results. In this chapter, we have rephrased a classic equation in fluid mechanics, i.e., the Falkner-Skan boundary-layer equation, in order to be used for nanofluid. This nonlinear equation has been solved by Homotopy analysis method (HAM) which was presented by Liao. This method is very capable to solve wide range of nonlinear equations. The physical interpretation of results which are velocity and temperature profiles are explained in detail and they are parallel with experimental outcomes of previous researchers (*this section has been worked by D.D. Ganji, Y. Rostamiyan, F. Hedayati, S.M. Hamidi in nonlinear dynamics team in Mechanical Engineering Department, 2012-2013*).

Application of Nonlinear Systems in Nanomechanics and Nanofluids. http://dx.doi.org/10.1016/B978-0-323-35237-6.00005-4

5.1.1 INTRODUCTION

As technology improved, it was realized that devices have to be cooled in a more effective way and the conventional fluids such as water are not appropriate anymore; so, the idea of adding particles to a fluid was presented. These tiny particles have high thermal conductivity, so the mixed fluids have better thermal properties (Goldstein et al., 2010; Kakaç and Pramuanjaroenkij, 2009). The materials of these nanoscale particles are aluminum oxide (Al2O3), copper (Cu), copper oxide (CuO), gold (Au), silver (Ag), etc., which are suspending in base fluids such as water, oil, acetone, ethylene glycol, etc. Al_2O_3 and CuO are the most well-known nanoparticles used by many researchers in their experimental works (i.e., see Das et al., 2003a,b,c; Xuan and Li, 2000). They claimed different results due to the size and shape and also the contact surface of the particles. In addition, the base fluid characteristics were important. The main obstacle in this field was how to keep the particles suspended in static fluid which is discussed in Wang et al. (1999) and Das et al. (2003a,b,c). Fortunately, the results were in a same trace that the thermal conductivity of the nanofluids is higher than the conventional fluids and this term is modeled mathematically in Xue (2003) and Wen et al. (2009). In this work, we will rephrase the Falkner-Skan equation (see Falkner and Skan, 1931) for a nanofluid with semianalytical method, i.e., HAM which is presented by Liao in 1992. The HAM is one of the well-known methods to solve nonlinear equations. Falkner and Skan considered two-dimensional wedge flows. They developed a similarity solution method in which the partial differential boundary-layer equation was reduced to a nonlinear third-order ordinary differential equation which does not have exact solution besides the numerical solution for this equation is time-consuming and difficult.

5.1.2 GOVERNING EQUATIONS

In this chapter, we consider an incompressible viscous fluid which flows over a wedge, as shown in Figure 5.1. The wall temperature, i.e. T_w, is uniform and constant and is greater than the free stream temperature, T_1. It is assumed that the free stream velocity, U_1, is also uniform and constant. Further, assuming that the flow in the laminar boundary layer is two-dimensional, and that the temperature gradients resulting from viscous dissipation are small, the continuity, momentum, and energy equations can be expressed as:

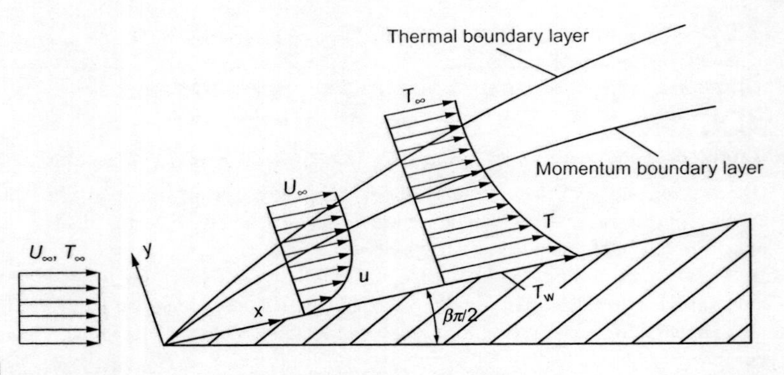

FIGURE 5.1

Velocity and thermal boundary layers for the Falkner-Skan wedge flow.

$$\frac{\partial u}{\partial x}+\frac{\partial u}{\partial y}=0 \tag{5.1}$$

$$u\frac{\partial u}{\partial x}+v\frac{\partial u}{\partial y}=U\frac{\mathrm{d}U}{\mathrm{d}x}+v\frac{\partial^2 u}{\partial^2 x} \tag{5.2}$$

$$u\frac{\partial T}{\partial x}+v\frac{\partial T}{\partial y}=\alpha\frac{\partial^2 T}{\partial^2 x} \tag{5.3}$$

where u and v are the respective velocity components in the x and y direction of the fluid flow, v is the kinematic viscosity of the fluid, and U is the reference velocity at the edge of the boundary layer and is a function of x. α is the thermal diffusivity of the fluid, T is the temperature in the vicinity of the wedge, and the boundary conditions are given by

$$\text{At } y=0: u=v=0 \text{ and } T=T_{\mathrm{w}} \tag{5.4}$$

$$\text{When } y\rightarrow\infty: u\rightarrow U(x)=U_{\infty}\left(\frac{x}{L}\right)^m \text{ and } T=T_{\infty} \tag{5.5}$$

$$\text{At } x=0: u=u_{\infty} \text{ and } T=T_{\infty} \tag{5.6}$$

where U_{∞} is the mean stream velocity, L is the length of the wedge, and m is the Falkner-Skan power-law parameter, and x is measured from the tip of the wedge. A stream function, $\Psi(x, y)$ is introduced such that:

$$u=\frac{\partial\Psi}{\partial y} \text{ and } v=-\frac{\partial\Psi}{\partial x} \tag{5.7}$$

By substituting Equation (5.7) into Equation (5.2), the momentum equation will be obtained as follows:

$$\frac{\partial\Psi}{\partial y}\frac{\partial^2\Psi}{\partial x\partial y}-\frac{\partial\Psi}{\partial x}\frac{\partial^2\Psi}{\partial y^2}=U\frac{\mathrm{d}U}{\mathrm{d}x}v\frac{\partial^3\Psi}{\partial y^3} \tag{5.8}$$

Integrating equation and using these similarity variable yields:

$$f(\eta)=\sqrt{\left(\frac{m+1}{2}\right)\left(\frac{L^m}{v_{\mathrm{f}}u_{\infty}}\right)}\frac{\Psi}{x^{(m+1)/2}} \tag{5.9}$$

$$\eta=\sqrt{\frac{m+1}{2}\frac{vu_{\infty}}{L^m}}\frac{y}{x^{(m+1)/2}} \tag{5.10}$$

In which v_{f} is the kinematic viscosity of the fluid. Substituting Equations (5.9) and (5.10) into Equation (5.8) gives:

$$\frac{\partial^3 f(\eta)}{\partial\eta^3}+f(\eta)\frac{\partial^2 f(\eta)}{\partial\eta^2}+\left(\frac{2m}{m+1}\right)\left(1-\left(\frac{\mathrm{d}f(\eta)}{\eta}\right)^2\right)=0 \tag{5.11}$$

which is known as the Falkner-Skan boundary-layer equation. The boundary conditions of $f(\eta)$ are

$$\text{At } \eta=0: f(0)=\frac{\mathrm{d}f(\eta)}{\eta}=0 \tag{5.12}$$

$$\text{When } \eta \to \infty : \frac{df(\infty)}{\eta} = 1 \tag{5.13}$$

Note that in the equations above, parameters β and m are related through the expression $\beta = 2m/(m+1)$. A dimensionless temperature is defined as follows:

$$\theta = \frac{T - T_w}{T_\infty - T_w} \tag{5.14}$$

If Equation (5.14) is substituted into Equation (5.3), the boundary-layer energy equation then becomes:

$$\frac{\partial^2 \theta(\eta)}{\partial \eta^2} + Prf(\eta)\frac{df(\eta)}{\eta} = 0 \tag{5.15}$$

With the following boundary conditions:

$$\text{At } \eta = 0 : \theta(0) = 1 \tag{5.16}$$

$$\text{When } \theta \to \infty : \theta(\infty) = 0 \tag{5.17}$$

The following equations can be obtained by this parameter changing to rephrase this classic equation and use it for nanofluids:

$$\mu_{nf} = \frac{\mu_f}{(1-\varphi)^{2.5}} \tag{5.18}$$

$$\rho_{nf} = (1-\varphi)\rho_f + \rho_p \tag{5.19}$$

$$\psi = \sqrt{\frac{2v_f x u_s(x)}{m+1}} f(\eta) \tag{5.20}$$

$$\eta = \sqrt{\frac{(m+1)u_s(x)}{2v_f x}} y \tag{5.21}$$

$$(\rho c_p)_{nf} = (1-\varphi)(\rho c_p)_f + (\varphi \rho c_p)_p \tag{5.22}$$

$$\frac{k_{nf}}{k_f} = \frac{(k_p + 2k_f) - 2\varphi(k_f - k_p)}{(k_p + 2k_f) + \varphi(k_f - k_p)} \tag{5.23}$$

Finally:

$$\frac{1}{(1-\varphi)^{2.5}(1-\varphi+\varphi(\rho_s/\rho_f))} f''' + ff'' + \left(\frac{2m}{m+1}\right)(1 - f'^2) = 0 \tag{5.24}$$

With these boundary conditions:

$$\text{At } \eta = 0 : f(0) = \frac{df(\eta)}{\eta} = 0 \tag{5.25}$$

$$\text{When } \eta \to \infty := \frac{df(\infty)}{\eta} = 1 \tag{5.26}$$

And also:

$$\frac{1}{Pr} \frac{k_{nf}/k_f}{\left(1-\varphi+\varphi\left((\rho c_p)_s/(\rho c_p)_f\right)\right)} \theta'' + f\theta' = 0 \tag{5.27}$$

With these boundary conditions:

$$\text{At } \eta = 0 : \theta(0) = 1 \tag{5.28}$$

$$\text{When } \theta \to \infty : \theta(\infty) = 0 \tag{5.29}$$

5.1.3 BASIC IDEA OF HAM

Let us assume the following nonlinear differential equation in form of:

$$N[u(\tau)] = 0 \tag{5.30}$$

where N is a nonlinear operator, τ is an independent variable and $u(\tau)$ is the solution of equation. We define the function, $\varphi(\tau, p)$ as follows:

$$\lim_{p \to 0} \varphi(\tau, p) = u_0(\tau) \tag{5.31}$$

where, $p \in [0, 1]$ and $u_0(\tau)$ is the initial guess which satisfies the initial or boundary conditions and if

$$\lim_{p \to 1} \varphi(\tau, p) = u(\tau) \tag{5.32}$$

And by using the generalized homotopy method, Liao's so-called zero-order deformation Equation (5.30) will be:

$$(1-P)L[\varphi(\tau, p) - u_0(\tau)] = p\hbar H(\tau) N[\varphi(\tau, p)] \tag{5.33}$$

where $\hbar$ is the auxiliary parameter which helps us increase the results convergence, $H(\tau)$ is the auxiliary function and L is the linear operator. It should be noted that there is a great freedom to choose the auxiliary parameter $\hbar$, the auxiliary function $H(\tau)$, the initial guess $u_0(\tau)$ and the auxiliary linear operator L.

Thus, when p increases from 0 to 1 the solution $\varphi(\tau, p)$ changes between the initial guess $u_0(\tau)$ and the solution $u(\tau)$. The Taylor series expansion of $\varphi(\tau, p)$ with respect to p is

$$\varphi(\tau, p) = u_0(\tau) + \sum_{m=1}^{\infty} u_m(\tau) p^m \tag{5.34}$$

and

$$u_0^{[m]}(\tau) = \left. \frac{\partial^m (\varphi(\tau; p))}{\partial p^m} \right|_{p=0} \tag{5.35}$$

where $u_0^{[m]}(\tau)$ for brevity is called the mth order of deformation derivation which reads:

$$u_m(\tau)=\frac{u_0^{[m]}}{m!}=\frac{1}{m!}\frac{\partial^m(\varphi(\tau;p))}{\partial p^m}\bigg|_{p=0} \tag{5.36}$$

According to the definition in Equation (5.36), the governing equation and the corresponding initial conditions of $u_m(\tau)$ can be deduced from zero-order deformation Equation (5.30). Differentiating Equation (5.30) for m times with respect to the embedding parameter p and setting $p=0$ and finally dividing by $m!$, we will have the so-called mth order deformation equation in the form:

$$L[u_m(\tau)-\chi_m u_{m-1}(\tau)]=\hbar H(\tau)R(\vec{u}_{m-1}) \tag{5.37}$$

where:

$$R(\vec{u}_{m-1})=\frac{1}{(m-1)!}\frac{\partial^{m-1}N[\varphi(\tau;p)]}{\partial p^{m-1}} \tag{5.38}$$

and

$$\chi_m=\begin{cases}0 & m\leq 1\\ 1 & m>1\end{cases} \tag{5.39}$$

So, by applying inverse linear operator to both sides of the linear equation, Equation (5.30), we can easily solve the equation and compute the generation constant by applying the initial or boundary condition.

5.1.4 APPLICATION OF HAM TO FALKNER-SKAN PROBLEM

As mentioned by Liao in 1992, a solution may be expressed with different base functions, among which some converge to the exact solution of the problem faster than others. Such base functions are obviously better suited for the final solution to be expressed. Noting these facts, we have decided to express $f(\eta)$ and $\theta(\eta)$ by a set of base functions of the form:

$$f(\eta)=\sum_{m=0}^{\infty}\sum_{n=0}^{\infty}\sum_{k=0}^{\infty}a_{m,n}^{k}\eta^{k}\exp(-n\eta) \tag{5.40}$$

$$\theta(\eta)=\sum_{m=0}^{\infty}\sum_{n=0}^{\infty}\sum_{k=0}^{\infty}b_{m,n}^{k}\eta^{k}\exp(-n\eta) \tag{5.41}$$

The rule of solution expression provides us with a starting point. It is under the rule of solution expression that initial approximations, auxiliary linear operators, and the auxiliary functions are determined. So, according to the rule of solution expression, we choose the initial guess and auxiliary linear operator in the following form:

$$f_0(\eta)=\exp(-\eta)+\eta-1 \tag{5.42}$$

$$\theta_0(\eta)=\exp(-\eta) \tag{5.43}$$

$$L_1(f) = f''' + f'' \tag{5.44}$$

$$L_2(\theta) = \theta'' + \theta \tag{5.45}$$

$$L_1(C_1 + C_1\eta + C_3\exp(-\eta)) = 0 \tag{5.46}$$

$$L_2(c_4 + c_5\exp(-\eta)) = 0 \tag{5.47}$$

And $c_i (i = 1\_5)$ are constants. Let $p \in [0, 1]$ denotes the embedding parameter and $\hbar$ indicates nonzero auxiliary parameters. We construct the following equations:

5.1.4.1 Zeroth-Order Deformation Equations

$$(1-P)L_1[f(\eta;p) - f_0(\eta)] = p\,\hbar N_1[f(\eta;p), \theta(\eta;p)] \tag{5.48}$$

$$f(0;p) = 0;\ f(0;p) = 0;\ f(\infty;p) = 1 \tag{5.49}$$

$$(1-P)L_2[\theta(\eta;p) - \theta_0(\eta)] = p\,\hbar N_2[f(\eta;p), \theta(\eta;p)] \tag{5.50}$$

$$\theta(0;p) = 1;\ \theta(\infty;p) = 0 \tag{5.51}$$

$$N_1[F(\eta;p), G(\eta;p)] = \frac{1}{(1-\varphi)^{2.5}(1-\varphi+\varphi(\rho_s/\rho_f))}\frac{\partial^3 f(\eta;p)}{\partial\eta^3} + f(\eta;p)\frac{\partial^2 f(\eta;p)}{\partial\eta^2} + \left(\frac{2m}{m+1}\right)\left(1 - \left(\frac{\partial f(\eta;p)}{\partial\eta}\right)^2\right) \tag{5.52}$$

$$N_2[F(\eta;p), G(\eta;p)] = \frac{1}{Pr}\frac{(k_{nf}/k_f)}{\left(1-\varphi+\varphi\frac{(\rho c_p)_s}{(\rho c_p)_f}\right)}\frac{\partial^2\theta(\eta;p)}{\partial\eta^2} + f(\eta;p)\frac{\partial\theta(\eta;p)}{\partial\eta} \tag{5.53}$$

For $p = 0$ and $p = 1$ we have:

$$f(\eta;0) = f_0(\eta) f(\eta;1) = f(\eta) \tag{5.54}$$

$$\theta(\eta;0) = \theta_0(\eta)\theta(\eta;1) = \theta(\eta) \tag{5.55}$$

When p increases from 0 to 1 then $f(\eta;p)$ and $\theta(\eta;p)$ vary from $f_0(\eta)$ and $\theta_0(\eta)$ to $f(\eta)$ and $\theta(\eta)$. Using Taylor's theorem for Equations (5.54) and (5.55), $f(\eta;p)$ and $\theta(\eta;p)$ can be expanded in a power series of p as follows:

$$f(\eta;0) = f_0(\eta) + \sum_{m-1}^{\infty} f_m(\eta)p^m,\ f_m(\eta) = \frac{1}{m!}\frac{\partial^m(f(\eta;p))}{\partial p^m} \tag{5.56}$$

$$\theta(\eta;0) = \theta_0(\eta) + \sum_{m-1}^{\infty} \theta_m(\eta)p^m,\ \theta_m(\eta) = \frac{1}{m!}\frac{\partial^m(\theta(\eta;p))}{\partial p^m} \tag{5.57}$$

In which $\hbar$ is chosen in such a way that these two series are convergent at $p = 1$, therefore we have through Equations (5.56) and (5.57) that

$$f(\eta;0)=f_0(\eta)+\sum_{m-1}^{\infty}f_m(\eta),\quad \theta(\eta;0)=\theta_0(\eta)+\sum_{m-1}^{\infty}\theta_m(\eta) \tag{5.58}$$

5.1.4.2 mth-Order Deformation Equations

$$L_1[f_m(\eta)-\chi_m f_{m-1}(\eta)]=\hbar_1 H_{\mathrm{f}}(\eta)R_m^{\mathrm{f}}(\eta)] \tag{5.59}$$

$$f_m(0)=0;\ f_m(0)=0;\ f_m(\infty)=0 \tag{5.60}$$

$$L_2[\theta_m(\eta)-\chi_m\theta_{m-1}(\eta)]=\hbar_2 H_\theta(\eta)R_m^\theta(\eta)] \tag{5.61}$$

$$\theta_m(0)=0;\ \theta_m(\infty)=0 \tag{5.62}$$

$$R_m^f(\eta)=\frac{1}{(1-\varphi)^{2.5}(1-\varphi+\varphi(\rho_{\mathrm{s}}/\rho_{\mathrm{f}}))}f'''_{m-1}+\sum_{n=0}^{m-1}f_{m-1-n}f''_n \tag{5.63}$$

$$R_m^\theta(\eta)=\frac{1}{Pr}\frac{k_{\mathrm{nf}}/k_{\mathrm{f}}}{\left(1-\varphi+\varphi\frac{(\rho c_p)_{\mathrm{s}}}{(\rho c_p)_{\mathrm{f}}}\right)}\theta''_{m-1}+\sum_{n=0}^{m-1}f_{m-1-n}\theta'_n \tag{5.64}$$

$$\chi_m=\begin{cases}1, & m\le 1\\ 0, & m>1\end{cases} \tag{5.65}$$

The general solutions will be:

$$f_m(\eta)-\chi_m f_{m-1}(\eta)=f_m^*(\eta)+\left(C_1^m+C_2^m\eta+C_3^m\exp(-\eta)\right) \tag{5.66}$$

$$\theta_m(\eta)-\chi_m\theta_{m-1}(\eta)=\theta_m^*(\eta)+\left(C_4^m+C_5^m\exp(-\eta)\right) \tag{5.67}$$

Where C_1^m to C_5^m are constants that can be obtained by applying the boundary conditions in Equations (5.60) and (5.62).

As discussed by Liao, the rule of coefficient ergodicity and the rule of solution existence play important roles in determining the auxiliary function and ensuring that the high-order deformation equations are closed and have solutions. In many cases, by means of the rule of solution expression and the rule of coefficient ergodicity, auxiliary functions can be uniquely determined. So, we define the auxiliary functions $H_f(\eta)$ and $H_\theta(\eta)$ in the following form:

$$H_f(\eta)=1 \tag{5.68}$$

$$H_\theta(\eta)=\exp(-\eta) \tag{5.69}$$

5.1.5 CONVERGENCE OF HAM SOLUTION

HAM provides us with great freedom in choosing the solution of a nonlinear problem by different base functions. This has a great effect on the convergence region because the convergence region and the rate of a series are chiefly determined by the base functions used to express the solution. Therefore, more accurate approximation of a nonlinear problem can be obtained by choosing a proper set of base functions and ensuring about its convergency. On the other hand, as pointed out by Liao, the convergence and rate of approximation for the HAM solution strongly depends on the value of auxiliary parameters $\hbar$. Even if the initial approximations $f_0(\eta)$ and $\theta_0(\eta)$, the auxiliary linear operator $\mathcal{L}$, and the auxiliary functions $H_f(\eta)$ and $H_\theta(\eta)$are given, we still have great freedom to choose the value of the auxiliary parameters $\hbar_1$ and $\hbar_2$. So, the auxiliary parameters provide us with an additional way to conveniently adjust and control the convergence region and rate of solution series. By means of the so-called $\hbar$ curves, it is easy to find out the so-called valid regions of auxiliary parameters to gain a convergent solution series. When the valid region of auxiliary parameters is a horizontal line segment, then the solution is converged. In our case study, suitable range of $\hbar_1$ and $\hbar_2$ for CuO and Al_2O_3 can be obtained from Figures 5.2–5.5.

5.1.6 RESULTS AND DISCUSSION

Figure 5.6 shows the relation between $f'(\eta)$ andη which represents the velocity of the fluid. Adding nanoparticles to a base fluid will cause a raise in fluid viscosity; therefore, the momentum exchange between the nanofluid layers is higher. It can be clearly seen that the pure fluid reaches to 1 approximately in $\eta=5$, however, nanofluids approach to 1 in $\eta=9$. Considering Equation (5.18), we have expected this trend.

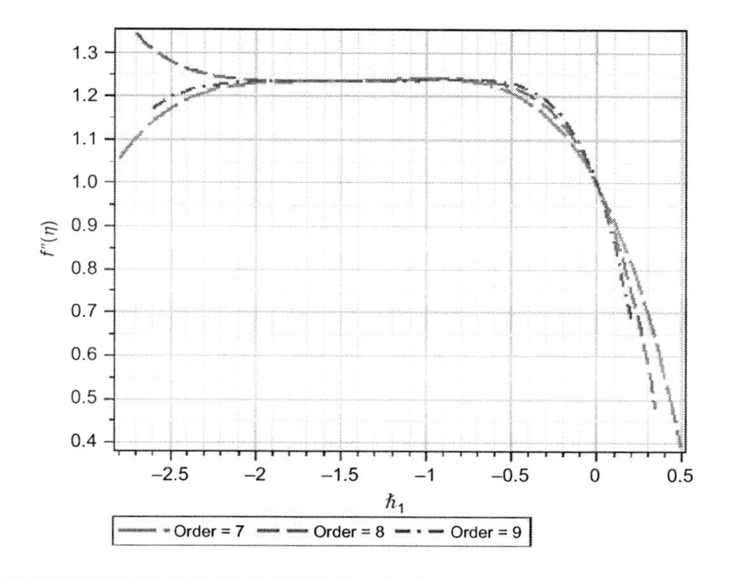

FIGURE 5.2

The $\hbar_1$-validity for CuO for $m=1$.

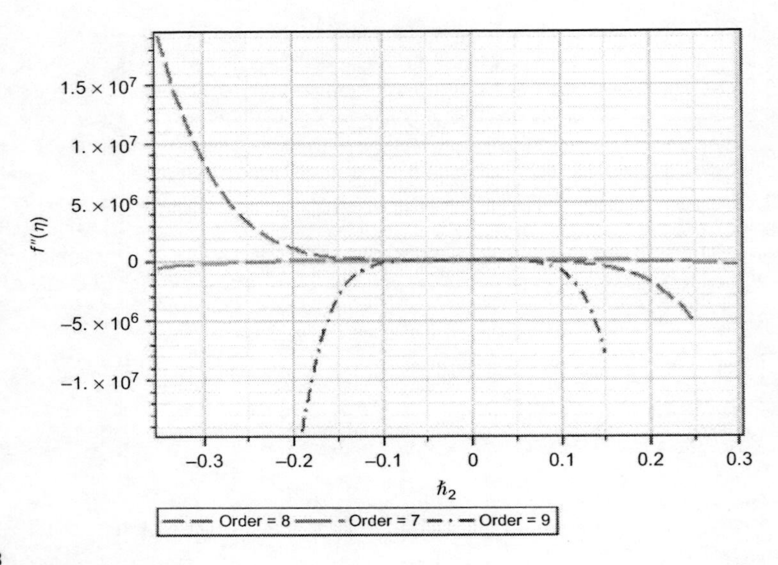

FIGURE 5.3

The $\hbar_2$-validity for CuO for $m=1$.

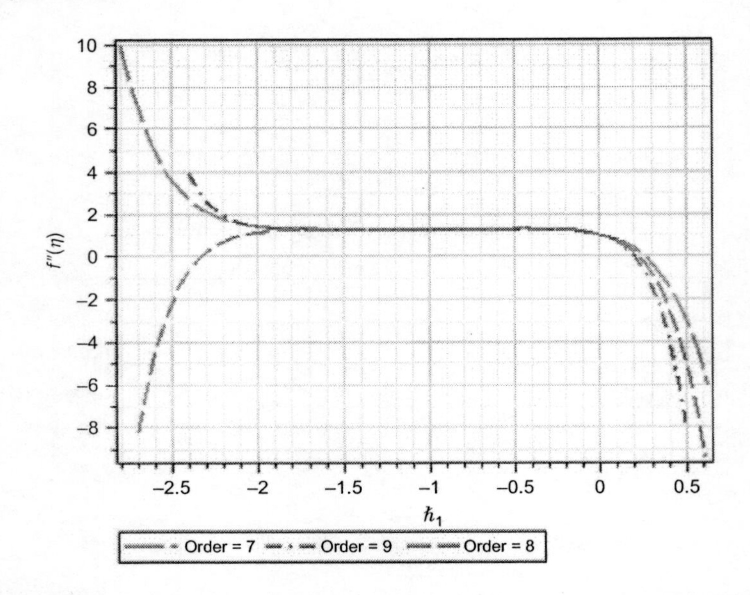

FIGURE 5.4

The $\hbar_1$-validity for Al_2O_3 for $m=1$.

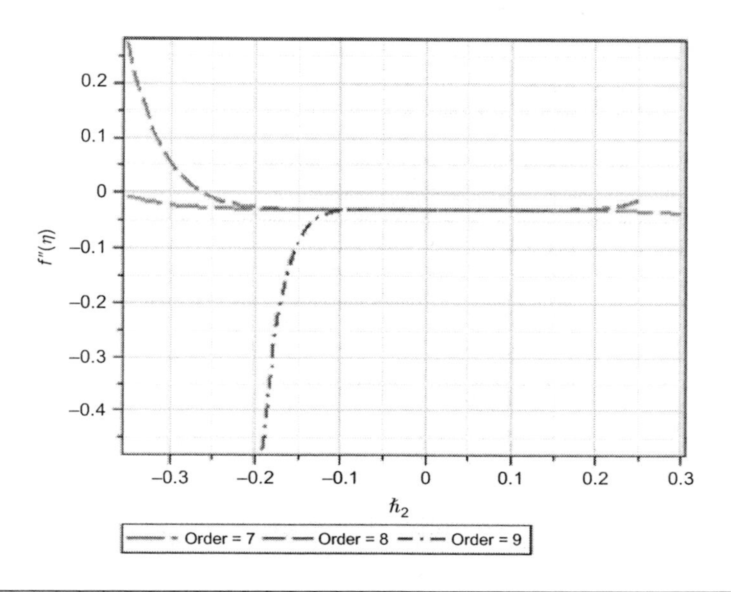

FIGURE 5.5

The $\hbar_2$-validity for Al_2O_3 for $m=1$.

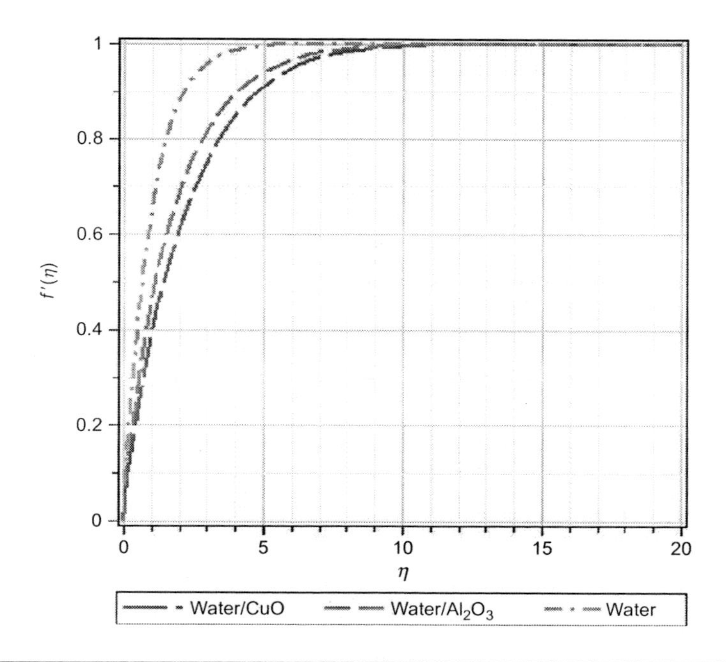

FIGURE 5.6

Velocity profile for different types of fluids for $m=1$.

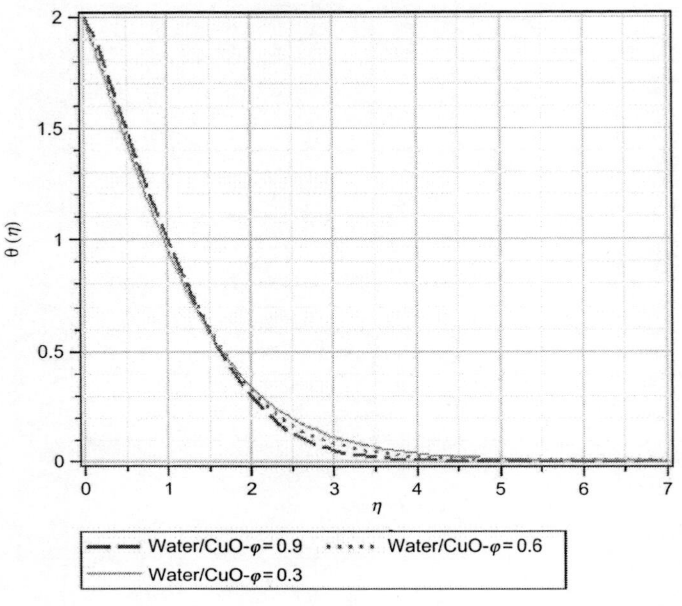

FIGURE 5.7

Temperature profile for CuO for different values of φ and $m=1$.

Figure 5.7 shows temperature curve variation with φ which represents the particle volume fraction of the suspension. When φ climbs up from 0.3% to 0.9%, as we expected, there is a noticeable enhancement in heat transfer rate. This can be justified by turbulence which the nanoscale particles cause. In addition, with increasing the amount of φ, the effects of the nanoparticle conductivity appear in the heat transfer rate as well. As we can see at about $\eta=4.5$, the nanofluid curve with $\varphi=0.9\%$ has dropped to 0 but when $\varphi=0.3\%$, this happens at $\eta=7$. This trend was observed before in experimental studies by others (see Kakaç and Pramuanjaroenkij, 2009).

Figure 5.8 shows the bulk temperature as a function of the fluid types. In agreement with previous experimental study results, our solution shows that nanofluids have better thermal properties and the temperature difference between the fluid and the wedge has disappeared sooner. Due to the addition of the nanoparticles in the base fluid, we can see a noticeable enhancement in heat transfer rate. In other words, the nanofluid temperatures are equal with the wedge surface at 4.4 for "Cuo/water" and 5.1 for "Al_2O_3/water"; however, water experience this at 6.5. To see this effect better, we have magnified the region between $3<\eta<8$ in related figure.

Figure 5.9 shows the relation between fluid's bulk temperature, i.e., $\theta(\eta)$ and m. We can observe a marked change in the η at which profiles drop to 0. As m grows, effects of the stagnation point will be more important and the conductivity of the fluid will play more important role. Because of high heat conduction in nanofluids, the curve approaches to 0 sooner with increase in m.

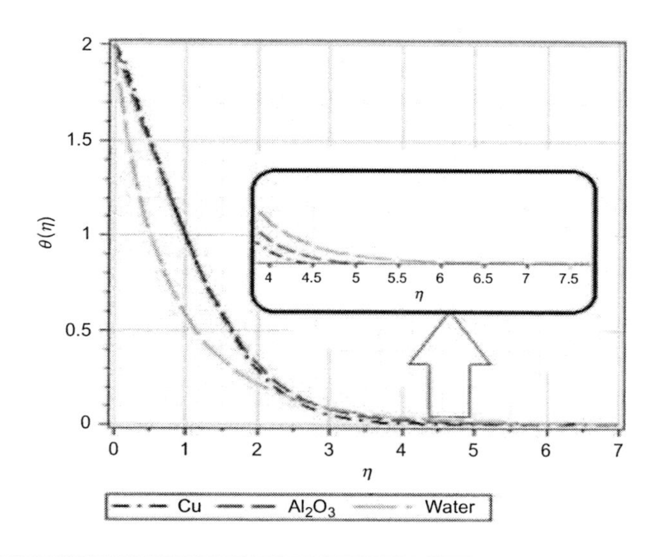

FIGURE 5.8

Temperature profile for different types of fluids for $\varphi=0.1\%$ and $m=1$.

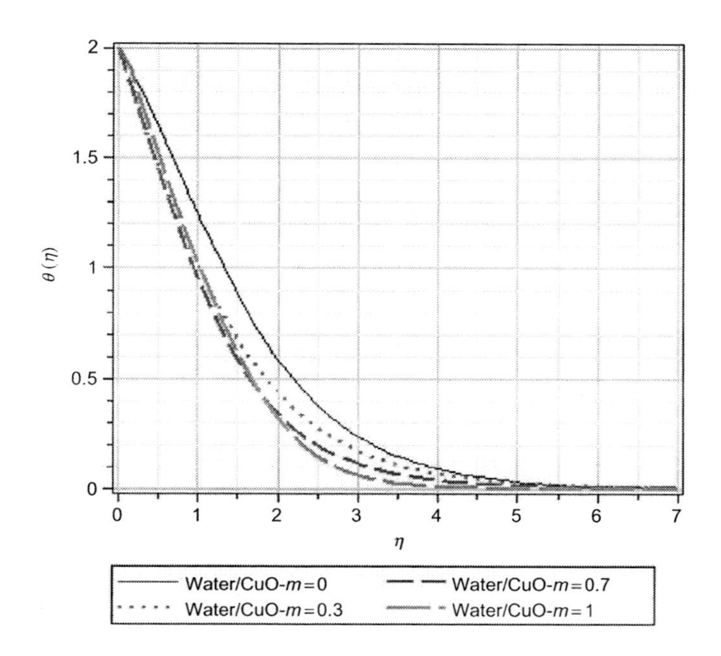

FIGURE 5.9

Temperature profile for CuO for different values of m and $\varphi=0.1\%$.

5.1.7 CONCLUSION

To make it brief, in this chapter we derived the Falkner-Skan equation for a nanofluid and solved it with HAM. The convergence region of HAM is discussed, i.e., auxiliary parameters $\hbar_1$ and $\hbar_2$. The physical interpretation of the results was explained. Temperature and velocity diagrams have been discussed in detail which shows that Cuo as a nanoparticle has more efficiency in comparison with Al_2O_3. Both of these nanoparticles are suspended in water as the base fluid. In addition, we observed that with adding nanoscale particles to a fluid, the viscosity of the fluid will increase and this will affect the momentum exchange in the fluid layers.

5.2 TEMPERATURE VARIATION ANALYSIS FOR NANOPARTICLE'S COMBUSTION

In this section, combustion process for iron nanoparticles involves the conversion of gaseous material aluminum, in the presence of carrier gas, to solid nanoparticles and their subsequent reaction with oxygen gas to form the iron oxide coating. Initial mixture is considered to be an aerosol. The solutions obtained by using Boubacker polynomials expansion scheme (BPES) technique are compared with those of the recent related method in literature (*this section has been worked by M. Hatami, D. D. Ganji, K. Boubaker in nonlinear dynamics team in Mechanical Engineering Department, 2012-2013*).

5.2.1 INTRODUCTION

Combustion of metallic particles is one of the most challenging issues in industries that manufacture, process, generate, or use combustible dusts, and an accurate knowledge of their explosion hazards is essential. Many studies have been done for estimating and modeling the particle and dust combustion. Haghiri and Bidabadi (2011) investigated the dynamic behavior of particles across flame propagation through a two-phase mixture consisting of micro-iron particles and air. They assumed three zones for flame structure: preheat, reaction, and postflame (burned). Liu et al. (2007) analyzed the flame propagation through hybrid mixture of coal dust and methane in a combustion chamber. A one-dimensional, steady-state theoretical analysis of flame propagation mechanism through micro-iron dust particles based on dust particles' behavior with special remark on the thermophoretic force in small Knudsen numbers is presented by Bidabadi et al. (2010). Haghiri and Bidabadi (2010) performed a mathematical model to analyze the structure of flame propagating through a two-phase mixture consisting of organic fuel particles and air. In contrast to previous analytical studies, they take thermal radiation effect into consideration, which has not been attempted before. Recently, Bidabadi et al. (2010) solved the non-linear energy equation that resulted from particle combustion modeling by using homotopy perturbation method, and they presented equations for calculating the convective heat transfer coefficient and burning time for iron particles.

Polynomial expansion methods are extensively used in many mathematical and engineering fields to yield meaningful results for both numerical and analytical analyses. Among the most frequently used polynomials, the Boubaker polynomials expansion scheme, which is firstly introduced by Boubaker (2008), is one of the interesting tools which were associated with several applied physical and mathematical problems (i.e., see Agida and Kumar, 2010; Fridjine et al., 2009; Tabatabaei et al., 2009; Rahmanov, 2011).

The differential transformation method (DTM) is an alternative procedure for obtaining an analytic Taylor series solution of differential equations. The main advantage of this method is that it can be applied directly to nonlinear differential equations without requiring linearization and discretization, and therefore, it is not affected by errors associated with discretization. The concept of DTM was first introduced by Zhou (1986), who solved linear and nonlinear problems in electrical circuits. Chen and Ho (1999) developed this method for partial differential equations. This method was successfully applied to various application problems (see Joneidi et al., 2009; Momeni et al., 2011).

Motivated by previously mentioned works, this work aims to introduce two analytical methods for obtaining the temperature of iron particle during combustion. So BPES and DTM are presented. These methods have an excellent agreement with numerical Runge-Kutta method; they also have very low errors without any need for perturbation or discretization compared to previous analytical methods in the literature.

5.2.2 PROBLEM DESCRIPTION

Consider a spherical particle which, due to high reaction with oxygen, will be combusted. Since the thermal diffusivity of substance is large and Biot number is small ($Bi_H \ll 0.1$), it is assumed that the particle is isothermal. In this state, a lumped system analysis is applicable. When this criterion is satisfied, the variation of temperature with location within the particle will be slight and can be approximated as being uniform, so particle has a spatially uniform temperature; therefore, the temperature of particle is a function of time only, $T=T(t)$, and is not a function of radial coordinate, $T \neq T(r)$. The assumptions used in this modeling are:

- The spherical particle burns in a quiescent, infinite ambient medium and there are no interactions with other particles; also, the effects of forced convection are ignored.
- Thermophysical properties for the particle and ambient gaseous oxidizer are assumed to be constant.
- The particle radiates as a gray body to the surroundings without contribution of the intervening medium.

By these assumptions and considering the particle as a thermodynamic system, and by using the principle of conservation of energy (first law of thermodynamics), the energy balance equation for this particle can be written as

$$\dot{E}_{in} - \dot{E}_{out} + \dot{E}_{gen} = \left(\frac{dE}{dt}\right)_p \tag{5.70}$$

where $\dot{E}_{in}$ is the rate of energy entering the system which is owing to absorption of total radiation incident on the particle surface from the surrounding; $\dot{E}_{out}$ is the rate of energy leaving the system by mechanisms of convection on the particle surface and thermal radiation that emits from the outer surface of particle; $\dot{E}_{gen}$ is the rate of generation of energy inside the particle due to the combustion process and equals to the heat released from the chemical reaction; and $(dE/dt)_p$ is the rate of change in total energy of particle. These energy terms can be calculated by:

$$\dot{E}_{in} = \alpha_s \sigma A_s T_{surr}^4 \tag{5.71}$$

$$\dot{E}_{out} = h_{conv} A_s (T_s - T_\infty) + \varepsilon_s \sigma A_s T_s^4 \tag{5.72}$$

$$\dot{E}_{gen} = \dot{Q}_{comb} = \dot{R}_p A_s \Delta h_{comb}^o \tag{5.73}$$

$$\left(\frac{dE}{dt}\right)_p = \rho_p V_p c_p \frac{dT_s}{dt} \tag{5.74}$$

By substituting the Equations (5.71)–(5.74) in Equation (5.70),

$$\alpha_s \sigma A_s T_{surr}^4 - \left(h_{conv} A_s (T_s - T_\infty) + \varepsilon_s \sigma A_s T_s^4\right) + \dot{R}_p A_s \Delta h_{comb}^o = \rho_p V_p c_p \frac{dT_s}{dt} \tag{5.75}$$

Three reasonable assumptions used for improving Equation (5.75) are:

I. Both absorptivity and emissivity of the surface depend on the temperature and the wavelength of radiation. Kirchhoff's law of radiation states that the absorptivity and the emissivity of a surface at a given temperature and wavelength are equal. ($\varepsilon_s \simeq \alpha_s$).

II. The initial temperature of the particle at the beginning of the combustion can be regarded as the initial condition. This temperature is known as ignition temperature ($T(0) = T_{ig}$).

III. The density of particle is a function of particle temperature; so, it can be considered as a linear function. ($\rho_p = \rho_p(T) = \rho_{p,\infty}[1 + \beta(T - T_\infty)]$)

By applying these assumptions, Equation (5.75) will be converted into the following:

$$\rho_{p,\infty}[1 + \beta(T - T_\infty)] V_p c_p \frac{dT_s}{dt} + h_{conv} A_s (T_s - T_\infty) + \varepsilon_s \sigma A_s \left(T_s^4 - T_{surr}^4\right) - \dot{R}_p A_s \Delta h_{comb}^o = 0 \tag{5.76}$$

For solving this nonlinear differential equation, it is more suitable that all the terms be converted into the dimensionless form. The following set of dimensionless variables are defined as

$$\begin{cases} \theta = \dfrac{T}{T_{ig}}, \quad \theta_\infty = \dfrac{T_\infty}{T_{ig}}, \quad \theta_{surr} = \dfrac{T_{surr}}{T_{ig}}, \quad \varepsilon_1 = \beta T_{ig} \\ \tau = \dfrac{t}{\left(\dfrac{\rho_{p,\infty} V_p c_p}{h_{conv} A_s}\right)}, \quad \psi = \dfrac{\dot{Q}_{comb}}{h_{conv} A_s T_{ig}}, \quad \varepsilon_2 = \dfrac{\varepsilon_s \sigma T_{ig}^3}{h_{conv}} \end{cases} \tag{5.77}$$

Consequently, the nonlinear differential equation and its initial condition can be expressed in the dimensionless form

$$\varepsilon_1 \theta \frac{d\theta}{d\tau} + (1 - \varepsilon_1 \theta_\infty) \frac{d\theta}{d\tau} + \varepsilon_2 \left(\theta^4 - \theta_{surr}^4\right) + \theta - \psi - \theta_\infty = 0 \tag{5.78}$$

$$\theta(0) = 1 \tag{5.79}$$

5.2.3 APPLIED ANALYTICAL METHODS

In this section, two analytical methods called DTM and BPES with their application in the problem are presented.

5.2.3.1 Differential Transformation Method

For understanding this method's concept, suppose that $x(t)$ is an analytic function in domain D, and $t=t_i$ represents any point in the domain. The function $x(t)$ is then represented by one power series whose center is located at t_i. The Taylor series expansion function of $x(t)$ is in the form of:

$$x(t)=\sum_{k=0}^{\infty}\frac{(t-t_i)^k}{k!}\left[\frac{\mathrm{d}^k x(t)}{\mathrm{d}t^k}\right]_{t=t_i}\quad \forall t\in D \tag{5.80}$$

The Maclaurin series of $x(t)$ can be obtained by taking $t_i=0$ in Equation (5.80) expressed as

$$x(t)=\sum_{k=0}^{\infty}\frac{t^k}{k!}\left[\frac{\mathrm{d}^k x(t)}{\mathrm{d}t^k}\right]_{t=0}\quad \forall t\in D \tag{5.81}$$

the differential transformation of the function $x(t)$ is defined as follows

$$X(k)=\sum_{k=0}^{\infty}\frac{H^k}{k!}\left[\frac{\mathrm{d}^k x(t)}{\mathrm{d}t^k}\right]_{t=0} \tag{5.82}$$

where $X(k)$ represents the transformed function and $x(t)$ is the original function. The differential spectrum of $X(k)$ is confined within the interval $t\in[0,H]$, where H is a constant value. The differential inverse transform of $X(k)$ is defined as follows

$$x(t)=\sum_{k=0}^{\infty}\left(\frac{t}{H}\right)^k X(k) \tag{5.83}$$

It is clear that the concept of differential transformation is based on the Taylor series expansion. The values of function at values of argument are referred to as discrete; that is, they are known as the zero discrete, the first discrete, and so forth. The more discreteness available, the more precise it is possible to restore the unknown function. The function consists of T-function, and its value is given by the sum of T-functions with as its coefficient. In real applications, at the right choice of the constant, if the values of argument are large, the discrete of spectrum reduces rapidly.

It is clear that the concept of differential transformation is based upon the Taylor series expansion. The values of function $X(k)$ at values of argument k are referred to as discrete, i.e., $X(0)$ is known as the zero discrete, $X(1)$ as the first discrete, etc. The more discrete available, the more precise it is possible to restore the unknown function. The function $x(t)$ consists of the T-function $X(k)$, and its value is given by the sum of the T-function with $(t/H)k$ as its coefficient. In real applications, at the right choice of constant H, if the values of argument k are larger, the discrete of spectrum reduces rapidly. Some important mathematical operations performed by DTM are listed in Table 5.2.

By applying DTM from Table 5.1, transformed form of (5.78) will be

$$\begin{aligned}(k+1)\Theta(k+1)&+\varepsilon_1\sum_{l=0}^{k}\Theta(l)\cdot(k+1-l)\cdot\Theta(k+1-l)-\varepsilon_1\theta_\infty(k+1)\cdot\Theta(k+1)+\Theta(k)\\&+\varepsilon_2\sum_{k2=0}^{k}\sum_{k1=0}^{k2}\sum_{l=0}^{k1}\Theta(l)\cdot\Theta(k2-k1)\cdot\Theta(k-k2)\cdot\Theta(k1-l)-\delta(k)\cdot\left(\theta_\infty+\varepsilon_2\theta_{\text{surr}}^4+\psi\right)=0\end{aligned} \tag{5.84}$$

Table 5.1 Thermophysical Properties of the Base Fluid and the Nanoparticles (Oztop and Abu-Nada, 2008).

Physical Properties	Fluid Phase(Water)	Cu	Al_2O_3
c_p (J/kg K)	4179	385	765
ρ (kg/m^3)	997.1	8933	3970
k (W/m K)	0.613	400	40
$\alpha \times 10^{-7}$ (m^2/s)	1.47	1163.1	131.7

Table 5.2 Some Fundamental Operations of the Differential Transform Method

Origin Function	Transformed Function
$x(t) = \alpha f(x) \pm \beta g(t)$	$X(k) = \alpha F(k) \pm \beta G(k)$
$x(t) = \dfrac{d^m f(t)}{dt^m}$	$X(k) = \dfrac{(k+m)!F(k+m)}{k!}$
$x(t) = f(t)g(t)$	$X(k) = \sum_{l=0}^{k} F(l)G(k-l)$
$x(t) = t^m$	$X(k) = \delta(k-m) = \begin{cases} 1, & \text{if } k = m, \\ 0, & \text{if } k \neq m \end{cases}$
$x(t) = \exp(t)$	$X(k) = \dfrac{1}{k!}$
$x(t) = \sin(\omega t + \alpha)$	$X(k) = \dfrac{\omega^k}{k!} \sin\left(\dfrac{k\pi}{2} + \alpha\right)$
$x(t) = \cos(\omega t + \alpha)$	$X(k) = \dfrac{\omega^k}{k!} \cos\left(\dfrac{k\pi}{2} + \alpha\right)$

where Θ is the transformed form of θ and

$$\delta(k) = \begin{cases} 1 & k = 0 \\ 0 & k \neq 1 \end{cases} \tag{5.85}$$

Transformed form of initial condition Equation (5.79) will be,

$$\Theta(0) = 1 \tag{5.86}$$

For example, for an iron particle with 20 μm diameter (see Table 5.3), solving Equation (5.84) makes,

$$\begin{aligned} &\Theta(0) = 1, \quad \Theta(1) = 1.170326597, \quad \Theta(2) = -0.6324112173, \quad \Theta(3) = 0.2462459086, \\ &\Theta(4) = -0.08656001439, \ldots \end{aligned} \tag{5.87}$$

Table 5.3 Properties and Conditions for Combustion of Iron Particles

Particle Diameter (μm)	ε_1	ε_2	Ψ	θ_∞	θ_{surr}
20	0.051595	0.002630	0.98579	1.17647	0.35294
60	0.051595	0.007567	2.83636	1.17647	0.35294
100	0.051595	0.011493	4.30774	1.17647	0.35294

By substituting DTM transformed terms of T (see Equation (5.87)) into Equation (5.82), $\theta(\tau)$ can be determined as,

$$\theta(\tau) = 1 + 1.17033\tau - 0.632411\tau^2 + 0.246245\tau^3 - 0.08656\tau^4 + 0.0324505\tau^5 - 0.0130388\tau^6 + 0.00504942\tau^7 - 0.0017\tau^8 + 0.00042358\tau^9 - 0.114879\text{e} - 4\tau^{10} \tag{5.88}$$

5.2.3.2 Boubaker Polynomials Expansion Scheme

The resolution of system (Equations 5.78 and 5.79) along with boundary conditions has been achieved using the BPES. This scheme is a resolution protocol which has been successfully applied to several applied physics and mathematics problems. The BPES protocol ensures the validity of the related boundary conditions regardless of main equation features. The BPES is based on the Boubaker polynomials first derivatives properties:

$$\begin{aligned}
&\sum_{q=1}^{N} B_{4q}(x)\bigg|_{x=0} = -2N \neq 0,\\
&\sum_{q=1}^{N} B_{4q}(x)\bigg|_{x=r_q} = 0,\\
&\sum_{q=1}^{N} \frac{dB_{4q}(x)}{dx}\bigg|_{x=0} = 0,\\
&\sum_{q=1}^{N} \frac{dB_{4q}(x)}{dx}\bigg|_{x=r_q} = \sum_{q=1}^{N} H_q,
\end{aligned} \tag{5.89}$$

with:

$$H_n = B'_{4n}(r_n) = \left(\frac{4r_n\left[2 - r_n^2\right] \times \sum_{q=1}^{n} B_{4q}^2(r_n)}{B_{4(n+1)}(r_n)} + 4r_n^3 \right) \tag{5.90}$$

Several solutions have been proposed through the BPES in many fields such as numerical analysis (Agida and Kumar, 2010), theoretical physics (Slama et al., 2008a,b), mathematical algorithms (Fridjine et al., 2009), heat transfer (Khélia et al., 2009), homodynamic (Ben Mahmoud and Amlouk, 2009), etc.

The resolution protocol is based on setting $\theta(\tau)$ as an estimator to the τ-dependent variable:

$$\theta(\tau) = \frac{1}{2N_0} \sum_{k=1}^{N_0} \xi_k \times B_{4k}(\tau \times r_k), \tag{5.91}$$

where B_{4k} are the $4k$-order Boubaker polynomials (see Agida and Kumar, 2010; Yildirim et al., 2010), r_k are B_{4k} minimal positive roots, N_0 is a prefixed integer, and $\xi_k|_{k=1...N_0}$ are unknown pondering real coefficients.

The main advantage of this formulation is the verification of boundary conditions, expressed in Equation (5.78), in advance to resolution process. In fact, thanks to the properties expressed in Equations (5.90) and (5.91); these conditions are reduced to the inherently verified linear equations:

$$\sum_{k=1}^{N_0} \xi_k = -N_0. \tag{5.92}$$

The BPES solution for Equation (5.70) is obtained, according to the principles of the BPES, by determining the nonnull set of coefficients $\xi_k|_{k=1...N_0}$ that minimize the absolute difference between left and right sides of the following equations:

$$\varepsilon_1 \left(\frac{1}{2N_0}\right)^2 \sum_{k=1}^{N_0} \xi_k B_{4k}(\hat{t}) \sum_{k=1}^{N_0} \xi_k r_k \times \frac{dB_{4k}(\hat{t})}{d\tau} + \frac{1-\varepsilon_1\theta_\infty}{2N_0} \sum_{k=1}^{N_0} \xi_k r_k \times \frac{dB_{4k}(\hat{t})}{d\tau}$$
$$+ \frac{1}{2N_0} \sum_{k=1}^{N_0} \xi_k B_{4k}(\hat{t}) - \psi - \theta_\infty + \frac{4\varepsilon_2}{(2N_0)^4} \sum_{k=1}^{N_0} \xi_k B_{4k}(\hat{t}) - \varepsilon_2 \theta_{\text{surr}} = 0, \quad \hat{t} = \tau \times r_k. \tag{5.93}$$

5.2.4 RESULTS AND DISCUSSION

As described in Section 5.2.2, the equation of combustion for a particle was introduced as Equation (5.78) with boundary condition Equation (5.79). Some constant parameters for this equation are introduced in Table 5.2 for iron particle. For these three particle diameters, Equation (5.78) is solved by DTM and BPES whose results are presented through Figure 5.10a and b.

As seen in this figure, both of the two analytical methods have very excellent agreement with numerical fourth-order Runge-Kutta method, and also this figure reveals that when particle diameter increases, an increase in combustion temperature due to higher energy realized in combustion is

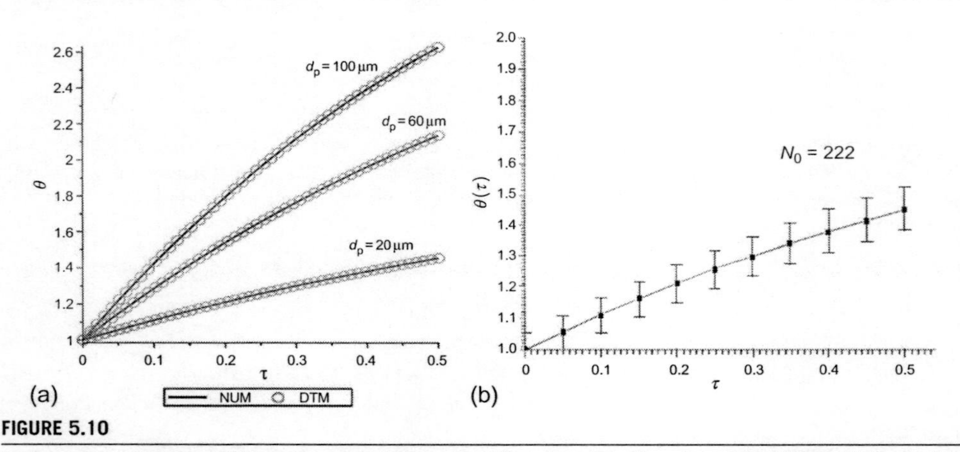

FIGURE 5.10

(a) Comparison between DTM and numerical method in different particle diameters. (b) BPES results for particle temperate for diameter 20 μm.

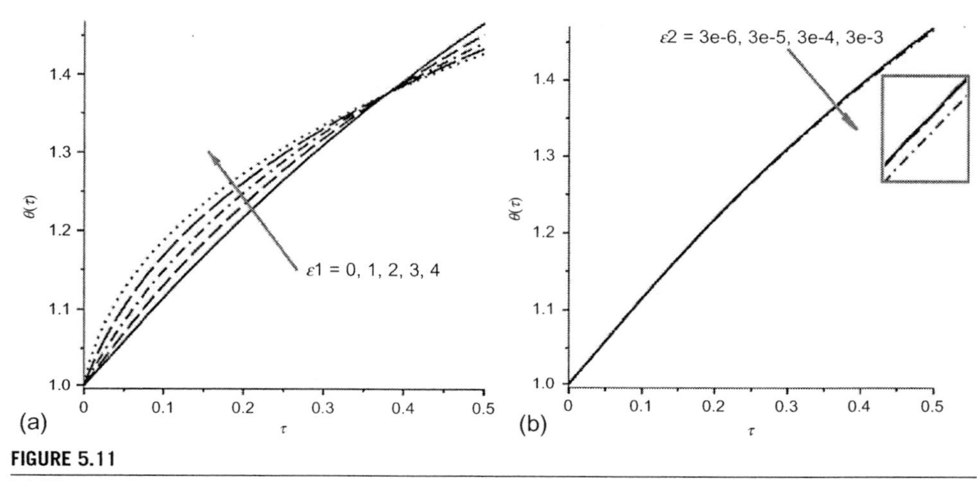

FIGURE 5.11

(a) Effect of $\varepsilon 1$ on nondimensional temperature profile for micro and nanoparticles. (b) Effect of $\varepsilon 2$ on nondimensional temperature profile for micro and nanoparticles.

observed. The figure indicates that the temperature of iron particle (for all sizes) reaches the maximum value at the end of the burning time, and the chemical reaction is finished, and larger particles have higher maximum temperature than that of smaller particles. This result is similar to the experimental result presented by Tang et al. (2011) and analytical method by Bidabadi and Mafi (2013).

This table confirms that BPES has lower errors compared to DTM, but both methods are convenient and accurate. Effects of $\varepsilon 1$ and $\varepsilon 2$ on nondimensional temperature profile for nanoparticles are shown in Figure 5.11a and b.

As seen, the increase in $\varepsilon 1$ makes an increase in temperature profile, but makes a decrease due to the increase in radiation heat transfer term in the particle.

An important point for particle combustion is the surrounding temperature in which particle is combusted. As Figure 5.11 reveals, energy inputs from the surrounding occur through radiation heat transfer, and when surrounding has higher temperature, input energy increases and consequently temperature in combustion of particle will increase. This effect is depicted in Figure 5.12a. Figure 5.12b demonstrates the effect of heat realized parameter (Ψ) on temperature profile versus time. It is completely evident that by increasing the generated heat in the combustion, temperature will increase significantly.

5.2.5 CONCLUSION

In this study, the equation of temperature variation in combustion process for particles is presented. The effects of thermal radiation from the external surface of burning particle and alterations of density of iron particle with temperature are considered. Due to nonlinearity of described equation, BPES and DTM have been presented in order to obtain analytical solutions. Results show that both methods have good agreement with numerical method, but BPES had lower errors than DTM. Also, the effects of

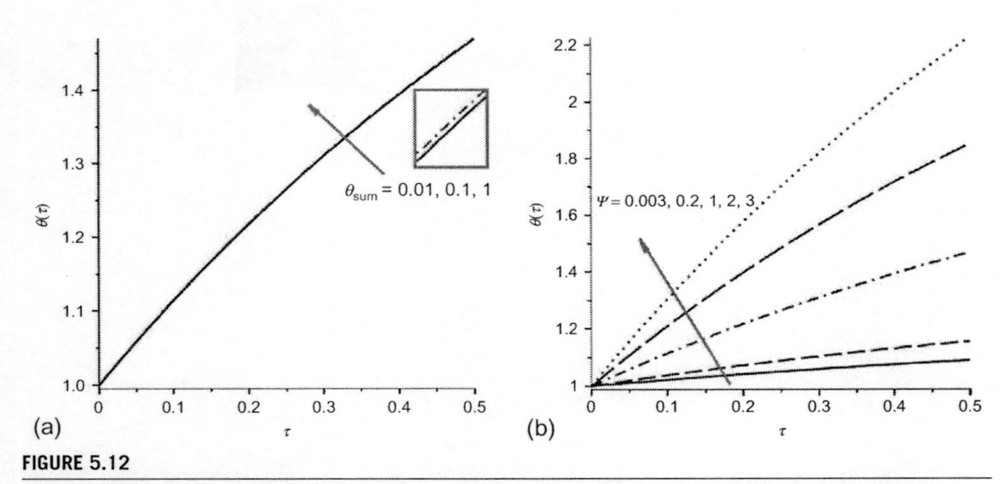

FIGURE 5.12

(a) Effect of surrounding temperature on nondimensional temperature profile. (b) Effect of heat realized parameter (Ψ) on nondimensional temperature profile.

surrounding temperature, realized heat parameter, particle diameter, and convection and radiation heat transfer parameter on combustion temperature have been investigated.

REFERENCES

Agida, M., Kumar, A.S., 2010. A Boubaker Polynomials Expansion Scheme solution to random Love's equation in the case of a rational kernel. Electron. J. Theor. Phys. 7 (24), 319–326.

Ben Mahmoud, K.B., Amlouk, M., 2009. The 3D Amlouk-Boubaker expansivity-energy gap-Vickers hardness abacus: a new tool for optimizing semiconductor thin film materials. Mater. Lett. 63 (12), 991–994.

Bidabadi, M., Mafi, M., 2013. Time variation of combustion temperature and burning time of a single iron particle. Int. J. Therm. Sci. 65, 136–147.

Bidabadi, M., Haghiri, A., Rahbari, A., 2010. Mathematical modeling of velocity and number density profiles of particles across the flame propagation through a micro-iron dust cloud. J. Hazard. Mater. 176 (1–3), 146–153.

Boubaker, K., 2008. The Boubaker polynomials, a new function class for solving bi-varied second order differential equations. Far East J. Appl. Math. 31, 299–320.

Chen, K., Ho, S.H., 1999. Solving partial differential equations by two-dimensional differential transform method. Appl. Math. Comput. 106 (2–3), 171–179.

Das, S.K., Putra, N., Roetzel, W., 2003a. Pool boiling characteristics of nano-fluids. Int. J. Heat Mass Transf. 46, 851–862.

Das, S.K., Putra, N., Roetzel, W., 2003b. Pool boiling of nano-fluids on horizontal narrow tubes. Int. J. Multiphase Flow 29, 1237–1247.

Das, S.K., Putra, N., Thiesen, P., Roetzel, W., 2003c. Temperature dependence of thermal conductivity enhancement for nanofluids. ASME J. Heat Transf. 125, 567–574.

Falkner, V.M., Skan, S.W., 1931. Solutions of the boundary-layer equations. Philos. Mag. 7 (12), 865–896.

Fridjine, S., Ben Mahmoud, K.B., Amlouk, M., Bouhafs, M., 2009. A study of sulfur/selenium substitution effects on physical and mechanical properties of vacuum-grown ZnS1-xSex compounds using Boubaker Polynomials Expansion Scheme (BPES). J. Alloys Compd. 479 (1–2), 457–461.

Goldstein, R.J., Ibele, W.E., Patankar, S.V., Simon, T.W., Kuehn, T.H., Strykowski, P.J., Tamma, K.K., Heberlein, J.V.R., Davidson, J.H., Bischof, J., Kulacki, F.A., Kortshagen, U., Garrick, S., Srinivasan, V., Ghosh, K., Mittal, R., 2010. Int. J. Heat Mass Transf. 53 (21-22), 4397–4447.

Haghiri, A., Bidabadi, M., 2010. Modeling of laminar flame propagation through organic dust cloud with thermal radiation effect. Int. J. Therm. Sci. 49 (8), 1446–1456.

Haghiri, A., Bidabadi, M., 2011. Dynamic behavior of particles across flame propagation through micro-iron dust cloud with thermal radiation effect. Fuel 90 (7), 2413–2421.

Joneidi, A., Ganji, D.D., Babaelahi, M., 2009. Differential transformation method to determine fin efficiency of convective straight fins with temperature dependent thermal conductivity. Int. Commun. Heat Mass Transfer 36 (7), 757–762.

Kakaç, Sadik, Pramuanjaroenkij, Anchasa, 2009. Review of convective heat transfer enhancement with nanofluids. Int. J. Heat Mass Transf. 52 (13-14), 3187–3196.

Khélia, C., Boubaker, K., Ben Nasrallah, T., Amlouk, M., Belgacem, S., 2009. Morphological and thermal properties of β-SnS2 sprayed thin films using Boubaker polynomials expansion. J. Alloys Compd. 477 (1-2), 461–467.

Liao, S.J., 1992. Proposed Homotopy Analysis Techniques for the Solution of Nonlinear Problems. Ph.D. Dissertation, Shanghai Liao Tong University, China.

Liu, Y., Sun, J., Chen, D., 2007. Flame propagation in hybrid mixture of coal dust and methane. J. Loss Prev. Process Ind. 20 (4–6), 691–697.

Momeni, M., Jamshidi, N., Barari, A., Domairry, G., 2011. Numerical analysis of flow and heat transfer of a viscoelastic fluid over a stretching sheet. Int. J. Numer. Methods Heat Fluid Flow 21 (2), 206–218.

Oztop, H.F., Abu-Nada, E., 2008. Numerical study of natural convection in partially heated rectangular enclosures filled with nanofluids. Int. J. Heat Fluid Flow 29, 1326–1336.

Rahmanov, H., 2011. A solution to the non linear Korteweg-De-Vries equation in the particular case dispersion-adsorption problem in porous media using the spectral Boubaker Polynomials Expansion Scheme (BPES). Stud. Nonlinear Sci. 2 (1), 46–49.

Slama, S., Bessrour, J., Boubaker, K., Bouhafs, M., 2008a. A dynamical model for investigation of A3 point maximal spatial evolution during resistance spot welding using Boubaker polynomials. Eur. Phys. J. Appl. Phys. 44 (3), 317–322.

Slama, S., Bouhafs, M., Ben Mahmoud, K.B., 2008b. A Boubaker polynomials solution to heat equation for monitoring A3 point evolution during resistance spot welding. Int. J. Heat Technol. 26 (2), 141–145.

Tabatabaei, S., Zhao, T., Awojoyogbe, O., Moses, F., 2009. Cut-off cooling velocity profiling inside a keyhole model using the Boubaker Polynomials Expansion Scheme. Int. J. Heat Mass Transf. 45 (10), 1247–1251.

Tang, F.D., Goroshin, S., Higgins, A.J., 2011. Modes of particle combustion in iron dust flames. Proc. Combust. Inst. 33 (2), 1975–1982.

Wang, X., Xu, X., Choi, S.U.S., 1999. Thermal conductivity of nanoparticle–fluid mixture. J. Thermophys. Heat Transf. 13, 474–480.

Wen, Dongsheng, Lin, Guiping, Vafaei, Saeid, Zhang, Kai, 2009. Review of nanofluids for heat transfer applications. Particuology 7, 141–150.

Xuan, Y., Li, Q., 2000. Heat transfer enhancement of nanofluids. Int. J. Heat Fluid Flow 21, 58–64.

Xue, Q.Z., 2003. Model for effective thermal conductivity of nanofluids. Phys. Lett. A 307, 313–317.

Yildirim, A., Mohyud-Din, S.T., Zhang, D.H., 2010. Analytical solutions to the pulsed Klein-Gordon equation using Modified Variational Iteration Method (MVIM) and Boubaker Polynomials Expansion Scheme (BPES). Comput. Math. Appl. 59 (8), 2473–2477.

Zhou, K., 1986. Differential Transformation and Its Applications for Electrical Circuits. Huazhong University Press, Wuhan, China.

This page intentionally left blank

CHAPTER

NATURAL, MIXED, AND FORCED CONVECTION IN NANOFLUID

6

CHAPTER CONTENTS

6.1 Natural Convection Flow of Nanofluid in a Concentric Annulus 206
6.1.1 Introduction 206
6.1.2 Problem Definition and Mathematical Model 207
6.1.2.1 Problem Statement 207
6.1.2.2 The Lattice Boltzmann Method 208
6.1.3 Boundary Conditions 209
6.1.3.1 Curved Boundary Treatment for Velocity 209
6.1.3.2 Curved Boundary Treatment for TEmperature 210
6.1.4 The Lattice Boltzmann Model for Nanofluid 210
6.1.5 Grid Testing and Code Validation 211
6.1.6 Results and Discussion 212
6.1.7 Conclusions 218
6.2 Mixed Convection Flow of a Nanofluid in a Horizontal Channel 219
6.2.1 Introduction 220
6.2.2 Describe Problem and Mathematical Formulation 220
6.2.3 Homotopy Perturbation Method Applied to the Problem 223
6.2.3.1 The HPM Applied to the Problem 223
6.2.4 Results and Discussion 224
6.2.5 Conclusion 226
6.3 The Effect of Nanofluid on the Forced Convection Heat Transfer 229
6.3.1 Introduction 229
6.3.2 Governing Equations 230
6.3.3 Solution Using the HAM 233
6.3.3.1 Zeroth-Order Deformation Equations 233
6.3.3.2 mth-Order Deformation Equations 234
6.3.4 Results and Discussion 235
6.3.5 Conclusion 241
6.4 Heat Transfer in Slip-Flow Boundary Condition of a Nanofluid in Microchannel 241
6.4.1 Introduction 242
6.4.2 Problem Statement and Governing Equation 243
6.4.3 Numerical Procedure and Validation 246

Application of Nonlinear Systems in Nanomechanics and Nanofluids. http://dx.doi.org/10.1016/B978-0-323-35237-6.00006-6

6.4.4 Results and Discussion 248
6.4.5 Conclusion 250
6.5 Forced Convection Analysis for Magnetohydrodynamics (MHD) Al_2O_3-Water Nanofluid Flow 253
6.5.1 Introduction 254
6.5.2 Description of the Problem 254
6.5.3 Basic Idea of HAM 257
6.5.3.1 Application in Problem 258
6.5.4 Numerical Method 260
6.5.5 Results and Discussion 260
6.5.6 Conclusion 266
References 266

6.1 NATURAL CONVECTION FLOW OF NANOFLUID IN A CONCENTRIC ANNULUS

In this section, lattice Boltzmann method (LBM) is applied to investigate natural convection flow of a nanofluid in a concentric annulus between a cold outer square cylinder and a heated inner circular cylinder. The fluid in the enclosure is a water-based nanofluid containing different types of nanoparticles: copper (Cu), silver (Ag), alumina (Al_2O_3), and titania (TiO_2). The effective thermal conductivity and viscosity of nanofluid are calculated by the Maxwell-Garnett (MG) model and Brinkman model, respectively. This investigation compared with other numerical methods is found to be in excellent agreement. Numerical results for the flow and heat transfer characteristics are obtained for various values of the nanoparticle volume fraction, Rayleigh numbers, and position of the inner circular cylinder. The results show that maximum value of enhancement is obtained at $\delta/L=0.4$ for $Ra=10^3$ and 10^4, whereas the minimum value of heat transfer enhancement at $Ra=10^5$ and 10^6 is obtained at $\delta/L=0.6$ and 0.4, respectively (*this section has been worked by M. Sheikholeslami, G. Domairry in nonlinear dynamics team in Mechanical Engineering Department, 2012-2013*).

6.1.1 INTRODUCTION

The LBM is a powerful numerical technique based on kinetic theory for simulating fluid flows and modeling the physics in fluids (see Succi, 2001; Yu et al., 2003). In comparison with the conventional CFD methods, the advantages of LBM include simple calculation procedure, simple and efficient implementation for parallel computation, easy and robust handling of complex geometries, and others. Various numerical simulations have been performed using different thermal LB models or Boltzmann-based schemes to investigate the natural convection problems (see Dixit and Babu, 2006; Kao and Yang, 2007).

Natural convection in horizontal annuli has attracted many attentions in recent years due to its wide applications such as in nuclear reactor design, cooling of electronic equipment, aircraft cabin insulation, and thermal storage system. A comprehensive review was presented by Kuehn and Goldstein (1976). Comparatively, little work has been done on natural convective heat transfer in more complex annuli such as the problem considered in this study. A few publications were involved in the experimental study (see Onyegegbu, 1986; Shaija and Narasimham, 2009).

Natural convection between a square outer cylinder and a heated elliptic inner cylinder has been studied numerically by Bararnia et al. (2011a,b).

Soleimani et al. (2012) studied natural convection heat transfer in a semiannulus enclosure filled with nanofluid using the control volume-based finite element method.

The LBM has been applied to investigate the nanofluid flow, and heat transfer characteristic has been studied in recent years. Bararnia et al. (2011a,b) studied the natural convection in a nanofluid-filled portion cavity with a heated built-in plate by LBM.

The aim of this chapter is to study natural convection of nanofluids in a concentric annulus between a cold outer square cylinder and a heated inner circular cylinder using LBM. Different types of nanoparticles such as Cu, Al_2O_3, Ag, and TiO_2 with water as their base fluid have been considered. Effects of nanoparticle volume fraction, types of nanofluids, Rayleigh numbers, and position of the inner circular cylinder on the flow and heat transfer characteristics have been examined.

6.1.2 PROBLEM DEFINITION AND MATHEMATICAL MODEL

6.1.2.1 Problem statement

The physical model used in this work is shown in Figure 6.1. In this figure, $\lambda = L/2r$ denotes aspect ratios and γ is measured counterclockwise from the upward vertical plane through the center of outer cylinders. The system consists of a square enclosure with sides of length L, within which a circular cylinder with an aspect ratio $\lambda = 3.5$ is located and moves along the vertical centerline in the range from $\delta/L = 0.2$ to 0.8. The walls of the square enclosure were kept at a constant low temperature of T_c, whereas the cylinder was kept at a constant high temperature of T_h ($T_h > T_c$).

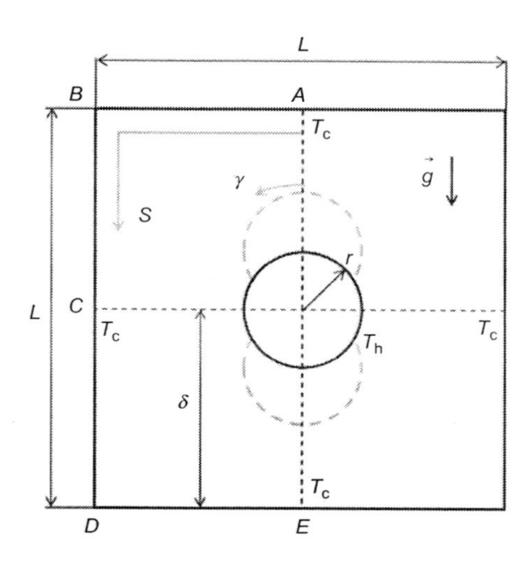

FIGURE 6.1

Geometry of the problem.

6.1.2.2 The lattice Boltzmann method

The LB model used here is the same as that employed in Kao and Yang (2007) and Barrios et al. (2005). The thermal LB model utilizes two distribution functions, f and g, for the flow and temperature fields, respectively. It uses modeling of movement of fluid particles to capture macroscopic fluid quantities such as velocity, pressure, and temperature. In this approach, the fluid domain is discretized to uniform Cartesian cells. Each cell holds a fixed number of distribution functions, which represent the number of fluid particles moving in these discrete directions. The D2Q9 model was used and values of $w_0 = 4/9$ for $|c_0| = 0$ (for the static particle), $w_{1-4} = 1/9$ for $|c_{1-4}| = 1$, and $w_{5-9} = 1/36$ for $|c_{5-9}| = \sqrt{2}$ are assigned in this model (Figure 6.2a).

The density and distribution functions, i.e., the f and g, are calculated by solving the lattice Boltzmann equation (LBE), which is a special discretization of the kinetic Boltzmann equation. After introducing BGK approximation, the general form of LBE with external force is as follows:

For the flow field:

$$f_i(x + c_i\Delta t, t + \Delta t) = f_i(x, t) + \frac{\Delta t}{\tau_v}\left[f_i^{eq}(x, t) - f_i(x, t)\right] + \Delta t c_i F_k \tag{6.1}$$

For the temperature field:

$$g_i(x + c_i\Delta t, t + \Delta t) = g_i(x, t) + \frac{\Delta t}{\tau_C}\left[g_i^{eq}(x, t) - g_i(x, t)\right] \tag{6.2}$$

where Δt denotes lattice time step, c_i is the discrete lattice velocity in direction i, F_k is the external force in the direction of lattice velocity, τ_v and τ_C denote the lattice relaxation time for the flow and temperature fields. The kinetic viscosity υ and the thermal diffusivity α are defined in terms of their respective relaxation times, i.e., $\upsilon = c_s^2(\tau_v - 1/2)$ and $\alpha = c_s^2(\tau_C - 1/2)$, respectively. Note that the limitation $0.5 < \tau$ should be satisfied for both relaxation times to ensure that viscosity and thermal diffusivity are positive. Furthermore, the local equilibrium distribution function determines the type of

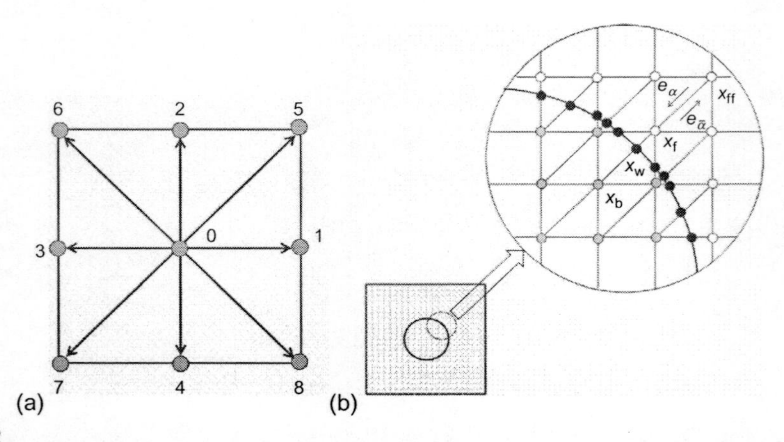

FIGURE 6.2

(a) Discrete velocity set of two-dimensional nine-velocity (D2Q9) model, (b) curved boundary and lattice nodes.

problem that needs to be solved. It also models the equilibrium distribution functions, which are calculated with Equations (6.3) and (6.4) for flow and temperature fields, respectively.

$$f_i^{eq} = w_i \rho \left[1 + \frac{c_i \cdot u}{c_s^2} + \frac{1}{2} \frac{(c_i \cdot u)^2}{c_s^4} - \frac{1}{2} \frac{u^2}{c_s^2} \right] \tag{6.3}$$

$$g_i^{eq} = w_i T \left[1 + \frac{c_i \cdot u}{c_s^2} \right] \tag{6.4}$$

where w_i is the weighting factor and ρ is the lattice fluid density.

In order to incorporate buoyancy forces in the model, the force term in the Equation (6.1) needs to calculated as below in vertical direction (y):

$$F = 3 w_i g_y \beta \theta \tag{6.5}$$

For natural convection, the Boussinesq approximation is applied and radiation heat transfer is negligible. To ensure that the code works in near-incompressible regime, the characteristic velocity of the flow for natural $\left(V_{\text{natural}} \equiv \sqrt{\beta g_y \Delta T H}\right)$ regime must be small when compared with the fluid speed of sound. In the present study, the characteristic velocity is selected as 0.1 of sound speed.

Finally, macroscopic variables are calculated with the following formula:

$$\begin{aligned} &\text{Flow density}: \ \rho = \sum_i f_i, \\ &\text{Momentum}: \ \rho u = \sum_i c_i f_i, \\ &\text{Temperature}: \ T = \sum_i g_i. \end{aligned} \tag{6.6}$$

6.1.3 BOUNDARY CONDITIONS

6.1.3.1 Curved boundary treatment for velocity

For treating velocity and temperature fields with curved boundaries, the method proposed in Yan and Zu (2008) has been used. An arbitrary curved wall separating solid region from fluid is shown in Figure 6.2b. The link between the fluid node x_f and the wall node x_w intersects the physical boundary at x_b. The fraction of the intersected link in the fluid region is $\Delta = |x_f - x_w| / |x_f - x_b|$. To calculate the postcollision distribution function $\tilde{f}_{\bar{\alpha}}(x_b, t)$ based upon the surrounding nodes information, a Chapman-Enskog expansion for the postcollision distribution function on the right-hand side of Equation (6.1) is conducted as:

$$\tilde{f}_{\bar{\alpha}}(x_b, t) = (1 - \chi) \tilde{f}_\alpha(x_f, t) + \chi f_\alpha^*(x_b, t) + 2 w_\alpha \rho \frac{3}{c^2} e_{\bar{\alpha}} \cdot u_w \tag{6.7}$$

where

$$\begin{aligned} &f_\alpha^*(x_b, t) = f_\alpha^{eq}(x_f, t) + w_\alpha \rho(x_f, t) \frac{3}{c^2} e_\alpha \cdot (u_{bf} - u_f), \\ &u_{bf} = u_{ff} = u(x_{ff}, t), \qquad \chi = \frac{(2\Delta - 1)}{\tau - 2}, \quad \text{if } 0 \le \Delta \le \frac{1}{2} \\ &u_{bf} = \frac{1}{2\Delta}(2\Delta - 3) u_f + \frac{3}{2\Delta} u_w, \quad \chi = \frac{(2\Delta - 1)}{\tau - 1/2}, \quad \text{if } \frac{1}{2} \le \Delta \le 1 \end{aligned} \tag{6.8}$$

In the above, $e_{\bar{\alpha}} \equiv -e_\alpha$; u_f is the fluid velocity near the wall; u_w is the velocity of solid wall; and u_{bf} is an imaginary velocity for interpolations.

6.1.3.2 Curved boundary treatment for temperature

Following the work of Yan and Zu (2008), the nonequilibrium parts of temperature distribution function can be defined as:

$$g_{\bar{\alpha}}(\mathbf{x}_b, t) = g_{\bar{\alpha}}^{eq}(\mathbf{x}_b, t) + g_{\bar{\alpha}}^{neq}(\mathbf{x}_b, t) \tag{6.9}$$

Substituting Equation (6.9) into Equation (6.2) yields:

$$\tilde{g}_{\bar{\alpha}}(\mathbf{x}_b, t+\Delta t) = g_{\bar{\alpha}}^{eq}(\mathbf{x}_b, t) + \left(1 - \frac{1}{\tau_s}\right) g_{\bar{\alpha}}^{neq}(\mathbf{x}_b, t) \tag{6.10}$$

Obviously, both $g_{\bar{\alpha}}^{eq}(\mathbf{x}_b, t)$ and $g_{\bar{\alpha}}^{neq}(\mathbf{x}_b, t)$ are needed to calculate the value of $\tilde{g}_{\bar{\alpha}}(\mathbf{x}_b, t+\Delta t)$. In Equation (6.10), the equilibrium part is defined as:

$$g_{\bar{\alpha}}^{eq}(\mathbf{x}_b, t) = w_{\bar{\alpha}} T_b{}^* \left(1 + \frac{3}{c^2} \mathbf{e}_{\bar{\alpha}} \cdot \mathbf{u}_b^*\right) \tag{6.11}$$

where T_b^* is defined as a function of $T_{b1} = [T_w + (\Delta - 1)T_f]/\Delta$ and $T_{b2} = [2T_w + (\Delta - 1)T_{ff}]/(1 + \Delta)$

$$\begin{aligned} T_b^* &= T_{b1}, && \text{if } \Delta \geq 0.75 \\ T_b^* &= T_{b1} + (1-\Delta)T_{b2}, && \text{if } \Delta \leq 0.75 \end{aligned} \tag{6.12}$$

and u_b^* is defined as a function of $u_{b1} = [u_w + (\Delta - 1)u_f]/\Delta$ and $u_{b2} = [2u_w + (\Delta - 1)u_{ff}]/(1 + \Delta)$

$$\begin{aligned} u_b^* &= u_{b1}, && \text{if } \Delta \geq 0.75 \\ u_b^* &= u_{b1} + (1-\Delta)u_{b2}, && \text{if } \Delta \leq 0.75 \end{aligned} \tag{6.13}$$

The nonequilibrium part in Equation (6.14) is defined as:

$$g_{\alpha}^{neq}(\mathbf{x}_b, t) = \Delta g_{\alpha}^{neq}(x_f, t) + (1-\Delta) g_{\alpha}^{neq}(x_{ff}, t) \tag{6.14}$$

6.1.4 THE LATTICE BOLTZMANN MODEL FOR NANOFLUID

In order to simulate the nanofluid by the LBM, because of the interparticle potentials and other forces on the nanoparticles, the nanofluid behaves differently from the pure liquid from the mesoscopic point of view and is of higher efficiency in energy transport as well as better stabilization than the common solid-liquid mixture. For pure fluid in the absence of nanoparticles in the enclosures, the governed equations are Equations (6.1)–(6.14). However, for modeling the nanofluid because of the change in the fluid thermal conductivity, density, heat capacitance, and thermal expansion, some of the governed equations should change.

The fluid is a water-based nanofluid containing different types of nanoparticles: Cu (copper), Al_2O_3 (alumina), Ag (silver), and TiO_2 (titanium oxide). The nanofluid is a two-component mixture with the following assumptions:

(i) incompressible;
(ii) no-chemical reaction;
(iii) negligible viscous dissipation;
(iv) negligible radiative heat transfer;
(v) nanosolid particles and the base fluid are in thermal equilibrium and no slip occurs between them.

The thermophysical properties of the nanofluid are given in Soleimani et al. (2012).

The effective density ρ_{nf}, the effective heat capacity $(\rho C_p)_{nf}$, and thermal expansion $(\rho\beta)_{nf}$ of the nanofluid are defined as:

$$\rho_{nf} = \rho_f(1-\phi) + \rho_s\phi \tag{6.15}$$

$$(\rho C_p)_{nf} = (\rho C_p)_f(1-\phi) + (\rho C_p)_s\phi \tag{6.16}$$

$$(\rho\beta)_{nf} = (\rho\beta)_f(1-\phi) + (\rho\beta)_s\phi \tag{6.17}$$

where ϕ is the solid volume fraction of the nanoparticles and subscripts f, nf, and s stand for base fluid, nanofluid and solid, respectively.

The viscosity of the nanofluid containing a dilute suspension of small rigid spherical particles is (Brinkman model (see Barrios et al., 2005)):

$$\mu_{nf} = \frac{\mu_f}{(1-\varphi)^{2.5}} \tag{6.18}$$

The effective thermal conductivity of the nanofluid can be approximated by the MG model as (Abu-Nada et al., 2008):

$$\frac{k_{nf}}{k_f} = \frac{k_s + 2k_f - 2\phi(k_f - k_s)}{k_s + 2k_f + \phi(k_f - k_s)} \tag{6.19}$$

In order to compare total heat transfer rate, Nusselt number is used. The local and average Nusselt numbers are defined as follows:

$$Nu = \frac{k_{nf}}{k_f}\frac{\partial T}{\partial n} \quad \text{and} \quad \overline{Nu} = \frac{1}{L}\int_0^L Nu\,dS \tag{6.20}$$

6.1.5 GRID TESTING AND CODE VALIDATION

To verify the grid independence of the solution scheme, numerical experiments are performed as shown in Soleimani et al. (2012). Different mesh sizes were used for the case of $Ra = 10^6$, $\lambda = 3.5$, $\delta/L = 0.8$, and $\phi = 0.06$.

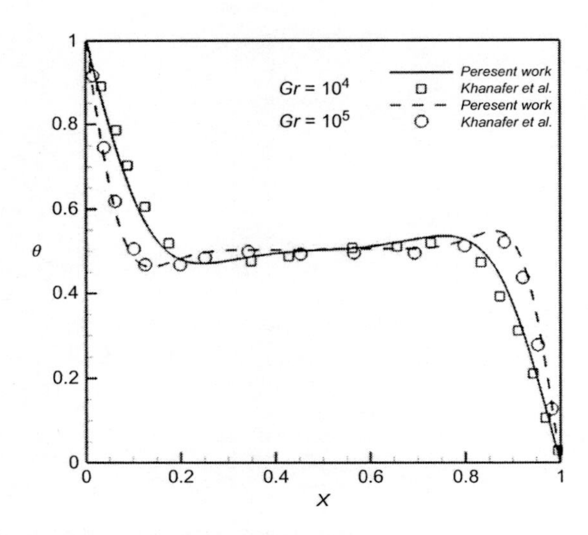

FIGURE 6.3

Comparison of the temperature on axial midline between the present results and numerical results by Khanafer et al. (2003). $\phi = 0.1$ and $Pr = 6.8$(Cu-water).

The present code is tested for grid independence by calculating the average Nusselt number on the inner wall. It is found that a grid size of 200 × 200 ensures the grid independent solution for the present case. The convergence criterion for the termination of all computations is:

$$\max_{\text{grid}}\left|\Gamma^{n+1} - \Gamma^{n}\right| \leq 10^{-7} \tag{6.21}$$

where n is the iteration number and Γ stands for the independent variables (U, V, T).

In this study, numerical solution is validated by comparing the present code results against the results of Moukalled and Acharya (1996) for viscous flow ($\phi = 0$).

Furthermore, another validation test was carried for natural convection in an enclosure filled with Cu-water for different Grashof numbers with the results of Khanafer et al. (2003) in Figure 6.3. All of the previous comparisons indicate the accuracy of the present LBM code.

6.1.6 RESULTS AND DISCUSSION

Natural convection in a concentric annulus between a cold outer square cylinder and a heated inner circular cylinder is investigated numerically using the LBM. Calculations are made for different types of nanoparticles Cu, Al_2O_3, Ag, and TiO_2; various values of volume fraction of nanoparticle ($\phi = 0, 0.02, 0.04$ and 0.06); Rayleigh number ($Ra = 10^3, 10^4, 10^5$ and 10^6) and position of the inner circular cylinder ($\delta/L = 0.2, 0.4, 0.6$ to 0.8) when Prandtl number is fixed ($Pr = 6.8$); and aspect ratios ($\lambda = 3.5$). The enhancement of heat transfer between the case of $\phi = 0.06$ and the pure fluid (base fluid) case is defined as:

$$En = \frac{\overline{Nu_h}(\phi = 0.06) - \overline{Nu_h}\,(\text{basefluid})}{\overline{Nu_h}\,(\text{basefluid})} \times 100 \tag{6.22}$$

The thermophysical characteristics of nanoparticles are key factors for heat transfer enhancement. Hence, at first a comparison among different types of nanoparticles is done to find which one of them gives the highest heat transfer enhancement for this problem. Heat transfer enhancement due to addition of nanoparticles for different types of nanofluids is shown in Figure 6.4.

The figure shows that the heat transfer enhancement has the highest value for copper nanoparticles in most of the computations. In this work, the numerical calculations were carried out for various values of nanoparticle volume fraction, Rayleigh number, and location of the inner circular cylinder for Cu-water nanofluid. Effects of Cu nanoparticles on the streamlines and isotherms are shown in Figure 6.5.

As seen, the flow intensity increases with increase of nanoparticles volume fraction which enhances the energy transport within the fluid. Hence, the absolute values of stream function which represent the strength of flow increase with increasing the volume fraction of nanofluid (Figure 6.5). The sensitivity of thermal boundary-layer thickness to volume fraction of nanoparticles is related to the increased thermal conductivity of the nanofluid. In fact, higher values of thermal conductivity are accompanied by higher values of thermal diffusivity. The high value of thermal diffusivity causes a drop in the temperature gradients and accordingly increases the thermal boundary-layer thickness. This increase in thermal boundary-layer thickness reduces the Nusselt number; however, according to Equation (6.20), the Nusselt number is a multiplication of temperature gradient and the thermal conductivity ratio (conductivity of the nanofluid to the conductivity of the base fluid). Since the reduction in temperature gradient

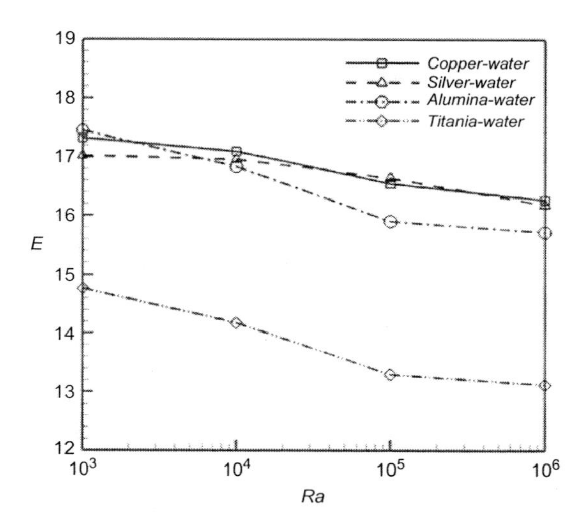

FIGURE 6.4

Ratio of enhancement of heat transfer due to addition of nanoparticles for different types of nanofluids when $\lambda = 3.5$, $\delta/L = 0$, and $Pr = 6.8$.

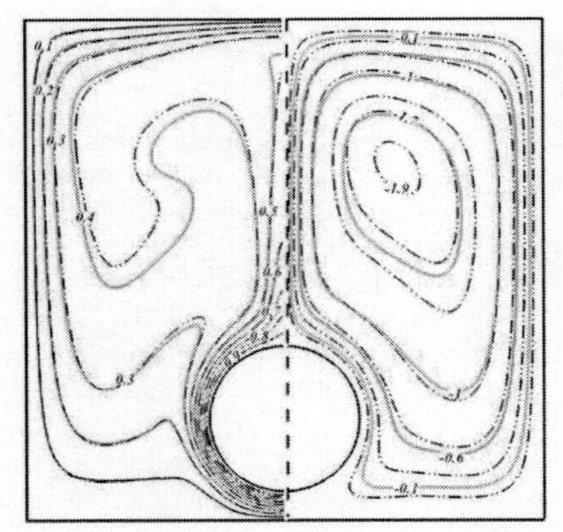

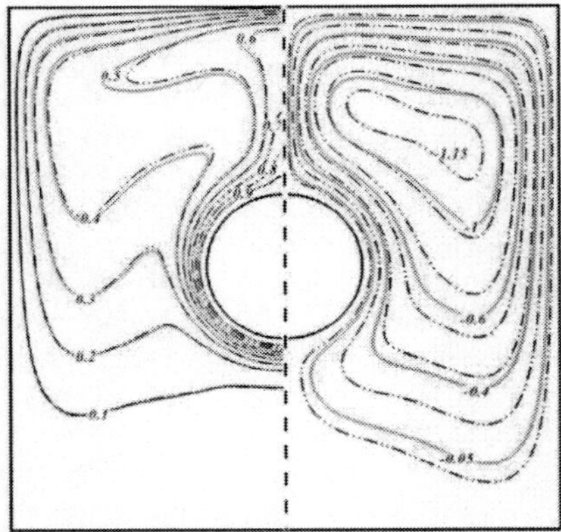

$|\psi_{max}|_f = 1.188, |\psi_{max}|_{nf} = 1.937$ $|\psi_{max}|_f = 0.698, |\psi_{max}|_{nf} = 1.200$

$\delta/L = 0.8$

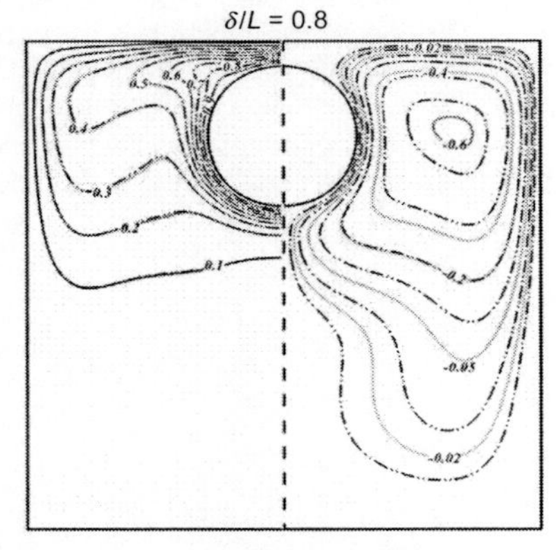

$|\psi_{max}|_f = 0.408, |\psi_{max}|_{nf} = 0.674$

FIGURE 6.5

Comparison of the isotherms (left) and streamline (right) contours between nanofluid ($\phi = 0.06$) (– –) and pure fluid ($\phi = 0$) (–···–) for different values of δ/L at $Ra = 10^6$ and $Pr = 6.8$.

due to the presence of nanoparticles is much smaller than the thermal conductivity ratio, an enhancement in Nusselt takes place by increasing the volume fraction of nanoparticles (Figures 6.7 and 6.8).

Figure 6.6 shows the effects of Ra and δ/L on isotherms (left) and streamlines (right) contours for Cu-water case ($\phi = 0.06$). At $Ra = 10^3$ and 10^4, increasing δ/L up to 0.6 leads to a decrease in the maximum value of stream function while as δ/L enhances furthermore, $|\psi_{max}|_{nf}$ increases.

At high Rayleigh number (e.g., $Ra = 10^5$ and 10^6), the values of $|\psi_{max}|_{nf}$ continuously decrease by increasing δ/L. At $Ra = 10^3$, because conduction is the dominant mode of heat transfer at this low

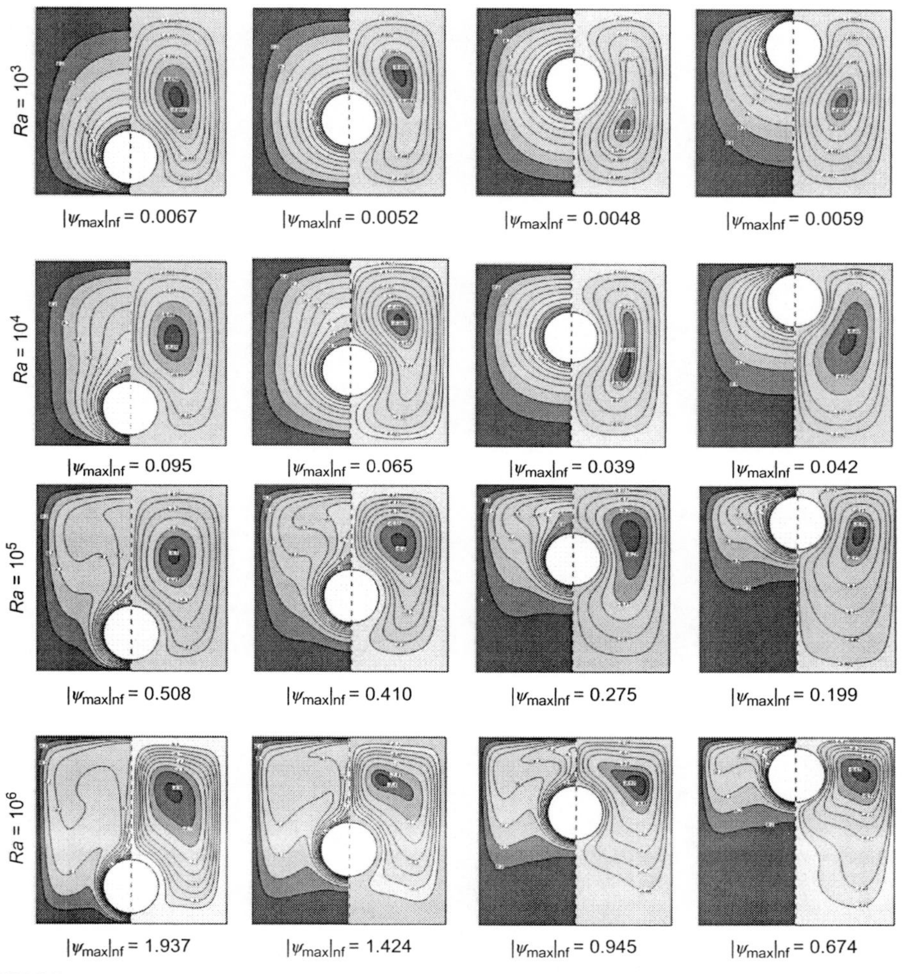

FIGURE 6.6

Effects of *Ra* and δ/L on isotherms (left) and streamline (right) contours for Cu-water case ($\phi = 0.06$).

Rayleigh number, the distribution of the flow and thermal fields for $\delta/L = 0.2$ and 0.4 shows the symmetric shapes about the horizontal center line, compared with that of corresponding $\delta/L = 0.8$ and 0.6, respectively. The circulation of flow shows one rotating eddy for the streamlines. When $\delta/L < 0.5$, the core of the main vortex is located in the upper half of the enclosure but when $\delta/L > 0.5$, the main vortex moves downward. Also, it can be seen that isotherms are parallel to each other which is the

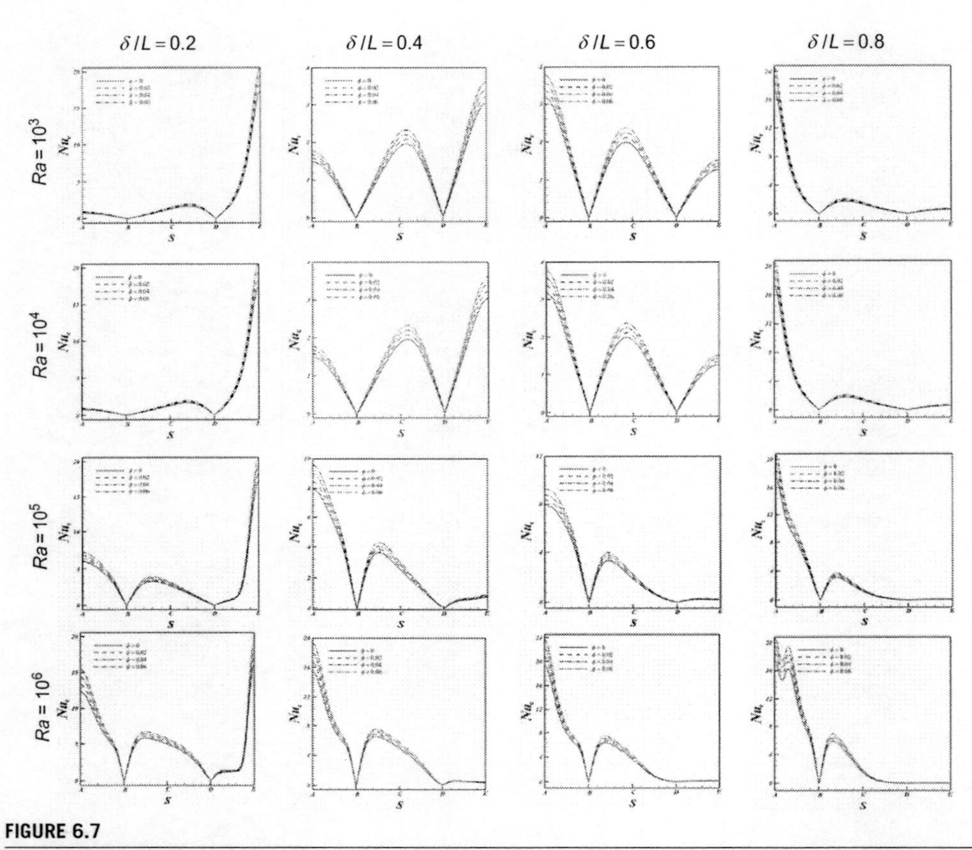

FIGURE 6.7

Effects of *Ra*, *δ/L*, and *ϕ* on local Nusselt number along the walls of the enclosure (Nu_C) for Cu-water case.

characteristics of conduction domination heat transfer mechanism at this Rayleigh number. As Rayleigh number increases up to 10^4, the effect of convection on heat transfer becomes more pronounced than that at $Ra = 10^3$. Thus, the distribution of flow and thermal fields for $\delta/L = 0.2$ and 0.4 shows the asymmetric shapes about the horizontal center line, compared with that of corresponding $\delta/L = 0.8$ and 0.6, respectively. Also, a thermal plume starts to appear on the top of the inner cylinder and as a result the isotherms move upward. The thermal plume diminishes as the inner cylinder moves upward. Streamline pattern for this Rayleigh number is similar to that of $Ra = 10^3$. As the Rayleigh number increases up to 10^4, the role of convection in heat transfer becomes more significant and consequently the thermal boundary-layer thickness on the surface of the inner cylinder becomes thinner. It is interesting to notice that at $Ra = 10^5$ the core of main vortex is located in upper half of the enclosure for all positions of inner cylinder. Also, this figure shows that at $\delta/L = 0.8$, the thermal

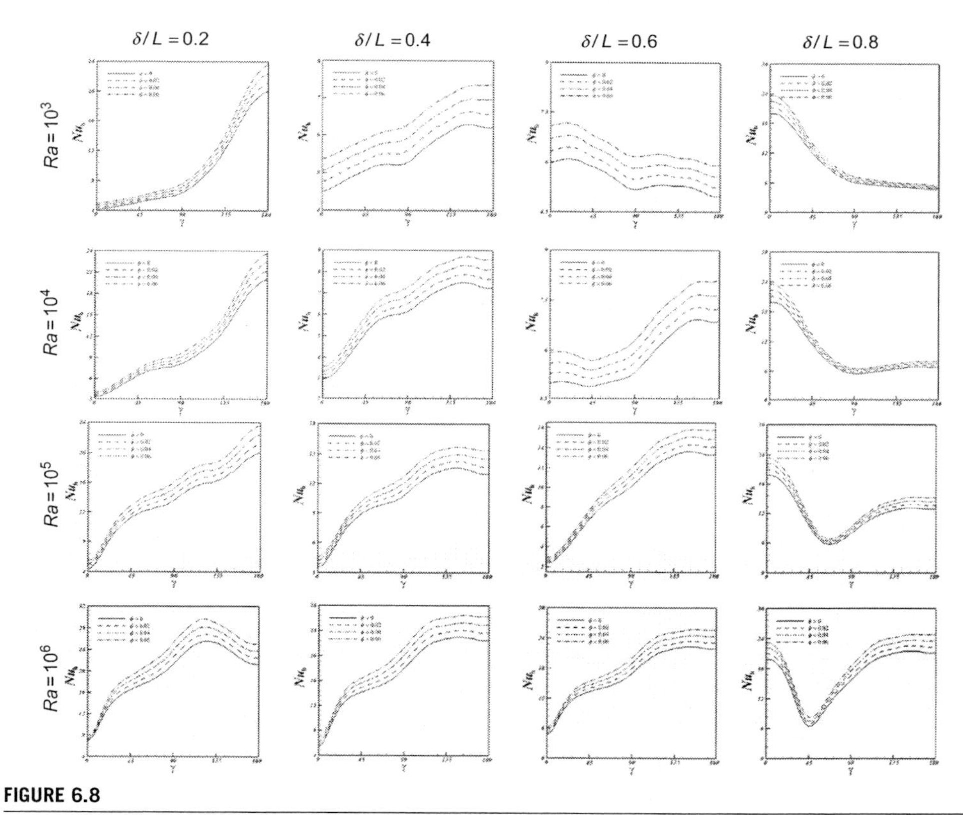

FIGURE 6.8

Effects of *Ra*, δ/L, and ϕ on local Nusselt number along the surface of the inner cylinder (Nu_h) for Cu-water case.

plume formed at the top of left cylinder. A similar flow pattern is observed at $Ra = 10^6$ but the value of $|\psi_{max}|_{nf}$ and thermal gradient near the hot inner cylinder are greater at this Rayleigh number. Effects of *Ra*, δ/L, and ϕ on local Nusselt number along the surface of the inner cylinder (Nu_h) for Cu-water case is shown in Figure 6.8.

At $Ra = 10^3$, when the center of inner cylinder located in the lower half of the enclosure minimum value of Nu_h is obtained at $\gamma = 0°$ and its maximum value occurs at $\gamma = 180°$, but opposite trend is observed for $\delta/L > 0.5$. The local Nusselt number over the hot surface shows similar tendency for $Ra = 10^4$ except at $\delta/L = 0.6$. At this position of inner cylinder, the maximum and minimum of Nu_h are obtained at $\gamma = 180°$ and 0°, respectively, which is in contrast with the case of $Ra = 10^3$ at this position of inner cylinder. At $Ra = 10^5$ and 10^6, the Nu_h profile shows two different behaviors observed at δ/L greater and smaller than 0.6. This means that for $\delta/L < 0.6$, local Nusselt number increases with augment of γ, whereas for $\delta/L > 0.6$ at first Nu_h decreases and then increases. The minimum value of Nu_h

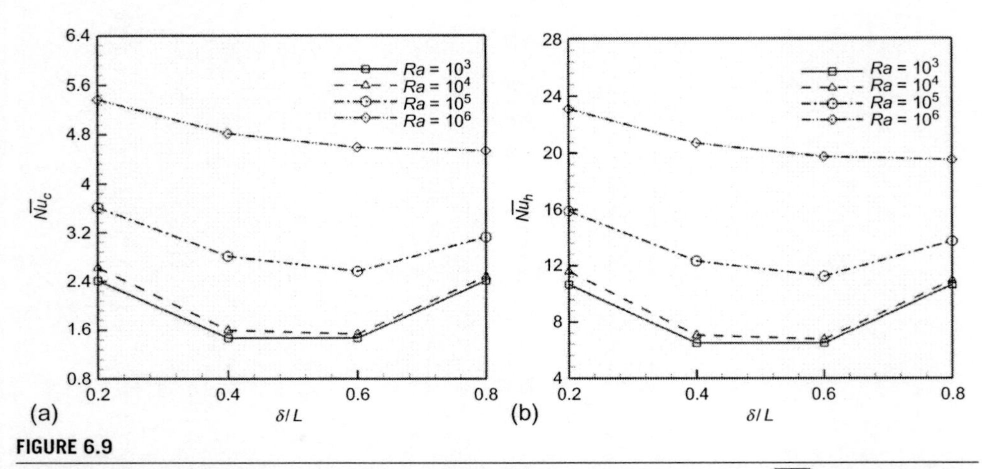

FIGURE 6.9

Effects of Ra and δ/L on average Nusselt number along (a) the walls of the enclosure $(\overline{Nu_c})$, (b) the surface of the inner cylinder $(\overline{Nu_h})$ for Cu-water case when $\phi = 0.06$.

for these cases is corresponding to the existence of thermal plume over the heated cylinder. Effects of Ra and δ/L on the average Nusselt number along the walls of the enclosure and the surface of the inner cylinder are depicted in Figure 6.9.

Variation of the average Nusselt number over the hot surface is similar to that over the cold surface. In addition, the value of Nusselt number over the inner cylinder is considerably greater than the cold one. Since the conduction heat transfer mechanism is dominant at $Ra = 10^3$, the profile of average Nusselt number is almost symmetric about $\delta/L = 0.5$. A similar trend is observed for $Ra = 10^4$. It can be found that the minimum value of average Nusselt number is obtained at $\delta/L = 0.6$ when $Ra = 10^5$. For $Ra = 10^6$, the average Nusselt number decreases continuously with increases of δ/L.

Figure 6.10 depicts the effects of Rayleigh number and position of inner cylinder on the enhancement of heat transfer. As seen, the effect of nanoparticles is more pronounced at low Rayleigh number than at high Rayleigh number because of greater amount of rate of enhancement, and increasing Rayleigh number leads to decrease in the ratio of enhancement of heat transfer. This observation can be explained by noting that at low Rayleigh number the heat transfer is dominant by conduction. Therefore, the addition of high thermal conductivity nanoparticles will increase the conduction and therefore make the enhancement more effective. Also, it can be seen that maximum value of En is obtained at $\delta/L = 0.4$ for $Ra = 10^3$ and 10^4, whereas the minimum value of heat transfer enhancement at $Ra = 10^5$ and 10^6 is obtained at $\delta/L = 0.6$ and 0.4, respectively.

6.1.7 CONCLUSIONS

In this study, natural convection heat transfer in a concentric annulus between a cold outer square cylinder and a heated inner circular cylinder filled with nanofluid is investigated numerically using LBM scheme.

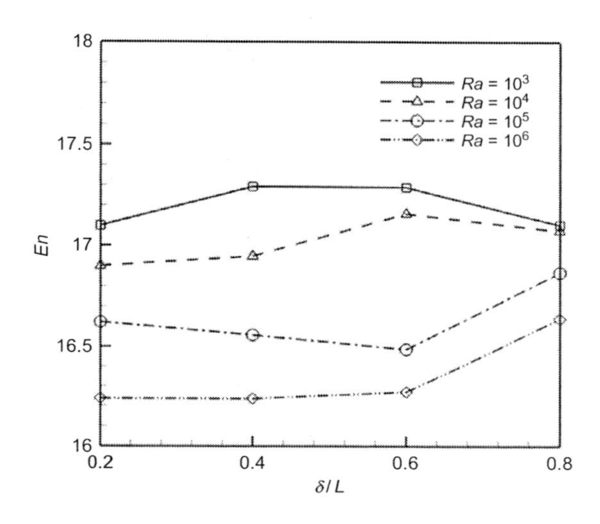

FIGURE 6.10

Effects of *Ra* and δ/L on ratio of enhancement of heat transfer due to addition of nanoparticles for Cu-water case.

Effects of nanoparticle volume fraction, types of nanofluids, Rayleigh numbers, and the position of inner cylinder on the flow and heat transfer characteristics have been examined. From this investigation, some conclusions were summarized as follows:

(a) The type of nanofluid is a key factor for heat transfer enhancement. The highest values of percentage of heat transfer enhancement are obtained when using copper nanoparticle.

(b) At $Ra = 10^3$ and 10^4, increasing δ/L up to 0.6 leads to increase in the maximum value of stream function where as δ/L enhances furthermore, $|\psi_{\max}|_{nf}$ increases. At high Rayleigh number (e.g., $Ra = 10^5$ and 10^6), the values of $|\psi_{\max}|_{nf}$ continuously decrease by increasing δ/L.

(c) The effect of nanoparticles is more pronounced at low Rayleigh number than at high Rayleigh number because of conduction domination at low Rayleigh number.

(d) Maximum value of enhancement is obtained at $\delta/L = 0.4$ for $Ra = 10^3$ and 10^4, whereas the minimum value of heat transfer enhancement at $Ra = 10^5$ and 10^6 is obtained at $\delta/L = 0.6$ and 0.4, respectively.

6.2 MIXED CONVECTION FLOW OF A NANOFLUID IN A HORIZONTAL CHANNEL

In this section, the laminar, fully developed mixed convection flow between two parallel horizontal flat plates filled by a nanofluid is investigated. Highly accurate solutions for the temperature and nanoparticle concentration distributions are obtained, which are solved along with the corresponding boundary conditions and the mass flux conservation relation by the homotopy perturbation method (HPM). The effects of the Brownian motion parameter (N_b), the thermophoresis parameter (N_t), and the Lewis number (L_e) on the temperature and nanoparticle concentration distributions are discussed. The current analysis shows that the nanoparticles can improve the heat transfer characteristics significantly for this flow problem (*this section has been worked by D.D. Ganji, M. Fakour, A. Bakhshi, A. Vahabzadeh in nonlinear dynamics team in Mechanical Engineering Department, 2012-2013*).

6.2.1 INTRODUCTION

Mixed convection flows or combined free and forced convection flows occur in many technological and industrial applications in nature, e.g., solar receivers exposed to wind currents, electronic devices cooled by fans, nuclear reactors cooled during emergency shutdown, heat exchangers placed in a low-velocity environment, flows in the ocean and in the atmosphere, and so on. The comprehensive reviews of convective flows are given in the monograph by Gebhart et al. (1988) and Martynenko and Khramtsov (2005). There are many numerical studies about heat and mass transfer of nanofluids in enclosures. In contrary, the number of studies on natural and mixed convection of nanofluids in vertical and horizontal channels is very small (see Cimpean and Pop, 2012). However, several researches on nanofluid flows in pipes and bends have been available in the literature, as shown by Lin et al. (2009) and Lin and Lin (2009). We mention to this end the very interesting paper by Lavine (1988) on steady fully developed opposing mixed convection between inclined parallel plates filled by a viscous and incompressible fluid (a regular fluid).

This chapter considers the steady fully developed mixed convection flow in a horizontal channel filled with nanofluids, which is driven by an external pressure gradient and also by a buoyancy force using the mathematical nanofluid model proposed by Buongiorno (2006). The work may be regarded as the extension of the problem considered by Chen and Chung (1996) on the mixed convection of a viscous (Newtonian) fluid in a vertical channel with linear variation of the wall temperature. Using similarity variables, the governing partial differential equations are transformed to ordinary differential equations, which are solved along with the corresponding boundary conditions and the mass flux conservation relation by the HPM (He, 2006).

6.2.2 DESCRIBE PROBLEM AND MATHEMATICAL FORMULATION

Consider the laminar steady mixed convective flow between two infinite horizontal parallel plane walls filled with the incompressible viscous nanofluid and separated by a spacing $2L$. The Cartesian coordinate system (x, y) is chosen with the origin at the center of the channel (see Figure 6.11). The x-axis is aligned parallel to the channel walls, and the y-axis is normal to them. Following Chen and Chung (1996), we assume that the temperature and the nanoparticle volume fraction of both the walls increase or decrease linearly with x, namely, $T_w(x) = T_0 + A_1 x$ and $C_w(x) = C_0 + A_2 x$, where T_0 and C_0 are, respectively, the reference temperature and the reference nanoparticle volume fraction at the channel entrance, A_1 is a constant, which is positive for the hot walls and negative for the cooled walls, and A_2 is a positive constant.

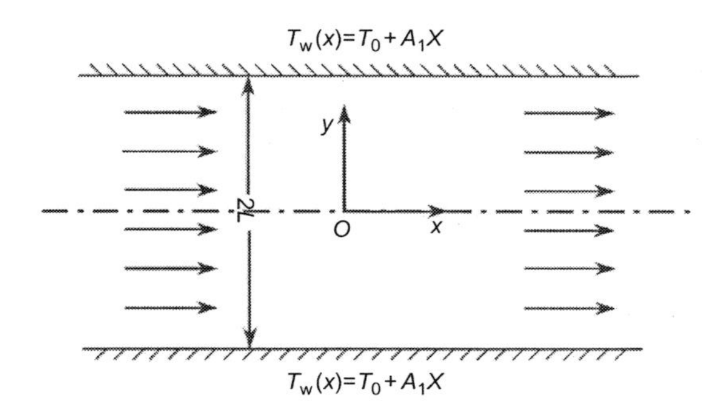

FIGURE 6.11

Schematic of the geometry and coordinate system.

Assuming that the nanofluid is dilute and then using the mathematical model suggested by Buongiorno (2006), the following four field equations including the conservation of the total mass, the momentum, the thermal energy, and the nanoparticle volume fraction can be simplified as

$$\frac{\partial u}{\partial x}+\frac{\partial v}{\partial y}=0, \tag{6.23}$$

$$u\frac{\partial u}{\partial x}+v\frac{\partial u}{\partial y}=-\frac{1}{\rho_f}\frac{\partial p}{\partial x}+\upsilon\left(\frac{\partial^2 u}{\partial x^2}+\frac{\partial^2 u}{\partial y^2}\right), \tag{6.24}$$

$$u\frac{\partial T}{\partial x}+v\frac{\partial T}{\partial y}=\alpha\left(\frac{\partial^2 T}{\partial x^2}+\frac{\partial^2 T}{\partial y^2}\right)+\tau\left(D_b\left(\frac{\partial C}{\partial x}\frac{\partial T}{\partial x}+\frac{\partial C}{\partial y}\frac{\partial T}{\partial y}\right)+\frac{D_t}{T_0}\left(\left(\frac{\partial T}{\partial x}\right)^2+\left(\frac{\partial T}{\partial y}\right)^2\right)\right), \tag{6.25}$$

$$u\frac{\partial C}{\partial x}+v\frac{\partial C}{\partial y}=D_b\left(\frac{\partial^2 C}{\partial x^2}+\frac{\partial^2 C}{\partial y^2}\right)+\frac{D_t}{T_0}\left(\frac{\partial^2 T}{\partial x^2}+\frac{\partial^2 T}{\partial y^2}\right), \tag{6.26}$$

subject to the boundary conditions

$$u(\pm L)=0,\;\; T_w(\pm L)=T_0+A_1x,\;\; C_w(\pm L)=C_0+A_2x. \tag{6.27}$$

Here, u and v are the velocity components along the x-axis and the y-axis, respectively. T is the temperature of the nanofluid. C is the nanoparticle volume fraction. p is the pressure. ρ_f is the density of the base fluid. ν is the kinematic viscosity. α is the thermal diffusivity. $\tau=(\rho C)_p/(\rho C)_f$, where $(\rho C)_p$ is the heat capacity of the nanoparticle material, and $(\rho C)_f$ is the heat capacity of the fluid. D_b is the Brownian diffusion coefficient. D_t is the thermophoretic diffusion coefficient.

It is a common practice in channel flow studies to assume the mass flow rate as a prescribed quantity. We thus obtain the following equation:

$$U_m=\frac{1}{2L}\int_{-L}^{+L}u(y)\,dy=\frac{1}{L}\int_0^{+L}u(y)\,dy, \tag{6.28}$$

where U_m is the average fluid velocity in the channel section.

With the assumption that the parallel flow is fully developed (far from the channel entrance), it is readily understood $v=0$. Thus, the continuity equation becomes $\partial u/\partial x=0$, which indicates $u=u(y)$. Therefore, Equation (6.24) is reduced to

$$\frac{1}{\rho_{\mathrm{f}}}\frac{\mathrm{d}p}{\mathrm{d}x}=\upsilon\frac{\mathrm{d}^2u}{\mathrm{d}y^2}. \tag{6.29}$$

We introduce now the following dimensionless variables

$$\begin{gathered} X=\frac{x}{L},\ Y=\frac{y}{L},\ U(Y)=\frac{RePr}{U_{\mathrm{m}}}u,\ \theta(Y)=\frac{T-T_{\mathrm{w}}}{A_1L}, \\ \varphi(Y)=\frac{C-C_{\mathrm{w}}}{A_2L},\ P(X)=\frac{p}{\rho_{\mathrm{f}}U_{\mathrm{m}}^2}. \end{gathered} \tag{6.30}$$

where $Re=U_{\mathrm{m}}L/\nu$ is the Reynolds number and $Pr=\nu/\alpha$ is the Prandtl number. Substituting similarity variables (6.30) into Equations (6.29), (6.25), (6.26), and (6.28), we obtain

$$U''(Y)-\sigma=0, \tag{6.31}$$

$$\theta''(Y)-N_{\mathrm{b}}(1+\theta'(Y)\phi'(Y))-N_{\mathrm{t}}\left(1+\theta'(Y)^2\right)-U(Y)=0, \tag{6.32}$$

$$\phi''(Y)-\frac{N_{\mathrm{t}}}{N_{\mathrm{b}}}\theta''(Y)-L_{\mathrm{e}}U(Y)=0 \tag{6.33}$$

subject to the boundary conditions

$$U(1)=U(-1)=0,\ \ \theta(1)=\theta(-1)=0,\ \ \varphi(1)=\varphi(-1)=0, \tag{6.34}$$

along with the integral constraint

$$\int_0^1 U\,\mathrm{d}y=RePr, \tag{6.35}$$

Here σ, N_{b}, N_{t}, and L_{e} denote the pressure parameter, the Brownian motion parameter, the thermophoresis parameter, and the Lewis number, respectively.

$$\sigma=\frac{U_{\mathrm{m}}^2L^4}{\upsilon\alpha}\frac{\mathrm{d}P}{\mathrm{d}X},\ \ N_{\mathrm{b}}=\frac{(\rho c)_{\mathrm{p}}}{(\rho c)_{\mathrm{f}}}D_{\mathrm{b}}\frac{A_2L}{\alpha},\ \ N_{\mathrm{t}}=\frac{(\rho c)_{\mathrm{p}}D_{\mathrm{t}}A_1L}{(\rho c)_{\mathrm{f}}T_0\ \alpha},\ \ L_{\mathrm{e}}=\frac{\alpha}{D_b}.$$

It is worth mentioning that when $N_{\mathrm{b}}=N_{\mathrm{t}}=0$, Equations (6.31)–(6.33) correspond to the channel flow in Newtonian fluids.

From Equation (6.31) and its boundary conditions given in Equation (6.34), we are able to obtain the following analytical solution:

$$U(Y)=\frac{3}{2}PrRe\left(1-Y^2\right). \tag{6.36}$$

6.2.3 HOMOTOPY PERTURBATION METHOD APPLIED TO THE PROBLEM

6.2.3.1 The HPM applied to the problem

A HPM can be constructed as follows:

$$H(\theta,p)=(1-p)(\theta'')+p\left(\theta''-N_b(1+\theta'\varphi')+N_t\left(1+\theta'^2\right)-U\right), \tag{6.37}$$

$$H(\varphi,p)=(1-p)(\varphi'')+p\left(\varphi''+\frac{N_t}{N_b}\theta''-L_eU\right), \tag{6.38}$$

One can now try to obtain a solution for Equations (6.37) and (6.38) in the form of:

$$\theta(Y)=\theta_0(Y)+p\theta_1(Y)+p^2\theta_2(Y)+\cdots \tag{6.39}$$

$$\varphi(Y)=\varphi_0(Y)+p\varphi_1(Y)+p^2\varphi_2(Y)+\cdots \tag{6.40}$$

where $v_i(Y), i=1, 2, 3,...$ are functions yet to be determined. According to Equations (6.37) and (6.38), the initial approximation to satisfy boundary condition is:

$$\theta_0(Y)=0, \tag{6.41}$$

$$\varphi_0(Y)=0, \tag{6.42}$$

Substituting Equations (6.39) and (6.40) into Equations (6.37) and (6.38) yields:

$$\frac{d^2\theta_1}{dy^2}+15y^2-15.2=0, \tag{6.43}$$

$$\frac{d^2\varphi_1}{dy^2}+150y^2-150=0, \tag{6.44}$$

The solutions of Equations (6.43) and (6.44) may be written as follow:

$$\theta_1(Y)=-1.25y^4+7.6y^2-6.35 \tag{6.45}$$

$$\varphi_1(Y)=-12.5y^4+75y^2-62.5 \tag{6.46}$$

where

$$\frac{d^2\theta_2}{dY^2}-0.1\left(-5Y^3+15.2Y\right)^2-0.1\left(-5Y^3+15.2Y\right)\left(-50Y^3+150Y\right)=0 \tag{6.47}$$

$$\frac{d^2\varphi_2}{dY^2}+15Y^2-15.2=0 \tag{6.48}$$

The solutions for Equations (6.47) and (6.48) may be written as follows:

$$\theta_2(Y)=0.49Y^8-5.54Y^6+20.92Y^4-15.87 \tag{6.49}$$

$$\varphi_2(Y)=-1.25Y^4+7.6Y^2-6.35 \tag{6.50}$$

In the same manner, the rest of the components were obtained by using the Maple package. According to the HPM, we can conclude:

$$\theta(Y) = 0.53Y^8 - 6.047Y^6 + 21.6Y^4 + 7.6Y^2 - 23.69 \tag{6.51}$$

$$\varphi(Y) = 0.49Y^8 - 5.54Y^6 + 7.175Y^4 + 82.6Y^2 - 84.72 \tag{6.52}$$

and so on. In the same manner, the rest of the components of the iteration formula can be obtained.

6.2.4 RESULTS AND DISCUSSION

To get better understanding of flow and heat transfer characteristics in the nanofluid, we thereafter make a detailed analysis on the problem of the fully developed nanofluid flow and heat transfer in a horizontal channel. Comparison between the results of HPM and exact solution is shown in Figure 6.12. It can be seen that the HPM method is very close to exact solution. Also, Table 6.1 exhibits the numerical magnitude of U, θ, and φ.

Since the analytical solution to $U(Y)$ given in Equation (6.36) possesses the same solution form and exhibits the similar behaviors as compared with the one obtained in Newtonian fluids, we neglect the discussion on the velocity field and analyze the temperature distribution and the nanoparticle concentration distribution. We first investigate the influence of the physical parameters N_b, N_t, and L_e on the temperature distribution and the nanoparticle concentration distribution, respectively. In this case,

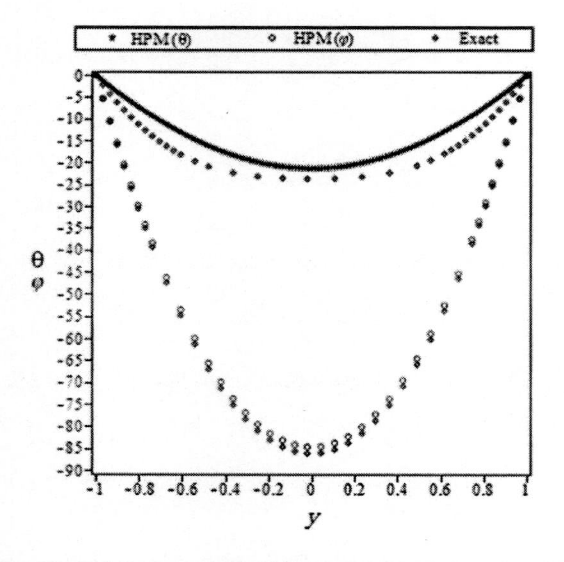

FIGURE 6.12

Comparison of θ and φ versus the exact solution with the result obtained by HPM at $N_t = N_b = 0.1$, $PrRe = 10$, $L_e = 10$.

Table 6.1 The Result of HPM and Exact Solution for θ and φ

y	HPM(θ)	Exact(θ)	HPM(φ)	Exact(φ)
−1.0	−0.000000001	0.0000000000	−0.0000000001	0.000000000
−0.8	−11.47330066	−12.00330066	−30.293277880	−31.09330066
−0.6	−18.42738322	−19.02738322	−54.310707690	−55.54738322
−0.4	−21.94515327	−22.54515327	−71.349086240	−72.76515327
−0.2	−23.35153883	−24.05153883	−81.411277530	−82.87153883
0.0	−23.68971428	−24.38971428	−84.726404760	−86.18971428
0.2	−23.35153883	−24.05153883	−81.411277530	−82.87153883
0.4	−21.94515327	−22.54515327	−71.349086240	−72.76515327
0.6	−18.42738322	−19.02738322	−54.310707690	−55.54738322
0.8	−11.47330066	−12.00330066	−30.293277880	−31.09330066
1.0	−0.000000001	0.0000000000	−0.0000000001	0.0000000000

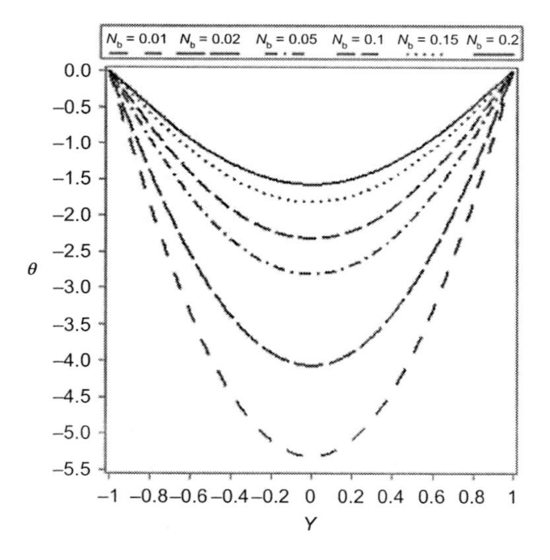

FIGURE 6.13

The dimensionless temperature profiles for various values of N_b with $N_t = 0.1$, $Pr \cdot Re = 10$, $L_e = 10$.

the flow problem is simplified to the channel flow in Newtonian fluids. As shown in Figure 6.13, the absolute amplitude of the temperature profile $\theta(Y)$ decreases monotonously with the increase in N_b, and the maximum value of the absolute amplitude of $\theta(Y)$ can reduce about five times as N_b evolves from 0.01 to 0.20. From Figure 6.14, it is found that N_b has a significant effect on the nanoparticle concentration distribution $\varphi(Y)$ as well. The increase in N_b leads to the enlargement of the absolute amplitude of the nanoparticle concentration profiles, while its effect becomes smaller and smaller.

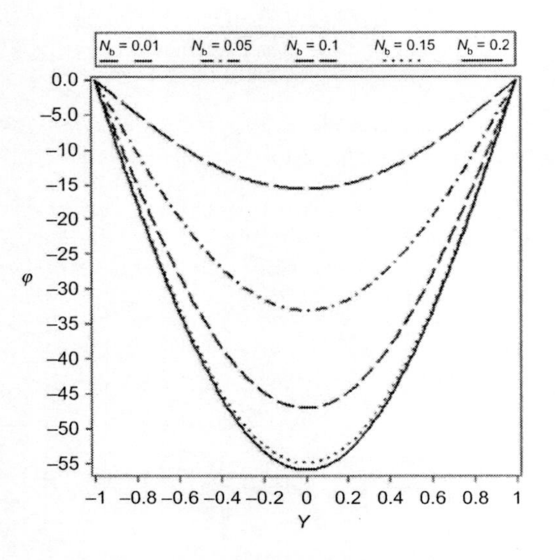

FIGURE 6.14

The dimensionless concentration distribution for various values of N_b with $N_t=0.1$, $Pr{\cdot}Re=10$, $L_e=10$.

The effects of N_t on the temperature distribution $\theta(Y)$ and the nanoparticle concentration distribution $\varphi(Y)$ are shown in Figures 6.15 and 6.16, respectively. It is found from Figure 6.15 that the absolute amplitude of the temperature distribution $\theta(Y)$ decreases gradually as N_t increases, while the changing difference is quite limited. In the same case, we notice that N_t plays a key role in the variation of the nanoparticle concentration distribution $\varphi(Y)$, as shown in Figure 6.16. From this figure, we notice that the absolute amplitude of $\varphi(Y)$ decreases continuously as N_t increases from 0.01 to about 0.11, while as N_t evolves consecutively from 0.11 to 0.20, $\varphi(Y)$ changes its sign and enlarges gradually.

The variations of the temperature distribution $\theta(Y)$ and the nanoparticle concentration distribution $\varphi(Y)$ as functions of L_e with $N_b=0.1$, $N_t=0.1$, and $PrRe=10$ are illustrated in Figures 6.17 and 6.18, respectively. It is shown from the figures that L_e has a huge effect on them. As shown in Figure 6.17, the absolute amplitude of $\theta(Y)$ diminishes successively as L_e enlarges. Different from the changing trend of $\theta(Y)$, the absolute value of the nanoparticle concentration $\varphi(Y)$ increases monotonically as L_e increases, as shown in Figure 6.18.

6.2.5 CONCLUSION

In this section, the heat transfer characteristics of the fully developed nanofluid flow in a horizontal channel are examined. Highly accurate solutions for the temperature and nanoparticle concentration distributions are obtained, which are solved along with the corresponding boundary conditions and the mass flux conservation relation by the HPM. The computational results for the temperature

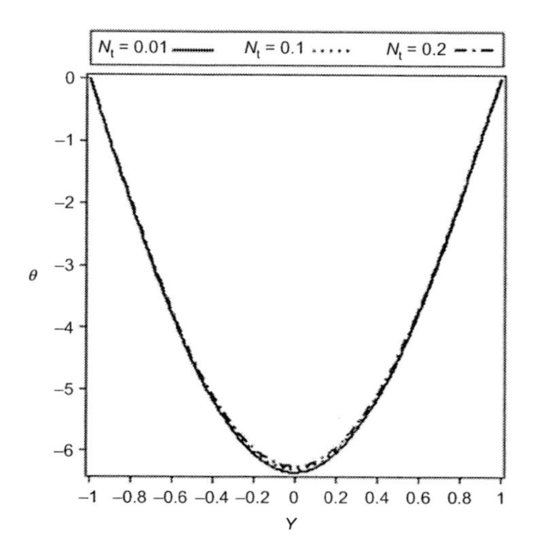

FIGURE 6.15

The dimensionless temperature profiles for various values of N_t with $N_b = 0.01$, $Pr \cdot Re = 10$, $L_e = 10$.

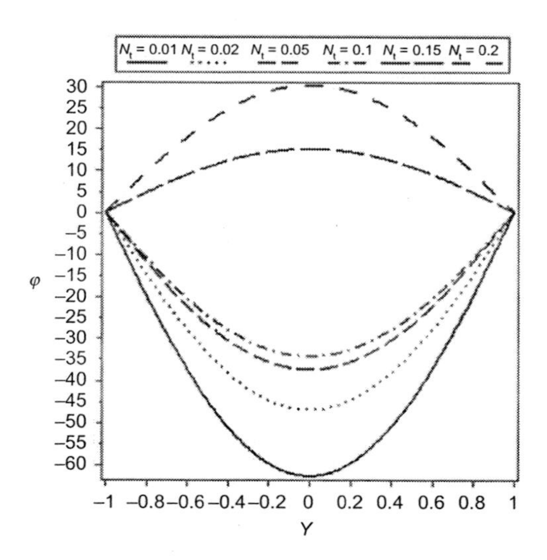

FIGURE 6.16

The dimensionless concentration distribution for various values of N_t with $N_t = 0.01$, $Pr \cdot Re = 10$, $L_e = 10$.

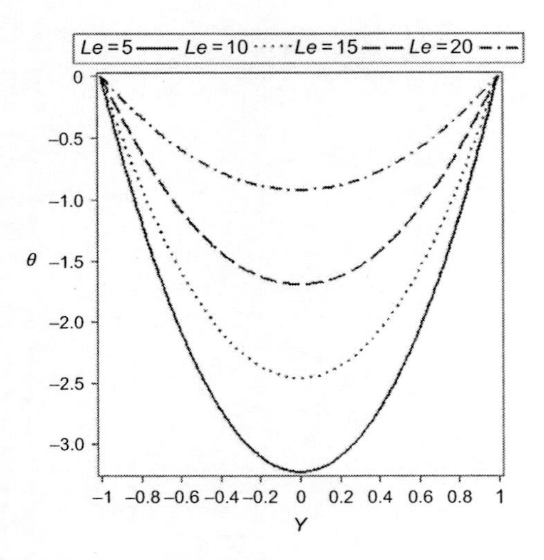

FIGURE 6.17

The dimensionless temperature profiles for various values of L_e with $N_b = N_t = 0.1$, $Pr{\cdot}Re = 10$.

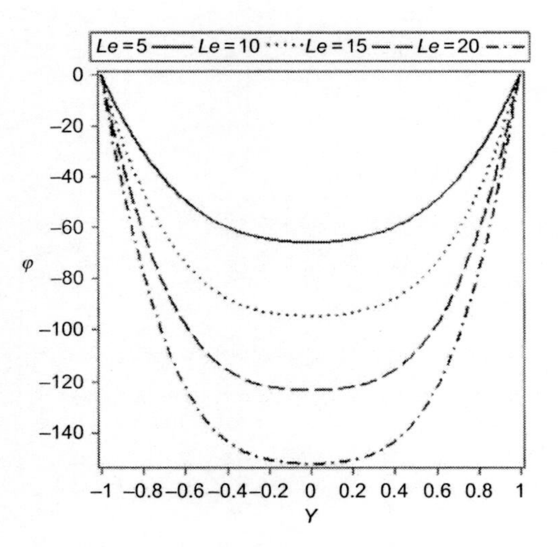

FIGURE 6.18

The dimensionless concentration distribution for various values of L_e with $N_b = N_t = 0.1$, $Pr{\cdot}Re = 10$.

$\theta(Y)$ and the nanoparticle concentration $\varphi(Y)$ show that the physical parameters N_b and N_t play important roles on the distributions of these profiles. The increase in N_b leads to the decrease in the absolute amplitude of the temperature. However, it can result in the enlargement in the absolute amplitude of the nanoparticle concentration. Moreover, for a given value of N_b, the temperature profiles decrease gradually as N_t increases. The absolute amplitude of the nanoparticle concentration decreases monotonously as N_t increases from 0.01 to 0.11 and then changes its sign and increases continuously as N_t evolves from 0.11 to 0.20. L_e has a significant effect on both the temperature and the nanoparticle concentration distributions. The temperature profiles diminish as L_e increases, while the nanoparticle concentration profiles enlarge as L_e increases. This analysis shows that the nanoparticles can effectively improve the heat transfer characteristics of the flow in the horizontal channel.

6.3 THE EFFECT OF NANOFLUID ON THE FORCED CONVECTION HEAT TRANSFER

This section deals with the problem of forced convection heat transfer of a nanofluid over a vertical flat plate in a porous medium. Various types of nanoparticles comprising Cuo, Al_2O_3, and TiO_2, mixing in ethylene glycol-water, have been examined. The governing partial differential equations subjected to boundary conditions are transformed to ordinary differential equations by local similarity solutions and scaling transformations, and then solved through efficient analytical technique "homotopy analysis method (HAM)." To clarify the problem, profiles of the dimensionless velocity and temperature together with the Nusselt number and skin-friction coefficient are shown graphically for various values of significant parameters. Comparisons between results of various methods, namely, Burgmann, Patel, Xuan, and Maxwell for describing nanoparticle effects are also illustrated in detail (*this section has been worked by S. Tavakoli, D.D. Ganji, A. Vosoughi, S. Naeejee, A. Rasekh, H. Jahani in nonlinear dynamics team in Mechanical Engineering Department, 2012-2013*).

6.3.1 INTRODUCTION

Convective flows in porous media have been extensively studied during the last several decades and they have included several different physical effects. This interest is due to the many practical applications which can be modeled or approximated as transport phenomena in porous media. These flows appear in a wide variety of industrial applications, as well as in many natural circumstances such as geothermal extraction, storage of nuclear waste material, ground water flows, industrial and agricultural water distribution, oil recovery processes, thermal insulation engineering, pollutant dispersion in aquifers, cooling of electronic components, packed-bed reactors, food processing, casting and welding of manufacturing processes, the dispersion of chemical contaminants in various processes in the chemical industry and in the environment, soil pollution, fibrous insulation, and even for obtaining approximate solutions for flow through turbo machinery. An excellent review of existing theoretical and experimental work on this subject can be found in the recent books and monographs by Ingham and Pop (1998, 2002, 2005), Vafai (2000), and Bejan et al. (2004).

This section aims to study the forced convection flow past a vertical flat plate embedded in a porous medium filled with different types of nanoparticles containing CuO, Al_2O_3, and TiO_2 as long as the basic fluid is ethylene glycol-water.

In this work, an analytical solution has been employed for solving the governing equations. The flow that has been reduced to a system and temperature of coupled nonlinear ordinary differential equations. The scope of the current research is to implement more appropriate models for nanofluid properties and to study the effect of these models on heat transfer. The MG model for the effective thermal conductivity of the nanofluid has been taken into account. The governing nonlinear differential equations are solved analytically using the HAM (Liao, 1992, 2003) to give such an explicit analytic solution.

Effects of dispersion of distinct types of metallic nanoparticles and the nanoparticle volume fraction on steady flow and heat characteristics are presented and discussed. The dependency of velocity and temperature on the dimensionless parameter, ζ, and the Nusselt number on the nanoparticle volume fraction between zero to 0.6% vol have been studied.

6.3.2 GOVERNING EQUATIONS

The problem under investigation is forced convection flow past a vertical flat plate embedded in a porous medium while accompanied by nanofluid effects. The schematic diagram is shown in Figure 6.19.

The surface of the plate is maintained at a uniform constant temperature T_w, which is adequately different from ambient temperature T_∞. It is also assumed that the free stream velocity U_∞, parallel to the vertical plate, is constant. The fluid is an ethylene glycol-water-based nanofluid containing different nanoparticles such as Al_2O_3, CuO, and TiO_2 which they are incompressible. The thermophysical properties of the nanofluid are given in Table 6.2.

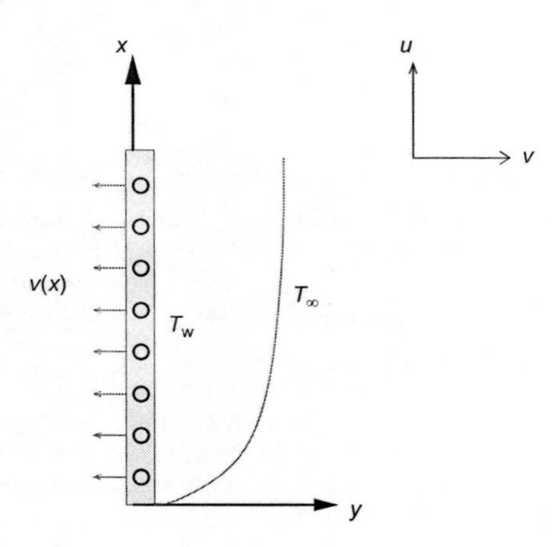

FIGURE 6.19

Physical configuration.

Table 6.2 Thermophysical Properties of Water and Nanoparticles

Physical Properties	E-gl Water	CuO	Al_2O_3	TiO_2
ρ (kg/m^3)	3970	6510	1041	4250
C_p (J/kg K)	3526	540	765	686.2
K (W/m K)	0.4621	18	40	8.9538
β (1/K)	3.2987×10^{-4}	8.5×10^{-6}	8.5×10^{-6}	0.9×10^{-5}

Thermophysical properties of the nanofluid are assumed to be constant. The governing equations for the flow and heat transfer under the boundary-layer approximations can be written in two-dimensional Cartesian coordinates (x, y) as:

Continuity equation:

$$\frac{\partial u}{\partial x} + \frac{\partial v}{\partial y} = 0 \tag{6.53}$$

Momentum equation:

$$u\frac{\partial u}{\partial x} + v\frac{\partial v}{\partial y} = -\frac{\partial^2 u}{\partial y^2}v + g\beta(T - T_w) - \frac{\nu u}{K} \tag{6.54}$$

Energy equation:

$$(\rho C_p)_{nf}\left(u\frac{\partial T}{\partial x} + v\frac{\partial T}{\partial y}\right) = k_{nf}\left(\frac{\partial^2 T}{\partial y^2}\right) \tag{6.55}$$

And the boundary conditions are in the following forms:

$$y \to 0 : u = 0, \ T \to T_w, \ v = \pm v \tag{6.56}$$

$$y \to \infty : u = u_\infty, \ T \to T_\infty \tag{6.57}$$

The density of the nanofluid is given by:

$$\rho_{nf} = \rho_f(1 - \phi) + \rho_s\phi \tag{6.58}$$

Whereas the heat capacitance of the nanofluid and thermal expansion coefficient of the nanofluid can be determined by:

$$(\rho C_p)_{nf} = (\rho C_p)_f(1 - \phi) + (\rho C_p)_s\phi \tag{6.59}$$

$$(\rho\beta)_{nf} = (\rho\beta)_f(1 - \phi) + (\rho\beta)_s\phi \tag{6.60}$$

where φ being the volume fraction of the solid particles and subscripts f, nf, and s stand for base fluid, nanofluid, and solid, respectively. The effective dynamic viscosity of the nanofluid given by Brinkman (1952) is:

$$\mu_{nf} = \frac{\mu_f}{(1 - \phi)^{2.5}} \tag{6.61}$$

The thermal conductivity of the stagnant nanofluid and the effective thermal conductivity of the nanofluid for spherical nanoparticles, according to Maxwell (1904), are:

$$\frac{k_{\mathrm{nf}}}{k_{\mathrm{f}}}=\frac{k_{\mathrm{s}}+2k_{\mathrm{f}}-2\varphi(k_{\mathrm{f}}-k_{\mathrm{s}})}{k_{\mathrm{s}}+2k_{\mathrm{f}}+\varphi(k_{\mathrm{f}}-k_{\mathrm{s}})} \tag{6.62}$$

The following dimensionless groups are introduced:

$$\eta=y\sqrt{\frac{u\infty}{\nu\cdot x}},\quad \psi=\sqrt{(\nu\cdot x\cdot u\infty)}f(\eta),\quad \theta(\eta)=\frac{T-T_{\infty}}{T_{\mathrm{w}}-T_{\infty}} \tag{6.63}$$

where ψ is the stream function that satisfies the continuity equation (6.53) and is defined in the usual manner such that $u=\frac{\partial\psi}{\partial y}$ and $v=-\frac{\partial\psi}{\partial x}$ and f, f', and θ are the dimensionless stream function, vertical velocity, and temperature field, respectively.

After transformation, we have:

$$\frac{1}{(1-\varphi)^{2.5}}\frac{1}{(1-\varphi)+\frac{\rho_{\mathrm{s}}}{\rho_{\mathrm{f}}}\varphi}f'''+\frac{1}{2}ff''+\frac{Gr}{Re^2}\theta-\frac{1}{(1-\varphi)^{2.5}}\frac{1}{(1-\varphi)+\frac{\rho_{\mathrm{s}}}{\rho_{\mathrm{f}}}\varphi}\frac{1}{ReDa}f'=0 \tag{6.64}$$

$$\theta''+\frac{1}{2}Pr\frac{(1-\varphi)+\frac{(\rho C_p)_{\mathrm{s}}}{(\rho C_p)_{\mathrm{f}}}\varphi}{\frac{k_{\mathrm{s}}+2k_{\mathrm{f}}-2\varphi(k_{\mathrm{f}}-k_{\mathrm{s}})}{k_{\mathrm{s}}+2k_{\mathrm{f}}+2\varphi(k_{\mathrm{f}}-k_{\mathrm{s}})}}f\theta'=0 \tag{6.65}$$

The boundary conditions are now transformed to:

$$\eta\to 0:\ f=f_w,\ f'=0,\ \theta=1 \tag{6.66}$$

$$\eta\to\infty:\ f'=1,\ \theta=0 \tag{6.67}$$

where

$$Gr=\frac{g\beta_T(T_{\mathrm{w}}-T_{\infty})x^3}{\nu^2} \tag{6.68}$$

$$Re=\frac{u_{\infty}x}{\upsilon} \tag{6.69}$$

$$Pr=\frac{\nu}{\alpha} \tag{6.70}$$

$$f_{\mathrm{w}}=\frac{2xv(x)}{\nu}Re^{-1/2} \tag{6.71}$$

are the Grashof number, Reynolds number, Prandtl number, and the dimensionless suction velocity, respectively. The nondimensional heat transfer coefficient can be written as

$$\frac{Nu}{Ra_x^{1/2}}=-\frac{k_{\mathrm{nf}}}{k_{\mathrm{f}}}\theta'(0) \tag{6.72}$$

It should be mentioned that the use of the approximation for k_{nf} is restricted to spherical nanoparticles and does not account for other shapes of nanoparticles. We also consider the three different models for comparison with the Maxwell model in order to predict effective thermal conductivity of the nanofluid. These models, namely, Patel model (2005), Xuan model (2003), and Bruggeman model (1935) are indicated respectively as follows.

$$\frac{k_{\mathrm{nf}}}{k_{\mathrm{f}}}=1+(1+C.pe)\left(\frac{k_{\mathrm{f}}}{k_{\mathrm{p}}}\right)\left(\frac{d_{\mathrm{f}}}{d_{\mathrm{s}}}\right)\left(\frac{\varphi}{1-\varphi}\right) \tag{6.73}$$

$$\frac{K_{\mathrm{nf}}}{K_{\mathrm{f}}}=\frac{k_{\mathrm{s}}+2k_{\mathrm{f}}-2\varphi(k_{\mathrm{f}}-k_{\mathrm{s}})(1+B^3)}{k_{\mathrm{s}}+2k_{\mathrm{f}}+\varphi(k_{\mathrm{f}}-k_{\mathrm{s}})(1+B^3)} \tag{6.74}$$

$$\frac{K_{\mathrm{nf}}}{K_{\mathrm{f}}}=\frac{k_{\mathrm{s}}+2k_{\mathrm{f}}-2\varphi(k_{\mathrm{f}}-k_{\mathrm{s}})}{k_{\mathrm{s}}+2k_{\mathrm{f}}+\varphi(k_{\mathrm{f}}-k_{\mathrm{s}})}+\left(\frac{\rho_s.\varphi.c_{\mathrm{ps}}}{2k_{\mathrm{f}}}\right)\left(\sqrt{\frac{k_{\mathrm{B}}T}{3\pi r\mu_{\mathrm{f}}}}\right) \tag{6.75}$$

where C and B are experimental constants, k_{B} is Boltzmann constant, Re is Reynolds number, and Pe is obtained by

$$Pe=\frac{u_{\mathrm{s}}d_{\mathrm{s}}}{\alpha_{\mathrm{f}}} \tag{6.76}$$

6.3.3 SOLUTION USING THE HAM

Here, the HAM has been employed to solve Equations (6.64) and (6.65) as the initial guess approximation for $f(\eta)$ and $\theta(\eta)$:

$$L_1(f)=f'''-f',\ \ L_2(\theta)=\theta''-\theta' \tag{6.77}$$

And, we choose linear operator as below:

$$L_1(C_1+C_2e^{-\eta}+C_3e^{\eta})=0,\ \ L_2(C_4+C_5e^{\eta})=0, \tag{6.78}$$

that C_1, C_2, C_3, C_4, and C_5 are constant.

Where $P\in[0,1]$, the embedding parameter, $\hbar=0$ is a nonzero auxiliary parameter. As the embedding parameter increases from 0 to 1, $f(\eta,q)$ varies from the initial guess $f_0(\eta)$ to the exact solution $f(\eta)$.

According to the discussed limitation and under the rule of solution expression and initial conditions, we choose initial guess in the form:

$$f_0(\eta)=f_{\mathrm{w}}+\eta-1+e^{\eta},\ \ \theta_0(\eta)=e^{-\eta} \tag{6.79}$$

6.3.3.1 *Zeroth-order deformation equations*

$$(1-p)L_1[f(\eta;p)-f_0(\eta)]=p\hbar_1N_1[f(\eta;p)] \tag{6.80}$$

$$(1-p)L_2[\theta(\eta;p)-\theta_0(\eta)]=p\hbar_2N_2[\theta(\eta;p)] \tag{6.81}$$

$$f(0;p)=f_{\mathrm{w}},\ \ \theta(0;p)=1 \tag{6.82}$$

$$f'(\infty;p)=1,\ \ \theta(\infty;p)=0 \tag{6.83}$$

$$N_1[f(\eta;p)] = \frac{1}{(1-\varphi)^{2.5}} \frac{1}{(1-\varphi)+\frac{\rho_s}{\rho_f}\varphi} \frac{d^3 f(\eta;p)}{d\eta^3} + \frac{1}{2} f \frac{d^2 f(\eta;p)}{d\eta^2} + \frac{Gr}{Re^2}\theta - \frac{1}{(1-\varphi)^{2.5}} \frac{1}{(1-\varphi)+\frac{\rho_s}{\rho_f}\varphi} \frac{1}{ReDa} \frac{df(\eta;p)}{d\eta} \tag{6.84}$$

$$N_2[\theta(\eta;p)] = \frac{d^2\theta(\eta;p)}{d\eta^2} + \frac{1}{2} pr \left[\frac{(1-\phi)+\frac{\rho_s c_{p_s}}{\rho_f c_{p_f}}\phi}{\left(\frac{k_s+2k_f-2\phi(k_f-k_s)}{k_s+2k_f+\phi(k_f-k_s)}\right)} \right] f \frac{d\theta(\eta;p)}{d\eta} = 0 \tag{6.85}$$

For $p=0$ and $p=1$, we have

$$f(\eta;0) = f_0(\eta), \quad (\eta;1) = f(\eta), \quad \theta(\eta;0) = \theta_0(\eta), \quad \theta(\eta;1) = \theta_0(\eta) \tag{6.86}$$

By Taylor's theorem, $f(\eta; p)$ and $\theta(\eta; p)$ can be expanded in a power series of p as follows:

$$f(\eta;p) = f_0(\eta) + \sum_{m-1}^{\infty} f_m(\eta) p^m, \quad f_m(\eta) = \frac{1}{m!} \frac{\partial^m [f(\eta;p)]}{\partial p^m} \tag{6.87}$$

$$\theta(\eta;p) = \theta_0(\eta) + \sum_{m-1}^{\infty} \theta_m(\eta) p^m, \quad \theta_m(\eta) = \frac{1}{m!} \frac{\partial^m [\theta(\eta;p)]}{\partial p^m} \tag{6.88}$$

Let us choose $\hbar_1 = \hbar_2 = \hbar$, where $\hbar$ is chosen in such a way that these two series are convergent at $p=1$.

Then:

$$f(\eta) = f_0(\eta) + \sum_{m-1}^{\infty} f_m(\eta) \tag{6.89}$$

$$\theta(\eta) = \theta_0(\eta) + \sum_{m-1}^{\infty} \theta_m(\eta) \tag{6.90}$$

6.3.3.2 mth-order deformation equations

$$L[f_m(\eta) - \chi_m f_{m-1}(\eta)] = \hbar R_m^f(\eta)$$

$$f_m(0) = f_m'(\infty) = f_m'' = 0$$

$$R_m^f(\eta) = \frac{1}{(1-\varphi)^{2.5}} \frac{1}{(1-\varphi)+\frac{\rho_s}{\rho_f}\varphi} f_{m-1}'''(\eta) + \frac{1}{2} \sum_{n=0}^{m-1} f_{m-1} f_n'' + \theta_{m-1} - \frac{1}{(1-\varphi)^{2.5}} \frac{1}{(1-\varphi)+\frac{\rho_s}{\rho_f}\varphi} \frac{1}{ReDa} f_{m-1}'$$

$$L[\theta_m(\eta) - \chi_m \theta_{m-1}(\eta)] = \hbar R_m^\theta(\eta)$$

$$\theta_m(0) = \theta_m(\infty) = 0$$

$$R_m^f(\eta) = \theta''_{m-1}(\eta) + \frac{1}{2}pr\sum_{n=0}^{m-1}\left[\frac{(1-\phi)+\dfrac{\rho_s c_{p_s}}{\rho_f c_{p_f}}\phi}{\left(\dfrac{k_s+2k_f-2\phi(k_f-k_s)}{k_s+2k_f+\phi(k_f-k_s)}\right)}\right]f_{m-1-n}\theta'_n$$

$$\chi_m = \begin{cases} 0, & m \le 1 \\ 1, & m > 1 \end{cases}$$

6.3.4 RESULTS AND DISCUSSION

We should ensure that the solution converges; as pointed out by Liao (1992), the convergence and rate of approximation for the HAM solution strongly depend on the value of the auxiliary parameter; this means that we are free to choose the auxiliary parameter $\hbar = 0.1$. The $\hbar$ region for this problem is shown in Figures 6.20 and 6.21. Figure 6.22 shows the profiles of temperature for various values of f_w for CuO-EG water when $\varphi = 0.03$.

It is seen that by increasing the suction parameter of CuO-EG water, from 0 to positive (suction) values, leads to increase in solid particles with high thermal conductivity take place in boundary layer. So, the heat transfer rate at the wall increases as thickness of thermal boundary layer decreases. The effect of f_w on vertical velocity is shown in Figure 6.23. Figures 6.24 and 6.25 show the variation of the skin-friction coefficient C_f and the heat transfer coefficient with the nanoparticle volume fraction parameter φ for different values of suction parameter f_w.

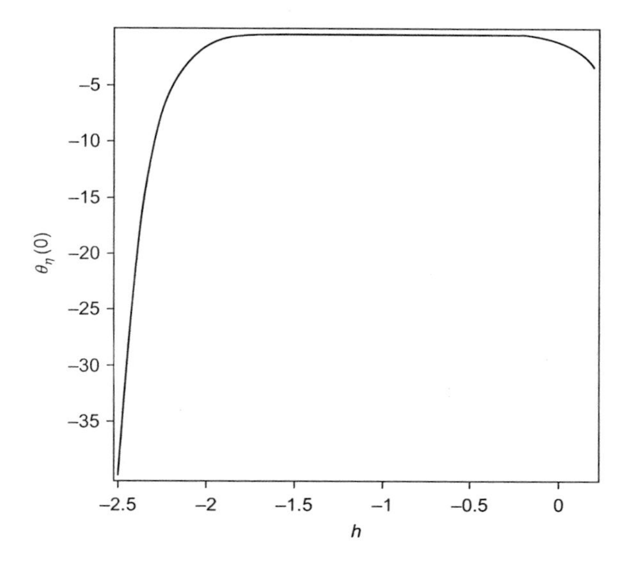

FIGURE 6.20

The $\hbar$ curve for temperature profile.

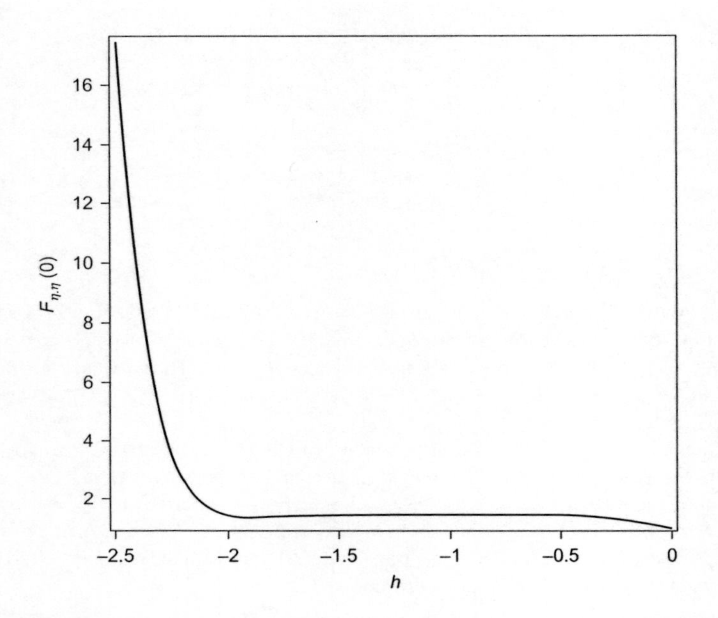

FIGURE 6.21

The $\hbar$ curve for $F''(\eta)$ profile.

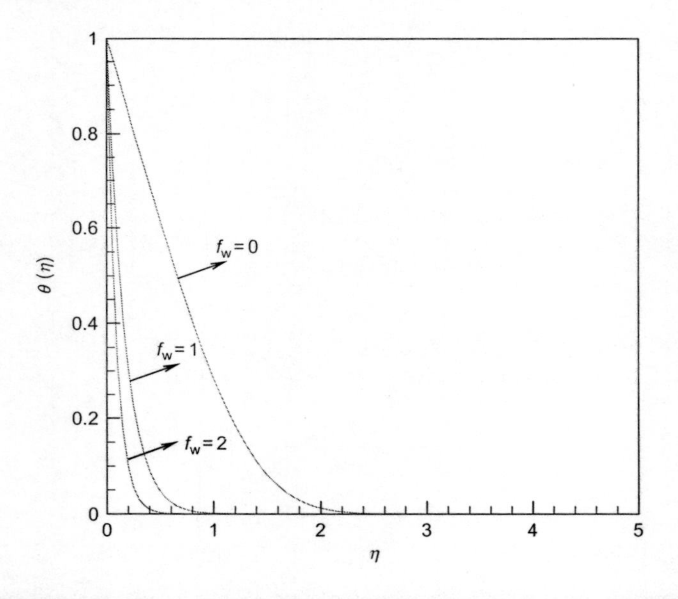

FIGURE 6.22

Temperature profiles $\theta(\eta)$ for various values of f_w with CuO-EG water when $\varphi = 0.03$.

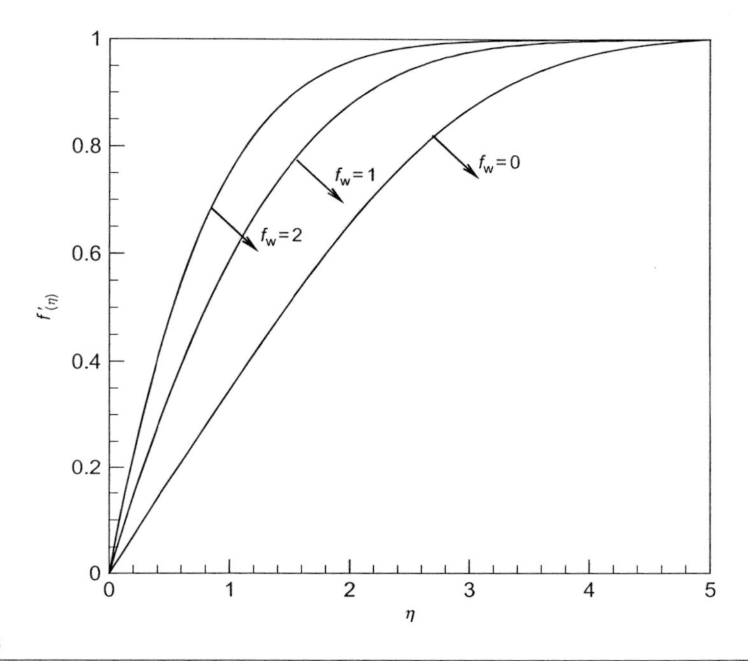

FIGURE 6.23

The effect of f_w on $f'(\eta)$ for CuO-EG water when $\varphi=0.03$.

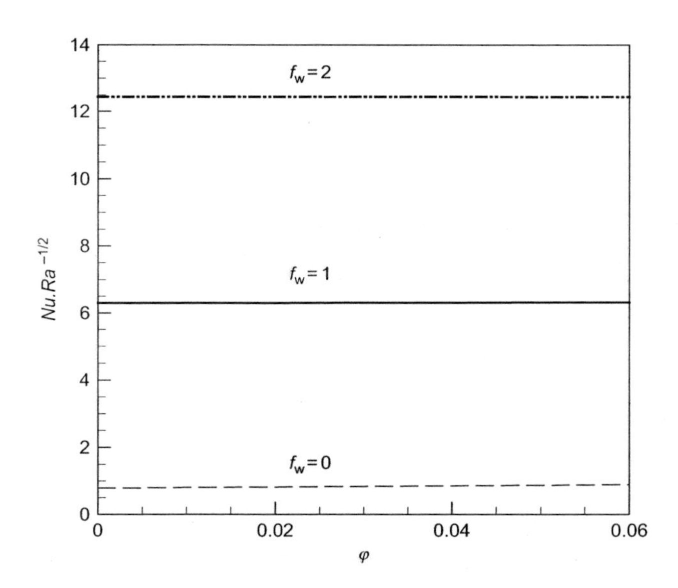

FIGURE 6.24

Variation of $Nu/Ra^{1/2}$ with φ for different values of f_w for CuO-EG water.

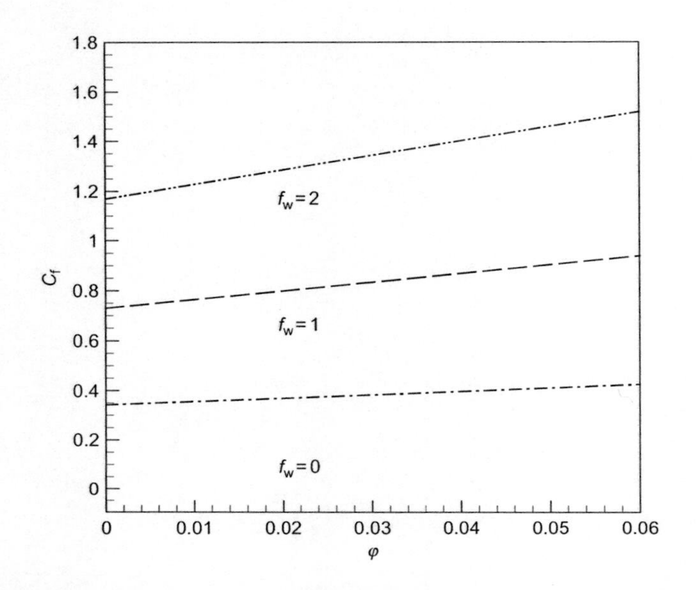

FIGURE 6.25

Variation of C_f with φ for different values of f_w for CuO-EG water.

By increasing the suction parameter, the nanoparticles increase too. These nanoparticles with high thermal conductivity cause more heat transfer rate as shown in Figure 6.24. Figures 6.26 and 6.27 illustrate the variations of the skin-friction coefficient and the local Nusselt number with the nanoparticle volume fraction parameter φ for three different of nanoparticles: copper oxide (CuO), alumina (Al_2O_3), and titania (TiO_2).

These figures show that these quantities increase almost linearly with φ. The presence of the nanoparticles in the fluids increases appreciably the effective thermal conductivity of the fluid and consequently enhances the heat transfer characteristics as seen in Figure 6.26. Nanofluids have a distinctive characteristic, which is quite different from those of traditional solid-liquid mixtures in which particles are involved. In addition, it is noted that the lowest heat transfer rate is obtained for the TiO_2 nanoparticles due to the domination of conduction mode of heat transfer. This is because TiO_2 has the lowest thermal conductivity compared to CuO and Al_2O_3. However, the difference in the values for CuO and Al_2O_3 is negligible. The thermal conductivity of Al_2O_3 is approximately half of CuO, as shown in Table 6.3.

A unique property of Al_2O_3 is its low thermal diffusivity. The reduced value of thermal diffusivity leads to higher temperature gradients and therefore, higher enhancements in heat transfer. The CuO nanoparticles have high values of thermal diffusivity, so this reduces the temperature gradient which affects the performance of CuO nanoparticles. In Figures 6.28 and 6.29, four different methods for variation of $Nu/Ra^{1/2}$ and C_f with volume fraction have been represented, respectively.

The utilized CuO nanoparticle is used and in basic EG water fluid. The comparisons between these methods, namely, Burgmann, Patel, Xuan, and Maxwell show that among all of these methods, the Burgmann method is not reliable, as shown in Figure 6.28. The other methods, however, have good

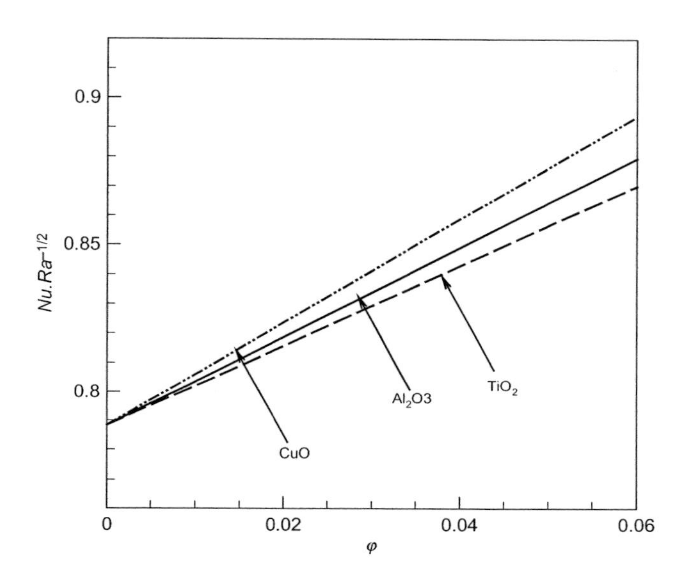

FIGURE 6.26

Variation of $Nu/Ra^{1/2}$ with φ for different nanoparticles in basic EG water fluid.

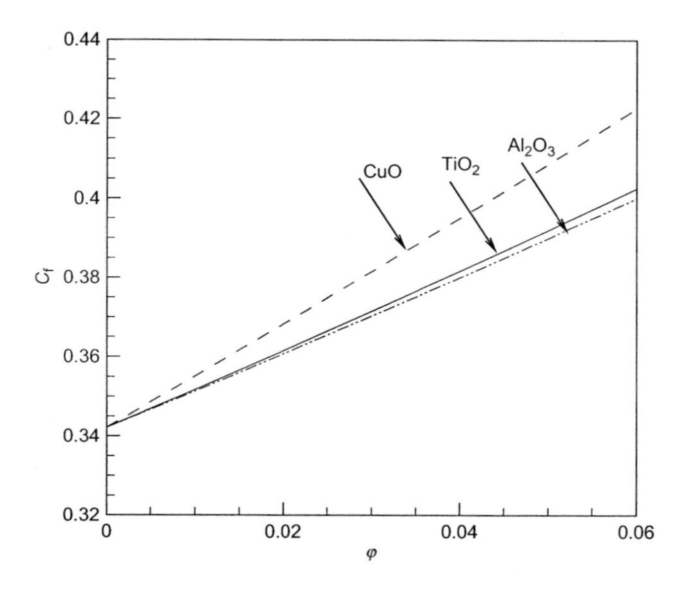

FIGURE 6.27

Variation of C_f with φ for different nanoparticles in basic EG water fluid.

Table 6.3 Comparison Between HAM Solution and Present Numerical Solution for θ (η) with $\varphi = 0$ and $\varphi = 0.03$ for CuO-Water as Working Fluid with 40th-Order Approximation and with $f_w = 0$

	$\varphi = 0.00$, $\theta(\eta)$			$\varphi = 0.03$, $\theta(\eta)$		
η	HAM Solution	Numerical Solution	Relative Error	HAM Solution	Numerical Solution	Relative Error
0	1.0000000000	1.0000000000	0.0000000000	1.0000000000	1.0000000000	0.0000000000
0.5	0.6114644445	0.6101095882	0.0013548563	0.6181421659	0.6166301445	0.0015120214
1	0.2754995548	0.2744993010	0.0010002538	0.2855881648	0.2842127123	0.0013754525
1.5	0.0763125601	0.0753539104	0.0009586497	0.0831814084	0.0819361489	0.0012452595
2	0.0109160793	0.0101618943	0.0007541850	0.0130045686	0.0120623402	0.0009422284
2.5	0.0005450622	0.0005419200	0.0000031422	0.0007457944	0.0007412697	0.0000045247
3	0.0000096202	0.0000094211	0.0000001991	0.0000162106	0.0000159284	0.0000002822
3.5	0.0000000459	0.0000000457	0.0000000002	0.0000001047	0.0000001042	0.0000000005
4	0.0000000001	0.0000000001	0.0000000000	0.0000000001	0.0000000001	0.0000000000
4.5	0.0000000040	0.0000000000	0.0000000000	0.0000000040	0.0000000000	0.0000000000
5	0	0	0	0	0	0

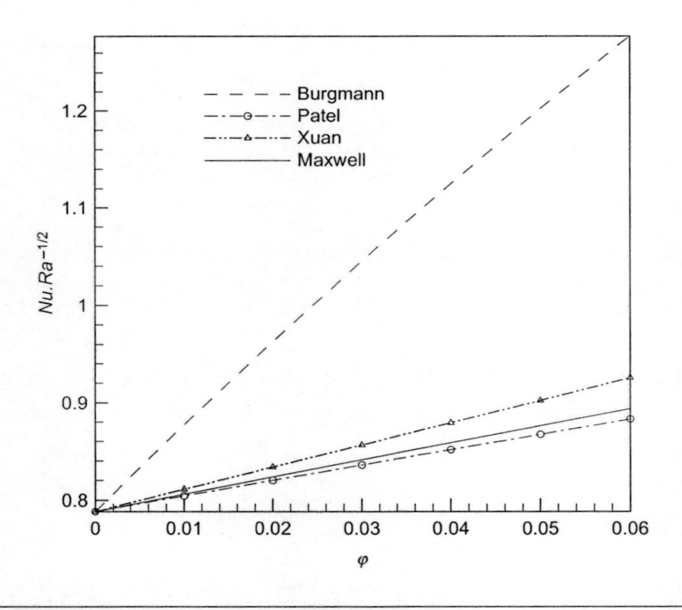

FIGURE 6.28

Variation of $Nu/Ra^{1/2}$ with φ for different methods for CuO nanoparticle.

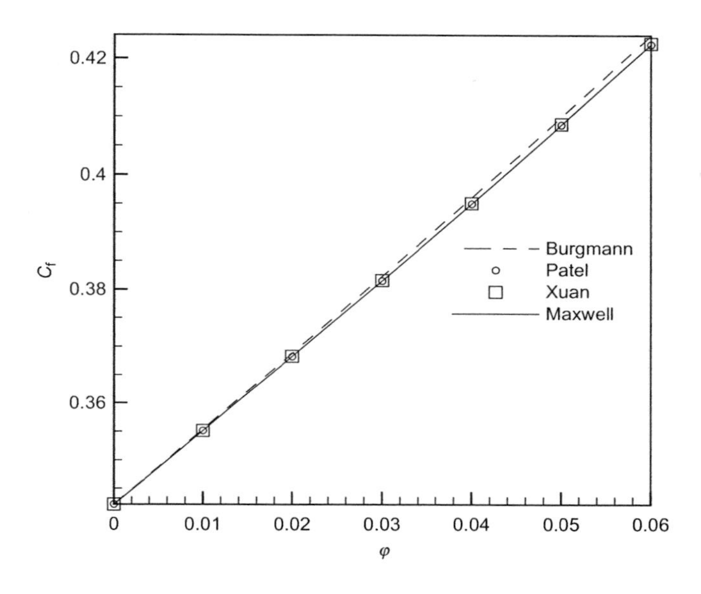

FIGURE 6.29

Variation of C_f with φ for different methods for CuO nanoparticle.

convergence for lower volume fraction than 0.06. Moreover, Figure 6.29 has approximately good results on variation of C_f with φ for the methods named above.

6.3.5 CONCLUSION

The analysis of forced convection flow past over a vertical porous flat plate embedded in an ethylene glycol-water-based nanofluid containing CuO, Al_2O_3, and TiO_2 nanoparticles in a porous medium has been investigated by HAM. The results have been compared with numerical methods that show good accuracy. It is found that for CuO-EG water by increasing the suction parameter f_w, the heat transfer rate at the wall increases, and the thickness of boundary layer decreases. An increase in f_w results in an increase in the skin-friction coefficient and a decrease in the vertical velocity. Utilizing nanoparticles with higher thermal conductivity, increases the heat transfer rate, so, CuO brings about higher heat transfer rate when compared to Al_2O_3 and TiO_2. As a result, it is more appropriate for enhancing the forced convection heat transfer at the wall.

6.4 HEAT TRANSFER IN SLIP-FLOW BOUNDARY CONDITION OF A NANOFLUID IN MICROCHANNEL

This section focuses on analysis of forced convection heat transfer over the horizontal surface with embedded open parallel microchannels at constant heat flux boundary condition by using three types of water-based nanofluids containing Cu, Al_2O_3, and TiO_2. The governing partial differential equations subjected to slip-flow boundary conditions are transformed into a set of ordinary differential equations. These equations were then solved numerically by using a forth-order Rung-Kutta and shooting method. Different thermal conductivity models of the nanofluid, namely, the Maxwell model, Jang and Choi model, Chon model, and Patel model are considered (*this section has been worked by A.R. Yousefi, G. Domairry, A. Rafiei in nonlinear dynamics team in Mechanical Engineering Department, 2012-2013*).

6.4.1 INTRODUCTION

In response to the ever-increasing demand for smaller and lighter high-performance cooling devices, the problem of slip-flow laminar-boundary layer in microchannels is of considerable practical interest. Microscale cooling devices, such as microchannel heat sinks (MCHS), are increasingly important in current and future heat removal applications. They microchannels are widely being considered for cooling of electronic devices, microheat exchanger systems, etc. Key is the very large heat transfer surface-to-volume ratio of the devices, leading to high compactness and effectiveness of heat removal. Slip-flow happens if the characteristic size of the flow system is small or the flow pressure is very low. Continuum physics is no longer suitable if the characteristic size of the flow system tends to the molecular mean free path. In no-slip-flow, as a requirement of continuum physics, the flow velocity is zero at a solid-fluid interface and the fluid temperature instantly closest to the solid walls is equal to that of the solid walls. In the existence of slip-flow, the flow velocity at the solid walls is nonzero and there is a temperature jump. Convective heat transfer in microchannel has been proved to be a very effective method for the thermal control microelectronic device (Arkilic et al., 1997; Beskok and Karniadakis, 1999; Gad-el-Hak, 1999; Ngoma and Erchiqui, 2007). The most frequently used coolants in the MCHS study were air, water, and fluorochemicals. The heat transfer capability is limited by the working fluid transport properties. Another approach to enhance the convective heat transfer coefficient in the microchannel may be utilizing nanofluids as working fluids. This can be possible because nanofluids having unprecedented stability of suspended nanoparticles were proven to be having anomalous thermal conductivity even with small volume fraction of the nanoparticles (see Martin and Boyd, 2006; Naterer, 2005).

In this section, the nanofluid is to be used as the working fluid in the microchannel. Yazdi et al. (2008), investigated the forced convection heat transfer of liquid flow over the horizontal surface with embedded open parallel microchannels at constant heat flux. Yang and Lai (2011) performed a mathematical modeling to simulate forced convection flow of Al_2O_3/water nanofluid in a microchannel using the LBM. Mohammed et al. (2010) investigated the effect of using nanofluid on heat transfer and fluid flow characteristics in rectangular-shaped MCHS for Reynolds number range of 100-1000.

Chein and Huang (2005) analyzed the silicon MCHS performance using nanofluid as coolants. Koo and Kleinstreuer (2005) numerically simulated the steady laminar liquid nanofluid flow in microchannels considering two types of nanofluids. Jung et al. (2009) measured a convective heat transfer coefficient and friction factor of nanofluid in rectangular microchannels.

In this work, we employ numerical solution for solving the equations governing the flow and temperature that have been reduced to a system of coupled nonlinear ordinary differential equations. The scope of the current research is to investigate the effects of nanoparticle dispersion on heat transfer performance of microchannel systems and also to implement and compare different effective thermal conductivity models for nanofluid and study the effect of these models on heat transfer.

6.4.2 PROBLEM STATEMENT AND GOVERNING EQUATION

The physical configuration of microchannel is shown in Figure 6.30. The heat is supplied to the microchannel through a top plate.

A coolant nanofluid passes through the microchannel and takes heat away. The specified dimensions of rectangular-shaped microchannel are given in Figure 6.30. The thermophysical properties for the nanoparticles and base fluid (water) are listed in Table 6.4.

It is assumed that the base fluid (i.e., water) and the nanoparticles are in thermal equilibrium and no slip occurs between them. The thermophysical properties of the nanofluid are given in Table 6.4.

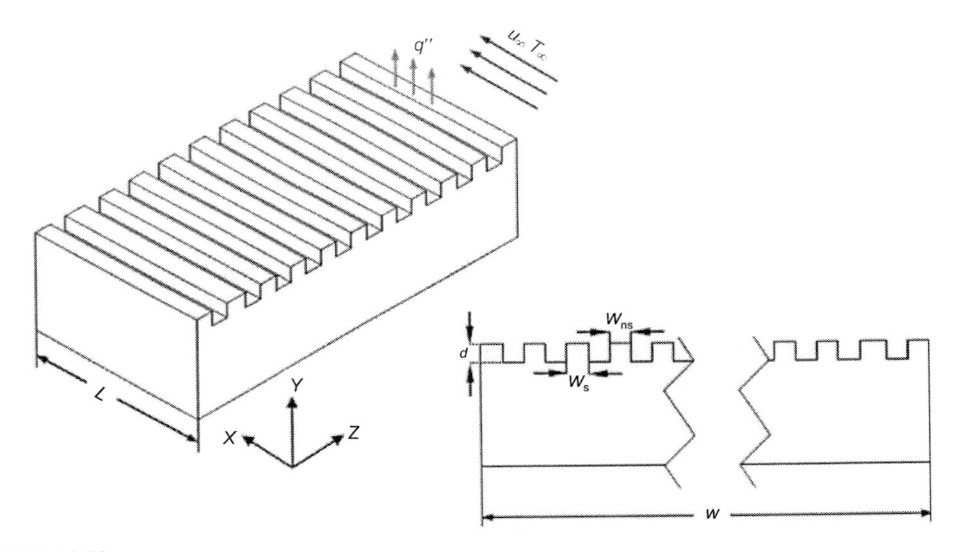

FIGURE 6.30

Schematic diagram of the embedded open microchannels within the surface.

Table 6.4 Thermophysical Properties of the Cu Nanoparticles and Base Fluid (Water)

Physical Property	Cu Nanoparticles	TiO_2 Nanoparticles	Al_2O_3 Nanoparticles	Base Fluid (Water)
ρ (kg/m^3)	8954	4250	3970	997.1
μ (Pa s)	–	–	–	8.9E−4
C_p (J/kg K)	383	686.2	765	4179
K (W/m K)	400	8.9538	40	0.6
Pr	–	–	–	6.2
d_p (m)	10^{-9}	10^{-9}	10^{-9}	–

Thermophysical properties of the nanofluid are assumed to be constant. For the nanoparticles-water nanofluid mixture, the density, thermal conductivity, and dynamic viscosity are significantly increased while the specific heat of the nanofluid is decreased with the increase of its particle volume fraction compared to pure water. In this work, the effect of particle volume fraction on heat transfer and fluid flow characteristics of rectangular-shaped MC was considered. Both fluid flow and heat transfer are in steady-state and two-dimensional nanofluid is in single phase, incompressible, and the flow is laminar.

The continuity and momentum equations for the two-dimensional steady-state flows using the usual boundary-layer approximations are expressed as below:

$$\frac{\partial u}{\partial x}+\frac{\partial v}{\partial y}=0 \tag{6.91}$$

$$\rho_{\text{nf}}\left(u\frac{\partial u}{\partial x}+v\frac{\partial v}{\partial y}\right)=\mu_{\text{nf}}\left(\frac{\partial^2 u}{\partial y^2}\right) \tag{6.92}$$

$$\left(\rho C_p\right)_{\text{nf}}\left(u\frac{\partial T}{\partial x}+v\frac{\partial T}{\partial y}\right)=k_{\text{nf}}\left(\frac{\partial^2 T}{\partial y^2}\right) \tag{6.93}$$

With associated boundary conditions:

$$\text{At } y=0\rightarrow\begin{cases} u=0 & \text{no slip}\\ u=u_{\text{s}} & \text{slip}\\ \dfrac{\partial T}{\partial y}=\dfrac{-q_{\text{w}}}{k_\infty} & \\ v=0 & \end{cases} \tag{6.94}$$

$$\text{At } y=\infty\rightarrow\begin{cases} u=u_\infty\\ T=T_\infty \end{cases} \tag{6.95}$$

In the above equations, ρ_{nf}, the density of the nanofluid is given by:

$$\rho_{\text{nf}}=(1-\varphi)\rho_{\text{f}}+\varphi\rho_{\text{s}} \tag{6.96}$$

Whereas, the heat capacitance of the nanofluid and thermal expansion coefficient of the nanofluid can be determined by:

$$\left(\rho C_p\right)_{\text{nf}}=(1-\varphi)\left(\rho C_p\right)_{\text{f}}+\varphi\left(\rho C_p\right)_s \tag{6.97}$$

$$(\rho\beta)_{\text{nf}}=(1-\varphi)(\rho\beta)_{\text{f}}+\varphi(\rho\beta)_{\text{s}} \tag{6.98}$$

With φ being the volume fraction of the solid particles and subscripts f, nf and s stand for base fluid, nanofluid, and solid, respectively. The effective dynamic viscosity of the nanofluid given by Brinkman (1952) is:

$$\mu_{\text{nf}}=\frac{\mu_{\text{f}}}{(1-\phi)^{2.5}} \tag{6.99}$$

Different models for predicting effective thermal conductivity of the nanofluid are given in Table 6.5.

Table 6.5 Different Models of Thermal Conductivity

Models	Equation	
Maxwell (1904)	$\frac{k_{nf0}}{k_f} = \frac{k_s + 2k_f - 2\phi(k_f - k_s)}{k_s + 2k_f + \phi(k_f - k_s)}$	(6.100a)
Jang and Choi (2007)	$\frac{k_{nf0}}{k_f} = (1-\varphi) + \beta\varphi\left(\frac{k_s}{k_f}\right) + 3C\varphi\left(\frac{d_f}{d_s}\right)Pr_T Re^2$ $C_{RM} = \frac{D_o}{L_f},\ D_o = \frac{K_B T}{3\pi d_s \mu_f},\ C = 1.8 \times 10^7$	(6.100b)
Patel et al. (2005)	$\frac{k_{nf0}}{k_f} = 1 + (1 + c \cdot Pe)\frac{k_s A_s}{k_f A_f}$ $Pe = \frac{u_s d_s}{\alpha_f},\ \frac{A_s}{A_f} = \frac{\varphi}{1-\varphi}\frac{d_f}{d_s}$ $u_s = \frac{2k_b T}{\pi \mu_f d_s^2},\ d_f = 2 \times 10^{-10}\ \text{m},\ c = 25{,}000$	(6.100c)
Chon et al. (2005)	$\frac{k_{nf0}}{k_f} = 1 + 64.7\varphi^{0.7640}\left(\frac{d_f}{d_s}\right)^{0.3690}\left(\frac{k_f}{k_s}\right)^{0.7476} Pr^{0.9955} Re^{1.2321}$ $Pr = \frac{\mu_f}{\alpha_f \rho_f},\ Re = \frac{\rho_f k_b T}{3\pi \mu_f^2 L_f},\ L_f = 0.17\,\text{nm}$	(6.100d)

Equations (6.91)-(6.93) can be converted to nondimensional forms, using the following nondimensional parameters and the similarity variable η

$$\psi = \sqrt{\nu_f u_\infty x} f(\eta)$$
$$u = \frac{\partial \psi}{\partial y} = f'(\eta),\quad v = -\frac{\partial \psi}{\partial x} = f(\eta) \tag{6.101}$$
$$\theta(\eta) = \frac{T_f - T_\infty}{q''_w x / k_f} Re_x^{1/2},\quad Pr = \frac{\nu_f}{\alpha_f},\quad \eta = \left(\frac{u_\infty}{x\nu_f}\right)^{1/2} y$$

where q''_w is the heat flux and k_f is the thermal conductivity. The velocity components u and v in Equation (6.101) automatically satisfy the continuity Equation (6.91). By using the transformation Equation (6.101), nondimensional velocity and temperature becomes identical.

For the case of a slip-flow condition at the wall, the boundary condition involves the nonzero wall velocity and the spatial gradient of velocity so that

$$f'(0) = Kf''(0) \tag{6.102}$$

where K is the slip coefficient defined for liquids by

$$K = \frac{\beta}{x} Re^{0.5} \tag{6.103}$$

β is the slip length for the liquid.

The governing Equations (6.92) and (6.93) and the boundary conditions Equations (6.94) and (6.95) are written as follows

$$\frac{2}{(1-\varphi)^{2.5}\left((1-\varphi)+\frac{\rho_s}{\rho_f}\varphi\right)}\frac{d^3f}{d\eta^3}+f\frac{d^2f}{d\eta^2}=0 \tag{6.104}$$

$$\rightarrow \theta''+\frac{Pr}{2}(\theta' f-\theta f')\left(\frac{\left((1-\varphi)+\varphi\frac{(\rho C_p)_s}{(\rho C_p)_f}\right)}{\frac{(k_s+2k_f)-2\varphi(k_f-k_s)}{(k_s+2k_f)+2\varphi(k_f-k_s)}}\right)=0 \tag{6.105}$$

And, the dimensionless boundary conditions used to solve Equations (6.104) and (6.105) are as follows:

$$f(0)=0,\ f'(0)=Kf''(0),\ \theta(0)=-1 \tag{6.106}$$

$$f'(\infty)=1,\ \theta(\infty)=0 \tag{6.107}$$

6.4.3 NUMERICAL PROCEDURE AND VALIDATION

The numerical solution of nonlinear differential Equations (6.104) and (6.105) with boundary conditions given in Equations (6.106) and (6.107) is accomplished by shooting technique with fourth-order Runge-Kutta algorithm (Gerald and Weatly, 1989). The nonlinear differential Equations (6.104) and (6.105) are first decomposed to a system of first-order differential equations in the form

$$\begin{aligned}
&f_0=f,\ \frac{df_0}{d\eta}=f_1,\ \frac{df_1}{d\eta}=f_2\\
&\frac{df_2}{d\eta}\frac{2}{(1-\varphi)^{2.5}\left((1-\varphi)+\frac{\rho_s}{\rho_f}\varphi\right)}+f_0f_2=0\\
&\theta_0=\theta,\ \frac{d\theta_0}{d\eta}=\theta_1\\
&\frac{d\theta_1}{d\eta}+\frac{Pr\left((1-\varphi)+\frac{(\rho C_p)_s}{(\rho C_p)_f}\right)}{\left(\frac{k_s+2k_f-2\varphi(k_f-k_s)}{k_s+2k_f+2\varphi(k_f-k_s)}\right)}\left(\frac{f_0\theta_1}{2}-\frac{\theta_0 f_1}{2}\right)=0
\end{aligned} \tag{6.108}$$

Corresponding boundary conditions take the form

$$\begin{aligned}
&f_0(0)=0,\ f_1(0)=Kf_2(0),\ f_1(\infty)=1\\
&\theta_1(0)=-1,\ \theta_0(\infty)=0
\end{aligned} \tag{6.109}$$

Appropriately guessing the missing slopes $f_2(0)$ and $\theta_1(0)$, the boundary value problem (BVP) is first converted into an initial value problem which is solved by a fourth-order Runge-Kutta method. Then the values of $f_2(0)$ and $\theta_1(0)$ are updated using shooting method. The convergence criterion largely depends on fairly good guesses of the initial conditions in the shooting technique.

After solving this slip-flow problem numerically, the wall stress and Nusselt number at the wall exhibit a dependence on the slip coefficient, K.

$$\frac{\tau_w}{\rho_f u_\infty^2} = f''(0) Re_x^{-0.5}, \quad \tau_w = \mu_{nf}\left(\frac{\partial u}{\partial y}\right)_{y=0} \tag{6.110}$$

It can be expressed as:

$$c_f \sqrt{Re} = \frac{F''(0)}{(1-\varphi)^{2.5}} \tag{6.111}$$

The local heat transfer coefficient h can be expressed in dimensionless form of the local Nusselt number Nu_x, which represents the heat transfer rate at the surface, and can be given in dimensionless form as:

$$\frac{Nu_x}{\sqrt{Re}} = \frac{k_{nf}}{k_f} \frac{1}{\theta(0)} \tag{6.112}$$

Present results are compared with some of the earlier published results by Yazdi et al. (2008) which are depicted in Figure 6.31.

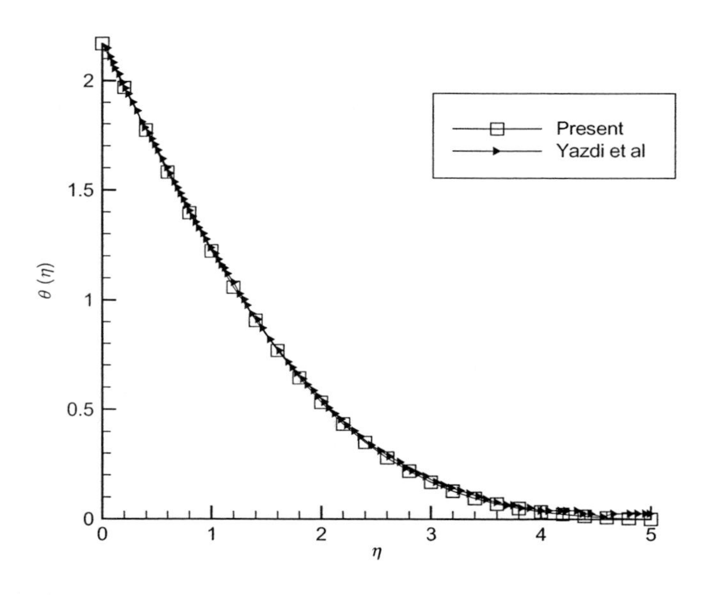

FIGURE 6.31

Comparison of temperature profile between present work with Yazdi et al. (2008) results.

6.4.4 RESULTS AND DISCUSSION

In this section, three types of water-based nanoparticles were investigated, namely, copper (Cu), alumina (Al_2O_3), and titania (TiO_2). The value of the Prandtl number is constant and is taken as 6.2 (for water), and the effect of the volume fraction parameter (φ), variation, and slip coefficient (K) for this geometry were studied. Figure 6.32 illustrates the effect of increasing volume fractions (φ) on behavior of temperature profile and it can clearly be seen that due to rising Cu nanoparticle concentration, thermal boundary-layer thickness grows which is in good agreement with physical behavior because as volume fractions (φ) increase nanofluid thermal conductance increases and consequently leads to an increase in thermal boundary-layer thickness.

Figure 6.33 depicts variation of velocity gradient $f''(0)$ at the surface as a function of slip coefficient (K). At and $f''(0) = 0.332$, the value of K equal to zero and no-slip condition occurs. By increasing volume fraction (φ), velocity gradient rises and due to improving velocity by increasing slip coefficient K, the shear stress at the surface increases.

Figure 6.34 shows the effect of nanoparticle type on temperature profile and heat transfer rate. Thermal boundary-layer thickness grows from Cu to TiO_2 and to Al_2O_3, and as a result, decreases wall temperature-local temperature difference.

Figure 6.35 shows that increasing volume fractions (φ) of Cu nanoparticles in water increases local Nusselt number.

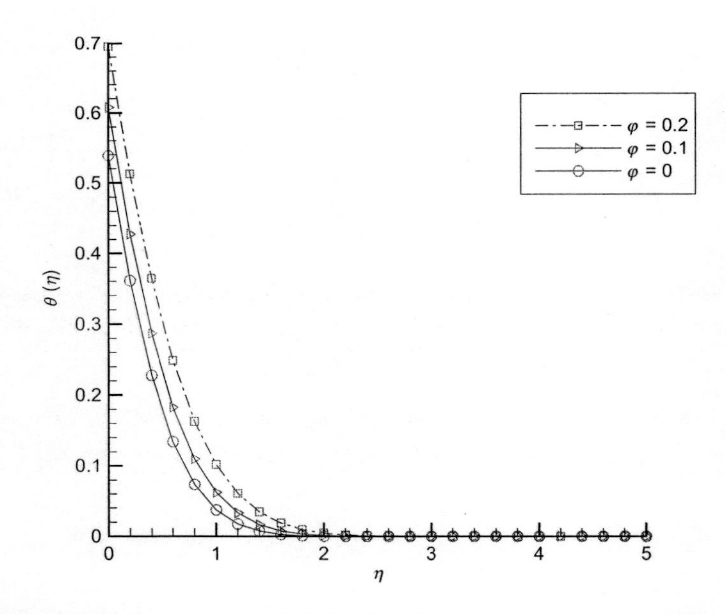

FIGURE 6.32

Temperature profiles for various variations of η for different values of nanoparticle volume fraction (φ) for Cu.

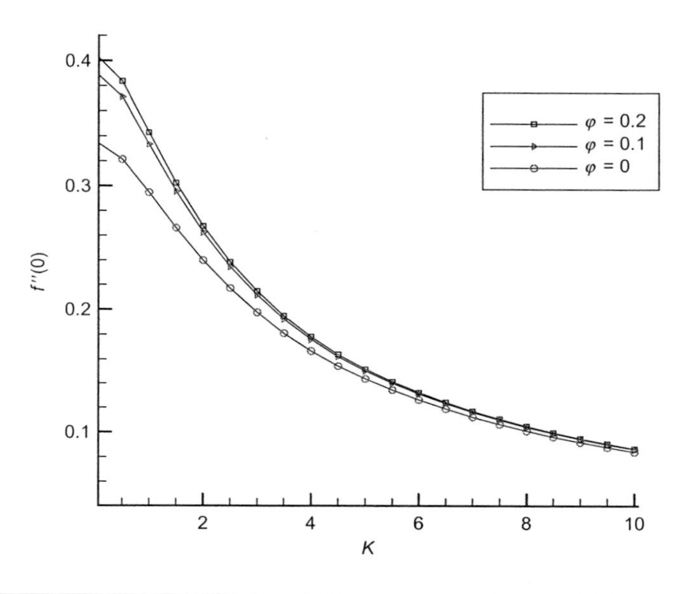

FIGURE 6.33

Variation of the velocity gradient at the surface as a function of slip coefficient (K) for different values of nanoparticle volume fraction (φ) for Cu.

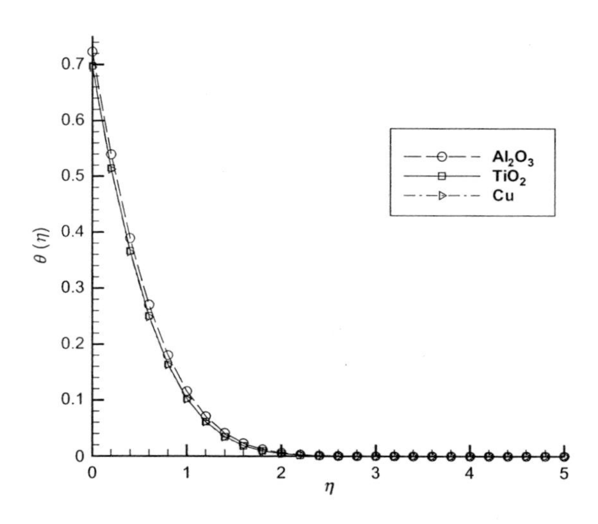

FIGURE 6.34

Temperature profiles for various values of η for different nanoparticles when $Pr=6.2$, $\Phi=0.2$.

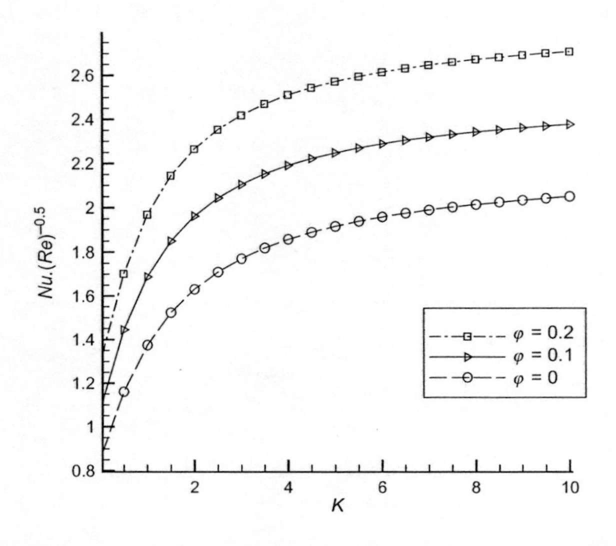

FIGURE 6.35

Variation of the local Nusselt number with K for different value of nanoparticle volume fraction (φ) for Cu.

Figure 6.36 illustrates variation of the Nusselt number with volume fractions (φ) for different nanoparticles. Moreover, it is obvious that for $\varphi = 0.2$ Cu-water nanofluid has the highest rate of heat transfer and Al_2O_3 and TiO_2 takes the second and third places, respectively. It is seen that rate of heat transfer increases by increase of slip coefficient (K).

Figure 6.37 shows the variation of the Nusselt number on increasing slip coefficient (K) for different nanoparticles and is in agreement with results of Figures 6.35 and 6.36. Dimensionless shear stress distributions for different nanoparticles observed from Figure 6.38 shows that by increase of slip coefficient (K) slip velocity and shear stress at the surface increases and decreases, respectively.

Also, Figure 6.39 shows that increasing volume fractions (φ) of Cu nanoparticles in water increases shear stress at the surface.

Figure 6.40 shows different models of thermal conductivity for local Nusselt number considering estimation of effect of distributed nanoparticles size for Cu-water nanofluid and for volume fractions (φ) up to 0.2. Among all these stated models, Jang and Choi seem less rational. This model is acceptable for volume fraction less than 6% because the maximum amount available for volume fraction by this model is 6%.

6.4.5 CONCLUSION

The problem of forced convection heat transfer over the horizontal surface with embedded open parallel microchannels at constant heat flux boundary condition was investigated numerically. The governing partial differential equations were transformed into a system of nonlinear ordinary differential equations using a similarity transformation, before being solved numerically by Rung-Kutta shooting method. Three different types of nanoparticles, namely, copper (Cu), alumina (Al_2O_3), and titania

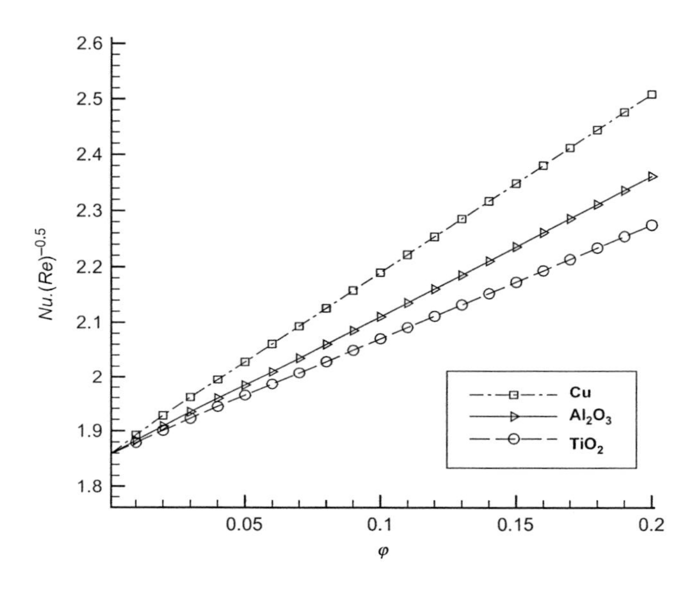

FIGURE 6.36

Variation of the Nusselt number with Φ for different nanoparticles with $Pr=6.2$ with $K=4$.

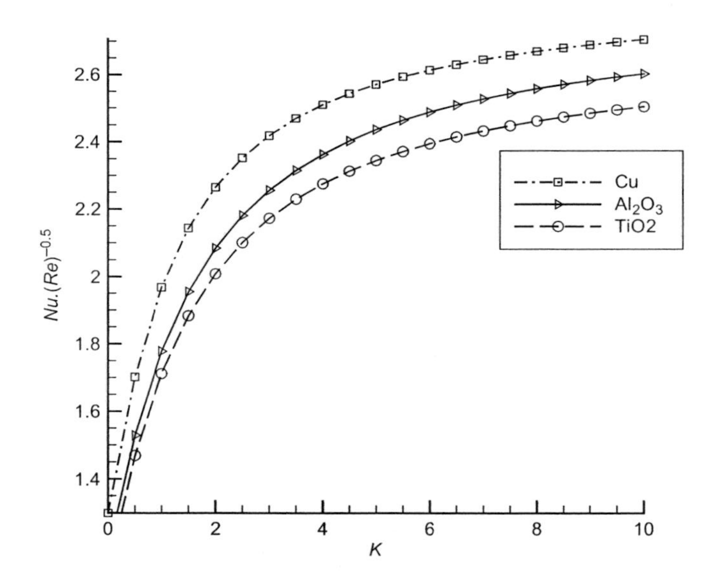

FIGURE 6.37

Variation of the Nusselt number as a function of slip coefficient (K) for different nanoparticles with $Pr=6.2$, $\Phi=0.2$.

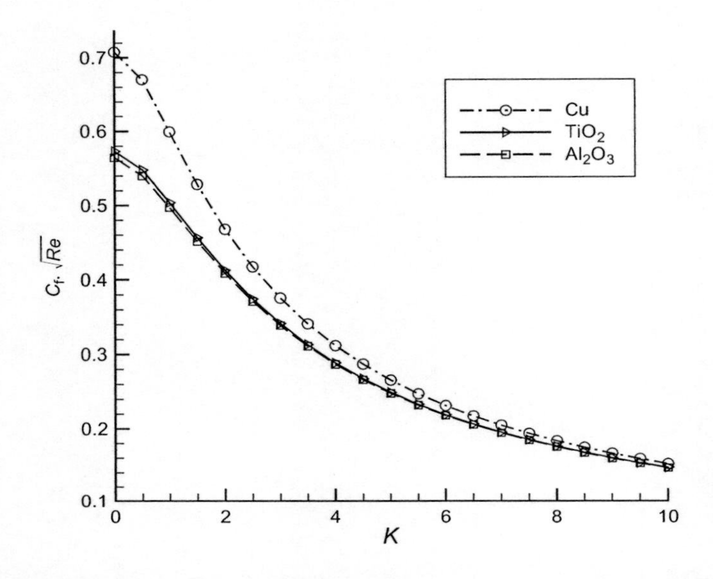

FIGURE 6.38

Dimensionless shear stress distributions for different nanoparticle at $\varphi=0.2$ with $Pr=6.2$, $\Phi=0.2$.

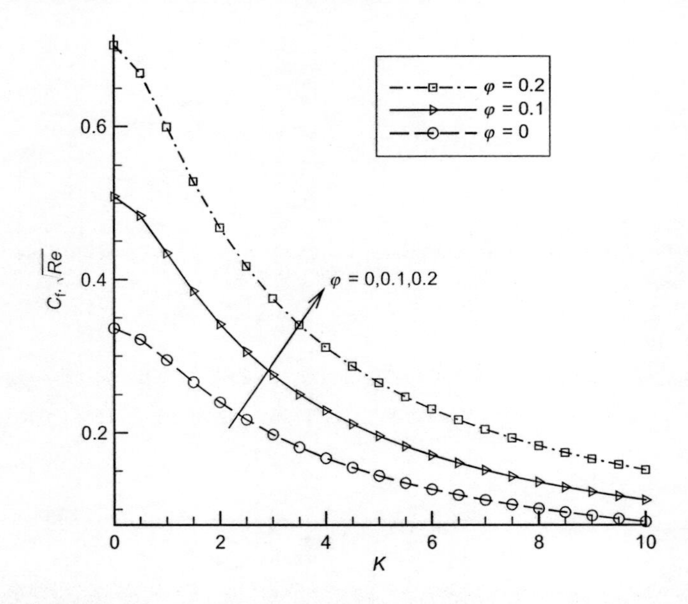

FIGURE 6.39

Dimensionless shear stress distributions at the surface for different values of nanoparticle volume fractions φ for Cu.

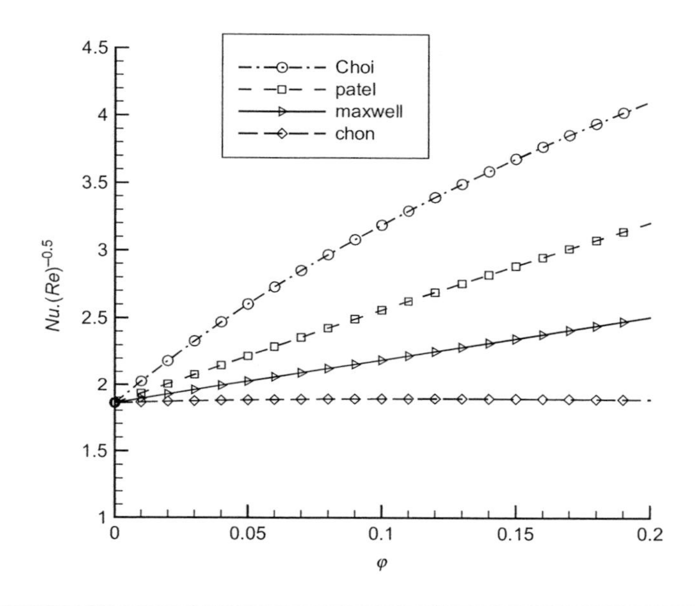

FIGURE 6.40

Comparison between different models of thermal conductivity for local Nusselt number for Cu.

(TiO_2) with water as the base fluid were taken into account. It is found that for each and all three nanoparticle cases, the thermal boundary-layer thickness increases as volume fraction φ increases. Moreover, four different models, namely, Maxwell model, Jang and Choi model, Chon model, and Patel model have been compared in this study. It is concluded, the Jang and Choi model does not give logical results for high volume fraction. Meanwhile, it can be seen that increase in the value of nanoparticle volume fraction φ led to an increase of local heat transfer coefficient.

6.5 FORCED CONVECTION ANALYSIS FOR MAGNETOHYDRODYNAMICS (MHD) AL_2O_3-WATER NANOFLUID FLOW

In this section, forced-convection boundary-layer of MHD Al_2O_3-water nanofluid flow over a horizontal stretching flat plate is investigated using HAM and fourth-order Runge-Kutta numerical method. The influence of the nanofluid volume fraction (φ) and magnetic parameter (Mn) on nondimensional temperature and velocity profiles is investigated. As an important outcome, by increasing Mn number, thermal boundary-layer thickness significantly increased but increasing the nanofluid volume fraction has not very sensible effect on it (*this section has been worked by M. Hatami, R. Nouri, D. D. Ganji in nonlinear dynamics team in Mechanical Engineering Department, 2012-2013*).

6.5.1 INTRODUCTION

Most scientific problems in fluid mechanics and heat transfer problems are inherently nonlinear. All these problems and phenomena are modeled by ordinary or partial nonlinear differential equations. Most of these described physical and mechanical problems are with a system of coupled nonlinear differential equations which can be solved by analytical methods such as HPM (He, 2003).

The term of MHD was first introduced by Alfvén (1942). The theory of MHD states that inducing current in a moving conductive fluid in the presence of magnetic field exerts force on the ions of the conductive fluid. The theoretical study of MHD channel has been a subject of great interest due to its extensive applications in designing cooling systems with liquid metals, MHD generators, accelerators, pumps, and flow meters.

The main motivation of this work is to solve a two-dimensional forced-convection boundary-layer MHD problem in the presence of an MHD field over a horizontal flat plate including the viscous dissipation term using the HAM. The main advantage of this study is using an analytical method which does not need any small parameter, and obtained results are compared with the numerical results which confirm the high accuracy of the applied method; furthermore, the effect of some parameters appeared in the mathematical formulation on stream function and temperature profile is investigated.

6.5.2 DESCRIPTION OF THE PROBLEM

Consider the two-dimensional forced-convection boundary-layer with variable MHD field of an incompressible flow including nanoparticles over a horizontal surface (see Figure 6.41).

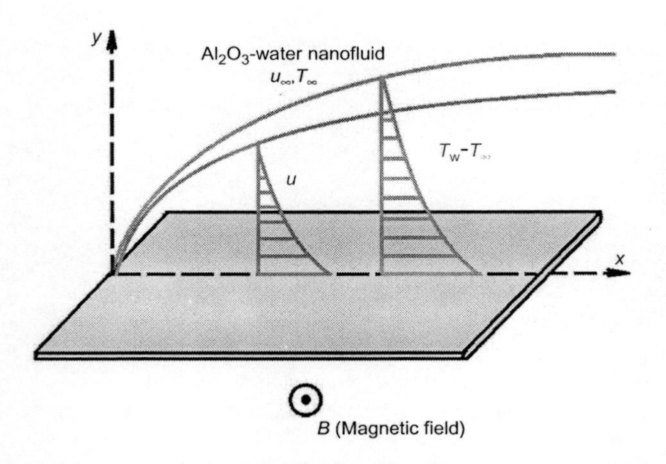

FIGURE 6.41

Schematic of the nanofluid boundary layer on a horizontal plate in presence of magnetic field.

By using boundary-layer approximation and considering viscous dissipation term, the simplified two-dimensional equations governing the flow in the boundary layer of a steady, laminar, and incompressible flow are (Nourazar et al., 2011):

$$\frac{\partial u}{\partial x}+\frac{\partial v}{\partial y}=0, \tag{6.113}$$

$$u\frac{\partial u}{\partial x}+v\frac{\partial u}{\partial y}=\frac{1}{\rho_{\mathrm{nf}}}\left(\mu_{\mathrm{nf}}\frac{\partial^2 u}{\partial y^2}-\sigma B(x)^2 u\right) \tag{6.114}$$

$$u\frac{\partial T}{\partial x}+v\frac{\partial T}{\partial y}=\alpha_{\mathrm{nf}}\frac{\partial^2 T}{\partial y^2}+\frac{\mu_{\mathrm{nf}}}{\left(\rho C_p\right)_{\mathrm{nf}}}\left(\frac{\partial u}{\partial y}\right)^2 \tag{6.115}$$

where u and v are the x and y components of velocity, respectively, σ is the electrical conductivity, $B(x)$ is the variable magnetic field acting in the perpendicular direction to the horizontal flat plate, μ_{nf} and ρ_{nf} are the viscosity and the density of the nanofluid, respectively, α_{nf} is the thermal diffusivity, and $(\rho C_p)_{\mathrm{nf}}$ is the heat capacitance of the nanofluid. The appropriate physical boundary conditions are defined as below:

$$u=u_{\mathrm{w}}=bx^m,\quad v=0,\quad T=T_{\mathrm{w}} \tag{6.116}$$

$$u\to 0,\quad T\to T_\infty \tag{6.117}$$

where u_{w} is the x component of velocity on the horizontal flat plate, b and m are constants, T_{w} and T_∞ are the plate and ambient temperatures, respectively. The nanofluid properties such as the density, ρ_{nf}, the dynamic viscosity, μ_{nf}, the heat capacitance, $(\rho C_p)_{\mathrm{nf}}$, and the thermal conductivity, k_{nf}, are defined in terms of fluid and nanoparticle properties as (Aminossadati and Ghasemi, 2009).

$$\rho_{\mathrm{nf}}=(1-\phi)\rho_{\mathrm{f}}+\phi\rho_{\mathrm{s}} \tag{6.118}$$

$$\mu_{\mathrm{nf}}=\frac{\mu_{\mathrm{f}}}{(1-\phi)^{2.5}} \tag{6.119}$$

$$\frac{k_{\mathrm{nf}}}{k_{\mathrm{f}}}=\frac{k_{\mathrm{s}}+2k_{\mathrm{f}}-2\phi(k_{\mathrm{f}}-k_{\mathrm{s}})}{k_{\mathrm{s}}+2k_{\mathrm{f}}+2\phi(k_{\mathrm{f}}-k_{\mathrm{s}})} \tag{6.120}$$

$$\left(\rho C_p\right)_{\mathrm{nf}}=(1-\phi)\left(\rho C_p\right)_{\mathrm{f}}+\phi\left(\rho C_p\right)_{\mathrm{s}} \tag{6.121}$$

$$\alpha_{\mathrm{nf}}=\frac{k_{\mathrm{nf}}}{\left(\rho C_p\right)_{\mathrm{nf}}} \tag{6.122}$$

where ρ_{f} is the density of fluid, ρ_{s} is the density of nanoparticles, ϕ is defined as the volume fraction of the nanoparticles, μ_{f} is the dynamic viscosity of fluid, $(\rho C_p)_{\mathrm{f}}$ is the thermal capacitance of fluid, $(\rho C_p)_{\mathrm{s}}$, is the thermal capacitance of nanoparticles, and k_{f} and k_{s} are the thermal conductivities of fluid and nanoparticles, respectively. The variable magnetic field is defined as (Chiam, 1995; Devi and Thiyagarajan, 2006):

$$B(x)=B_0\left(x^{\frac{m-1}{2}}\right) \tag{6.123}$$

where B_0 and m are constants. The following dimensionless similarity variable is used to transform the governing equations into the ordinary differential equations;

$$\eta = \frac{y}{x} Re_x^{1/2} \tag{6.124}$$

$$Re_x = \frac{\rho_f u_w(x)}{\mu_f} x \tag{6.125}$$

The dimensionless stream function and dimensionless temperature are defined as:

$$f(\eta) = \frac{\psi(x, y)(Re_x)^{1/2}}{u_w(x)} \tag{6.126}$$

$$\theta(\eta) = \frac{T - T_\infty}{T_w - T_\infty} \tag{6.127}$$

where the stream function $\psi(x, y)$ is defined as:

$$u = \frac{\partial \psi}{\partial y}, \quad v = -\frac{\partial \psi}{\partial x} \tag{6.128}$$

By applying the similarity transformation parameters, the momentum equation (Equation (6.114)) and the energy equation (Equation (6.115)) can be rewritten as (Nourazar et al., 2011):

$$f''' + \left((1-\phi) + \phi\left(\frac{\rho_s}{\rho_f}\right)\right)(1-\phi)^{2.5}\left(\frac{m+1}{2}\right)^2 ff'' - \left((1-\phi) + \phi\left(\frac{\rho_s}{\rho_f}\right)\right)(1-\phi)^{2.5}(m)f'^2 - \left((1-\phi)^{2.5} Mn\right)f' = 0 \tag{6.129}$$

$$\theta'' + \left((1-\phi) + \phi\frac{(\rho C_p)_s}{(\rho C_p)_f}\right) Prf\theta' + \frac{EcPr}{(1-\phi)^{2.5}} f''^2 = 0 \tag{6.130}$$

Therefore, the transformed boundary conditions are:

$$\begin{gathered} f'(0) = 1, \; f(0) = 0, \; f'(\infty) = 1 \\ \theta(0) = 1, \; \theta(\infty) = 0 \end{gathered} \tag{6.131}$$

The dimensionless parameters of Mn, Pr, Ec, and Re_x are the magnetic parameter, Prandtl, Eckert, and Reynolds numbers, respectively. They are defined as:

$$Mn = \frac{\sigma B_0^2}{\rho_f b}, \quad Pr = \frac{(\rho C_p)_f}{k_{ef}} \nu_f, \quad Ec = \frac{u_w(x)^2}{C_p \Delta T}, \quad Re_x = \frac{\rho_f u_w(x)}{\mu_f} x \tag{6.132}$$

The Equations (6.129) and (6.130) are rewritten as:

$$f''' + Aff'' - Bf'^2 - Cf' = 0 \tag{6.133}$$

$$\theta'' + Df\theta' + Ef''^2 = 0 \tag{6.134}$$

where coefficients A, B, C, D, and E are written as:

$$A=\left((1-\phi)+\phi\left(\frac{\rho_s}{\rho_f}\right)\right)(1-\phi)^{2.5}\left(\frac{m+1}{2}\right)^2 \tag{6.135}$$

$$B=\left((1-\phi)+\phi\left(\frac{\rho_s}{\rho_f}\right)\right)(1-\phi)^{2.5}(m) \tag{6.136}$$

$$C=\left((1-\phi)^{2.5}Mn\right) \tag{6.137}$$

$$D=\left((1-\phi)+\phi\frac{(\rho C_p)_s}{(\rho C_p)_f}\right)Pr \tag{6.138}$$

6.5.3 BASIC IDEA OF HAM

Let us assume the following nonlinear differential equation in the form of (Liao, 1992, 2003):

$$N[u(\tau)]=0, \tag{6.139}$$

where N is a nonlinear operator, τ is an independent variable, and $u(\tau)$ is the solution of the equation. We define the function $\omega(\tau,p)$ as follows:

$$\lim_{p\to 0}\omega(\tau,p)=u_0(\tau), \tag{6.140}$$

where p: [0,1] and $u_0(\tau)$ is the initial guess which satisfies initial or boundary conditions and

$$\lim_{p\to 1}\omega(\tau,p)=u(\tau), \tag{6.141}$$

And by using the generalized homotopy method, Liao's so-called zero-order deformation (Equation (6.139)) is

$$(1-p)L[\omega(\tau,p)-u_0(\tau)]=p\hbar H(\tau)N[\omega(\tau,p)] \tag{6.142}$$

where $\hbar$ is the auxiliary parameter which helps us to increase the convergence results, $H(\tau)$ is the auxiliary function, and L is the linear operator. It should be noted that there is a great freedom to choose the auxiliary parameter $\hbar$, the auxiliary function $H(\tau)$, the initial guess $u_0(\tau)$, and the auxiliary linear operator L. This freedom plays an important role in establishing the keystone of validity and flexibility of HAM as shown in this work.

Thus, when p increases from 0 to 1, the solution $\omega(\tau,p)$ changes between the initial guess $u_0(\tau)$ and the solution $u(\tau)$ The Taylor series expansion of $\omega(\tau,p)$ with respect to p is

$$\omega(\tau,p)=u_0(\tau)+\sum_{m=1}^{+\infty}u_m(\tau)p^m \tag{6.143}$$

and

$$u_0^{[m]}(\tau)=\left.\frac{\partial^m\omega(\tau;p)}{\partial p^m}\right|_{p=0} \tag{6.144}$$

where $u_0^{[m]}(\tau)$ briefly is called the mth-order of deformation derivation which reads

$$u_m(\tau)=\frac{u_0^{[m]}(\tau)}{m!}=\frac{1}{m!}\frac{\partial^m\omega(\tau;p)}{\partial p^m}\bigg|_{p=0} \tag{6.145}$$

It is clear that if the auxiliary parameter $\hbar=-1$ and auxiliary function $H(\tau)=1$, then Equation (6.139) will become:

$$(1-p)L[\omega(\tau,p)-u_0(\tau)]+p(\tau)N[\omega(\tau,p)]=0 \tag{6.146}$$

This statement is commonly used in HPM procedure. Indeed, in HPM, we solve the nonlinear differential equation by separating any Taylor expansion term. Now, we define the vector of

$$\vec{u}_m=\left\{\vec{u}_1,\vec{u}_2,\vec{u}_3,\ldots,\vec{u}_n\right\} \tag{6.147}$$

According to the definition in Equation (6.145), the governing equation and corresponding initial conditions of $u_m(\tau)$ can be deduced from zero-order deformation Equation (6.139). Differentiating Equation (6.139) m times with respect to the embedding parameter p and setting $p=0$ and finally dividing by $m!$, we will have the so-called mth-order deformation equation in the form:

$$L[u_m(\tau)-\chi_m u_{m-1}(\tau)]=\hbar H(\tau)R\left(\vec{u}_{m-1}\right), \tag{6.148}$$

where

$$R_m\left(\vec{u}_{m-1}\right)=\frac{1}{(m-1)!}\frac{\partial^{m-1}N[\omega(\tau;p)]}{\partial p^{m-1}}\bigg|_{p=0}, \tag{6.149}$$

and

$$\chi_m=\begin{cases}0 & m\leq 1\\ 1 & m>1\end{cases} \tag{6.150}$$

So by applying inverse linear operator to both sides of the linear equation, Equation (6.139), we can easily solve the equation and compute the generation constant by applying the initial or boundary condition.

6.5.3.1 Application in problem

For HAM solution, we choose the initial guesses and auxiliary linear operator in the following form which Ziabakhsh et al. (2009) reveal that this form of auxiliary has the best results for these kind of problems.

$$f_0(\eta)=1-\exp(-\eta),\quad \theta_0(\eta)=\exp(-\eta), \tag{6.151}$$

$$L_1(f)=f'''+f'',\quad L_2(\theta)=\theta''+\theta' \tag{6.152}$$

$$L_1(c_1\exp(-\eta)+c_2\eta+c_3)=0,\quad L_2(c_4\exp(-\eta)+c_5)=0, \tag{6.153}$$

And $c_i\,(i=1-5)$ are constants. Let $P\in[0,1]$ denote the embedding parameter and $\hbar$ indicate nonzero auxiliary parameters. We then construct the following equations:

6.5.3.1.1 Zeroth-order deformation equations

$$(1-P)L_1[f(\eta;p)-f_0(\eta)]=p\hbar N_1[f(\eta;p),\theta(\eta;p)] \tag{6.154}$$

$$f(0;p)=0;\ f'(0;p)=1,\ f'(\infty;p)=0. \tag{6.155}$$

$$(1-P)L_2[\theta(\eta;p)-\theta_0(\eta)]=p\hbar N_2[f(\eta;p),\theta(\eta;p)] \tag{6.156}$$

$$\theta(0;p)=1;\ \theta(\infty;p)=0. \tag{6.157}$$

$$N_1[f(\eta;p),\theta(\eta;p)]=\frac{\partial^3 f(\eta;p)}{\partial\eta^3}+Af(\eta;p)\frac{\partial^2 f(\eta;p)}{\partial\eta^2}-B\frac{\partial f(\eta;p)}{\partial\eta}\frac{\partial f(\eta;p)}{\partial\eta}-C\frac{\partial f(\eta;p)}{\partial\eta} \tag{6.158}$$

$$N_2[f(\eta;p),\theta(\eta;p)]=\frac{\partial^2\theta(\eta;p)}{\partial\eta^2}+Df(\eta;p)\frac{\partial\theta(\eta;p)}{\partial\eta}+E\theta(\eta;p) \tag{6.159}$$

For $p=0$ and $p=1$ we have

$$f(\eta;0)=f_0(\eta),\ f(\eta;1)=f(\eta) \tag{6.160}$$

$$\theta(\eta;0)=\theta_0(\eta),\ \theta(\eta;1)=\theta(\eta) \tag{6.161}$$

When p increases from 0 to 1 then $f(\eta;p)$ and $\theta(\eta;p)$ vary from $f_0(\eta)$ and $\theta_0(\eta)$ to $f(\eta)$ and $\theta(\eta)$. By Taylor's theorem and using Equations (6.160) and (6.161), $f(\eta;p)$ and $\theta(\eta;p)$ can be expanded in a power series of p as follows:

$$f(\eta;p)=f_0(\eta)+\sum_{m-1}^{\infty}f_m(\eta)p^m,\ f_m(\eta)=\frac{1}{m!}\frac{\partial^m(f(\eta;p))}{\partial p^m} \tag{6.162}$$

$$\theta(\eta;p)=\theta_0(\eta)+\sum_{m-1}^{\infty}\theta_m(\eta)p^m,\ \theta_m(\eta)=\frac{1}{m!}\frac{\partial^m(\theta(\eta;p))}{\partial p^m} \tag{6.163}$$

In which $\hbar$ is chosen in such a way that these two series are convergent at $p=1$; therefore, we have through Equations (6.162) and (6.163) that

$$f(\eta)=f_0(\eta)+\sum_{m-1}^{\infty}f_m(\eta),\ \theta(\eta)=\theta_0(\eta)+\sum_{m-1}^{\infty}\theta_m(\eta), \tag{6.164}$$

6.5.3.1.2 *m*th-order deformation equations

$$L_1[f_m(\eta)-\chi_m f_{m-1}(\eta)]=\hbar_1 H(\eta)R_m^f(\eta) \tag{6.165}$$

$$f_m(0)=0;\ f_m'(0)=0,\ f_m'(\infty)=0 \tag{6.166}$$

$$L_2[\theta_m(\eta)-\chi_m\theta_{m-1}(\eta)]=\hbar_2 H(\eta)R_m^{\theta}(\eta) \tag{6.167}$$

$$\theta_m(0)=0;\ \theta_m(\infty)=0. \tag{6.168}$$

$$R_m^f(\eta)=f'''_{m-1}+A\sum_{n=0}^{m-1} f_{m-1-n}f''_n-B\sum_{n=0}^{m-1} f'_n f'_{m-1-n}-Cf'_{m-1} \tag{6.169}$$

$$R_m^{\theta}(\eta)=\theta''_{m-1}+D\sum_{n=0}^{m-1} f_{m-1-n}\theta'_n+E\sum_{n=0}^{m-1} f''_{m-1-n}f''_n \tag{6.170}$$

$$\chi_m=\begin{cases}0 & m\le 1\\ 1 & m>1\end{cases} \tag{6.171}$$

The general solutions then become:

$$\begin{aligned} f_m(\eta)-\chi_m f_{m-1}(\eta)&=f_m^*(\eta)+C_1^m\exp(-\eta)+C_2^m\eta+C_3^m,\\ \theta_m(\eta)-\chi_m\theta_{m-1}(\eta)&=\theta_m^*(\eta)+C_4^m\exp(-\eta)+C_5^m, \end{aligned} \tag{6.172}$$

where C_1^m to C_5^m are constants that can be obtained by applying the boundary conditions in Equations (6.166) and (6.168). We define the auxiliary function $H(\eta)$ in the following form:

$$H(\eta)=\exp(-\eta) \tag{6.173}$$

6.5.4 NUMERICAL METHOD

The numerical solution is performed using the algebra package Maple 15.0 to solve the present case. The package uses a fourth-order Runge-Kutta procedure for solving nonlinear boundary value (B-V) problem. The algorithm is proved to be precise and accurate in solving a wide range of mathematical and engineering problems, especially, heat transfer cases.

As mentioned, the type of the current problem is BVP and the appropriate method needs to be selected. The available submethods in the Maple 15.0 are a combination of the base schemes, trapezoid or midpoint method. There are two major considerations when choosing a method for a problem. The trapezoid method is generally efficient for typical problems, but the midpoint method is so capable of handling harmless end-point singularities that the trapezoid method cannot. The midpoint method, also known as the fourth-order Runge-Kutta-Fehlberg method, improves the Euler method by adding a midpoint in the step which increases the accuracy by one order. Thus, the midpoint method is used as a suitable numerical technique (see Aziz, 2006).

6.5.5 RESULTS AND DISCUSSION

In this study, governing equations for forced convection of nanofluid over a horizontal plate (Figure 6.41) is solved by an analytical method.

HAM is used for this aim due to many advantages; some of them are described by Liao as follows:

(a) Unlike all other analytic techniques, HAM provides us with great freedom to express solutions of a given nonlinear problem by means of different base functions.

(b) HAM always provides us with a family of solution expressions in the auxiliary parameter $\hbar_1$, even if a nonlinear problem has a unique solution; so, the auxiliary parameter $\hbar_1$ provides us with an additional way to conveniently adjust and control the convergence region and rate of solution series.

(c) Unlike perturbation techniques, the HAM is independent of any small or large quantities. So, the HAM can be applied no matter if governing equations and boundary/initial conditions of a given nonlinear problem contain small or large quantities or not.

As pointed out by Liao, the convergence and rate of approximation for the HAM solution strongly depend on the values of auxiliary parameters $\hbar_1$ and $\hbar_2$. Note that the Equations (6.154) and (6.156) contain the auxiliary parameter $\hbar$ which is not yet defined. This parameter plays an important role in the framework of HAM. In fact, this parameter controls the rate of convergence and the convergence region of the series. Proper values for this auxiliary parameter can be found by plotting the so-called $\hbar$ curves when $\eta=0$ for Equations (6.154) and (6.156) for θ and f' functions, respectively. For f', the acceptable ranges for values of $\hbar_1$ are $-2<\hbar_1<-0.5$ (Figure 6.42a) and $-1<\hbar_1<-0.1$ (Figure 6.42b).

For θ, the acceptable ranges for values of $\hbar_2$ are $-1.8<\hbar_2<-0.7$ (Figure 6.43a) and $-0.8<\hbar_2<-0.2$ (Figure 6.43b).

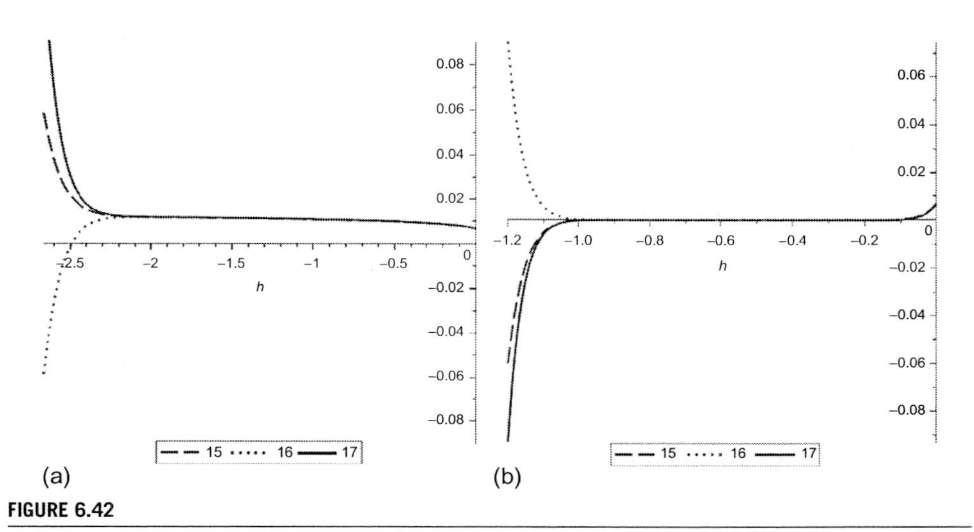

FIGURE 6.42

The $\hbar_1$ validity for different orders (15th, 16th, and 17th) of approximation (a) $Ec=0.1$, $m=0.1$, $Pr=6.2$, $\phi=0.05$, and $Mn=0.5$; (b) $Ec=0.1$, $m=0.1$, $Pr=6.2$, $\phi=0.05$, and $Mn=5$.

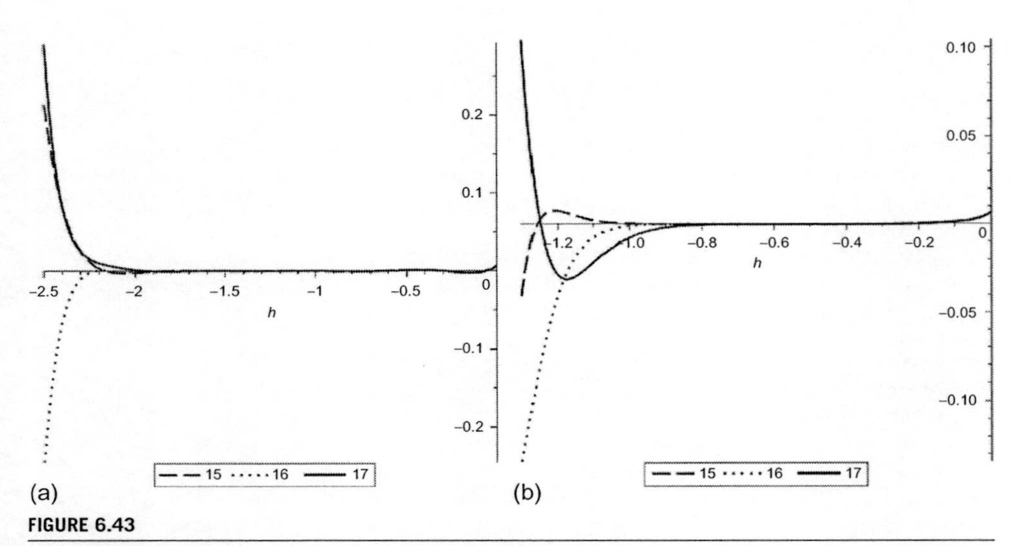

FIGURE 6.43

The $\hbar_2$ validity for different orders (15th, 16th, and 17th) of approximation (a) $Ec=0.1$, $m=0.1$, $Pr=6.2$, $\phi=0.05$, and $Mn=0.5$; (b) $Ec=0.1$, $m=0.1$, $Pr=6.2$, $\phi=0.05$, and $Mn=5$.

Table 6.6 Physical Properties of Fluid (Water) and Nanoparticles (Aluminum Oxide)

Properties	Water	Aluminum oxide (Al_2O_3)
$\rho\,(\mathrm{kgm^{-3}})$	997.1	3890
$C_p\,(\mathrm{Jkg^{-1}K^{-1}})$	4179	880
$K\,(\mathrm{Wm^{-1}K^{-1}})$	0.613	35

For showing the accuracy of the HAM method, by using Table 6.6 properties, this problem is solved for water with Al_2O_3 nanoparticles in different Mn numbers (1, 2, and 5) when $Ec=0.1$, $m=0.1$, $Pr=6.2$, and $\phi=0.05$, and results are compared with the numerical solution in Figure 6.44.

Figure 6.44a and b illustrates the nondimensional temperature and velocity profiles. As seen in Figure 6.44, the HAM has an excellent compatibility with numerical procedure in both temperature and velocity profiles.

Table 6.7 data confirm the high accuracy of HAM, and Table 6.8 compares HPM and HAM results and it is obvious that HAM has lower errors and is more accurate than HPM. Because the traditional perturbation techniques are the special cases of HAM when $\hbar=-1$, these techniques cannot give convergent results in all cases versus HAM.

The effect of nanoparticle volume fraction (φ) on temperature and velocity profiles is depicted in Figure 6.45 for a special case when $Ec=0.1$, $m=0.1$, $Pr=6.2$, and $Mn=0.5$.

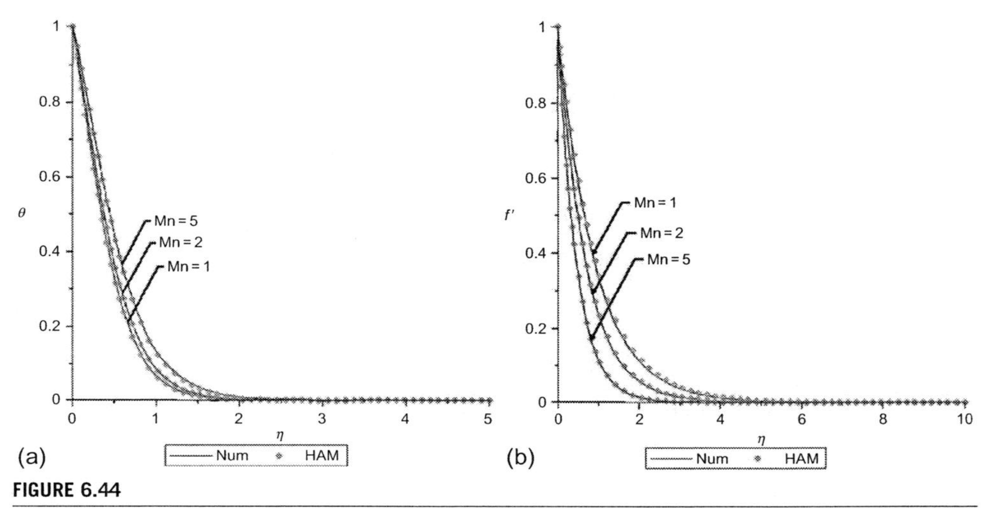

FIGURE 6.44

Comparison of HAM and numerical solution in different *Mn* numbers when $Ec=0.1$, $m=0.1$, $Pr=6.2$, and $\phi=0.05$ for (a) temperature profile, (b) velocity profile.

Table 6.7 Comparison Between Num. and HAM Results When $Ec=0.1$, $m=0.1$, $Pr=6.2$, $\phi=0.05$, and $Mn=2$

	f'			θ		
η	Num	HAM	Relative Errors	Num	HAM	Relative Errors
0	1.00000000	0.99999829	1.7E−06	1.00000000	1.00008099	0.000499
0.2	0.75660464	0.75659825	6.4E−06	0.72420677	0.72470547	0.000654
0.4	0.57125079	0.57123916	1.2E−05	0.47175314	0.47240716	0.000724
0.6	0.43061741	0.43059937	1.8E−05	0.28201760	0.28274178	0.000745
0.8	0.32421347	0.32418678	2.7E−05	0.15774431	0.15848938	0.000738
1	0.24387842	0.24383966	3.9E−05	0.08397983	0.08471762	0.000719
1.2	0.18332261	0.18326907	5.4E−05	0.04317549	0.04389474	0.0007
1.4	0.13773137	0.13766635	6.5E−05	0.02169578	0.02239620	0.000684
1.6	0.10343793	0.10337658	6.1E−05	0.01076203	0.01144609	0.000667
1.8	0.07766028	0.07763097	2.9E−05	0.00531243	0.00597920	0.000644
2	0.05829376	0.05833281	3.9E−05	0.00262640	0.00326993	0.000611
2.2	0.04374949	0.04389195	0.000142	0.00130685	0.00191804	0.000569
2.4	0.03282992	0.03310081	0.000271	0.00065676	0.00122622	0.00052

Continued

Table 6.7 Comparison Between Num. and HAM Results When $Ec=0.1$, $m=0.1$, $Pr=6.2$, $\phi=0.05$, and $Mn=2$—cont'd

	f'			θ		
η	Num	HAM	Relative Errors	Num	HAM	Relative Errors
2.6	0.02463349	0.02504282	0.000409	0.00033408	0.00085443	0.000467
2.8	0.01848211	0.01902429	0.000542	0.00017218	0.00063909	0.000412
3	0.01386609	0.01452277	0.000657	0.00008992	0.00050221	0.000359
3.2	0.01040255	0.01114706	0.000745	0.00004754	0.00040665	0.000309
3.4	0.00780391	0.00860604	0.000802	0.00002542	0.00033463	0.000264
3.6	0.00585430	0.00668413	0.00083	0.00001373	0.00027747	0.000223
3.8	0.00439168	0.00522233	0.000831	0.00000748	0.00023069	0.000188
4	0.00329444	0.00410363	0.000809	0.00000410	0.00019183	0.000157
4.2	0.00247131	0.00324195	0.000771	0.00000226	0.00015934	0.000131
4.4	0.00185383	0.00257390	0.00072	0.00000125	0.00013215	0.000109
4.6	0.00139063	0.00205266	0.000662	0.00000070	0.00010941	9E−05
4.8	0.00104316	0.00164348	0.0006	0.00000039	0.00009043	7.44E−05

Table 6.8 Comparison Between HAM and HPM Results When $Ec=0.1$, $m=0.1$, $Pr=6.2$, $\phi=0.2$, and $Mn=0.2$

	θ			% Error	
η	Num	HAM	HPM [1]	HAM	HPM [1]
0	1.000000	1.000000	1.000000	0	0
0.2	0.913421	0.913231	0.914671	0.00021	0.001368
0.4	0.827645	0.827531	0.830011	0.00014	0.002859
0.6	0.744232	0.744124	0.747524	0.00015	0.004423
0.8	0.664324	0.664267	0.668484	8.6E−05	0.006262
1	0.588935	0.588649	0.593911	0.00049	0.008449
1.2	0.518546	0.518416	0.524578	0.00025	0.011633
1.4	0.453628	0.453129	0.461022	0.0011	0.0163
1.6	0.394431	0.393298	0.403573	0.00287	0.023178
1.8	0.340898	0.346982	0.352385	0.017847	0.033696
2	0.292854	0.291236	0.307456	0.00552	0.049861
2.2	0.249931	0.248621	0.268638	0.00524	0.074849
2.4	0.212125	0.211698	0.235634	0.00201	0.110826
2.6	0.178646	0.177564	0.207977	0.00606	0.164185
2.8	0.149359	0.148131	0.185003	0.00822	0.238646
3	0.123748	0.122692	0.165822	0.00853	0.339997
3.2	0.101569	0.100691	0.149317	0.00864	0.470104
3.4	0.082226	0.081223	0.134198	0.0122	0.632063

Table 6.8 Comparison Between HAM and HPM Results When $Ec=0.1$, $m=0.1$, $Pr=6.2$, $\phi=0.2$, and $Mn=0.2$—cont'd

	θ			% Error	
η	Num	HAM	HPM [1]	HAM	HPM [1]
3.6	0.065648	0.056982	0.119118	0.13201	0.814495
3.8	0.051214	0.046982	0.102891	0.08263	1.00904
4	0.038856	0.029631	0.084773	0.23742	1.181722
4.2	0.028149	0.012396	0.064787	0.55963	1.301574
4.4	0.013087	0.008913	0.043964	0.31894	2.359364
4.6	0.011226	0.009235	0.024391	0.17736	1.172724
4.8	0.004432	0.003797	0.008861	0.14328	0.999323

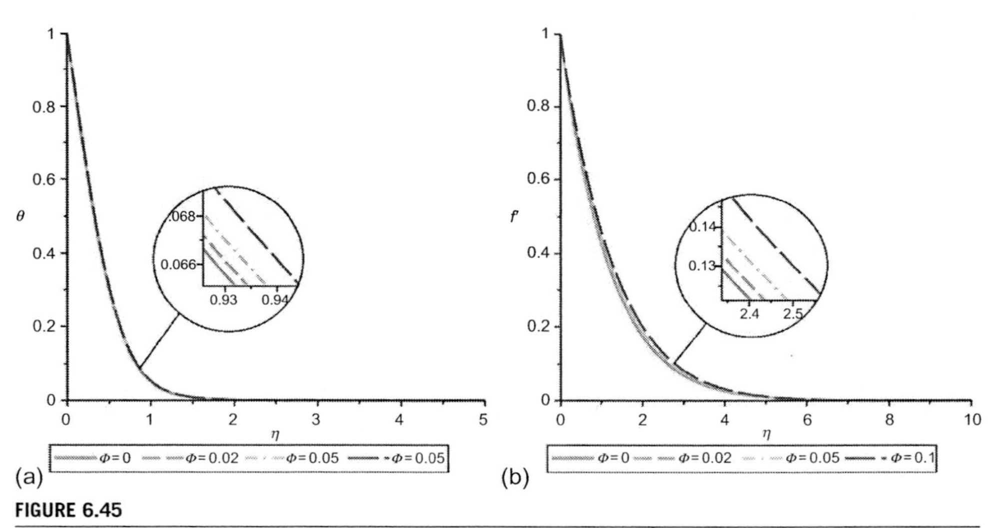

FIGURE 6.45

Effect of nanoparticles volume fraction when $Ec=0.1$, $m=0.1$, $Pr=6.2$, and $Mn=0.5$ on nondimensional (a) temperature profile, (b) velocity profile.

It can be concluded from Figure 6.45 that increase in nanoparticle volume fraction makes an increase in temperature boundary layer; also, it causes an increase in velocity profile and velocity boundary layer. Figure 6.46a and b shows the magnetic field effect (Mn number) on temperature and velocity profiles when $Ec=0.1$, $m=0.1$, $Pr=6.2$, and $\phi=0.05$. As seen by increasing the Mn number, thermal boundary layer is increased but velocity boundary layer, due to magnetic effect on nanofluid flow, is decreased.

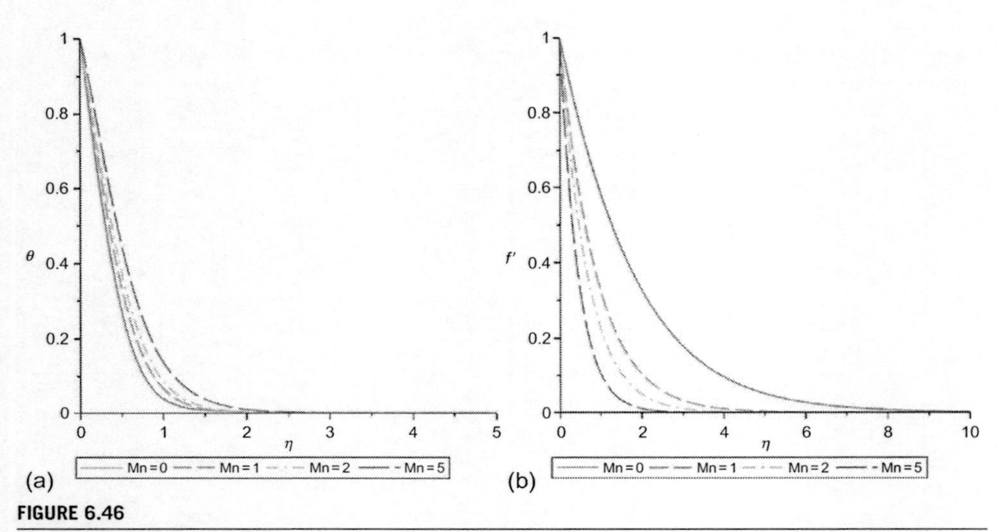

FIGURE 6.46

Effect of magnetic parameter (*Mn*) when $Ec = 0.1$, $m = 0.1$, $Pr = 6.2$, and $\phi = 0.05$ on nondimensional (a) temperature profile (b) velocity profile.

6.5.6 CONCLUSION

In this section, HAM has been successfully applied to find the solution to forced convection of MHD nanofluid flow over a horizontal plate. The proper range of auxiliary parameter $\hbar$ to ensure the convergency of the solution series was obtained through the so-called $\hbar$ curves. The following points can be concluded from the present study:

- The obtained solutions revealed that HAM can be a simple, powerful, and efficient technique for finding analytical solutions in science and engineering nonlinear differential equations.
- Calculated errors revealed that HAM, due to its accuracy, led to high appropriate results compared to HPM.
- Velocity boundary-layer thickness decreases with increase in magnetic field effect (*Mn* number) and it increases with nanoparticle volume fraction (φ) increasing.
- Thermal boundary-layer thickness is increased by increasing the *Mn* number.

REFERENCES

Abu-Nada, E., Masoud, Z., Hijazi, A., 2008. Natural convection heat transfer enhancement in horizontal concentric annuli using nanofluids. Int. Commun. Heat Mass Transfer 35, 657–665.

Alfvén, H., 1942. Existence of electromagnetic-hydrodynamic waves. Nature 150, 405.

Aminossadati, S.M., Ghasemi, B., 2009. Natural convection cooling of a localised heat source at the bottom of a nanofluid-filled enclosure. Eur. J. Mech. B. Fluids 28, 630–640.

Arkilic, E.B., Schmidt, M.A., Breuer, K.S., 1997. Gaseous slip flow in long microchannels. J. Microelectromech. Syst. 6, 167–178.

Aziz, A., 2006. Heat Conduction with Maple. R.T. Edwards, Philadelphia, PA.

Bararnia, H., Soleimani, Soheil, Ganji, D.D., 2011a. Lattice Boltzmann simulation of natural convection around a horizontal elliptic cylinder inside a square enclosure. Int. Commun. Heat Mass Transfer 38, 1436–1442.

Bararnia, H., Hooman, K., Ganji, D.D., 2011b. Natural convection in a nanofluid filled portion cavity; the lattice-Boltzmann method. Numer. Heat Transfer, Part A 59, 487–502.

Barrios, G., Rechtman, R., Rojas, J., Tovar, R., 2005. The lattice Boltzmann equation for natural convection in a two-dimensional cavity with a partially heated wall. J. Fluid Mech. 522, 91–100.

Bejan, A., Dincer, I., Lorente, S., Miguel, A.F., Rei, A.H., 2004. Porous and Complex Flow Structures in Modern Technologies. Springer, New York.

Beskok, A., Karniadakis, G.E., 1999. A model for flows in channels, pipes and ducts at micro and nano scales. Microscale Thermophys. Eng. 3 (1), 43–47.

Brinkman, H.C., 1952. The viscosity of concentrated suspensions and solution. J. Chem. Phys. 20, 571–581.

Bruggeman, D.A.G., 1935. Berechnung verschiedener physikalischer konstanten von heterogenen substanzen, I. Dielektrizitatskonstanten und leitfahigkeiten der mischkorper aus isotropen substanzen. Ann. Phys. (Leipzig) 24, 636–679.

Buongiorno, J., 2006. Convective transport in nanofluids. ASME J. Heat Transfer 128, 240–250.

Chein, Reiyu, Huang, Guanming, 2005. Analysis of microchannel heat sink performance using nanofluids. Appl. Therm. Eng. 25 (17–18), 3104–3114.

Chen, Y.-C., Chung, J.N., 1996. The linear stability of mixed convection in a vertical channel flow. J. Fluid Mech. 325, 29–51.

Chiam, T.C., 1995. Hydromagnetic flow over a surface stretching with a power-law velocity. Int. J. Eng. Sci. 33 (3), 429–435.

Chon, C.H., Kihm, K.D., Lee, S.P., Choi, S.U.S., 2005. Empirical correlation finding the role of temperature and particle size for nanofluid (Al_2O_3) thermal conductivity enhancement. Appl. Phys. Lett. 89(97).

Cimpean, D.S., Pop, I., 2012. Fully developed mixed convection flow of a nanofluid through an inclined channel filled with a porous medium. Int. J. Heat Mass Transfer 55, 907–914.

Devi, S.P.A., Thiyagarajan, M., 2006. Steady nonlinear hydromagnetic flow and heat transfer over a stretching surface of variable temperature. Heat Mass Transfer 42 (8), 671–677.

Dixit, H.N., Babu, V., 2006. Simulations of high Rayleigh number natural convection in a square cavity using the lattice Boltzmann method. Int. J. Heat Mass Transfer 49, 727–739.

Gad-el-Hak, M., 1999. The fluid mechanics of microdevices: the Freeman scholar lecture. ASME J. Fluid Eng. 121, 5–23.

Gebhart, B., Jaluria, Y., Mahajan, R.L., Sammakia, B., 1988. Buoyancy-Induced Flows and Transport. Hemisphere Publishing, New York.

Gerald, C.F., Weatly, P.O., 1989. Applied Numerical Analysis. Addison Wesley Publishing Company, New York.

He, J.H., 2003. Homotopy perturbation method: a new nonlinear analytical technique. Appl. Math. Comput. 135, 73–79.

He, J.-H., 2006. Homotopy perturbation method for solving boundary value problems. Phys. Lett. A 350 (1–2), 87–88.

Ingham, D.B., Pop, I. (Eds.), 1998. Transport Phenomena in Porous Media. Pergamon, Oxford, vol. II 2002, vol. III 2005.

Jang, S.P., Choi, S.V.S., 2007. Effect of various parameters on nanofluid thermal conductivity. ASME J. Heat Transfer 129, 617–623.

Jung, Jung-Yeul, Oh, Hoo-Suk, Kwak, Ho-Young, 2009. Forced convective heat transfer of nanofluids in microchannels. Int. J. Heat Mass Transfer 52 (1–2), 466–472.

Kao, P.H., Yang, R.J., 2007. Simulating oscillatory flows in Rayleigh-Benard convection using the lattice Boltzmann method. Int. J. Heat Mass Transfer 50, 3315–3328.

Khanafer, K., Vafai, K., Lightstone, M., 2003. Buoyancy-driven heat transfer enhancement in a two-dimensional enclosure utilizing nanofluids. Int. J. Heat Mass Transfer 46, 3639–3653.

Koo, J., Kleinstreuer, C., 2005. Laminar nanofluid flow in microheat-sinks. Int. J. Heat Mass Transfer 48 (13), 2652–2661.

Kuehn, T.H., Goldstein, R.J., 1976. An experimental and theoretical study of natural convection in the annulus between horizontal concentric cylinders. J. Fluid Mech. 74, 695–719.

Laving, A.S., 1988. Analysis of fully developed opposing mixed convection between inclined parallel plates. Warme Stoffubertragung 23, 249–257.

Liao, S.J., 1992. The Proposed Homotopy Analysis Technique for the Solution of Nonlinear Problems. Ph.D. thesis, Shanghai Jiao Tong University.

Liao, S.J., 2003. Beyond Perturbation: Introduction to Homotopy Analysis Method. Chapman & Hall/CRC Press, Boca Raton, FL.

Lin, P.F., Lin, J.Z., 2009. Prediction of nanoparticle transport and deposition in bends. Appl. Math. Mech. 30 (8), 957–968. http://dx.doi.org/10.1007/s10483-009-0802-z.

Lin, J., Lin, P., Chen, H., 2009. Research on the transport and deposition of nanoparticles in a rotating curved pipe. Phys. Fluids 21, 122001.

Martin, M.J., Boyd, I.D., 2006. Momentum and heat transfer in a laminar boundary layer with slip flow. J. Thermophys Heat Transfer 20 (4), 710–719.

Martynenko, O.G., Khramtsov, P.P., 2005. Free-Convective Heat Transfer. Springer, Berlin.

Maxwell, J., 1904. A Treatise on Electricity and Magnetism, second ed. Oxford University Press, Cambridge, UK.

Mohammed, H.A., Gunasegaram, P., Shuaib, N.H., 2010. Heat transfer in rectangular microchannels heat sink using nanofluids. Int. Commun. Heat Mass Transfer 37 (10), 1496–1503.

Moukalled, F., Acharya, S., 1996. Natural convection in the annulus between concentric horizontal circular and square cylinders. J. Thermophys Heat Transfer 10 (3), 524–531.

Naterer, G.F., 2005. Embedded converging surface microchannels for minimized friction and thermal irreversibilities. Int. J. Heat Mass Transfer 48, 1225–1235.

Ngoma, G.D., Erchiqui, F., 2007. Heat flux and slip effects on liquid flow in a microchannel. Int. J. Therm. Sci. 46, 1076–1083.

Nourazar, S.S., Habibi Matin, M., Simiari, M., 2011. The HPM applied to MHD nanofluid flow over a horizontal stretching plate. J. Appl. Math. 2011. http://dx.doi.org/10.1155/2011/876437, Article ID 876437.

Onyegegbu, S.O., 1986. Heat transfer inside a horizontal cylindrical annulus in the presence of thermal radiation and buoyancy. Int. J. Heat Mass Transfer 29, 659–671.

Patel, H.E., Sundararajan, T., Pradeep, T., Dasgupta, A., Dasgupta, N., Das, S.K., 2005. A micro-convection model for thermal conductivity of nanofluids. Pramana J. Phys. 65, 863–869.

Shaija, A., Narasimham, G.S.V.L., 2009. Effect of surface radiation on conjugate natural convection in a horizontal annulus driven by inner heat generating solid cylinder. Int. J. Heat Mass Transfer 52, 5759–5769.

Soleimani, Soheil, Sheikholeslami, M., Ganji, D.D., Gorji-Bandpay, M., 2012. Natural convection heat transfer in a nanofluid filled semi-annulus enclosure. Int. Commun. Heat Mass Transfer 39 (4), 565–574.

Succi, S., 2001. The Lattice Boltzmann Equation for Fluid Dynamics and Beyond. Clarendon Press, Oxford.

Vafai, K. (Ed.), 2000. Handbook of Porous Media. Marcel Dekker, New York.

Xuan, Y., Li, Q., Hu, W., 2003. Aggregation structure and thermal conductivity of nanofluids. AIChE J. 49 (4), 1038–1043.

Yan, Y.Y., Zu, Y.Q., 2008. Numerical simulation of heat transfer and fluid flow past a rotating isothermal cylinder—a LBM approach. Int. J. Heat Mass Transfer 51, 2519–2536.

Yang, Y.T., Lai, F.H., 2011. Numerical study of flow and heat transfer characteristics of alumina-water nanofluids in a microchannel using the lattice Boltzmann method. Int. Commun. Heat Mass Transfer 38 (5), 607–614.

Yazdi, M.H., Abdullah, S., Hashim, I., Sopian, K., 2008. Friction and heat transfer in slip flow boundary layer at constant heat flux boundary conditions. In: Presented at 10th WSEAS International Conference on Mathematical Methods, Computational Techniques and Intelligent Systems, Greece.

Yu, D., Mei, R., Luo, L.S., Shyy, W., 2003. Viscous flow computations with the method of lattice Boltzmann equation. Prog. Aerosp. Sci. 39, 329–367.

Ziabakhsh, Z., Domairry, G., Bararnia, H., 2009. Analytical solution of non-Newtonian micropolar fluid flow with uniform suction/blowing and heat generation. J. Taiwan Inst. Chem. Eng. 40, 443–451.

CHAPTER

7 NANOFLUID FLOW IN POROUS MEDIUM

CHAPTER CONTENTS

7.1 Introduction of Porous Medium ... 272
7.2 Stagnation Point Flow of Nanofluids in a Porous Medium ... 272
7.2.1 Introduction ... 272
7.2.2 Mathematical Formulation ... 273
7.2.3 Numerical Procedure and Validation ... 276
7.2.4 Results and Discussions ... 277
7.2.5 Conclusion ... 284
7.3 Flow and Heat Transfer of Nanofluids in a Porous Medium ... 286
7.3.1 Introduction ... 286
7.3.2 Problem Statement ... 287
7.3.3 Flow Analysis and Mathematical Formulation ... 288
7.3.3.1 Hydrodynamics ... 288
7.3.3.2 Thermal Analysis ... 289
7.3.4 The HAM Solution of the Problem ... 291
7.3.5 Convergence of the HAM Solution ... 293
7.3.6 Results and Discussions ... 293
7.3.7 Conclusions ... 300
7.4 Natural Convection in a Non-Darcy Porous Medium of Nanofluids ... 301
7.4.1 Introduction ... 303
7.4.2 Governing Equations ... 303
7.4.3 Solution Using HAM ... 306
7.4.3.1 Zeroth-Order Deformation Equations ... 306
7.4.3.2 mth-Order Deformation Equations ... 307
7.4.4 Convergence of HAM Solution ... 307
7.4.5 Results and Discussions ... 308
7.4.6 Conclusion ... 314
References ... 314

Application of Nonlinear Systems in Nanomechanics and Nanofluids. http://dx.doi.org/10.1016/B978-0-323-35237-6.00007-8

7.1 INTRODUCTION OF POROUS MEDIUM

A porous medium (or a porous material) is a material containing pores (voids). The skeletal portion of the material is often called the "matrix" or "frame." The pores are typically filled with a fluid (liquid or gas). The skeletal material is usually a solid, but structures like foams are often also usefully analyzed using concept of porous media.

A porous medium is most often characterized by its porosity. Other properties of the medium (e.g., permeability, tensile strength, electrical conductivity) can sometimes be derived from the respective properties of its constituents (solid matrix and fluid) and the media porosity and pores structure, but such a derivation is usually complex. Even the concept of porosity is only straightforward for a poroelastic medium.

Often both the solid matrix and the pore network (also known as the pore space) are continuous, so as to form two interpenetrating continua such as in a sponge. However, there is also a concept of closed porosity and effective porosity, i.e., the pore space accessible to flow.

Many natural substances such as rocks and soil (e.g., aquifers, petroleum reservoirs), zeolites, biological tissues (e.g., bones, wood, cork), and man-made materials such as cements and ceramics can be considered as porous media. Many of their important properties can only be rationalized by considering them to be porous media.

The concept of porous media is used in many areas of applied science and engineering: filtration, mechanics (acoustics, geomechanics, soil mechanics, rock mechanics), engineering (petroleum engineering, bio-remediation, construction engineering), geosciences (hydrogeology, petroleum geology, geophysics), biology and biophysics, material science, etc. Fluid flow through porous media is a subject of most common interest and has emerged a separate field of study. The study of more general behavior of porous media involving deformation of the solid frame is called poromechanics.

7.2 STAGNATION POINT FLOW OF NANOFLUIDS IN A POROUS MEDIUM

In this section, the steady two-dimensional stagnation point flow of three types of nanofluids, namely, Cu-water, Al_2O_3-water, and TiO_2-water, toward a permeable stretching surface with heat generation is investigated. The governing system of partial differential equations is transformed into ordinary differential equations, which are then solved numerically by shooting technique along with the fourth-order Runge-Kutta method. Comparisons with previously published works show the accuracy of the obtained results. Effects of dispersion of different nanoparticles along with changes in nanoparticles volume fraction on flow and heat characteristics are presented through temperature and velocity profiles. Also reduced Nusselt number and skin-friction coefficients are at the surface and are determined numerically for various values of nanoparticle volume fractions. It was found that increasing the volume fraction of nanoparticles enhances the heat transfer rate and increases the skin-friction coefficient at the surface when $C > 1$. Moreover, Cu-water has the highest heat transfer rate and the highest skin-friction coefficient at the surface compared with the others (this section has been worked by N. Ghadimi, D.D. Ganji, M. Abdollahzadeh in nonlinear dynamics team in Mechanical Engineering Department, 2012-2013).

7.2.1 INTRODUCTION

The problem of heat transfer in fluid-saturated porous media is frequently encountered in various possible applications of this type of flow in many processes such as cooling of electronic devices by fans, food processing, and hydrodynamic processes. Due to these applications, in recent years, the flow and heat transfer in porous media have attended by many researchers in different conditions (see Attia, 2006; Layek et al., 2007; Ishak et al., 2009).

The practical interest in enhancement of convective heat transfer in porous media was greatly increased over the last several years, and this led to the utilization of nanofluids in industrial instruments. The utility of a particular nanofluid for a heat transfer application can be established by suitably modeling the convective transport in the nanofluid (Kumar et al., 2010).

Aminossadati and Ghasemi (2011) numerically studied the homogenous and Newtonian natural convection of water-CuO nanofluid in a two-dimensional square cavity with two pairs of heat source-sink covering the entire length of the bottom wall of the cavity. The results showed that regardless of the position of the pairs of source-sink, the heat transfer rate increases with an increase of the Rayleigh number and the solid volume fraction. Talebi et al. (2010) executed a numerical investigation of laminar mixed convection flow of copper-water nanofluid in a square lid-driven cavity. In the present study, the top and bottom horizontal walls are insulated, while the vertical walls are maintained at constant but different temperatures. It was found that at the fixed Reynolds number, the solid concentration affects on the flow pattern and thermal behavior particularly for a higher Rayleigh number.

Transient natural convection heat transfer of aqueous nanofluids in a differentially heated square cavity is investigated numerically by Yu et al. (2011). Kuznetsov and Nield (2010) studied the influence of nanoparticles on natural convection boundary-layer flow past a vertical plate, using a model in which Brownian motion and thermophoresis are accounted for. Further, Nield and Kuznetsov (2009) have studied the Cheng and Minkowycz (1977) problem of natural convection past a vertical plate, in a porous medium saturated by a nanofluid. The model used for the nanofluid incorporates the effects of Brownian motion and thermophoresis. For the porous medium, the Darcy model has been employed.

The present work investigates the thermal behavior of the stagnation point flow of nanofluids in a porous medium on a stretching sheet with heat generation. The resulting governing nonlinear ordinary differential equations are integrated numerically using a Runge-Kutta shooting method. Effects of dispersion of different types of metallic nanoparticles such as Al_2O_3, TiO_2, and Cu along with changes in nanoparticles volume fraction on flow and heat characteristics are presented through temperature and velocity profiles. Results are provided for representative values nanoparticle volume fraction ($\varphi = 0, 0.1, 0.2$). Furthermore, the variation of the Nusselt number and skin-friction coefficient due to nanoparticles dispersion is presented in graphical forms.

7.2.2 MATHEMATICAL FORMULATION

We consider the steady two-dimensional stagnation flow of a water-based nanofluid containing nanoparticle on a permeable stretching surface. The flow situation and the coordinate system with the origin being the stagnation point and the y-axis is normal to the plane are shown in Figure 7.1. The stretching velocity $U_w(x)$ and the free-stream velocity $U_\infty(x)$ are assumed to vary proportional to the distance x from the stagnation point, i.e., $U_w(x) = cx$ and $U_\infty(x) = ax$, where c and a are constants with $c > 0$ and $a \geq 0$. Also, the wall temperature and the free-stream temperature are re-assumed to be constants.

The nanofluid and the flow are assumed to be incompressible and laminar, respectively. It is assumed that the base fluid (i.e., water) and the nanoparticles are in thermal equilibrium and no slip occurs between them. The thermophysical properties of the nanofluid are given in Table 7.1. Thermophysical properties of the nanofluid are assumed to be constant.

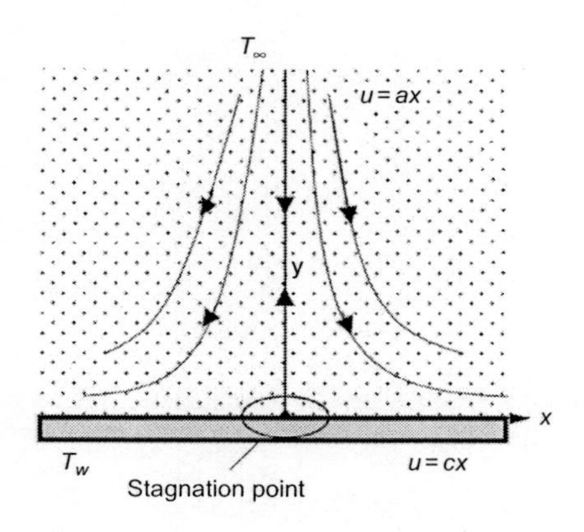

FIGURE 7.1

Schematic representation of problem.

Table 7.1 Thermophysical Properties of the Cu Nanoparticles and Base Fluid (Water)

Physical Property	Cu Nanoparticle	TiO_2 Nanoparticle	Al_2O_3 Nanoparticle	Base Fluid (Water)
ρ[kg/m^3]	8954	4250	3970	997.1
μ[Pa. s]	–	–	–	8.9×10^{-4}
c_p[J/kg K]	383	686.2	765	4179
k[W/m K]	400	8.9538	40	0.6
Pr	–	–	–	6.2
$d_p[m]$	10^{-9}	10^{-9}	10^{-9}	–

The continuity and momentum equations for the two-dimensional steady state flows, using the usual boundary-layer approximations, are expressed as below:
Continuity:

$$\frac{\partial u}{\partial x}+\frac{\partial v}{\partial y}=0 \tag{7.1}$$

Momentum:

$$\rho_{\mathrm{nf}}\left(u\frac{\partial u}{\partial x}+v\frac{\partial u}{\partial y}\right)=\rho_{\mathrm{nf}}U_\infty\frac{\mathrm{d}U_\infty}{\mathrm{d}x}+\mu_{\mathrm{nf}}\left(\frac{\partial^2 u}{\partial y^2}\right)+\frac{\mu}{K}(U_\infty-u) \tag{7.2}$$

Energy:

$$(\rho C_\mathrm{p})_{\mathrm{nf}}\left(u\frac{\partial T}{\partial x}+v\frac{\partial T}{\partial y}\right)=k_{\mathrm{nf}}\left(\frac{\partial^2 T}{\partial y^2}\right)+Q(T-T_\infty) \tag{7.3}$$

and the boundary conditions are in the following form:

$$u(x,0)=U_w=cx, \quad u(x,\infty)=U_\infty=ax \tag{7.4a}$$

$$v(x,0)=0, \tag{7.4b}$$

$$T(x,0)=T_w, \quad T(x,\infty)=T_\infty \tag{7.4c}$$

In the above equations, ρ_{nf}, the density of the nanofluid is given by

$$\rho_{nf}=(1-\varphi)\rho_f+\varphi\rho_s \tag{7.5}$$

whereas the heat capacitance of the nanofluid and thermal expansion coefficient of the nanofluid can be determined by

$$(\rho C_p)_{nf}=(1-\varphi)(\rho C_p)_f+\varphi(\rho C_p)_s \tag{7.6}$$

$$(\rho\beta)_{nf}=(1-\varphi)(\rho\beta)_f+\varphi(\rho\beta)_s \tag{7.7}$$

with φ being the volume fraction of the solid particles and subscripts f, nf, and s stand for base fluid, nanofluid, and solid, respectively. The effective dynamic viscosity of the nanofluid given by Brinkman (1952a, 1952b) is

$$\mu_{nf}=\frac{\mu_f}{(1-\phi)^{2.5}} \tag{7.8}$$

The thermal conductivity of the stagnant (subscript 0) nanofluid and the effective thermal conductivity of the nanofluid for spherical nanoparticles, according to Maxwell (1904), are

$$\frac{k_{nf0}}{k_f}=\frac{k_s+2k_f-2\phi(k_f-k_s)}{k_s+2k_f+\phi(k_f-k_s)} \tag{7.9}$$

Equations (7.1)–(7.3) can be converted to nondimensional forms, using the following nondimensional parameters by introducing a stream function ψ and the similarity variable η

$$\begin{gathered} u=\frac{\partial\psi}{\partial y} \quad v=-\frac{\partial\psi}{\partial x} \\ \psi=\sqrt{c\upsilon_f}\,xf(\eta) \\ \eta=\sqrt{\frac{c}{\upsilon_f}}y, \quad \theta=\frac{T-T_\infty}{T_w-T_\infty} \\ M=\frac{\upsilon_f}{cK}, \quad C=\frac{a}{c}, \quad B=\frac{Q}{c(\rho C_p)_f} \end{gathered} \tag{7.10}$$

The continuity Equation (7.1) is automatically satisfied by introducing a stream function. By using the transformation (7.10), the governing Equations (7.2) and (7.3), and the boundary condition (7.4a)–(7.4c) are written as follows

$$f'^2(\eta)+f(\eta)f''(\eta)=C^2-\frac{1}{(1-\varphi)^{2.5}\left((1-\varphi)+\frac{\rho_s}{\rho_f}\varphi\right)}f'''(\eta)+\frac{M}{(1-\varphi)^{2.5}\left((1-\varphi)+\frac{\rho_s}{\rho_f}\varphi\right)}(C-f'(\eta)) \tag{7.11}$$

$$\theta'' + \frac{\mathrm{Pr}B}{\left(\frac{k_s+2k_f-2\varphi(k_f-k_s)}{k_s+2k_f+2\varphi(k_f-k_s)}\right)}\theta + \frac{\mathrm{Pr}\left((1-\varphi)+\frac{(\rho C_p)_s}{(\rho C_p)_f}\right)}{\left(\frac{k_s+2k_f-2\varphi(k_f-k_s)}{k_s+2k_f+2\varphi(k_f-k_s)}\right)}f\theta' = 0 \tag{7.12}$$

And the dimensionless boundary conditions, used to solve the above equations, are as follows:

$$f(0)=0,\ f'(0)=1,\ f'(\infty)=C \tag{7.13}$$

$$\theta(0)=1,\ \theta(\infty)=0, \tag{7.14}$$

The physical quantities of interest are the skin-friction coefficient C_f and the local Nusselt number Nu_x. The skin friction at the sheet is given by

$$C_f = \frac{\tau_w}{\rho_f U_w^2/2} = \frac{\mu_{nf}\left(\frac{\partial u}{\partial y}+\frac{\partial v}{\partial x}\right)\Big|_{y=0}}{\rho_f U_w^2/2} = \frac{1}{2}\frac{\mu_{nf}}{\mu_f}\sqrt{\frac{\upsilon_f}{U_w x}}f''(0) = \frac{1}{2}\frac{1}{(1-\varphi)^{2.5}}Re_x^{-1/2}f''(0) \tag{7.15}$$

and the rate of heat transfer in terms of the Nusselt number at the sheet is given by

$$Nu = \frac{xq_w}{k_f(T_w-T_\infty)} = \frac{-xk_{nf}\frac{\partial T}{\partial y}\Big|_{y=0}}{k_f(T_w-T_\infty)} = -\frac{k_{nf}}{k_f}Re_x^{1/2}\theta'(0) \tag{7.16}$$

where $Re_x = \frac{U_w x}{\upsilon_f}$ is the local Reynolds number.

7.2.3 NUMERICAL PROCEDURE AND VALIDATION

The numerical solution of nonlinear differential Equations (7.11) and (7.12) with boundary conditions given in Equations (7.13) and (7.14) is accomplished by shooting technique with fourth-order Runge-Kutta algorithm (1989). The nonlinear differential Equations (7.12) and (7.13) are first decomposed to a system of first-order differential equations in the form

$$\begin{aligned}
&f_0 = f\frac{df_0}{d\eta} = f_1\\
&\frac{df_1}{d\eta} = f_2\\
&f_1^2 + f_0 f_2 = C^2 - \frac{1}{(1-\varphi)^{2.5}\left((1-\varphi)+\frac{\rho_s}{\rho_f}\varphi\right)}\frac{df_2}{d\eta} - \frac{M}{(1-\varphi)^{2.5}\left((1-\varphi)+\frac{\rho_s}{\rho_f}\varphi\right)}(C-f_1)\\
&\theta_0 = \theta\frac{d\theta_0}{d\eta} = \theta_1\\
&\frac{d\theta_1}{d\eta} + \frac{\mathrm{Pr}B}{\left(\frac{k_s+2k_f-2\varphi(k_f-k_s)}{k_s+2k_f+2\varphi(k_f-k_s)}\right)}\theta_0 + \frac{\mathrm{Pr}\left((1-\varphi)+\frac{(\rho C_p)_s}{(\rho C_p)_f}\right)}{\left(\frac{k_s+2k_f-2\varphi(k_f-k_s)}{k_s+2k_f+2\varphi(k_f-k_s)}\right)}f_0\theta_1 = 0
\end{aligned} \tag{7.17}$$

Corresponding boundary conditions take the form,

$$\begin{gathered} f_0(0)=0,\ f_1(0)=1,\ f_1(\infty)=C \\ \theta_0(0)=1,\ \theta_0(\infty)=0 \end{gathered} \tag{7.18}$$

Appropriately guessing the missing slopes $f_2(0)$ and $\theta_1(0)$, the boundary value problem is first converted into an initial value problem, which is solved by a fourth-order Runge-Kutta method. Then the values of $f_2(0)$ and $\theta_1(0)$ are updated using shooting method. The convergence criterion largely depends on fairly good guesses of the initial conditions in the shooting technique. Present results are compared with some of the earlier published results Attia (2006), which are depicted in Figure 7.2.

7.2.4 RESULTS AND DISCUSSIONS

Stagnation point flow and heat transfer of three types of nanofluids, namely, Cu-water, Al_2O_3-water, and TiO_2-water, toward a permeable stretching surface with heat generation are studied numerically for over a range of particle volume fraction. Adopting appropriate similarity transformation, the governing partial differential equations of flow and heat transfer are transferred into a system of nonlinear ordinary differential equations and then are solved using shooting technique along with the fourth-order Runge-Kutta method. The numerical results are obtained for $C<1$ and $C>1$ as there exist a distinct behavior for these two ranges of values of velocity ratio parameter. Further, the effects of porosity parameter M and surface heat generation B are not affected by the presence of nanoparticles. Hence, we omit the discussion on the results of M and B as they are extensively studied by Attia (2006).

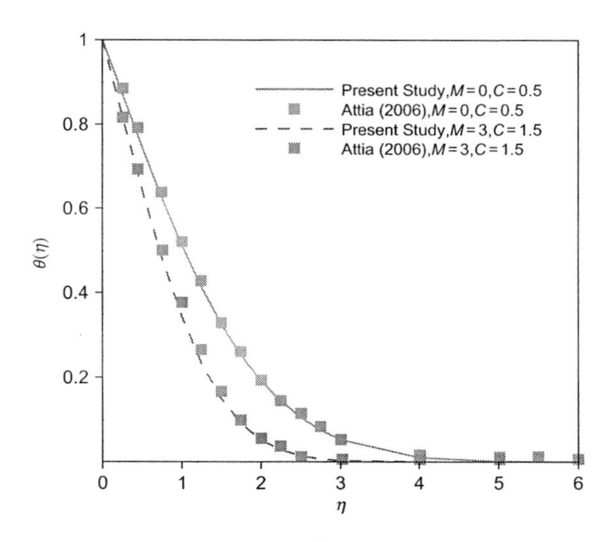

FIGURE 7.2

Comparison of the results of the present work and numerical results of Attia (2006).

The effects of nanoparticle volume fraction on various flow and heat transfer quantities are shown in Figures 7.3–7.8 for different parameter. In all cases, the default value is Cu nanoparticle, unless otherwise stated.

The variation of horizontal velocity profiles $f'(\eta)$ with respect to the nanoparticle volume fraction φ is projected in Figure 7.3a for two values of velocity ratio parameter.

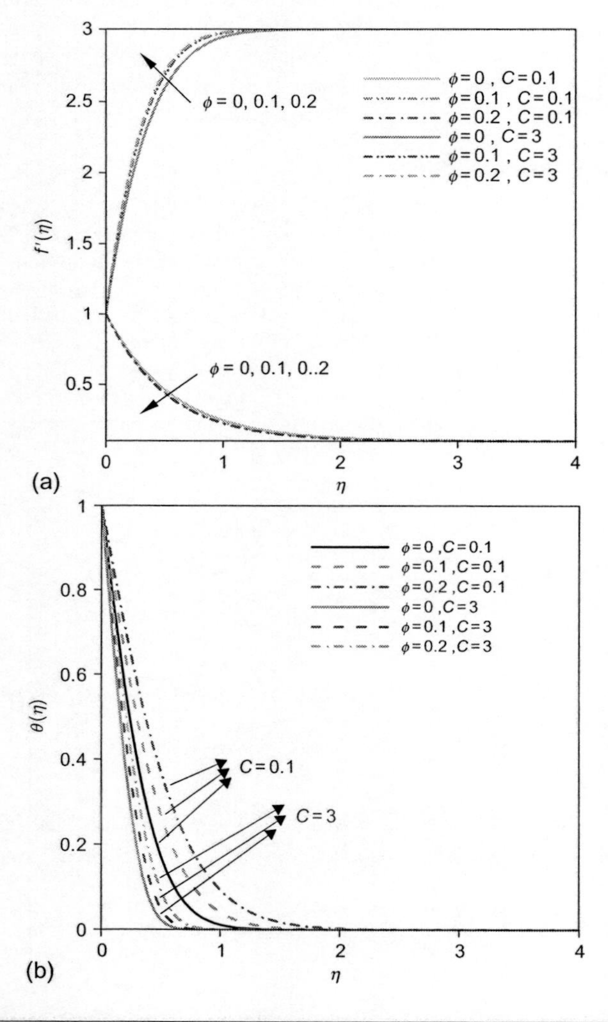

FIGURE 7.3

Variation of (a) horizontal in the velocity profiles $f'(\eta)$ and (b) temperature profiles $\theta(\eta)$ for different values of nanoparticle volume fraction with $C=0.1$ and $C=3$.

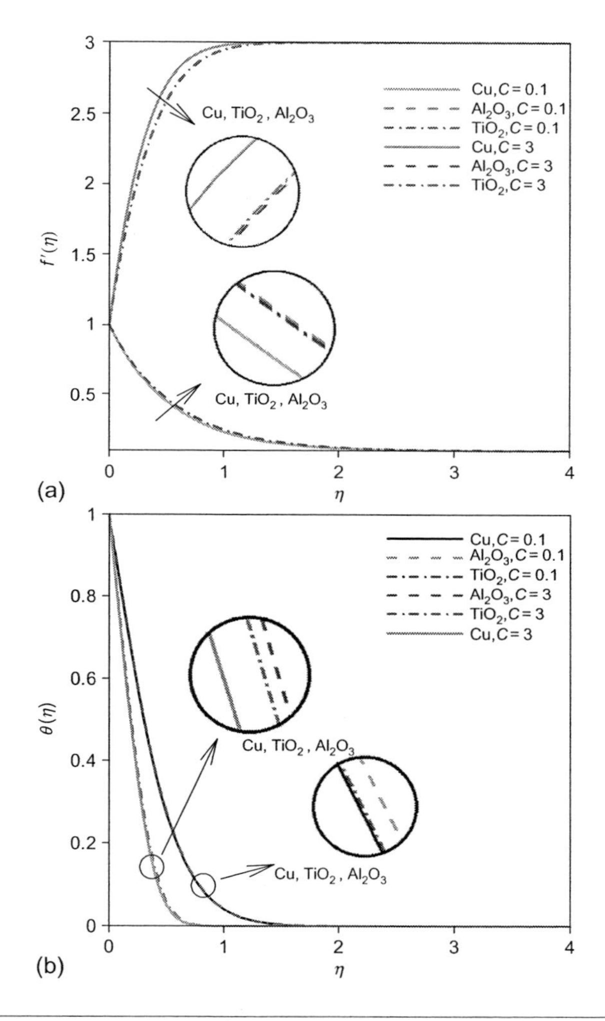

FIGURE 7.4

Variation in (a) the horizontal velocity profiles $f'(\eta)$ and (b) the temperature profiles $\theta(\eta)$ for different types of nanoparticle with $\varphi=0.1$, $C=0.1$, and $C=3$.

It is evident from this figure that when $C>1$, the flow has a boundary-layer structure, and the thickness of the boundary layer decreases with the increase in C. It can be explained as follows. For fixed value of C, $C>1$ implies increase in straining motion near the stagnation region. Due to this reason, the acceleration of the external stream is increased, and this leads to thinning of the boundary layer. On the other hand when $C<1$, the flow has an inverted boundary-layer structure. In this case, the stretching

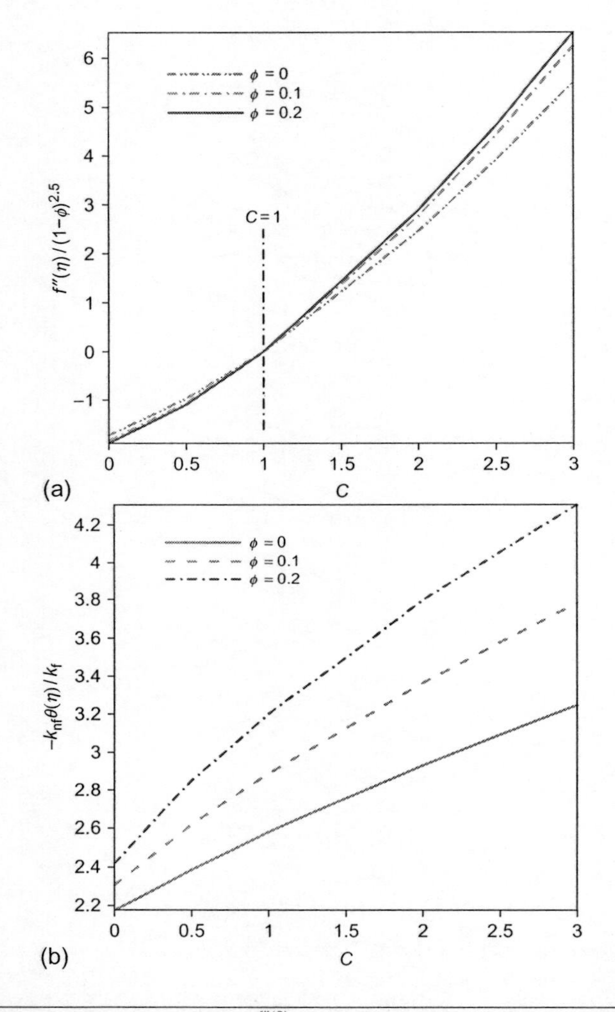

FIGURE 7.5

Variation of (a) reduced skin-friction coefficient $\frac{f''(0)}{(1-\varphi)^{2.5}}$ and (b) reduced Nusselt number $-\frac{k_{nf}}{k_f}\theta'(0)$ for different values of nanoparticle volume fraction and velocity ratio parameter.

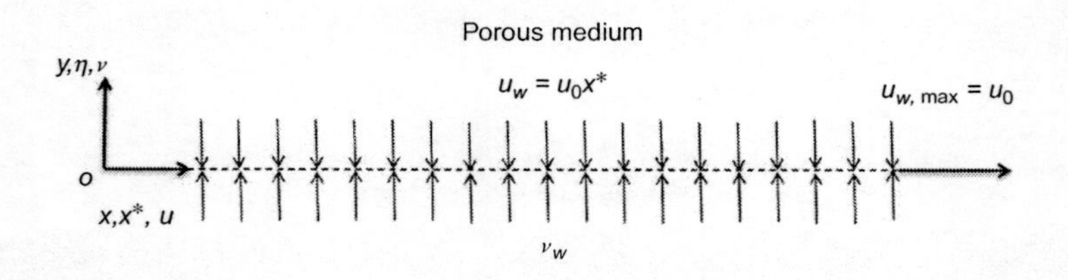

FIGURE 7.6

Schematic theme of the problem geometry.

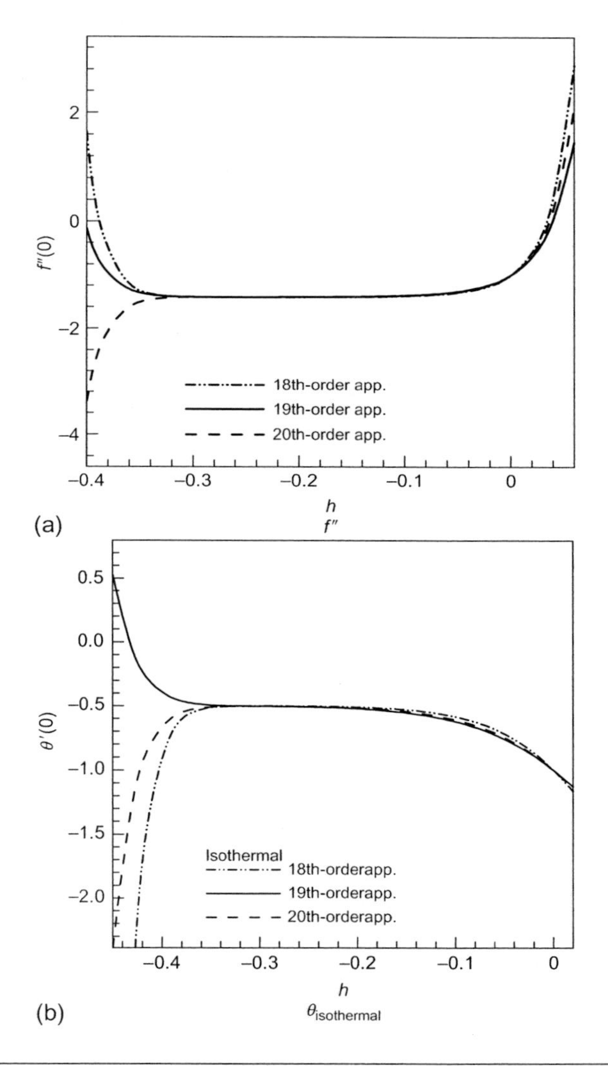

FIGURE 7.7

$\hbar$ curves in $Re = f_w = Pr = 1$, $\phi = 0.01$ and $n = 0$.

Continued

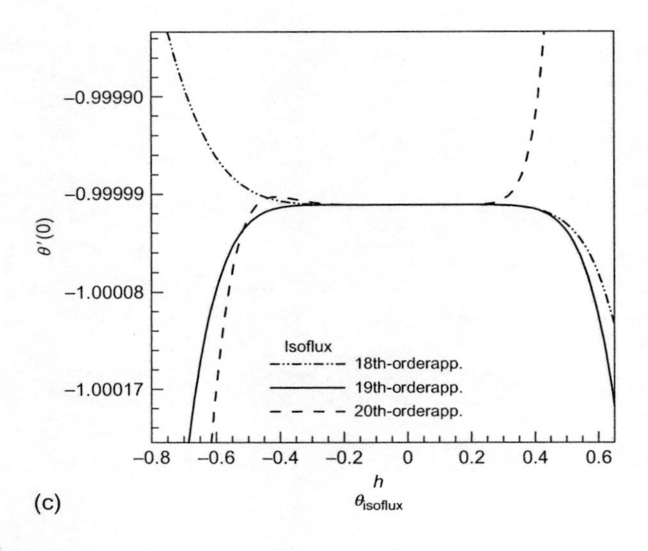

FIGURE 7.7, cont'd

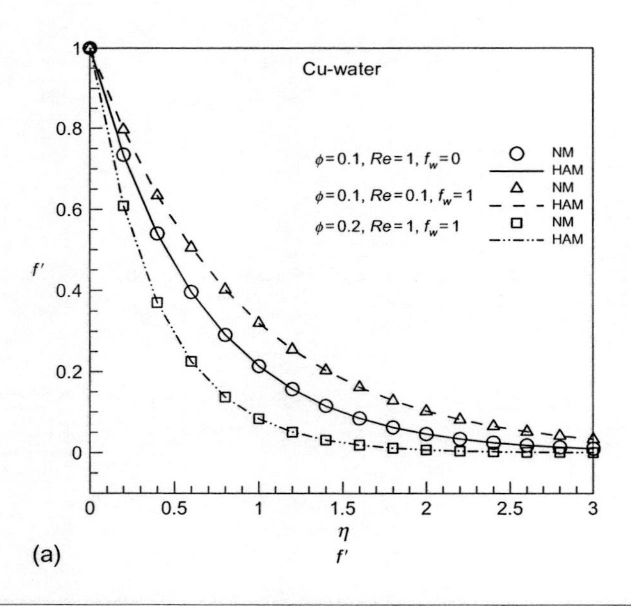

FIGURE 7.8

Comparison between numerical results and HAM solution results when $\hbar = -0.21$.

Continued

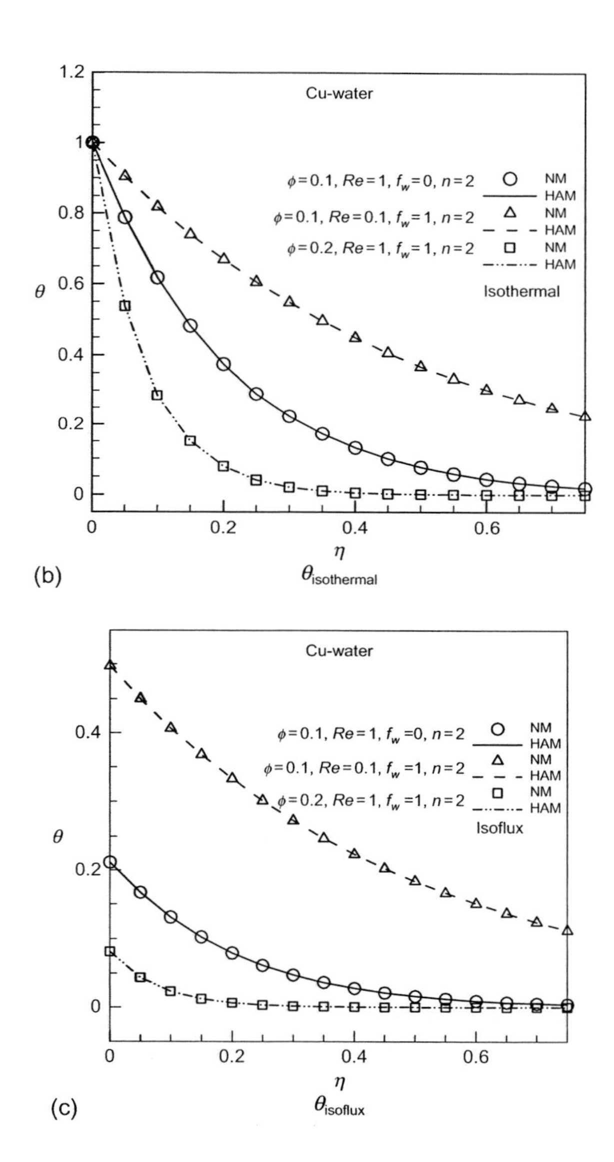

FIGURE 7.8, cont'd

velocity of the sheet exceeds the velocity of the external stream. It is to be that no boundary layer is formed when $C = 1$. Furthermore, it is clear from this plot that the increasing value of volume fraction decreases the boundary-layer thickness for $C > 1$ and decreases the velocity profiles. And the result is true for different values of velocity ratio parameter C. This behavior is ascribed to increase in density and viscosity of nanofluid mixture due to nanoparticle dispersion.

Figure 7.3b represents the effect of φ on temperature profiles for $C = 0.1,\ 3$. For all values of C considered, the temperature θ is found to decrease with the increase in η. Temperature at a point on the sheet decreases significantly with the increase in C. The results show that the thermal boundary-layer thickness increases with the increasing values of nanoparticle volume fraction φ. This is because solid particles have high thermal conductivity, so the thickness of the thermal boundary layer increases.

The effects of different types of nanoparticles on horizontal velocity and temperature profiles are depicted in Figure 7.4a and b. It is observed that the hydraulic boundary layer is thinner for heavier nanoparticles. In the present study, the effects of nanoparticle on hydraulic boundary are considered only with the changes in mixture density. Utilizing nanoparticle with higher thermal conductivity reduces the thermal boundary-layer thickness. Nevertheless, higher heat capacitance of nanoparticles increases the thickness of thermal boundary layer. Considering the opposite effects of nanoparticles properties of thermal boundary layer, temperature profiles are plotted for three nanoparticles (Cu, TiO_2, Al_2O_3). For both $C < 1$ and $C > 1$, Cu-water nanofluid has the thinner thermal boundary-layer thickness.

The effect of φ-reduced skin-friction coefficient $f''(0)/(1-\varphi)^{2.5}$ is illustrated in Figure 7.5a. Clearly, for $C > 1$, increasing values of φ result in increasing the skin-friction coefficient (values of $f''(0)/(1-\varphi)^{2.5}$ are positive). And for $C < 1$, the behavior is reverse, and the skin-friction coefficient is decreased by increasing nanoparticle volume fraction (values of $f''(0)/(1-\varphi)^{2.5}$ are negative). When $C > 1$, the increase in shear stress at the sheet is directly related to the phenomenon of thinning of boundary layer. But when $C < 1$, an inverted boundary layer is formed. So a result of opposite behavior is observed in this case.

Figure 7.5b highlights the effect of φ on the dimensionless wall heat flux $-\frac{k_{nf}}{k_f}\theta'(0)$ for different values of velocity ratio parameter. It is found from this plot that the dimensionless wall heat flux $-\frac{k_{nf}}{k_f}\theta'(0)$ increases with the increasing values of φ for all values of C. It should be noted that the reduced Nusselt number $-\frac{k_{nf}}{k_f}\theta'(0)$, increases with increase in C, which is mainly a result of thinning of thermal boundary layer.

Tables 7.2 and 7.3 show the variation of the reduced skin-friction coefficient and Nusselt number for different types of nanofluids. When $C < 1$, because of the inverted boundary-layer structure, it is observed that the skin-friction coefficient and Nusselt number are, respectively, lowest and highest for Cu compared to TiO_2 and Al_2O_3. This is because Cu nanoparticles have the highest thermal conductivity and density. When $C > 1$, the hydraulic boundary-layer thickness is highest and the thermal boundary-layer thickness is the smallest.

7.2.5 CONCLUSION

We have numerically studied flow and heat transfer characteristics of stagnation point flow of a nanofluid on a stretching sheet in a porous medium. The governing partial differential equations were transformed to a set of nonlinear ordinary differential equations using a similarity transformation, before

Table 7.2 The Variation of Reduced Skin-Friction Coefficient and Nusselt Number for Different Types of Nanofluids for Different Values of φ and M with $C=0.1$

$C=0.1$		$\frac{f''(0)}{(1-\varphi)^{2.5}}$			$-\frac{k_{nf}}{k_f}\theta'(0)$		
M	φ	Cu	Al_2O_3	TiO_2	Cu	Al_2O_3	TiO_2
0	0.05	−1.075573	−0.975144	−0.981084	2.450999	2.388654	2.341221
	0.1	−1.142395	−0.971267	−0.981716	2.541669	2.491175	2.419565
	0.15	−1.182156	−0.960153	−0.974026	2.633358	2.593561	2.497365
	0.2	−1.202261	−0.943533	−0.959987	2.727479	2.697195	2.575513
0.5	0.05	−1.248650	−1.163347	−1.168325	2.406333	2.340004	2.293691
	0.1	−1.307413	−1.161012	−1.169758	2.493003	2.435179	2.366049
	0.15	−1.343966	−1.153718	−1.165275	2.579229	2.528769	2.436688
	0.2	−1.364493	−1.143313	−1.156913	2.666314	2.621792	2.506246
1	0.05	−1.401037	−1.325673	−1.330037	2.367172	2.298211	2.252802
	0.1	−1.454326	−1.324428	−1.332091	2.449935	2.387208	2.320082
	0.15	−1.488777	−1.319785	−1.329886	2.531173	2.473563	2.384794
	0.2	−1.509878	−1.313663	−1.325499	2.612067	2.558056	2.447413
2	0.05	−1.664960	−1.602171	−1.605778	2.299960	2.227660	2.183696
	0.1	−1.711212	−1.602470	−1.608798	2.375518	2.306518	2.242585
	0.15	−1.743172	−1.601509	−1.609828	1.743172	2.381243	2.297722
	0.2	−1.765595	−1.601325	−1.611032	2.518456	2.452328	2.349382

Table 7.3 The Variation of Reduced Skin-Friction Coefficient and Nusselt Number for Different Types of Nanofluids for Different Values of φ and M with $C=3$

$C=3$		$\frac{f''(0)}{(1-\varphi)^{2.5}}$			$-\frac{k_{nf}}{k_f}\theta'(0)$		
M	φ	Cu	Al_2O_3	TiO_2	Cu	Al_2O_3	TiO_2
0	0.05	5.247329	4.757373	4.786350	3.465796	3.324795	3.264359
	0.1	5.573330	4.738459	4.789435	3.730849	3.535081	3.436126
	0.15	5.767309	4.684239	4.751919	3.991969	3.744675	3.604277
	0.2	5.865396	4.603154	4.683429	4.252379	3.954756	3.769367
0.5	0.05	5.432519	4.960916	4.988709	3.479320	3.340409	3.279535
	0.1	5.748914	4.943829	4.992706	3.744008	3.552062	3.452311
	0.15	5.939034	4.894174	4.958984	4.005332	3.763396	3.621816
	0.2	6.037455	4.820549	4.897258	4.266397	3.975676	3.788665
1	0.05	5.611821	5.156710	5.183449	3.492139	3.355059	3.293782
	0.1	5.919481	5.141287	5.188296	3.756538	3.567972	3.467494
	0.15	6.106108	5.095767	5.158033	4.018080	3.780887	3.638234
	0.2	6.204919	5.028879	5.102442	4.279774	3.995144	3.806666
2	0.05	5.954755	5.528152	5.553090	3.515939	3.381906	3.319913
	0.1	6.247109	5.515665	5.559486	3.779939	3.597070	3.495308
	0.15	6.427664	5.477389	5.535333	4.041946	3.812769	3.668227
	0.2	6.527390	5.422269	5.490524	4.304817	4.030451	3.839407

being solved numerically by the shooting technique along with the fourth-order Runge-Kutta method. Three different types of nanoparticles, namely, copper (Cu), alumina (Al_2O_3), and titanium oxide (TiO_2) with water as the base fluid were considered.

It was found that increasing the volume fraction of nanoparticles angle enhanced the heat transfer rate and increases the skin-friction coefficient and at the surface at $C > 1$. Moreover, Cu-water has the highest heat transfer rate and the highest skin-friction coefficient at the surface compared with the others when $C > 1$.

7.3 FLOW AND HEAT TRANSFER OF NANOFLUIDS IN A POROUS MEDIUM

The steady boundary-layer flow passing through a permeable stretching wall embedded in a porous medium filled with nanofluids is studied using different types of nanoparticles such as copper (Cu), silver (Ag), alumina (Al_2O_3), and titanium oxide (TiO_2) with water as its base fluid. The basic partial equations are reduced to ordinary differential equations, which are solved with an analytical method named homotopy analysis method (HAM). Comparison between HAM and numerical solutions results showed an excellent agreement. The influence of pertinent parameters such as solid volume fraction of nanoparticles, the type of nanofluids on the flow, heat transfer, entropy generation, skin-friction coefficient, Nusselt number, and Bejan number is discussed.

The results indicate that increase in the nanoparticle volume fraction will decrease momentum boundary-layer thickness and entropy generation rate, while this increases the thermal boundary-layer thickness. Such effects are found to be more noticeable in the Ag-water solution than in the other solutions (this section has been worked by M. Sheikholeslami, H.R. Ashorynejad, D.D. Ganji in nonlinear dynamics team in Mechanical Engineering Department, 2012-2013).

7.3.1 INTRODUCTION

The study of boundary layers over a stretching surface has lots of applications in several engineering processes such as liquid composite molding, extrusion of plastic sheets, paper production, glass blowing, metal spinning, wire drawing, and hot rolling (Al-Odat et al., 2006; Nazar et al., 2008). The flow field of a stretching surface with a power-law velocity variation was discussed by Banks (1983). Furthermore, a stretching surface subject to suction or injection was studied by Ali (1995) for uniform. For an engineering (real) system, the generated entropy is proportional to the destroyed exergy (which is always destroyed as a result of the Second Law (Bejan, 1995)), it makes perfect engineering sense to study and minimize the entropy generation of the system, which is commonly referred to as entropy generation minimization (EGM), as an optimization tool (Bejan et al., 2004).

The basic aim of this work is to study the boundary-layer flow passing through a permeable stretching wall in a porous medium filled with a nanofluid, where the basic fluid is water. The reduced coupled ordinary differential equations are solved with analytical method that is named as HAM (Liao, 1999). Also, entropy generation rate and the Bejan number variations are investigated. The effects of the parameters governing on the problem are studied and discussed.

7.3.2 PROBLEM STATEMENT

A steady, constant property, two-dimensional flow of an incompressible nanofluid through a homogenous porous medium with permeability of K, over a stretching surface with linear velocity distribution, i.e., $u_w = \frac{u_0 x}{L}$ is assumed (Figure 7.6).

The fluid is a water-based nanofluid containing different types of nanoparticles: Cu, Al_2O_3, Ag, and TiO_2. It is assumed that the base fluid and the nanoparticles are in thermal equilibrium and no slip occurs between them. The thermophysical properties of the nanofluid are given in Table 7.4 (see Oztop and Abu-Nada, 2008a, 2008b).

The transport properties of the medium can be considered independent from the temperature when the temperature difference between wall and ambient is not significant (see Kaviany, 1992). The origin is kept fixed while the wall is stretching and the y-axis is perpendicular to the surface. Using the abovementioned assumptions, the continuity equation is

$$\frac{\partial u}{\partial x} + \frac{\partial v}{\partial y} = 0 \tag{7.19}$$

where u and v are velocity components in the x and y directions, respectively. The Brinkman model x-momentum equation reads (Nield and Bejan, 2006a, 2006b):

$$\rho_{nf}\left(u\frac{\partial u}{\partial x} + v\frac{\partial v}{\partial y}\right) = \mu_{eff_{nf}}\frac{\partial^2 u}{\partial y^2} - \frac{\mu_{nf}}{K}u \tag{7.20}$$

$$u\frac{\partial T}{\partial x} + v\frac{\partial T}{\partial y} = \frac{k_{nf}}{(\rho C_p)_{nf}}\frac{\partial^2 T}{\partial y^2} \tag{7.21}$$

where $\mu_{eff_{nf}}$ is the effective viscosity, which for simplicity in the present study is considered to be identical to the dynamic viscosity, μ_{nf}. This assumption is reasonable for packed beds of particles (Starov and Zhdanov, 2001).

The effective density ρ_{nf}, the effective dynamic viscosity μ_{nf}, the heat capacitance $(\rho C_p)_{nf}$, and the thermal conductivity k_{nf} of the nanofluid are given as (see Aminossadati and Ghasemi, 2009):

$$\rho_{nf} = \rho_f(1-\phi) + \rho_s\phi \tag{7.22}$$

$$\mu_{nf} = \frac{\mu_f}{(1-\phi)^{2.5}} \tag{7.23}$$

$$(\rho C_p)_{nf} = (\rho C_p)_f(1-\phi) + (\rho C_p)_s\phi \tag{7.24}$$

Table 7.4 Thermo Physical Properties of Water and Nanoparticles [17]

	ρ(kg/m^3)	C_p(J/kg K)	k(W/mK)	$\beta \times 10^5$ (K^{-1})
Pure water	997.1	4179	0.613	21
Copper (Cu)	8933	385	401	1.67
Silver (Ag)	10 500	235	429	1.89
Alumina (Al_2O_3)	3970	765	40	0.85
Titanium oxide (TiO_2)	4250	686.2	8.9538	0.9

$$\frac{k_{nf}}{k_f} = \frac{k_s + 2k_f - 2\phi(k_f - k_s)}{k_s + 2k_f + 2\phi(k_f - k_s)} \tag{7.25}$$

Here, ϕ is the solid volume fraction.

The hydrodynamic boundary conditions are

$$u(x^*, 0) = u_0 x^*, \quad v(x^*, 0) = v_w, \quad (x^*, \infty) = 0 \tag{7.26}$$

where $x^* = \frac{x}{L}$ is the nondimensional x-coordinate, and L is the length of the porous plate.

The following thermal boundary conditions are considered:

$$T(x^*, 0) = T_\infty + T_0 (x^*)^n, \quad T(x^*, \infty) = T_\infty \tag{7.27}$$

$$-k_{nf} \left.\frac{\partial T}{\partial y}\right|_{(x^*, 0)} = q_0 (x^*)^n, \quad T(x^*, \infty) = T_\infty \tag{7.28}$$

The power-law temperature and heat flux distribution, described in Equations (7.27) and (7.28), resent a wider range of thermal boundary conditions including isoflux and isothermal cases. For example, by setting n equal to zero, Equations (7.27) and (7.28) yield isothermal and isoflux, respectively.

Second law of thermodynamics analysis of porous media is found to be more complicated compared to the clear-fluid counterpart due to increased number of variables in governing equations (Hooman et al., 2007). In the non-Darcian regime, there are three alternative models for the fluid friction term, which are the clear-fluid compatible model, the Darcy model, and the Nield model or the power of drag model. Following the entropy generation function introduced by (Hooman et al., 2007), the volumetric entropy generation rate, $S^{\bullet}_{gen}$, reads:

$$S_{gen} = \frac{k_{nf}}{T^2}\left[\left(\frac{\partial T}{\partial x}\right)^2 + \left(\frac{\partial T}{\partial y}\right)^2\right] + \frac{\mu_{nf}}{T}\left\{2\left[\left(\frac{\partial u}{\partial x}\right)^2 + \left(\frac{\partial u}{\partial y}\right)^2 + \left(\frac{\partial u}{\partial x} + \frac{\partial u}{\partial y}\right)^2\right]\right\} + \frac{\mu_{nf}}{TK}(u^2 + v^2) \tag{7.29}$$

Using boundary-layer approximations (Nield and Bejan, 2006a, 2006b), Equation (7.29) reduces to:

$$S^{\cdot}_{gen} = \frac{k_{nf}}{T^2}\left(\frac{\partial T}{\partial y}\right)^2 + \frac{\mu_{nf}}{T}\left(\frac{\partial u}{\partial y}\right)^2 + \frac{\mu_{nf}}{TK}u^2 \tag{7.30}$$

7.3.3 FLOW ANALYSIS AND MATHEMATICAL FORMULATION

7.3.3.1 Hydrodynamics

Using the stream function, $\psi(x,y)$, the continuity equation is satisfied:

$$u = \frac{\partial \psi}{\partial y}, \quad v = \frac{\partial \psi}{\partial x} \tag{7.31}$$

According to Nield and Kuznetsov (2005), the hydrodynamic boundary-layer thickness scales with $\sqrt{K}$. This can be found through a scale analysis between the first and the second terms on the right hand side of Equation (7.20), i.e., the viscous and the Darcy terms. Therefore, instead of the other similarity parameters reported in the literature, the following dimensionless similarity parameter is defined as

$$\eta=\frac{y}{\sqrt{K}} \tag{7.32}$$

The u-velocity is assumed to be correlated to $f(\eta)$, a dimensionless similarity function as:

$$u=u_0x^*f'(\eta) \tag{7.33}$$

where $f'(\eta)$ is $\frac{df}{d\eta}$. Using stream function definition, Equation (7.33), the stream function and the v velocity take the following form:

$$v=-\frac{u_0}{L}\sqrt{K}f(\eta),\quad v=u_0x^*\sqrt{K}f(\eta) \tag{7.34}$$

Substituting from u and v into Equations (7.20) and (7.22), one will find the following differential equation for the u momentum equation:

$$f'''+Re\cdot A_1\cdot\left(ff''-(f')^2\right)-f'=0,\quad Re=\frac{\rho_f u_0 K}{L\mu_f} \tag{7.35}$$

where $A_1=(1-\phi)+(\rho_s/\rho_f)\phi$ is a parameter.

where Re is the Reynolds number. Equation (7.35) should be solved and subjected to the following boundary conditions:

$$f(0)=\frac{-v_w L}{u_0\sqrt{K}}=f_w,\quad f'(0)=1,\quad f'(\infty)=0 \tag{7.36}$$

f_w is the injection parameter. Positive/negative values of f_w show suction/injection into/from the porous surface, respectively.

The wall shear stress is the driving force that drags fluid flow along the stretching wall. The wall shear stress term can then be found, in terms of the similarity function, as

$$\tau_w=-\mu_{nf}\left.\frac{\partial u}{\partial y}\right|_{y=0}=\frac{-\mu_{nf}u_0x^*f''(0)}{2\sqrt{K}} \tag{7.37}$$

7.3.3.2 Thermal analysis

Introducing a similarity function, θ, as

$$T-T_\infty=T_{ref}(x^*)^n\theta(\eta) \tag{7.38}$$

where T_{ref} is T_0 and $q_0\frac{\sqrt{K}}{k}$ for the power-law temperature and heat flux boundary conditions, respectively. The thermal energy equation reads:

$$\theta''+Re\cdot Pr\cdot\frac{A_1}{A_3}\cdot A_2\cdot(1-\phi)^{2.5}(f\theta'-nf'\theta))=0\quad Pr=\frac{\mu_f(\rho C_p)_f}{\rho_f k_f},\quad Re=\frac{\rho_f u_0 K}{L\mu_f} \tag{7.39}$$

where A_2, A_3 are parameters having the following form:

$$A_2 = (1-\phi) + \frac{(\rho C_p)_s}{(\rho C_p)_f}\phi$$
$$A_3 = \frac{k_{nf}}{k_f} = \frac{k_s + 2k_f - 2\varphi(k_f - k_s)}{k_s + 2k_f + 2\varphi(k_f - k_s)} \tag{7.40}$$

which are subjected to the following boundary conditions:

$$\begin{aligned} &\theta(0) = 1,\ \theta(\infty) = 0 \ \text{Power-law temperature} \\ &\theta'(0) = -1,\ \theta(\infty) = 0 \ \text{Power-law heat flux} \end{aligned} \tag{7.41}$$

For power-law temperature and heat flux boundary conditions, respectively. Employing the definition of convective heat transfer coefficient, the local Nusselt numbers, become

$$Nu_x = \frac{hx}{k} = \begin{cases} \dfrac{-\theta'(0)x}{\sqrt{K}} & \text{Power-law temperature} \\ \dfrac{q_w x}{K(T_w - T_\infty)} = \dfrac{x}{\theta(0)\sqrt{K}} & \text{Power-law heat flux} \end{cases} \tag{7.42}$$

Finally, the local volumetric entropy generation rate for the above cases, respectively, reads

$$S_{gen}^{\cdot} = \text{HTI} + \text{FFI} \tag{7.43}$$

where HTI is the heat transfer irreversibility due to heat transfer in the direction of finite temperature gradients. HTI is common in all types of thermal engineering applications.

The last term (FFI) is the contribution of fluid friction irreversibility to the total entropy generation. Not only the wall and fluid layer shear stress but also the momentum exchange at the solid boundaries (pore level) contribute to FFI. In terms of the primitive variables, HTI and FFI become

$$\text{HTI} = \begin{cases} \dfrac{A_3 \cdot k_f}{(\theta T_0 (x^*)^n + T_\infty)^2}\left(\dfrac{\theta' T_0 (x^*)^n}{\sqrt{K}}\right)^2 & \text{Power-law temperature} \\ \dfrac{A_3 \cdot k_f}{(\theta\sqrt{K} q_0 (x^*)^n / k + T_\infty)^2}\left(\dfrac{\theta' q_0 (x^*)^n}{A_3 \cdot k_f}\right)^2 & \text{Power-law heat flux} \end{cases} \tag{7.44}$$

$$\text{FFI} = \begin{cases} \dfrac{\mu_f \cdot (1-\phi)^{2.5}}{(\theta T_0 (x^*)^n + T_\infty)}\left[\left(\dfrac{f'' u_0 x^*}{\sqrt{K}}\right)^2 + \left(\dfrac{f' u_0 x^*}{\sqrt{K}}\right)^2\right] & \text{Power-law temperature} \\ \dfrac{\mu_f \cdot (1-\phi)^{2.5}}{(\theta\sqrt{K} q_0 (x^*)^n / k + T_\infty)}\left[\left(\dfrac{f'' u_0 x^*}{\sqrt{K}}\right)^2 + \left(\dfrac{f' u_0 x^*}{\sqrt{K}}\right)^2\right] & \text{Power-law heat flux} \end{cases} \tag{7.45}$$

where T_∞ and T_0 are measured in degrees of Kelvin.

One can also define the Bejan number, Be, as

$$Be = \frac{\text{HTI}}{\text{HTI} + \text{FFI}} \tag{7.46}$$

The Bejan number shows the ratio of entropy generation due to heat transfer irreversibility to the total entropy generation so that a Be value more/less than 0.5 shows that the contribution of HTI to the total entropy generation is higher/less than that of FFI.

The limiting value of $Be=1$ shows that the only active entropy generation mechanism is HTI, while $Be=0$ represents no HTI contribution to the total entropy production.

7.3.4 THE HAM SOLUTION OF THE PROBLEM

For HAM solutions of the governing equations, we choose the initial approximations of $f(\eta)$ and $\theta(\eta)$ as follow:

$$f_0(\eta)=-\exp(-\eta)+1+f_w \tag{7.47}$$

$$\theta_0(\eta)=\exp(-\eta) \tag{7.48}$$

and the auxiliary linear operators are:

$$L_1(f)=f'''+f'' \tag{7.49}$$

$$L_2(\theta)=\theta''+\theta' \tag{7.50}$$

These auxiliary linear operators satisfy:

$$L_1(C_1+C_2+C_3\exp(-\eta)) \tag{7.51}$$

$$L_2(C_4+C_5\exp(-\eta)) \tag{7.52}$$

where C_i ($i=1,2,3,4,5,6$) are constants. Introducing a nonzero auxiliary parameters $\hbar_1$ and $\hbar_2$, we develop the zeroth-order deformation problems as follow:

$$(1-p)L[f(\eta;p)-f_0(\eta)]=p\hbar_1 N[f(\eta;p)] \tag{7.53}$$

$$f(0;p)=f_w,\ \ f'(0;p)=1,\ \ f'(\infty;p)=0 \tag{7.54}$$

$$(1-p)L[\theta(\eta;p)-\theta_0(\eta)]=p\hbar_2 N[\theta(\eta;p)] \tag{7.55}$$

$$\theta(0;p)=1,\ \ \theta(\infty;p)=0 \ \ \text{Power-law temperature} \tag{7.56}$$

$$\theta'(0;p)=-1,\ \ \theta(\infty;p)=0 \ \ \text{Power-law heat flux} \tag{7.57}$$

where nonlinear operators, N_1 and N_2 are defined as:

$$N_1[f(\eta;p),\ \theta(\eta;p)]=\frac{\partial^3 f(\eta;p)}{\partial\eta^3}+Re\cdot A_1\cdot\left(f(\eta;p)\frac{\partial^2 f(\eta;p)}{\partial\eta^2}-\left(\frac{\partial f(\eta;p)}{\partial\eta}\right)^2\right)-\frac{\partial f(\eta;p)}{\partial\eta} \tag{7.58}$$

$$\begin{aligned}N_2[f(\eta;p),\ \theta(\eta;p)]=&\frac{\partial^2\theta(\eta;p)}{\partial\eta^2}+Re\cdot Pr\cdot\frac{A_1}{A_3}\cdot A_2\cdot\left(f(\eta;p)\frac{\partial\theta(\eta;p)}{\partial\eta}\right)\\&-n\cdot Re\cdot Pr\cdot\frac{A_1}{A_3}\cdot A_2\cdot\left(\theta(\eta;p)\frac{\partial f(\eta;p)}{\partial\eta}\right)\end{aligned} \tag{7.59}$$

For $p=0$ and $p=1$ we, respectively, have:

$$f(\eta;0)=f_0(\eta) \quad f(\eta;1)=f(\eta) \tag{7.60}$$

$$\theta(\eta;0)=\theta_0(\eta) \quad \theta(\eta;1)=\theta(\eta) \tag{7.61}$$

As p increases from 0 to 1, $f(\eta;p)$ and $\theta(\eta;p)$ vary, respectively, from $f_0(\eta)$ and $\theta_0(\eta)$ to $f(\eta)$ and $\theta(\eta)$. By Taylor's theorem and using Equations (7.60) and (7.61), $f(\eta)$ and $\theta(\eta)$ can be expanded in a power series of p as follows:

$$f(\eta;p)=f_0(\eta)+\sum_{m=1}^{\infty} f_m(\eta)p^m \tag{7.62a}$$

$$f_m(\tau)=\frac{1}{m!}\frac{\partial^m f(\eta,p)}{\partial p^m} \tag{7.62b}$$

and

$$\theta(\eta;p)=\theta_0(\eta)+\sum_{m=1}^{\infty} \theta_m(\eta)p^m \tag{7.63a}$$

$$\theta_m(\tau)=\frac{1}{m!}\frac{\partial^m \theta(\eta,p)}{\partial p^m} \tag{7.63b}$$

In which $\hbar_1$ and $\hbar_2$ are chosen in such a way that these series are convergent at $p=1$. Convergence of the series (7.62a) and (7.62b) depends on the auxiliary parameters $\hbar_1$ and $\hbar_2$.

Assume that $\hbar_1$ and $\hbar_2$ are selected such that the series (7.62a) and (7.63a) are convergent at $p=1$, using HAM, we have:

$$f(\eta)=f_0(\eta)+\sum_{m=1}^{\infty} f_m(\eta)o \tag{7.64}$$

$$\theta(\eta)=\theta_0(\eta)+\sum_{m=1}^{\infty} \theta_m(\eta) \tag{7.65}$$

Differentiating the zeroth-order deformation m times with respect to p and then dividing them by $m!$ and finally setting $p=0$, we have the following mth-order deformation problem:

$$L_1[f_m(\eta)-\chi_m f_{m-1}(\eta)]=\hbar_1 R_m^f(\eta) \tag{7.66}$$

$$f_m(0)=0,\ f'_m(0;p)=0,\ f'_m(\infty;p)=0 \tag{7.67}$$

$$L_2[\theta_m(\eta)-\chi_m \theta_{m-1}(\eta)]=\hbar_2 R_m^\theta(\eta) \tag{7.68}$$

$$\theta_m(0)=0,\ \theta_m(\infty;p)=0 \ \text{Power-law temperature} \tag{7.69}$$

$$\theta'_m(0;p)=0,\ \theta_m(\infty;p)=0 \ \text{Power-law heat flux} \tag{7.70}$$

$$R_m^f(\eta) = f_{m-1}''' + Re \cdot A_1 \cdot \left(\sum_{n=0}^{m-1} f_{m-1-n} f_n'' - \sum_{n=0}^{m-1} f_{m-1-n}' f_n' \right) - f_{m-1}' \tag{7.71}$$

$$R_m^\theta(\eta) = \theta_{m-1}'' + \mathrm{Re} \cdot \mathrm{Pr} \cdot \frac{A_1}{A_3} \cdot A_2 \cdot \left(\sum_{n=0}^{m-1} f_{m-1-n} \theta_n' \right) - n \cdot \mathrm{Re} \cdot \mathrm{Pr} \cdot \frac{A_1}{A_3} \cdot A_2 \cdot \left(\sum_{n=0}^{m-1} \theta_{m-1-n} f_n' \right) \tag{7.72}$$

We use MAPLE software to obtain the solution of these equations. Two first deformations of the coupled solutions are presented as follow:

$$f_1(\eta) = -\hbar \cdot \mathrm{Re} \cdot A_1 \cdot \exp(-\eta)[1 + f_w + \eta + f_w \eta - (1 + f_w)\exp(\eta)) \tag{7.73}$$

Power-law temperature

$$\begin{aligned} \theta_1(\eta) = & -0.5\hbar \cdot Pr \cdot Re \cdot \frac{A_1}{A_3} \cdot A_2 \cdot \exp(-2\eta) + \hbar \cdot Pr \cdot Re \cdot \frac{A_1}{A_3} \cdot A_2 \cdot \exp(-\eta) \\ & -0.5\hbar \cdot Pr \cdot Re \cdot \frac{A_1}{A_3} \cdot A_2 \cdot \exp(-\eta) + \hbar \cdot Pr \cdot Re \cdot f_w \cdot \frac{A_1}{A_3} \cdot A_2 \cdot \exp(-\eta) - \hbar \cdot \eta \cdot \exp(-\eta) \\ & -0.5n \cdot \hbar \cdot Pr \cdot Re \cdot \frac{A_1}{A_3} \cdot A_2 \cdot \exp(-2\eta) + 0.5n \cdot \hbar \cdot Pr \cdot Re \cdot \frac{A_1}{A_3} \cdot A_2 \cdot \exp(-\eta) \end{aligned} \tag{7.74}$$

Power-law heat flux

$$\begin{aligned} \theta_1(\eta) = & -0.5\hbar \cdot \frac{A_1}{A_3} \cdot A_2 \cdot [-Pr \cdot Re \cdot \exp(-\eta) - 2Pr \cdot Re \cdot \eta - 2Pr \cdot Re \cdot \eta \cdot f_w + 2\eta \\ & + 2 + n \cdot Pr \cdot Re \cdot \exp(-\eta) - 2n \cdot Pr \cdot Re] \end{aligned} \tag{7.75}$$

The solutions $f_2(\eta)$ and $\theta_2(\eta)$ were too long to be mentioned here, therefore, they are shown graphically.

7.3.5 CONVERGENCE OF THE HAM SOLUTION

As pointed out by Liao, the convergence and the rate of approximation for the HAM solution strongly depend on the values of auxiliary parameter $\hbar$. $\hbar$ curves for (a) f, (b) $\theta_{\text{isothermal}}$, and (c) θ_{isoflux} in $Re = f_w = Pr = 1$, $\phi = 0.01$ and $n = 0$ are shown in Figure 7.7. Using the $\hbar$-curve, we can easily choose the value of auxiliary parameter $\hbar$ to guarantee the convergence of series (7.64) and (7.65). For this problem $\hbar = -0.21$ in step 20 has good accuracy.

7.3.6 RESULTS AND DISCUSSIONS

The governing equations and their boundary conditions are transformed to ordinary differential equations, which are solved analytically using HAM and the results are compared with numerical method (fourth-order Runge-Kutta).

The results that obtained by HAM were well matched with the results carried out by the numerical solution obtained by fourth-order Runge-kutta method as shown in Figure 7.8.

After this validity, results are given for the velocity, temperature distribution, wall shear stress, Nusselt number, and entropy generation for different nondimensional numbers. Figures 7.9 and 7.10 are presented to show the effect of the volume fraction of nanoparticles (Cu) on velocity profiles

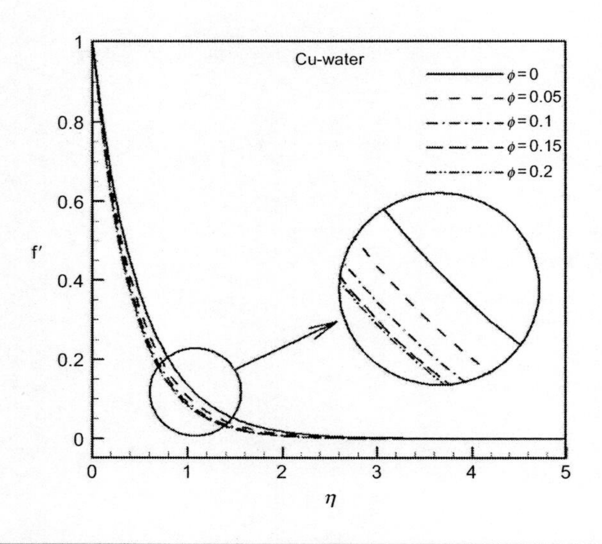

FIGURE 7.9

Effect of nanoparticle volume fraction on velocity profiles when $f_w = 1$, $Re = 1$.

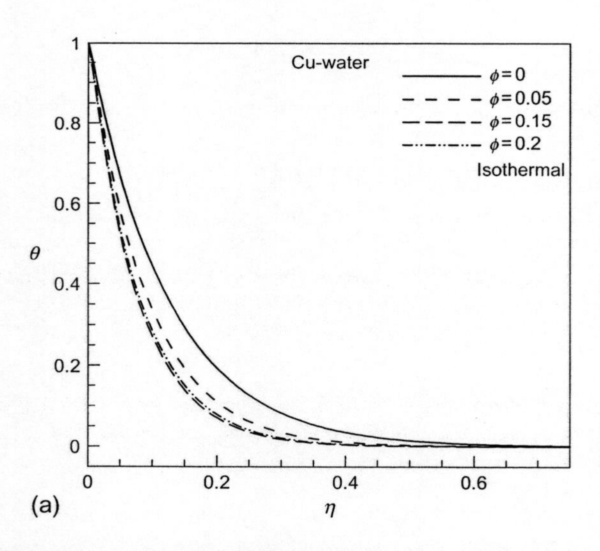

FIGURE 7.10

Effect of nanoparticle volume fraction on temperature distribution (a) power-law temperature and (b) power-law heat flux when $Pr = 6.2$, $f_w = 1$, $Re = 1$, $n = 2$.

Continued

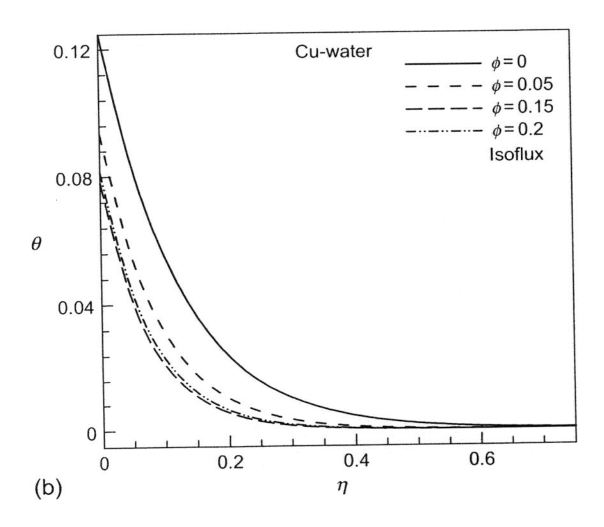

FIGURE 7.10, cont'd

and temperature distribution: (a) power-law temperature and (b) power-law heat flux, respectively when $Pr=6.2$, $f_w=1$, $Re=1$, $n=2$. When the volume of fraction for the nanoparticles increases from 0 to 0.2, all boundary-layers' thicknesses decrease. This agrees with the physical behavior, when the volume of copper nanoparticles increases the thermal conductivity increases and then the thermal boundary-layer thickness decreases.

Figures 7.11 and 7.12 display the behavior of the velocity and the temperature profiles using different nanofluids.

When $Pr=6.2$, $\phi=0.1$, $Re=1$, $f_w=1$, $n=2$, the tables show that by using different types of nanofluids, the values of the velocity and temperature change. When silver is chosen as the nanoparticle, the maximum amount of all boundary-layers' thicknesses observed, while minimum amount of those amounts observed by choosing alumina (Figures 7.11 and 7.12).

Figure 7.13 shows variation of skin-friction coefficients $(-f''(0))$ versus (a) Re and (b) f_w for selected values of the nanoparticles volume parameter in the case of Cu-water. As shown in Figure 7.9, for both suction and injection, it is observed that skin-friction increases as ϕ increases. Also this change occurs when Re or f_w increases. It should be noticed that the changes are more noticeable for higher values of ϕ when values of Re and f_w are greater.

Figure 7.14 shows variation of Nusselt number for power-law temperature case $(-\theta'(0))$ versus (a) Re, (b) f_w, and (c) n for selected values of the nanoparticles volume parameter in the case of Cu-water.

It is obvious from Figure 7.14 that the heat transfer rates increase with the increase of the nanoparticles volume fraction (ϕ), Re, f_w, and n. The change in the Nusselt number is found to be high for higher values of the parameter ϕ, and this change is more noticeable with the increase of Re, f_w, and n.

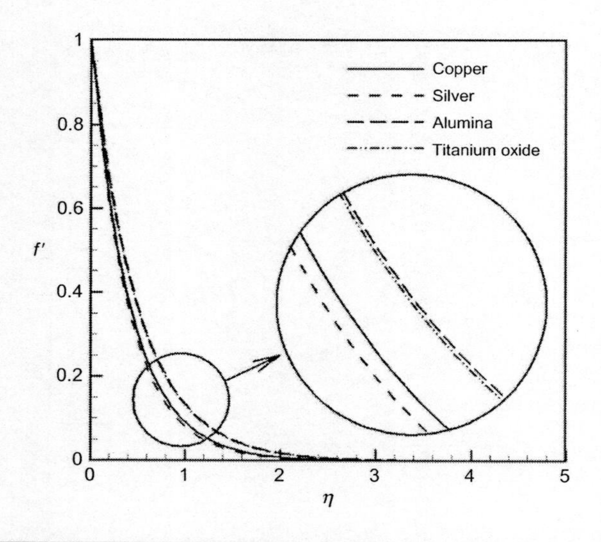

FIGURE 7.11

Velocity for different types of nanofluids when $\phi = 0.1$, $Re = 1$, $f_w = 1$.

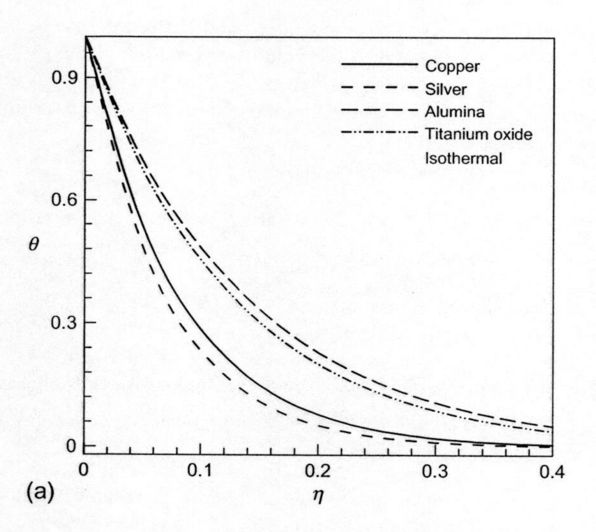

FIGURE 7.12

Temperature profiles (a) power-law temperature and (b) power-law heat flux for different types of nanofluids when $Pr = 6.2$, $\phi = 0.1$, $Re = 1$, $f_w = 1$, $n = 2$.

Continued

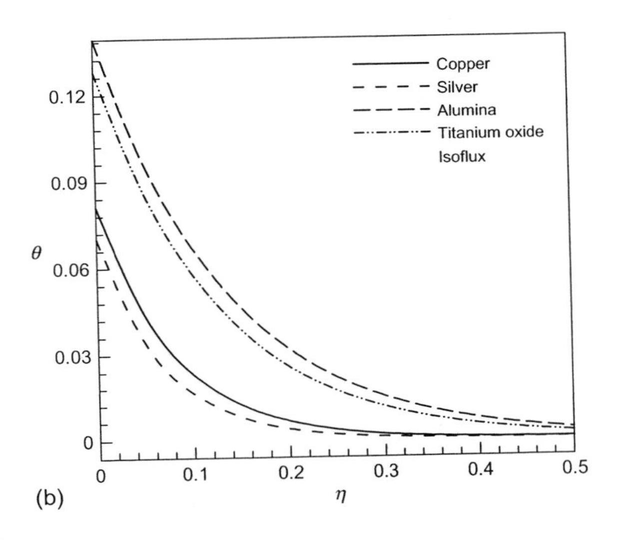

FIGURE 7.12, cont'd

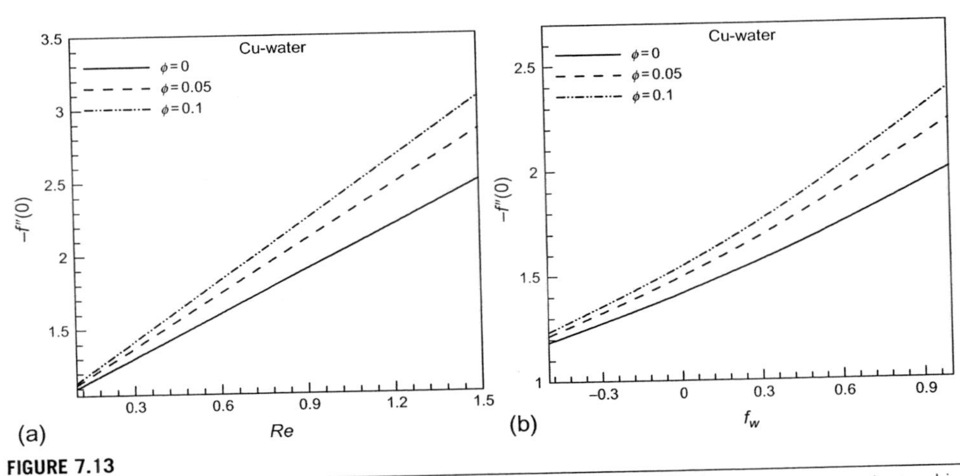

FIGURE 7.13

Effects of the nanoparticle volume fraction $f''(0)$, Reynolds number, and wall injection/suction parameter on skin-friction coefficient when (a) $f_w = 1$ and (b) $Re = 1$.

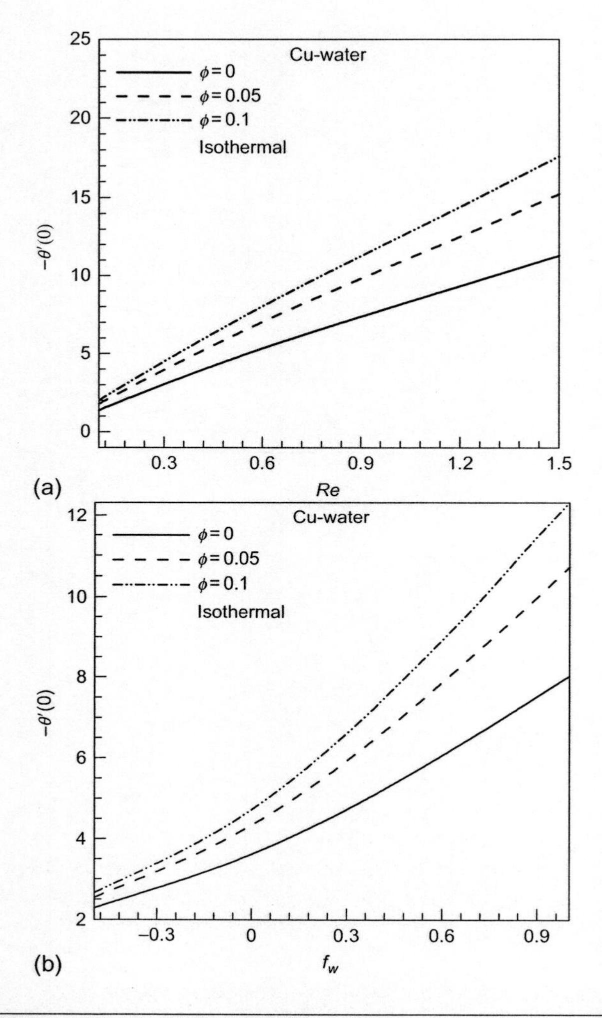

FIGURE 7.14

Effects of the nanoparticle volume fraction $\theta'(0)$, Reynolds number, wall injection/suction parameter, and power of temperature/heat flux distribution on Nusselt number (power-law temperature) when (a) $Pr = 6.2, \ f_w = 1, \ n = 2$, (b) $Pr = 6.2, \ Re = 1, \ n = 2$, and (c) $Pr = 6.2, \ f_w = 1, \ Re = 1$.

Continued

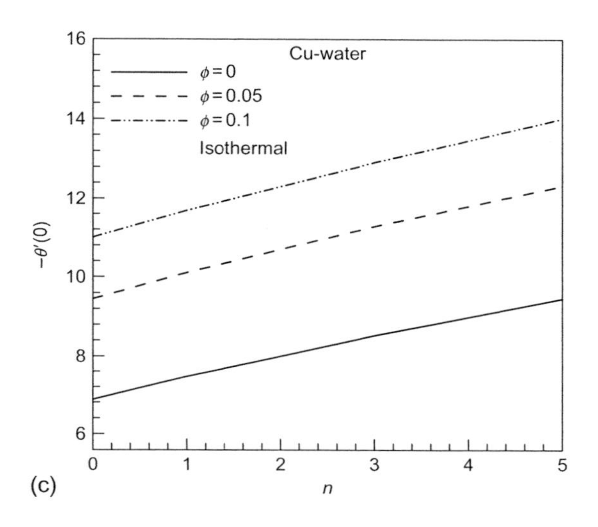

FIGURE 7.14, cont'd

Table 7.5 Effects of the Nanoparticle Volume Fraction for Different Types of Nanofluids on Skin-Friction Coefficient when $Re=1, f_w=1$

	Nanoparticles			
ϕ	Cu	Ag	Al_2O_3	TiO_2
0.05	2.229703	2.298824	2.010783788	2.023135
0.1	2.380028	2.500792	1.9975454	2.019124
0.2	2.483631	2.663553	1.913780762	1.94593

Table 7.5 shows the effects of the nanoparticle volume fraction ϕ for different types of nanofluids on skin-friction coefficient when $Re = 1,\ f_w = 1$.

Tables 7.6 and 7.7 show the effects of the nanoparticle volume fraction ϕ for different types of nanofluids on Nusselt number for power-law temperature case and power-law heat flux case, respectively, when $Pr = 6.2,\ n = 2,\ Re = 1,\ f_w = 1$.

Table 7.6 Effects of the Nanoparticle Volume Fraction for Different Types of Nanofluids on Nusselt Number (Power-Law Temperature) when $Pr=6.2, n=2, Re=1, f_w=1$

	Nanoprticles			
ϕ	Cu	Ag	Al_2O_3	TiO_2
0.05	10.71299	11.77269	7.741767	8.085111
0.1	12.31904	14.26799	7.201689	7.822867
0.2	12.30637	15.04481	5.630545	6.602276

Table 7.7 Effects of the Nanoparticle Volume Fraction for Different Types of Nanofluids on Nusselt Number (Power-Law Heat Flux) when $Pr=6.2$, $n=2$, $Re=1$, $f_w=1$

	Nanoprticles			
ϕ	Cu	Ag	Al_2O_3	TiO_2
0.05	10.71299	11.77268758	7.741767427	8.085111261
0.1	12.31904	14.26799218	7.201688959	7.822867028
0.2	12.30637	15.04480717	5.630544587	6.60227588

These tables show that the values of $-f''(0)$, $-\theta'(0)$ and $1/\theta(0)$ change with nanofluid changes, i.e., we can say that the shear stress and rate of hate transfer change by using different types of nanofluid. This means that the nanofluids will be important in the cooling and heating processes. Choosing silver as the nanoparticle leads to the maximum amount of skin-friction coefficient and rate of hate transfer, while selecting alumina leads to the minimum amount of those values (Tables 7.5–7.7 and Figures 7.15 and 7.16).

It is observed that, regardless of the boundary condition, increasing percentage of nanoparticles (ϕ) leads to the increase in the heat transfer irreversibility due to heat transfer in the direction of finite temperature gradients (HTI) and the contribution of fluid friction irreversibility to the total entropy generation (FFI), whereas the increase in HTI is more than increase in FFI. The entropy generation rate (S_{gen}) reduces while we get farther from the surface of the porous plate and HTI reduction in higher nanoparticles percentages occurs in farther distances.

By studying the Bejan number, we can see that HTI beats FFI near the surface of porous plate. Adding alumina nanoparticles leads to the minimum amount of heat loss, while the maximum amount of heat loss occurs when we use silver as the nanoparticles and there will be the same for HTI and FFI.

By choosing silver and copper as the nanoparticle HTI beats FFI close to the porous plate, while selecting alumina and titanium oxide as the nanoparticle causes this fact to occur farther from the porous plate.

7.3.7 CONCLUSIONS

In this study, HAM is applied to solve the problem of a system of partial differential equations for a nanofluid flow passing over a permeable stretching wall in a porous medium. Some conclusions were summarized as follows:

a. HAM is a powerful approach for solving this nonlinear equation problem field; also it can be observed that there is a good agreement between the present and numerical result.
b. The type of nanofluid is a key factor for heat transfer enhancement. The highest values of Nusselt numbers are obtained by using silver nanoparticles.
c. Nusselt number has direct relationship with nanoparticle volume fraction, Reynolds number, wall injection/suction parameter, and power of temperature/heat flux distribution.
d. Bejan number has a direct relationship with nanoparticle volume fraction, and the minimum amount of heat loss is obtained by using alumina as the nanoparticle.

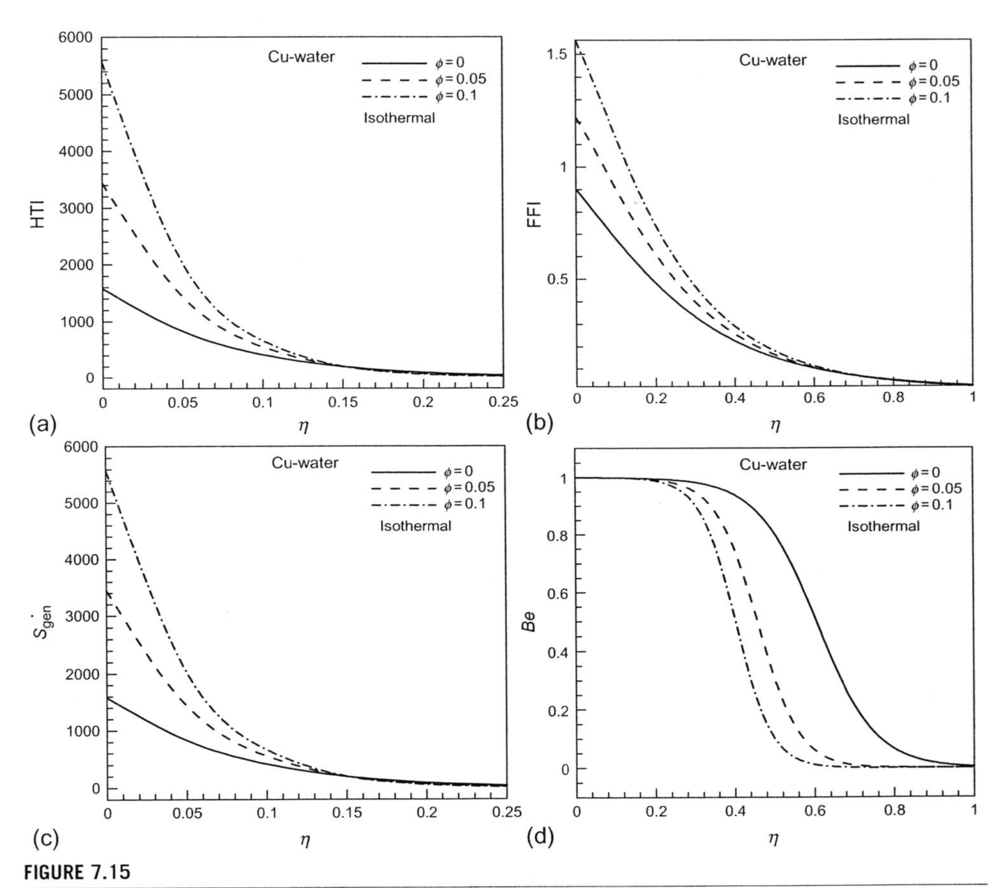

FIGURE 7.15

Effect of nanoparticle volume fraction on (a) HTI, (b) FFI, (c) S_{gen}, and (d) Be when $Pr=6.2$, $Re=1$, $n=2$, $x^*=0.5$, $f_w=1$ $u_0=1m/s$, $T_\infty=T_0=10\text{K}$ and $K=0.001$.

7.4 NATURAL CONVECTION IN A NON-DARCY POROUS MEDIUM OF NANOFLUIDS

Third section of this chapter focuses on the study of natural convection heat transfer characteristics in a vertical wall embedded in a non-Darcy porous medium filled with nanofluid. The governed partial differential equations are transformed into ordinary differential equations, which are obtained by similarity solution; then they have been solved through HAM. The obtained analytical solution in comparison with the numerical ones represents a remarkable accuracy. Considering Maxwell-Garnett model for the effective thermal conductivity of the nanofluid, influence of different nanoparticles along with changes

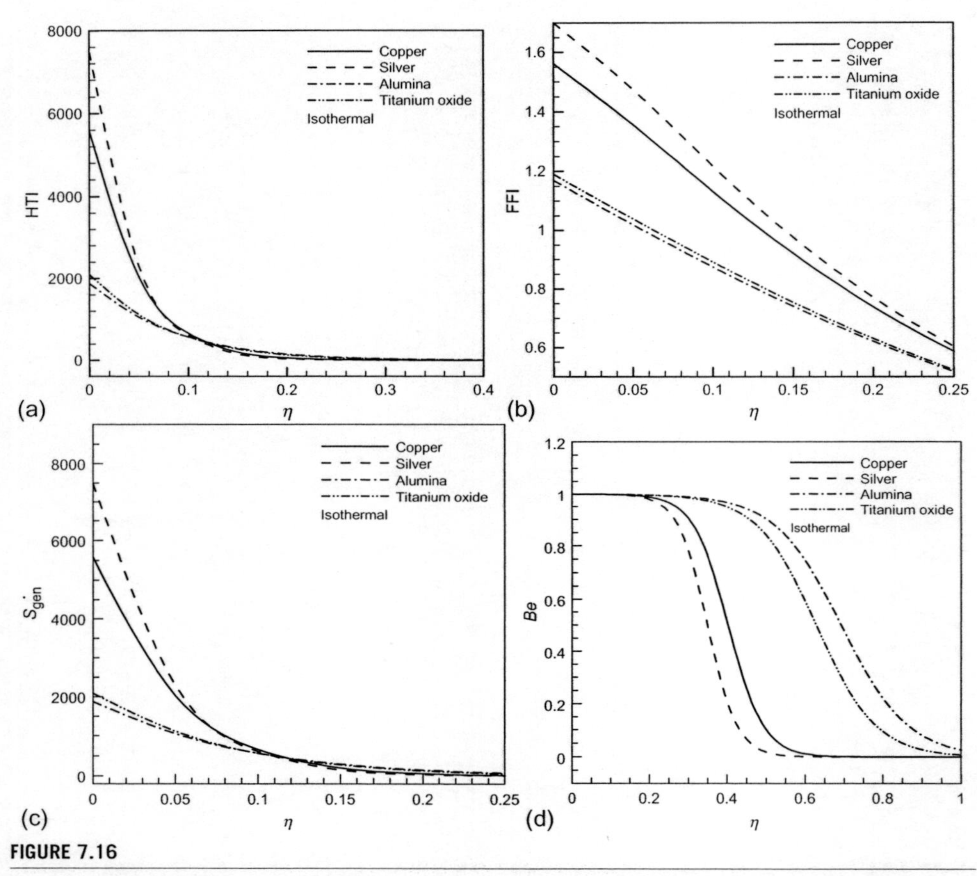

FIGURE 7.16

The effect of different types of nanofluids on (a) HTI, (b) FFI, (c) S_{gen}, and (d) Be when $Pr = 6.2$, $Re = 1$, $n = 2$, $\phi = 0.1$, $x^* = 0.5$, $f_w = 1$ $u_0 = 1m/s$, $T_\infty = T_0 = 10K$ and $K = 0.001$.

in nanoparticles volume fraction φ on flow and heat characteristics is presented through nondimensional temperature and velocity profiles and also Nusselt number at the wall. It is found that increasing the value of φ intensifies the heat transfer rate at the wall surface in the case of injection, whereas different behaviors are observed for the case of suction, i.e., addition of TiO_2 and Al_2O_3 as nanoparticles leads to decrease the rate of heat transfer at wall. Moreover, the results show that Cu-water as nanofluid enhances the local Nusselt number more than other nanofluids (this section has been worked by S. Tavakoli, D.D. Ganji, A. Rasekh, B. Haghighi, H. Jahani in nonlinear dynamics team in Mechanical Engineering Department, 2012-2013).

7.4.1 INTRODUCTION

Natural convection heat transfer in porous media has long been studied as it is a representative model problem for numerous engineering applications such as in electronics cooling, heat exchangers, and various thermal systems (see some works in Nield and Bejan, 1998 and Sohouli et al., 2008).

Note that studies on natural convection using nanofluids are very limited; Oztop and Abu-Nada (2009) presented that addition of nanoparticles leads to enhancement in heat transfers. On the contrary, Putra et al. (2003) showed experimentally using Al_2O_3-water and Cu-water nanofluids bring about decrement in heat transfer. Additionally, Hwang et al. (2007) reported an adverse effect of nanoparticles on heat transfer in natural convection regime in which rectangular cavity heated from below (Bénard convection) with nanofluids.

In accord with researches, the influence of nanofluids on heat transfer in natural convection is still debatable.

The scope of the current research is to study the natural convection boundary-layer flow past a vertical wall embedded in a non-Darcy porous medium filled with different types of nanoparticles including: Cu, Al_2O_3, and TiO_2, while the basic fluid is water. The Maxwell-Garnett model for the effective thermal conductivity of the nanofluid has been taken into account, and the governing nonlinear differential equations are solved analytically using the HAM to give such an explicit analytic solution. Effects of dispersion of distinct types of metallic nanoparticles and the nanoparticle volume fraction on steady flow and heat characteristics are presented and discussed.

7.4.2 GOVERNING EQUATIONS

The problem under investigation is natural convection heat transfer flow in vertical wall embedded in a non-Darcy porous medium while accompanied by nanofluid effects. The schematic diagram is shown in Figure 7.17.

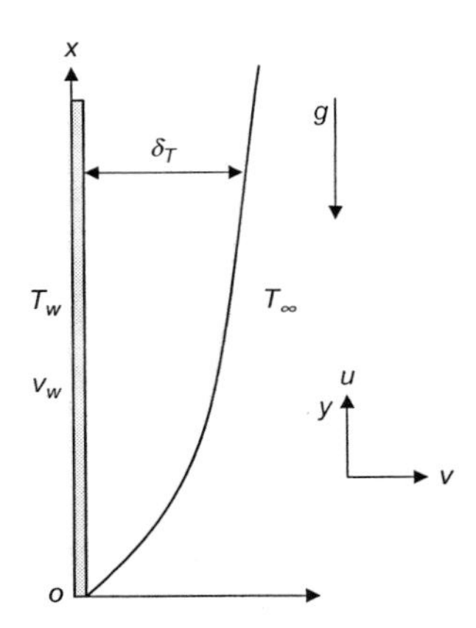

FIGURE 7.17

Physical configuration.

Table 7.8 Thermophysical Properties of Water and Nanoparticles

TiO_2	Al_2O_3	Cu	Water	Physical Properties
4250	3970	8933	997.1	ρ (kg/m^3)
686.2	765	385	4179	C_P (J/kg K)
8.9538	40	400	0.613	K (W/mK)
0.9×10^{-5}	8.5×10^{-6}	1.67×10^{-5}	2.1×10^{-4}	β (1/K)

The plate may be permeable ($\upsilon_w \neq 0$) or impermeable ($\upsilon_w = 0$). The surface temperature of the plate T_w is constant, which is sufficiently different from ambient temperature T_∞. The fluid is a water-based nanofluid containing different nanoparticles such as Al_2O_3, Cu, and TiO_2, in which they are incompressible. The thermophysical properties of the nanofluid are considered to be constant that is given in Table 7.8.

The flow is assumed to be laminar; also, water and nanoparticles are in thermal equilibrium and no slip occurs between them. The governing equations for the flow, heat transfer within the boundary layer near the vertical flat plate can be written in two-dimensional Cartesian coordinates (x,y) by Murthy and Singh (1999) as,

$$\frac{\partial u}{\partial x} + \frac{\partial v}{\partial y} = 0 \tag{7.76}$$

$$\frac{\partial u}{\partial y} + \frac{c\sqrt{K}}{\upsilon}\frac{\partial u^2}{\partial y} = \frac{Kg}{\upsilon}\left(\beta_{\mathrm{nf}}\frac{\partial T}{\partial y}\right) \tag{7.77}$$

$$(\rho C_{\mathrm{P}})_{\mathrm{nf}}\left(u\frac{\partial T}{\partial x} + v\frac{\partial T}{\partial y}\right) = k_{\mathrm{nf}}\left(\frac{\partial^2 T}{\partial y^2}\right) \tag{7.78}$$

The boundary conditions to be considered are:

$$y = 0 : \nu = \nu_w, \quad T = T_w \tag{7.79}$$

$$y \to \infty : u = 0 \quad T = T_\infty \tag{7.80}$$

where $\upsilon_w = Ex^{-1/2}$, E is a real constant. In the above equations, ρ_{nf}, the density of the nanofluid is given by:

$$\rho_{\mathrm{nf}} = \rho_{\mathrm{f}}(1-\phi) + \rho_{\mathrm{s}}\phi \tag{7.81}$$

whereas the heat capacitance of the nanofluid and thermal expansion coefficient of the nanofluid can be determined by:

$$(\rho C_{\mathrm{P}})_{\mathrm{nf}} = (\rho C_{\mathrm{P}})_{\mathrm{f}}(1-\phi) + (\rho C_{\mathrm{P}})_{\mathrm{s}}\phi \tag{7.82}$$

$$(\rho\beta)_{\mathrm{nf}} = (\rho\beta)_{\mathrm{f}}(1-\phi) + (\rho\beta)_{\mathrm{s}}\phi \tag{7.83}$$

where φ being the volume fraction of the solid particles and subscripts f, nf, and s stand for base fluid, nanofluid, and solid, respectively. The effective dynamic viscosity of the nanofluid given by Brinkman (1952a, 1952b) is:

$$\mu_{\mathrm{nf}} = \frac{\mu_{\mathrm{f}}}{(1-\phi)^{2.5}} \tag{7.84}$$

The effective thermal conductivity of the nanofluid calculated by Maxwell (1904) as,

$$\frac{k_{\mathrm{nf}}}{k_{\mathrm{f}}} = \frac{k_{\mathrm{s}} + 2k_{\mathrm{f}} - 2\varphi(k_{\mathrm{f}} - k_{\mathrm{s}})}{k_{\mathrm{s}} + 2k_{\mathrm{f}} + \varphi(k_{\mathrm{f}} - k_{\mathrm{s}})} \tag{7.85}$$

Then Equations (7.87)–(7.89) admit the following similarity solution:

$$\eta = \frac{y}{x}Ra_x^{1/2}, \quad u = \frac{\alpha}{x}Ra_x f'(\eta), \quad \upsilon = -\frac{\alpha}{2x}Ra_x^{1/2}(f - \eta f'), \quad \theta(\eta) = \frac{T - T_\infty}{T_w - T_\infty} \tag{7.86}$$

where f, f', and θ are the dimensionless stream function, vertical velocity, and temperature field, respectively, and Ra_x is the modified Rayleigh number defined by

$$Ra_x = \frac{Kg\beta_T(T_w - T_\infty)x}{\alpha \upsilon} Ra_x^{1/2} \tag{7.87}$$

After transformation we have:

$$f'' + 2Grf'f'' = \left[(1-\phi) + \frac{\rho_{\mathrm{s}}\beta_{\mathrm{s}}}{\rho_{\mathrm{f}}\beta_{\mathrm{f}}}\phi\right](1-\phi)^{2.5}\theta' \tag{7.88}$$

$$\theta'' + \left[\frac{\left(\frac{1-\phi}{2}\right) + \frac{\rho_{\mathrm{s}}c_{p_{\mathrm{s}}}}{2\rho_{\mathrm{f}}c_{p_{\mathrm{f}}}}\phi}{\left(\frac{k_{\mathrm{s}} + 2k_{\mathrm{f}} - 2\phi(k_{\mathrm{f}} - k_{\mathrm{s}})}{k_{\mathrm{s}} + 2k_{\mathrm{f}} + \phi(k_{\mathrm{f}} - k_{\mathrm{s}})}\right)}\right] f\theta' = 0 \tag{7.89}$$

which are subjected to the following boundary conditions:

$$f(0) = f_w, \quad \theta(0) = 1 \tag{7.90}$$

$$f'(\infty) = 0, \quad \theta(\infty) = 0 \tag{7.91}$$

where

$$Gr = \frac{c\sqrt{K}Kg\beta_T(T_w - T_\infty)}{\nu^2} \tag{7.92}$$

$$f_w = -\frac{2E}{\sqrt{\alpha Kg\beta_T(T_w - T_\infty)}} \tag{7.93}$$

are the Grash of number and the mass flux parameter, respectively. The nondimensional heat transfer coefficient can be written as,

$$\frac{Nu}{Ra_x^{1/2}} = -\frac{k_{\mathrm{nf}}}{k_{\mathrm{f}}}\theta'(0) \tag{7.94}$$

7.4.3 SOLUTION USING HAM

Here, the HAM has been employed to solve Equations (7.88) and (7.89) as the initial guess approximation for $f(\eta)$ and $\theta(\eta)$:

$$L_1(f) = f'', \quad L_2(\theta) = \theta'' + \theta' \tag{7.95}$$

And as the auxiliary linear operator, which has the property:

$$L_1(c_1 + c_2\eta) = 0, \quad L_2(c_3 + c_4 e^{-\eta}) = 0 \tag{7.96}$$

where c_1-c_4 are constants. Let $P \in [0, 1]$ denotes the embedding parameter and $\hbar$ indicates the nonzero auxiliary parameters. Let us choose the initial guesses using the auxiliary linear operators and boundary conditions in the following form,

$$f_0(\eta) = f_w, \quad \theta_0(\eta) = e^{-\eta} \tag{7.97}$$

Then, the following equations will be constructed.

7.4.3.1 Zeroth-order deformation equations

$$(1-p)L_1[f(\eta;p) - f_0(\eta)] = p\hbar_1 N_1[f(\eta;p)] \tag{7.98}$$

$$(1-p)L_2[\theta(\eta;p) - \theta_0(\eta)] = p\hbar_2 N_2[\theta(\eta;p)] \tag{7.99}$$

$$f(0;p) = f_w, \quad \theta(0;p) = 1 \tag{7.100}$$

$$f'(\infty;p) = \theta(\infty;p) = 0 \tag{7.101}$$

$$N_1[f(\eta;p)] = \frac{d^2 f(\eta;p)}{d\eta^2} + 2Gr\frac{df(\eta;p)}{d\eta}\frac{d^2 f(\eta;p)}{d\eta^2} - \left[(1-\phi) + \frac{\rho_s\beta_s}{\rho_f\beta_f}\phi\right](1-\phi)^{2.5}\frac{d\theta(\eta;p)}{d\eta} = 0 \tag{7.102}$$

$$N_2[\theta(\eta;p)] = \frac{d^2\theta(\eta;p)}{d\eta^2} + \left[\frac{\left(\frac{1-\phi}{2}\right) + \frac{\rho_s c_{p_s}}{2\rho_f c_{p_f}}\phi}{\left(\frac{k_s + 2k_f - 2\phi(k_f - k_s)}{k_s + 2k_f + \phi(k_f - k_s)}\right)}\right] f\frac{d\theta(\eta;p)}{d\eta} = 0 \tag{7.103}$$

For $p = 0$ and $p = 1$, we have

$$f(\eta;0) = f_0(\eta), \quad f(\eta;1) = f(\eta), \quad \theta(\eta;0) = \theta_0(\eta), \quad \theta(\eta;1) = \theta_0(\eta) \tag{7.104}$$

When p increases from 0 to 1, then $f(\eta;p)$ vary from $f_0(\eta)$ to $f(\eta)$, and $\theta(\eta;p)$ vary from $\theta_0(\eta)$ to $\theta(\eta)$. By Taylor's theorem and using Equations (7.102) and (7.103), $f(\eta;p)$ and $\theta(\eta;p)$ can be expanded in a power series of p as follows,

$$f(\eta;p) = f_0(\eta) + \sum_{m-1}^{\infty} f_m(\eta)p^m$$
$$f_m(\eta) = \frac{1}{m!}\frac{\partial^m[f(\eta;p)]}{\partial p^m} \tag{7.105}$$

$$\theta(\eta;p) = \theta_0(\eta) + \sum_{m-1}^{\infty} \theta_m(\eta)p^m$$
$$\theta_m(\eta) = \frac{1}{m!}\frac{\partial^m[\theta(\eta;p)]}{\partial p^m} \tag{7.106}$$

For simplicity, we suppose $\hbar_1 = \hbar_2 = \hbar$, where $\hbar$ is chosen in such a way that these two series are convergent at $p=1$. Therefore, we have Equations (7.105) and (7.106) in the following form:

$$f(\eta) = f_0(\eta) + \sum_{m-1}^{\infty} f_m(\eta) \tag{7.107}$$

$$\theta(\eta) = \theta_0(\eta) + \sum_{m-1}^{\infty} \theta_m(\eta) \tag{7.108}$$

7.4.3.2 mth-Order deformation equations

$$L[f_m(\eta) - \chi_m f_{m-1}(\eta)] = \hbar R_m^f(\eta) \tag{7.109}$$

$$f_m(0) = f_m'(\infty) = 0 \tag{7.110}$$

$$R_m^f(\eta) = f_{m-1}''(\eta) - \left[(1-\phi) + \frac{\rho_s\beta_s}{\rho_f\beta_f}\phi\right](1-\phi)^{2.5}\theta_{m-1}' + \sum_{n=0}^{m-1} 2f_{m-1-n}'f_n'' \tag{7.111}$$

$$L[\theta_m(\eta) - \chi_m\theta_{m-1}(\eta)] = \hbar R_m^\theta(\eta) \tag{7.112}$$

$$\theta_m(0) = \theta_m(\infty) = 0 \tag{7.113}$$

$$R_m^f(\eta) = \theta_{m-1}''(\eta) + \sum_{n=0}^{m-1}\left[\frac{\left(\frac{1-\phi}{2}\right) + \frac{\rho_s c_{p_s}}{2\rho_f c_{p_f}}\phi}{\left(\frac{k_s + 2k_f - 2\phi(k_f - k_s)}{k_s + 2k_f + \phi(k_f - k_s)}\right)}\right] f_{m-1-n}\theta_n' \tag{7.114}$$

$$\chi_m = \begin{cases} 0, & m \le 1 \\ 1, & m > 1 \end{cases} \tag{7.115}$$

7.4.4 CONVERGENCE OF HAM SOLUTION

The HAM provides us with a great freedom in choosing the solution of a nonlinear problem by different base functions. This has a great effect on the convergence region because the convergence region and rate of a series are chiefly determined by the base functions and its convergence is ensured. On the other hand, as pointed out by Liao (1999), the convergence and rate of approximation for the HAM solution strongly depends on the value of the auxiliary parameter $\hbar$. Even though the initial approximation and the auxiliary linear operator L are given, we still have great freedom to choose the value of the auxiliary parameter $\hbar$. Hence, the auxiliary parameter $\hbar$ provides us with an additional way to conveniently

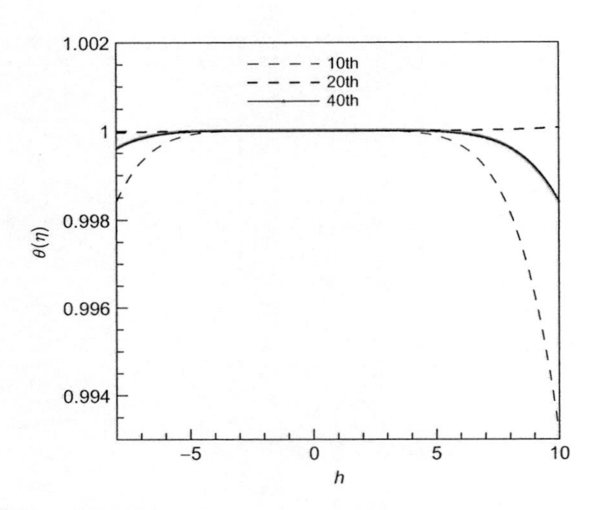

FIGURE 7.18

The ħ curve for $\theta(0)$ when $\varphi=0$ and $Gr=1$.

adjust and control the convergence region and the solution series rate. The $\hbar$ region for this problem is shown in Figure 7.18. It is easy to discover that $-4 \le \hbar \le 4$ is the valid region of $\hbar$, for all value of η. Note that we consider $\hbar$ as -0.01 in this study.

7.4.5 RESULTS AND DISCUSSIONS

The explicit analytic solution has been verified by the numerical integration using the fourth-order Runge-Kutta method, where the initial guesses of the numerical solutions are given by f_0 and θ_0 as defined in Equation (7.107). Note that our explicit analytic solution contains the auxiliary parameter $\hbar$ which we have great freedom to choose. In Table 7.9, we made comparison between present HAM solution and numerical solutions with various value of η. It is found that our analytic approximations agree well with the numerical ones.

The nondimensional temperature distribution $\theta(\eta)$ and velocity component $f'(\eta)$ in the x-direction for various parameters including volume fraction factor φ, different nanoparticles, and f_w are plotted in Figures 7.19–7.24 when $Gr=1$.

Figure 7.19 shows the temperature $\theta(\eta)$ profiles across the boundary layers for Cu-water as nanofluid for different values of the suction/injection parameter f_w when $\varphi=0.05$.

It should be mentioned that suction corresponds to $f_w>0$, injection to $f_w<0$, and impermeable plate to $f_w=0$; therefore, it is clear that suction reduces the boundary-layer thickness sharply. The effect of different volume fractions, φ, on temperature $\theta(\eta)$ profiles is indicated in Figure 7.20 for Cu-water as nanofluid.

It is seen that, increasing the volume fraction of nanoparticles, φ, tends to thicken thermal boundary layer. The reason is that solid particles have high thermal conductivity, so the thickness of the thermal boundary layer increases. Figure 7.21 demonstrates the influences of different nanoparticles such as

Table 7.9 Comparison Between HAM Solution and Present Numerical Solution for $\theta(\eta)$ with $\varphi=0$ and $\varphi=0.05$ for Cu-Water as Working Fluid with 40th-Order Approximation

	$\varphi=0,\ \theta(\eta)$			$\varphi=0.1,\ \theta(\eta)$		
η	HAM Solution	Numerical Solution	Relative Error	HAM Solution	Numerical Solution	Relative Error
0	1.000000	1.000000	0.000000	1.000000	1.000000	0.000000
0.5	0.685926	0.685279	0.000646	0.720575	0.719824	0.000751
1	0.455879	0.455511	0.000368	0.506208	0.505753	0.000456
1.5	0.296403	0.296052	0.000350	0.349082	0.348755	0.000326
2	0.189522	0.189351	0.000171	0.237396	0.237102	0.000294
2.5	0.119894	0.119743	0.000152	0.159646	0.159460	0.000186
3	0.075252	0.075104	0.000148	0.106475	0.106328	0.000147
3.5	0.046925	0.046801	0.000124	0.070499	0.070374	0.000125
4	0.029100	0.028986	0.000115	0.046342	0.046229	0.000113
4.5	0.017935	0.017824	0.000111	0.030202	0.030099	0.000102
5	0.010940	0.010851	0.000090	0.019455	0.019361	0.000094
5.5	0.065102	0.065026	0.000076	0.012303	0.012228	0.000075
6	0.003843	0.003794	0.000048	0.007549	0.007497	0.000053
6.5	0.002140	0.002108	0.000032	0.004394	0.004362	0.000033
7	0.001065	0.001059	0.000006	0.002288	0.002285	0.000003
7.5	0.000408	0.000406	0.000002	0.000910	0.000910	0.000000
8	0.000000	0.000000	0.000000	0.000000	0.000000	0.000000

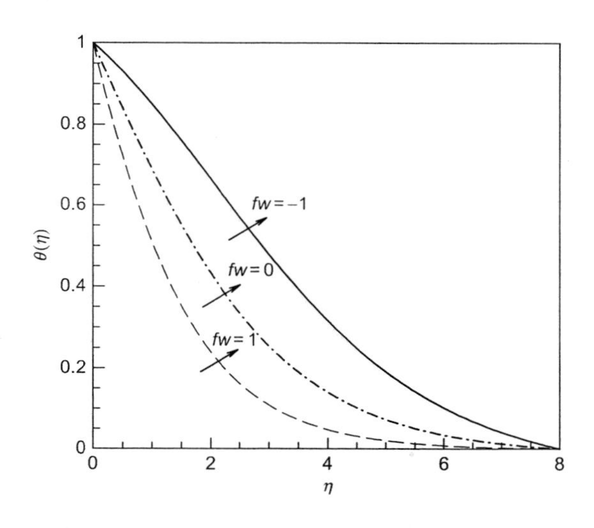

FIGURE 7.19

Temperature profiles $\theta(\eta)$ for various values of f_w with Cu-water when $\varphi=0.05$.

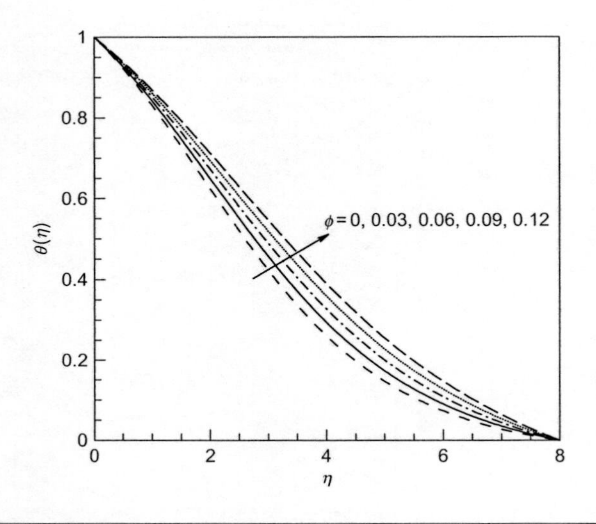

FIGURE 7.20

Temperature profiles $\theta(\eta)$ for various values of φ for Cu-water nanofluid when $f_w = -1$.

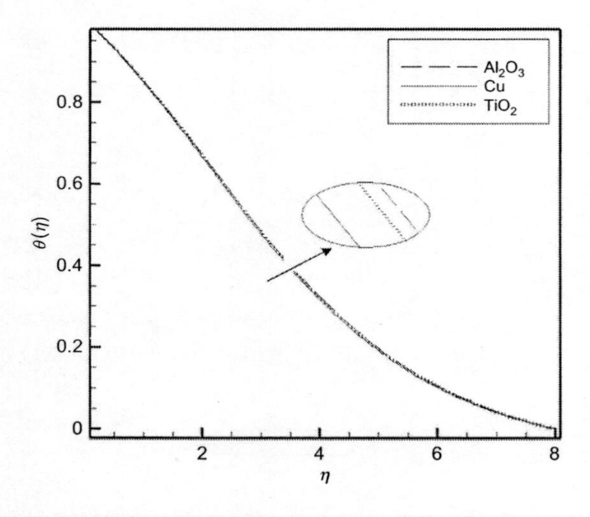

FIGURE 7.21

The effect of the different nanoparticles on temperature profiles $\theta(\eta)$ when $\varphi = 0.05$ and $f_w = -1$.

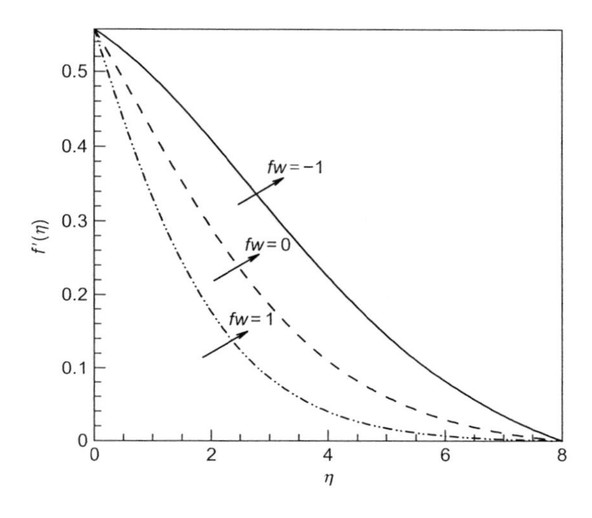

FIGURE 7.22

The effect of f_w on $f'(\eta)$ for Cu-water when $\varphi = 0.05$.

Al_2O_3, Cu, and TiO_2 on temperature profile $\theta(\eta)$ at $\varphi = 0.05$ in the case of injection. It is realized that, using Al_2O_3 eventuates increasing thermal boundary layer more in comparison with Cu and TiO_2.

The variation of nondimensional velocity, $f'(\eta)$, with f_w in the case of Cu-water nanofluid and $\varphi = 0.05$ is indicated in Figure 7.22.

It is seen that the magnitude of velocity profiles is increasing with decreasing values of f_w. Figure 7.23 illustrates the variation of nondimensional velocity $f'(\eta)$, with volume fraction for Cu-water when $f_w = -1$.

It is observed that addition of nanoparticle tends to increase the magnitude of $f'(\eta)$ profiles at far from of wall; however, nanofluid reveals opposite behavior at wall vicinity. The velocity profiles $f'(\eta)$, as shown in Figure 7.24, are found to increase with utilizing Cu-water as nanofluid. Besides, it is seen that these profiles satisfy the far field boundary condition, $f'(\infty) = 0$ asymptotically, which support the obtained numerical results.

Ultimately, the variations of heat transfer rate at the surface or the local Nusselt number $Nu\,Ra^{-1/2}$ are illustrated in Figures 7.25–7.27. As shown in Figure 7.25, the heat transfer rate at the wall surface increases for higher value of f_w for Cu-water nanofluid at the same value of volume fraction.

From Figure 7.26, it is observed that the local Nusselt number increases with increasing values of the nanoparticles volume fraction parameter φ in the case of injection ($f_w = -1$).

Additionally, it is found that among the considered nanoparticles, Cu has predominant effect on enhancing natural convection heat transfer rate at the vertical wall. In the case of suction, however, heat transfer shows an opposite trend (see Figure 7.27). As seen, the local Nusselt number is decreasing with the addition of nanoparticles except for Cu. Also, it should be mentioned that the lowest heat transfer rate is obtained for the nanoparticles TiO_2 due to the domination of conduction mode of heat

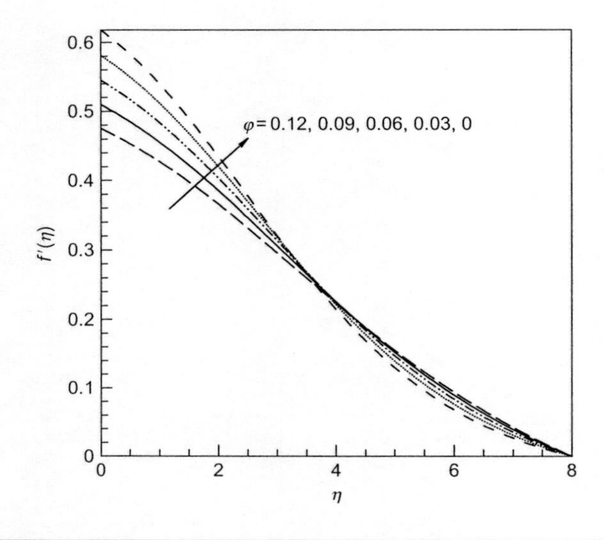

FIGURE 7.23

The effect of φ on $f'(\eta)$ for Cu-water when $f_w = -1$.

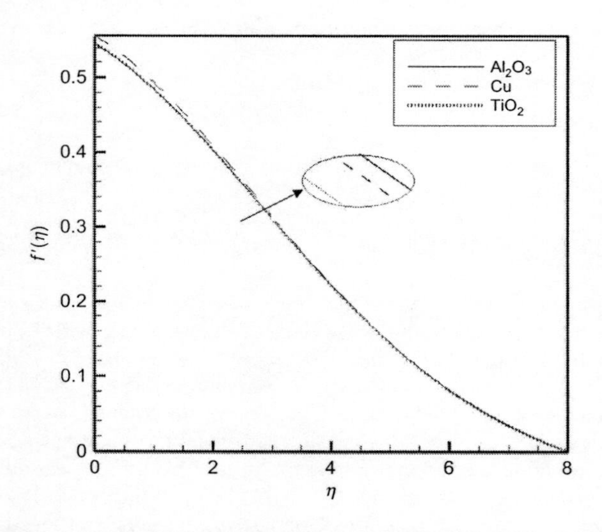

FIGURE 7.24

The effect of the different nanoparticles on $f'(\eta)$ when $\varphi = 0.05$ and $f_w = -1$.

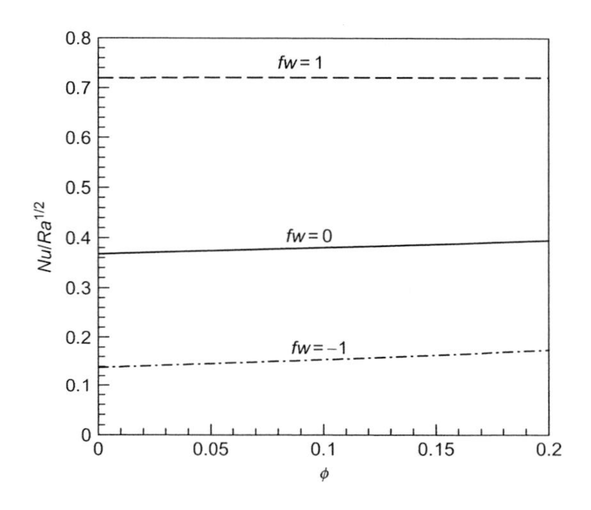

FIGURE 7.25

Variation of NU/Ra$^{1/2}$ with φ for different values of f_w for Cu-water.

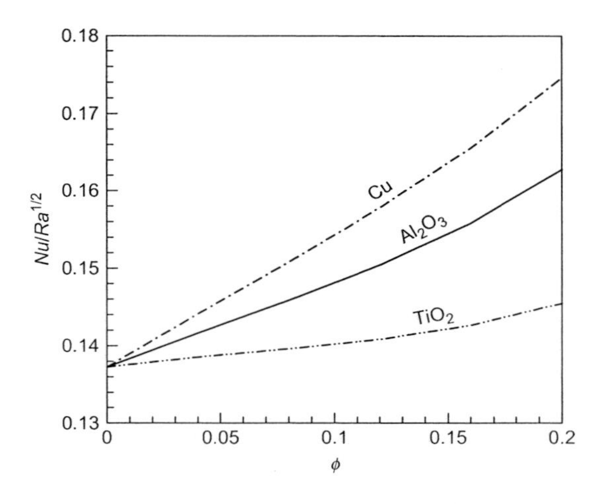

FIGURE 7.26

Variation of NU/Ra$^{1/2}$ with φ for different nanoparticles with $f_w = -1$.

transfer. The main reason behind the behavior is that TiO_2 has the lowest value of thermal conductivity compared to Cu and Al_2O_3, as can be seen from Table 7.8. On the contrary, the Cu nanoparticles have higher values of thermal diffusivity, and therefore reduce the temperature gradients, which increase Nusselt number accordingly.

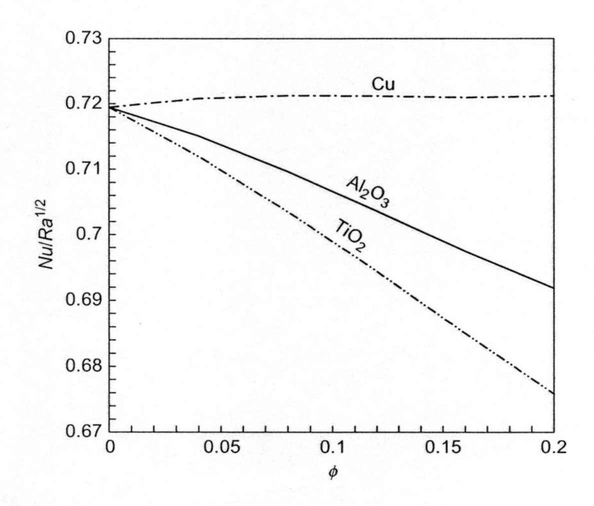

FIGURE 7.27

Variation of NU/Ra$^{1/2}$ with φ for different nanoparticles with f_w=1.

7.4.6 CONCLUSION

In this section, natural convection heat transfer from a vertical wall embedded in non-Darcy porous medium filled with nanofluid has been studied. The governing partial differential equations were transformed into a system of nonlinear ordinary differential equations using a similarity transformation, and then solved through HAM to obtain an explicit, totally analytic solution. The validity of our analytic solution was verified by numerical results. The effects of various parameters including the nanoparticle volume φ and suction/injection parameter f_w and also different types of nanoparticle, namely, Cu, Al_2O_3, and TiO_2 with water as the base fluid on heat transfer characteristics were analyzed. It is found that addition of nanoparticles enhances the rate of heat transfer in the case of injection; however, the opposite trend is observed for suction. Moreover, it can be concluded that utilizing Cu-water nanofluid increases the local Nusselt number more than other nanoparticles.

REFERENCES

Ali, M.E., 1995. On thermal boundary layer on a power-law stretched surface with suction or injection. Int. J. HeatFluid Flow 16, 280–290.

Al-Odat, M.Q., Damesh, R.A., Al-Azab, T.A., 2006. Thermal boundary layer on an exponentially stretching continuous surface in the presence of magnetic field effect. Int. J. Appl. Mech. Eng. 11, 289–299.

Aminossadati, S.M., Ghasemi, B., 2009. Natural convection cooling of a localized heat source at the bottom of a nanofluid-filled enclosure. European J. Mech. B/Fluids 28, 630–640.

Aminossadati, S.M., Ghasemi, B., 2011. Natural convection of water–CuO nanofluid in a cavity with two pairs of heat source–sink. Int. Commun. Heat and Mass Transfer 38 (5), 672–678.

Attia, H.A., 2006. Stagnation point flow towards a stretching surface through a porous medium with heat generation. Turk. J. Eng. Environ. Sci. 30, 299–306.

Banks, W.H.H., 1983. Similarity solutions of the boundary layer equations for a stretching wall. J. Mecan. Theor. Appl. 2, 375–392.

Bejan, A., 1995. Entropy Generation Minimization, the Method of Thermodynamic Optimization of Finite-Size Systems and Finite-Time Processes. CRC Press, Boca Rotan.

Bejan, A., Dincer, I., Lorenteh, S., Reyes, H., 2004. Porous and Complex Flow Structures in Modern Technologies. Springer, New York.

Brinkman, H.C., 1952a. The viscosity of concentrated suspensions and solution. J. Chem. Phys. 20, 571–581.

Brinkman, H.C., 1952b. The viscosity of concentrated suspensions and solution. J. Chem. Phys. 20, 571–581.

Cheng, P., Minkowycz, W.J., 1977. Free convection about a vertical flat plate embedded in a porous medium with application to heat transfer from a dike. J. Geophys. Res. 82, 2040–2044.

Gerald, C.F., Weatly, P.O., 1989. Applied Numerical Analysis. Addison Wesley Publishing Company, New York.

Hooman, K., Gurgenci, H., Merrikh, A.A., 2007. Heat transfer and entropy generation optimization of forced convection in porous-saturated ducts of rectangular cross-section. Int. J. Heat Mass Transfer 50, 2051–2059.

Hwang, K.S., Lee, J.H., Jang, S.P., 2007. Buoyancy-driven heat transfer of water-based Al_2O_3 nanofluids in a rectangular cavity. Int. J. Heat Mass Transfer 50, 4003–4010.

Ishak, A., Jafar, K., Nazar, R., Pop, I., 2009. MHD stagnation point flow towards a stretching sheet. Phys. A 388, 3377–3383.

Kaviany, M., 1992. Principles of Heat Transfer in Porous Media. Springer, New York.

Kumar, S., Prasad, S.K., Banerjee, J., 2010. Analysis of flow and thermal field in nanofluid using a single phase thermal dispersion model. Appl. Math. Model. 34 (3), 573–592.

Kuznetsov, A.V., Nield, D.A., 2010. Natural convective boundary-layer flow of a nanofluid past a vertical plate. Int. J. Therm. Sci. 49, 243–247.

Layek, G.C., Mukhopadhyay, S., Samad, Sk.A., 2007. Heat and mass transfer analysis for boundary layer stagnation-point flow towards a heated porous stretching sheet with heat absorption/generation and suction/blowing. Int. Commun. Heat and Mass Transfer 34, 347–356.

Liao, S.J., 1999. An explicit, totally analytic approximate solution for Blasius' viscous flow problems. J Non-Linear Mech 34, 759–782.

Maxwell, J., 1904. A Treatise on Electricity and Magnetism, 2nd ed. Oxford University Press, Cambridge, UK.

Murthy, P.V.S.N., Singh, P., 1999. Heat and mass transfer by natural convection in a non-Darcy porous medium. Acta Mech. 138, 243–254.

Nazar, R., Ishak, A., Pop, I., 2008. Unsteady boundary layer flow over a stretching sheet in a micropolar fluid. Int. J. Math. Phys. Eng. Sci. 2, 161–165.

Nield, D.A., Bejan, A., 1998. Convection in Porous Media, second ed. Springer, New York.

Nield, D.A., Bejan, A., 2006a. Convection in Porous Media. Springer, New York.

Nield, D.A., Bejan, A., 2006b. Convection in Porous Media. Springer, New York.

Nield, D.A., Kuznetsov, A.V., 2005. Forced Convection in Porous Media: Transverse Heterogeneity Effects and Thermaldevelopment, vol. 2. Taylor and Francis, New York, pp. 143–193.

Nield, D.A., Kuznetsov, A.V., 2009. The Cheng-Minkowycz problem for natural convective boundary layer flow in a porous medium saturated by a nanofluid. Int. J. Heat Mass Transfer 52, 5792–5795.

Oztop, H.F., Abu-Nada, E., 2008a. Numerical study of natural convection in partially heated rectangular enclosures filled with nanofluids. Int. J. Heat Fluid Flow 29, 1326–1336.

Oztop, H.F., Abu-Nada, E., 2008b. Numerical study of natural convection in partially heated rectangular enclosure filled with nanofluids. Int. J. Heat Fluid Flow 29, 1326–1336.

Putra, N., Roetzel, W., Das, S.K., 2003. Natural convection of nano-fluids. Heat Mass Transfer 39, 775–784.

Sohouli, A.R., Domairry, D., Famouri, M., Mohsenzadeh, A., 2008. Analytical solution of natural convection of Darcian fluid about a vertical full cone embedded in porous media prescribed wall temperature by means of HAM. Int. Commun. Heat and Mass Transfer 35, 1380–1384.

Starov, V.M., Zhdanov, V.G., 2001. Effective viscosity and permeability of porous media. Colloids Surf. A 192, 363–375.

Talebi, F., Mahmoudi, A.H., Shahi, M., 2010. Numerical study of mixed convection flows in a square lid-driven cavity utilizing nanofluid. Int. Commun. Heat and Mass Transfer 37 (1), 79–90.

Yu, Z.T., Wang, W., Xu, X., Fan, L.W., Hu, Y.C., Cen, K.F., 2011. A numerical investigation of transient natural convection heat transfer of aqueous nanofluids in a differentially heated square cavity. Int. Commun. Heat and Mass Transfer 38 (5), 585–589.

CHAPTER

NANOFLUID FLOW IN MAGNETIC FIELD

8

CHAPTER CONTENTS

8.1 MHD Nanofluid Flow Analysis in Divergent and Convergent Channels 318
8.1.1 Introduction 318
8.1.2 Problem Description 319
8.1.3 Weighted Residual Methods 320
8.1.3.1 Collocation Method 321
8.1.3.2 Least Square Method 322
8.1.3.3 Galerkin Method 323
8.1.4 Results and Discussions 324
8.1.5 Conclusion 327
8.2 MHD Stagnation-Point Flow of a Nanofluid and Heat Flux 330
8.2.1 Introduction 330
8.2.2 Mathematical Model 331
8.2.3 Methods of Solution 333
8.2.3.1 Review of HPM 333
8.2.3.2 Padé Approximants 333
8.2.4 Analytical Solution Statement 334
8.2.5 Results and Discussion 335
8.2.6 Conclusions 347
8.3 Jeffery-Hamel Flow with High Magnetic Field and Nanoparticle 349
8.3.1 Introduction 350
8.3.2 Governing Equations 351
8.3.3 Fundamentals of ADM 353
8.3.4 Application 354
8.3.5 Results and Discussion 355
8.3.6 Conclusion 359
8.4 The Transverse Magnetic Field on Jeffery-Hamel Problem with Cu-Water Nanofluid 359
8.4.1 Introduction 360
8.4.2 Problem Statement and Mathematical Formulation 360
8.4.3 Application of HAM on MHD Jeffery-Hamel Flow 362
8.4.3.1 Zeroth-Order Deformation Equations 362
8.4.3.2 mth-Order Deformation Equations 363

Application of Nonlinear Systems in Nanomechanics and Nanofluids. http://dx.doi.org/10.1016/B978-0-323-35237-6.00008-X

8.4.4 Convergence of the HAM Solution 363
8.4.5 Results and Discussions 365
8.4.6 Conclusion 369
8.5 Investigation of MHD Nanofluid Flow in a Semiporous Channel 369
8.5.1 Introduction 370
8.5.2 Problem Description 370
8.5.3 Weighted Residual Methods 373
8.5.3.1 Least Square Method 374
8.5.3.2 Galerkin Method 375
8.5.4 Results and Discussions 376
8.5.6 Conclusion 379
References 385

8.1 MHD NANOFLUID FLOW ANALYSIS IN DIVERGENT AND CONVERGENT CHANNELS

In this section, magnetohydrodynamic (MHD) nanofluid flow in divergent and convergent channels called, Jeffery-Hamel flow, is investigated using three weighted residual methods (WRMs) and numerical method. Different base fluids and nanoparticle are used. The effective thermal conductivity and viscosity of nanofluid are calculated by the Maxwell-Garnett (MG) and Brinkman models, respectively. Comparison between collocation method (CM), Galerkin method (GM), and least square method (LSM) show that LSM is more accurate than other methods. The influence of the nanofluid volume friction, Reynolds number, Hartmann number, and angle of the channel on velocity profiles are investigated. Also, it can be found that skin-friction coefficient is an increasing function of Reynolds number, opening angle, and nanoparticle volume friction, but decrease function of Hartmann number (*this section has been worked by D.D. Ganji, M. Hatami in nonlinear dynamics team in Mechanical Engineering Department, 2012-2013*).

8.1.1 INTRODUCTION

Internal flow between two plates is one of the most applicable cases in industrial, fluid mechanics, civil, environmental, and biomechanical engineering. The incompressible viscous fluid flow through convergent and divergent channels is one of the most applicable cases in flow between two inclined plates which first was introduced by Jeffery (1915) and Hamel (1916) and so, it is known as Jeffery-Hamel flow.

This section aims to apply WRMs to MHD Jeffery-Hamel nanofluid flow. CM, GM, and LSM were selected and after a brief introduction to the principles of the methods and problem, WRMs were applied to find the approximate solutions of both divergent and convergent channels. Obtained results were compared with those of the fourth-order Runge-Kutta numerical technique. By calculating the errors of methods, it was found that LSM is more acceptable than the other two WRMs. Also, the effects of the Reynolds number (Re), channel slope angle (α), and Hartmann number (Ha) for divergent and convergent channels are studied and treatment of the velocity profile near the centerline and walls is discussed in this chapter.

8.1.2 PROBLEM DESCRIPTION

Consider a system of cylindrical polar coordinates (r, θ, z), which steady two-dimensional flow of an incompressible conducting viscous fluid from a source or sink at channel walls, lie in planes, and intersect in the axis of z. Assuming purely radial motion means that there is no change in the flow parameter along the z-direction. The flow depends on r and θ and further assume that there is no magnetic field in the z-direction (see Figure 8.1).

The reduced form of continuity, Navier-Stokes, and Maxwell's equations are (Sheikholeslami et al., 2012a,b):

$$\frac{\rho_{\mathrm{nf}}}{r}\frac{\partial(ru(r,\theta))}{\partial r}(ru(r,\theta))=0 \tag{8.1}$$

$$u(r,\theta)\frac{\partial u(r,\theta)}{\partial r}=-\frac{1}{\rho_{\mathrm{nf}}}\frac{\partial P}{\partial r}+\upsilon_{\mathrm{nf}}\left[\frac{\partial^2 u(r,\theta)}{\partial r^2}+\frac{1}{r}\frac{\partial u(r,\theta)}{\partial r}+\frac{1}{r^2}\frac{\partial^2 u(r,\theta)}{\partial\theta^2}-\frac{u(r,\theta)}{r^2}\right]-\frac{\sigma B_0^2}{\rho_{\mathrm{nf}} r^2}u(r,\theta) \tag{8.2}$$

$$\frac{1}{\rho_{\mathrm{nf}} r}\frac{\partial P}{\partial\theta}-\frac{2\upsilon_{\mathrm{nf}}}{r^2}\frac{\partial u(r,\theta)}{\partial\theta}=0 \tag{8.3}$$

where B_0 is the electromagnetic induction, σ_{nf} is the conductivity of the fluid, $u(r)$ is the velocity along radial direction, P is the fluid pressure, υ_{nf} is the coefficient of kinematic viscosity and ρ_{nf} is the fluid density. The effective density ρ_{nf}, the effective dynamic viscosity μ_{nf}, and the kinematic viscosity υ_{nf} of the nanofluid are given as (Sheikholeslami et al., 2012a,b):

$$\rho_{\mathrm{nf}}=\rho_{\mathrm{f}}(1-\phi)+\rho_{\mathrm{s}}\phi,\ \mu_{\mathrm{nf}}=\frac{\mu_{\mathrm{f}}}{(1-\phi)^{2.5}},\ \upsilon_{\mathrm{nf}}=\frac{\mu_{\mathrm{f}}}{\rho_{\mathrm{nf}}},\ \frac{\sigma_{\mathrm{nf}}}{\sigma_{\mathrm{f}}}=1+\frac{3\left(\frac{\sigma_{\mathrm{s}}}{\sigma_{\mathrm{f}}}-1\right)\phi}{\left(\frac{\sigma_{\mathrm{s}}}{\sigma_{\mathrm{f}}}+2\right)-\left(\frac{\sigma_{\mathrm{s}}}{\sigma_{\mathrm{f}}}-1\right)\phi} \tag{8.4}$$

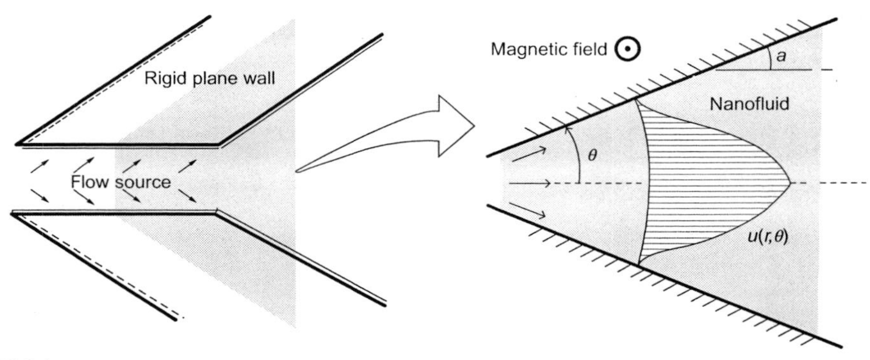

FIGURE 8.1

Schematic of the problem (MHD Jeffery-Hamel nanofluid flow).

Here, ϕ is the solid volume fraction.

Considering $u_\theta = 0$ for purely radial flow, one can define the velocity parameter as

$$f(\theta) = ru(r) \tag{8.5}$$

Introducing $x = \dfrac{\theta}{\alpha}$ as the dimensionless degree, the dimensionless form of the velocity parameter can be obtained by dividing that to its maximum value as

$$F(x) = \frac{f(\theta)}{f_{\max}} \tag{8.6}$$

Substituting Equation (8.5) into Equations (8.2) and (8.3), and eliminating P, one can obtain the ordinary differential equation for the normalized function profile as (Sheikholeslami et al., 2012a,b)

$$F'''(x) + 2\alpha Re{\cdot}A^*(1-\phi)^{2.5}F(x)F'(x) + \left(4 - (1-\phi)^{1.25}B^* Ha\right)\alpha^2 F'(x) = 0 \tag{8.7}$$

where A^* is a parameter. Reynolds number and Hartmann number based on the electromagnetic parameter are introduced as following:

$$A^* = (1-\phi) + \frac{\rho_s}{\rho_f}\phi \tag{8.8}$$

$$B^* = 1 + \frac{3\left(\frac{\sigma_s}{\sigma_f} - 1\right)\phi}{\left(\frac{\sigma_s}{\sigma_f} + 2\right) - \left(\frac{\sigma_s}{\sigma_f} - 1\right)\phi} \tag{8.9}$$

$$Re = \frac{f_{\max}\alpha}{\upsilon_f} = \frac{U_{\max} r\alpha}{\upsilon_f}\left(\begin{array}{l}\text{divergent-channel}: \alpha > 0,\ f_{\max} > 0\\ \text{convergent-channel}: \alpha < 0,\ f_{\max} < 0\end{array}\right) \tag{8.10}$$

$$Ha = \sqrt{\frac{\sigma_f B_0^2}{\rho_f \upsilon_f}} \tag{8.11}$$

With the following reduced form of boundary conditions

$$F(0) = 1,\ \ F'(0) = 0,\ \ F(1) = 0 \tag{8.12}$$

Physically, these boundary conditions mean that maximum values of velocity are observed at centerline $(x = 0)$ as shown in Figure 8.1, and we consider fully developed velocity profile, thus rate of velocity is zero at $(x = 0)$. Also, in fluid dynamics, the no-slip condition for fluid states that at a solid boundary, the fluid will have zero velocity relative to the boundary. The fluid velocity at all fluid-solid boundaries is equal to that of the solid boundary, so we can see that value of velocity is zero at $(x = 1)$.

8.1.3 WEIGHTED RESIDUAL METHODS

There existed an approximation technique for solving differential equations called the WRMs. Suppose a differential operator D is acted on a function u to produce a function p:

$$D(u(x)) = p(x) \tag{8.13}$$

It is considered that u is approximated by a function $\tilde{u}$, which is a linear combination of basic functions chosen from a linearly independent set. That is,

$$u \cong \tilde{u} = \sum_{i=1}^{n} c_i \varphi_i \tag{8.14}$$

Now, when substituted into the differential operator, D, the result of the operations generally is not $p(x)$. Hence, an error or residual will exist:

$$R(x) = D(\tilde{u}(x)) - p(x) \neq 0 \tag{8.15}$$

The notion in WRMs is to force the residual to zero in some average sense over the domain. That is:

$$\int_X R(x)\, W_i(x) = 0 \qquad i = 1, 2, \ldots, n \tag{8.16}$$

where the number of weight functions W_i is exactly equal to the number of unknown constants c_i in $\tilde{u}$. The result is a set of n algebraic equations for the unknown constants c_i. Three methods of WRMs are explained in the following subsections.

8.1.3.1 Collocation method

8.1.3.1.1 Mathematical formulation

In this method, the weighting functions are taken from the family of Dirac δ functions in the domain. That is, $W_i(x) = \delta(x - x_i)$. The Dirac δ function is defined as:

$$\delta(x - x_i) = \begin{cases} 1 & \text{if } x = x_i \\ 0 & \text{otherwise} \end{cases} \tag{8.17}$$

And residual function in Equation (8.15) must be forced to zero at specific points.

8.1.3.1.2 Application

In this problem, the trial function is considered as:

$$F(x) = (1 - x^2) + c_1(x^2 - x^3) + c_2(x^2 - x^4) + c_3(x^2 - x^5) + c_4(x^2 - x^6) \tag{8.18}$$

This trial function satisfies the boundary condition in Equation (8.12). By setting it into Equation (8.15) and using Equation (8.7), the residual function, $R(c_1, c_2, c_3, c_4, x)$, will be found as

$$\begin{aligned} R(c_1, c_2, c_3, c_4, x) &= -6c_1 - 24c_2 x - 60c_3 x^2 - 120c_4 x^3 \\ &\quad + 2\alpha Re\left(1 - \phi + \frac{\rho_s \phi}{\rho_f}\right)(1-\phi)^{2.5}\begin{pmatrix} 1 - x^2 + c_1(x^2 - x^3) + c_2(x^2 - x^4) \\ + c_3(x^2 - x^5) + c_4(x^2 - x^6) \end{pmatrix} \\ &\quad \times \left(-2x + c_1(2x - 3x^2) + c_2(2x - 4x^3) + c_3(2x - 5x^4) + c_4(2x - 6x^5)\right) \\ &\quad + \left(4 - (1-\phi)^{1.25} B^* Ha\right)\alpha^2\left(-2x + c_1(2x - 3x^2)\right. \\ &\quad \left. + c_2(2x - 4x^3) + c_3(2x - 5x^4) + c_4(2x - 6x^5)\right) \end{aligned} \tag{8.19}$$

On the other hand, the residual function must be close to zero. For reaching this aim, four specific points in the domain $t \in [0, 1]$ should be chosen. These points are:

$$X_1 = \frac{1}{5},\ X_2 = \frac{2}{5},\ X_3 = \frac{3}{5},\ X_4 = \frac{4}{5} \tag{8.20}$$

Finally, by substituting these points into the residual function, $R(c_1, c_2, c_3, c_4, x)$, a set of four equations and four unknown coefficients were obtained. After solving these unknown parameters (c_1, c_2, c_3, and c_4), the $F(x)$ equation will be determined (see Equation (8.14)). Using CM for a divergent channel with $\alpha = 3°$, $Re = 50$, and $Ha = 500$, $F(x)$ for a Cu-water nanofluid with $\phi = 0.04$ is as follows:

$$\begin{aligned} F(x) = {} & 1 - 1.310491199x^2 - 0.1312693419x^3 + 0.8786833168x^4 \\ & -0.4959297324x^5 + 0.05900695729x^6 \end{aligned} \tag{8.21}$$

With the same manner, for a convergent channel with $\alpha = -3°$, $Re = 50$, $Ha = 500$, and $\phi = 0.04$, $F(x)$ will be:

$$\begin{aligned} F(x) = {} & 1 - 0.553403033x^2 - 0.1917604785x^3 + 0.0933406758x^4 \\ & -0.5654355399x^5 + 0.2172583759x^6 \end{aligned} \tag{8.22}$$

8.1.3.2 Least square method

8.1.3.2.1 Mathematical formulation

If the continuous summation of all the squared residuals is minimized, the rationale behind the name can be seen. In other words, a minimum of

$$S = \int_X R(x)R(x)\,dx = \int_X R^2(x)\,dx \tag{8.23}$$

In order to achieve a minimum of this scalar function, the derivatives of S with respect to all the unknown parameters must be zero. That is,

$$\frac{\partial S}{\partial c_i} = 2\int_X R(x)\frac{\partial R}{\partial c_i}\,dx = 0 \tag{8.24}$$

Comparing with Equation (8.16), the weight functions are seen to be

$$W_i = 2\frac{\partial R}{\partial c_i} \tag{8.25}$$

However, the "2" coefficient can be dropped, since it cancels out in the equation. Therefore, the weight functions for the LSM are just the derivatives of the residual with respect to the unknown constants

$$W_i = \frac{\partial R}{\partial c_i} \tag{8.26}$$

8.1.3.2.2 Application

Because the trial function must satisfy the boundary condition in Equation (8.12), it will be considered as Equation (8.18) function, and residual will be as Equation (8.19). By substituting the residual function, $R(c_1, c_2, c_3, c_4, x)$, into Equation (8.24), a set of equations with four equations will appear and by solving this system of equations, coefficients c_1-c_4 will be determined. For example, using LSM for a divergent channel with $\alpha = 3°$, $Re = 50$, and $Ha = 500$, $F(x)$ for a Cu-water nanofluid with $\phi = 0.04$ is as follows:

$$F(x) = 1 - 1.334479475x^2 - 0.03418432046x^3 + 0.7250515693x^4 \\ -0.3857798169x^5 + 0.02939204338x^6 \quad (8.27)$$

With the same method, for a convergent channel with $\alpha = -3°$ and $Re = 50$ with $Ha = 500$, $F(x)$ will be:

$$F(x) = 1 - 0.5883323529x^2 - 0.1005355489x^3 + 0.057245669x^4 \\ -0.6407429523x^5 + 0.2723651847x^6 \quad (8.28)$$

8.1.3.3 Galerkin method

8.1.3.3.1 Mathematical formulation

This method may be viewed as a modification of the LSM. Rather than using the derivative of the residual with respect to the unknown c_i, the derivative of the approximating function or trial function is used. In this method, weight functions are:

$$W_i = \frac{\partial \tilde{u}}{\partial c_i} \quad i = 1, 2, \ldots, n \quad (8.29)$$

8.1.3.3.2 Application

Now, we apply GM for solving the $F(x)$ function as nondimensional velocity equation for MHD nanofluid Jeffery-Hamel flow. First, as already described, consider the trial function as Equation (8.18) which satisfies described boundary condition in Equation (8.12). Using Equation (8.29), weight functions will be obtained as

$$W_1 = x^2 - x^3, \quad W_2 = x^2 - x^4, \quad W_3 = x^2 - x^5, \quad W_4 = x^2 - x^6 \quad (8.30)$$

Applying Equation (8.24), a set of algebraic equations can be defined and solving this set of equations, c_1-c_4 coefficients and finally $F(x)$ function for a divergent channel with $\alpha = 3°$, $Re = 50$ and $Ha = 500$ for Cu-water nanofluid flow with $\phi = 0.04$ will be calculated as follows:

$$F(x) = 1 - 1.319461287x^2 - 0.1019978144x^3 + 0.8400824721x^4 \\ -0.4720277387x^5 + 0.05340436848x^6 \quad (8.31)$$

By repeating these steps for a convergent channel with $\alpha = -3°$, $Re = 50$, and $Ha = 500$, obtained $F(x)$ is:

$$F(x) = 1 - 0.5324936504x^2 - 0.3271451221x^3 + 0.411684785x^4 \\ -0.8861451438x^5 + 0.334099130x^6 \quad (8.32)$$

8.1.4 RESULTS AND DISCUSSIONS

In this section, the accuracy of three WRMs, namely, CM, GM, and LSM for obtaining the velocity profile of the MHD Jeffery-Hamel flow with nanofluid (Figure 8.1) is investigated. Figure 8.2 displays plots of $F(x)$ (nondimensional velocity profile) for MHD Jeffery-Hamel flow for different cases of α, Ha, and Re numbers for a divergent channel where Cu-water with $\phi=0.04$ is selected as nanofluid from Table 8.1.

This figure compares three described methods with those of the numerical method. The numerical solution which is applied to solve the present case is the fourth-order Runge-Kutta procedure. As in the diagrams of Figure 8.2, applied methods, especially LSM, show a good agreement with the numerical solution. Application of WRMs on MHD nanofluid flow analysis in convergent channels is shown in Figure 8.3 for Cu-water nanofluid.

Tables 8.2 and 8.3 show the values of $F(x)$ when $Re=50$, $Ha=500$, $\phi=0.04$, and $\alpha=3°$ for divergent and convergent channels, respectively which are derived from different applied methods for showing validity of them.

These tables confirmed that the WRMs were accurate and reliable methods for solving the divergent and convergent Jeffery-Hamel flow equation. Also, results reveal that the LSM has the lower error and higher accuracy among the other WRMs; so, it has been used for illustrating α, Reynolds effect on the physical treatment of the velocity profile, and Ha number for divergent and convergent channels in the following paragraph.

The effect of Hartmann number (Ha) for divergent and convergent channels is demonstrated in Figure 8.4a and b.

The velocity curves show that the rate of transport is considerably reduced with increase of Hartmann number. This clearly indicates that the transverse magnetic field opposes the transport phenomena, because the variation of Ha leads to the variation of the Lorentz force due to magnetic field and the Lorentz force produces more resistance to transport phenomena. As seen in this figure, increase in the Ha makes an increase in the velocity profile; so, by the increase of Hartmann number, the flow reversal disappears. Increasing Hartmann number leads to decrease in skin-friction coefficient.

Figure 8.5a and b displays the effect of Reynolds number (Re) for a divergent and a convergent channel with slope 5°, respectively.

These figures reveal that increase in Reynolds number make a decrease in velocity profile in divergent channels; also, for higher Reynolds number, the flow moves reversely and a region of back flow near the wall is observed (see Figure 8.4a for $Re=200$). As shown, in Figure 8.4b for convergent channel, results were inversed and by increasing Reynolds number, velocity profiles were increased and no back flow was observed. Also, for large Reynolds numbers, velocity profile was approximately constant near the centerline and suddenly reached to zero near the wall.

Figure 8.6 displays the effect of nanoparticle volume fraction, ϕ, when $Re=100$, $Ha=1000$ for a divergent and a convergent channel with 5° slope. It is assumed that the base fluid and the nanoparticles (Cu-water) are in thermal equilibrium and no slip occurs between them. It can be seen that increasing nanoparticle volume fraction in divergent channel leads to decrease in velocity profile and the back flow may be started at high Reynolds numbers.

Finally, we considered four different and common structures of nanofluid from Table 8.1 and their nondimensional velocity profiles, $F(x)$, are depicted in Figure 8.7a and b for divergent and convergent channels, respectively. As seen in this figure, for a divergent channel, when nanofluid includes copper (as nanoparticles) or ethylene glycol (as fluid phase) in its structure, $F(x)$ values are greater than the other structures, but this treatment of nanofluid structure is completely vice versa for convergent channels.

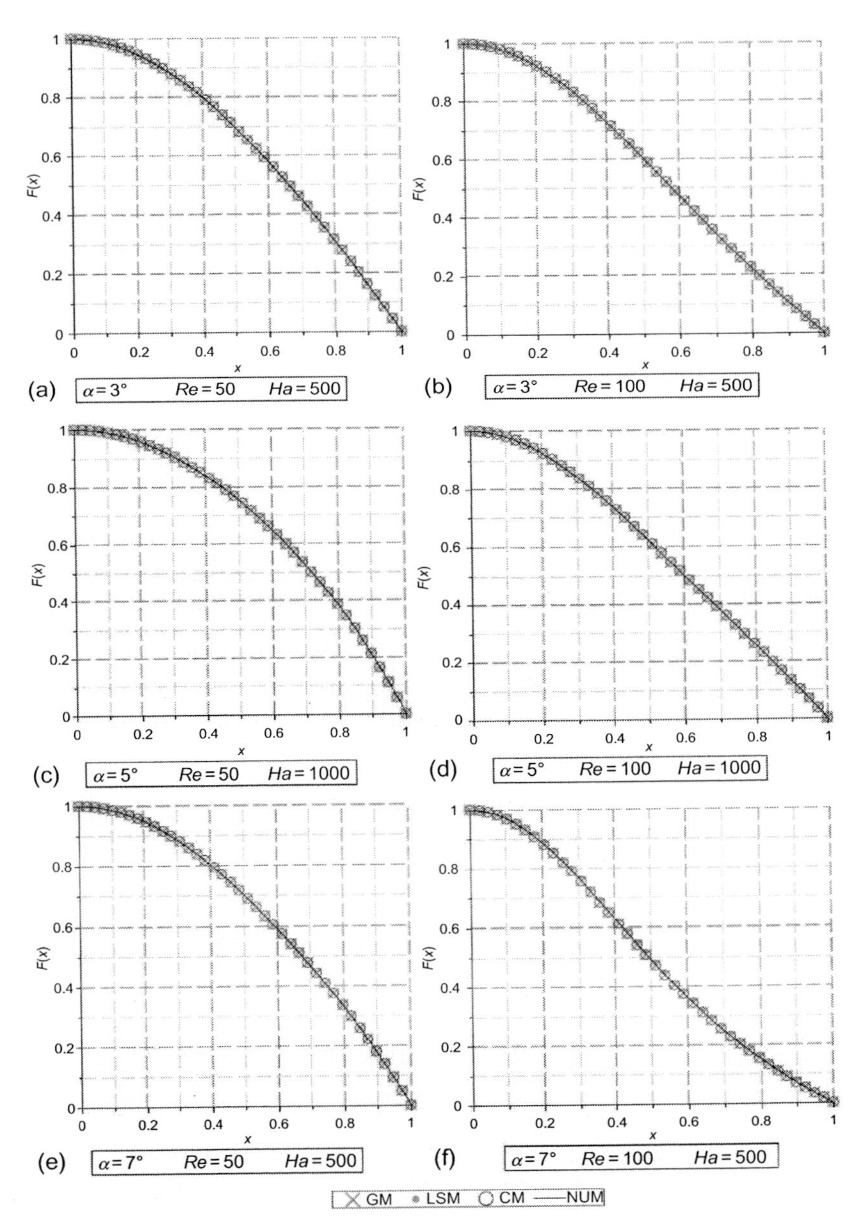

FIGURE 8.2

Comparison of WRMs (GM, CM, LSM) and numerical results for nondimensional velocity in different values of parameters for a divergent channel filled with Cu-water nanofluid with $\phi=0.04$.

Table 8.1 Thermo Physical Properties of Nanofluids and Nanoparticles

Material	Density (kg/m^3)	Electrical Conductivity, σ ((Ω m)$^{-1}$)
Silver	10,500	6.30×10^7
Copper	8933	5.96×10^7
Ethylene glycol	1113.2	1.07×10^{-4}
Drinking water	997.1	0.05

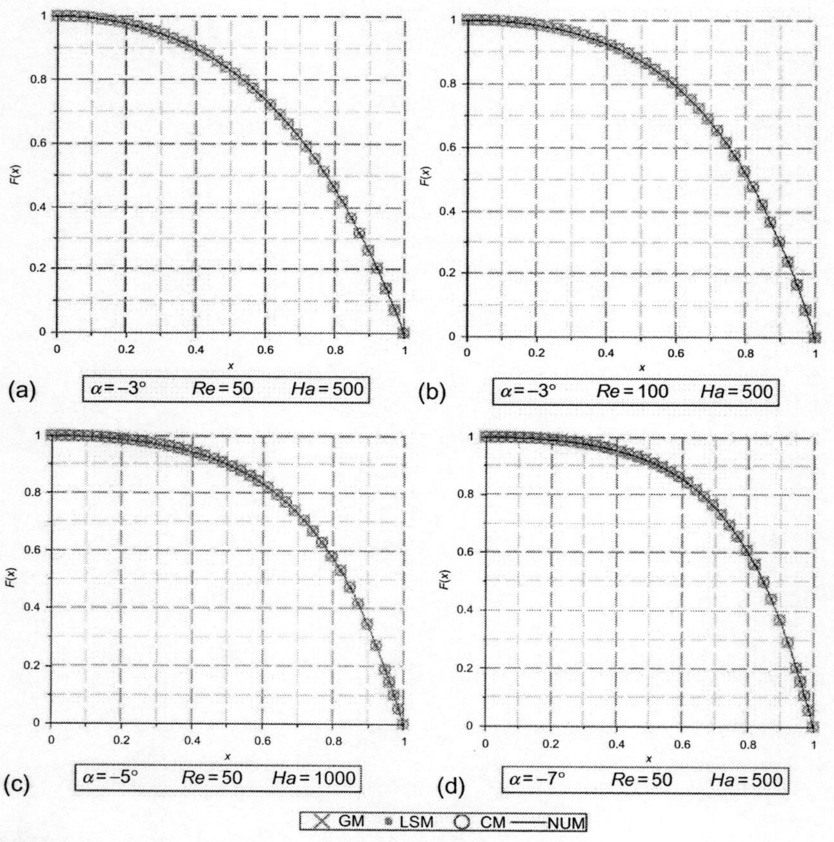

FIGURE 8.3

Comparison of WRMs (GM, CM, LSM) and numerical results for nondimensional velocity in different values of parameters for a convergent channel filled with Cu-water nanofluid with $\phi=0.04$.

Table 8.2 Comparison of $F(x)$ Values and Errors (%) of Applied Methods for a Divergent Channel When $Re = 50$, $Ha = 500$, $\phi=0.04$, and $\alpha=3°$ with Cu-Water Nanofluid

	Numerical	CM	GM	LSM	CM	GM	LSM
x	$F(x)$ Values				Errors (%)		
0.0	1.000	1.000	1.000	1.000	0.000	0.000	0.000
0.1	0.98670994	0.9868467	0.9867827	0.9866896	0.00014	7.37E−05	2.1E−05
0.2	0.94746148	0.9477811	0.9476020	0.9473858	0.00034	0.000148	8E−05
0.3	0.88402760	0.8844667	0.8841911	0.8839307	0.0005	0.000185	0.00011
0.4	0.79907449	0.7995778	0.7992496	0.7990268	0.00063	0.000219	6E−05
0.5	0.69578644	0.6963104	0.6959736	0.6958264	0.00075	0.000269	5.74E−05
0.6	0.57743773	0.5779358	0.5776238	0.5775433	0.00086	0.000322	0.000183
0.7	0.44697218	0.4473969	0.4471317	0.4470846	0.00095	0.000357	0.000252
0.8	0.30663219	0.3069464	0.3067452	0.3067045	0.00102	0.000369	0.000236
0.9	0.15765177	0.1578280	0.1577116	0.1576786	0.00112	0.00038	0.00017
1.0	0.000	7.9×10^{-10}	4.8×10^{-10}	2.8×10^{-10}	0.000	0.000	0.000

Table 8.3 Comparison of $F(x)$ Values and Errors (%) of Applied Methods for a Convergent Channel When $Re = 50$, $Ha = 500$, $\phi=0.04$, and $\alpha=-3°$ with Cu-Water Nanofluid

	Numerical	CM	GM	LSM	CM	GM	LSM
x	$F(x)$ Values				Errors (%)		
0.0	1.000	1.000	1.000	1.000	0.000	0.000	0.000
0.1	0.99407897	0.9942781	0.9943805	0.9940157	0.0002	0.000303	6.4E−05
0.2	0.97586003	0.9763121	0.9764796	0.9755664	0.00046	0.000635	0.0003
0.3	0.94395457	0.9445566	0.9446675	0.9434408	0.00064	0.000755	0.00054
0.4	0.89598228	0.8966721	0.8966972	0.8954524	0.00077	0.000798	0.00059
0.5	0.82848724	0.8292377	0.8292420	0.8281603	0.00091	0.000911	0.00039
0.6	0.73685715	0.7376197	0.7376743	0.7367870	0.00103	0.001109	9.5E−05
0.7	0.61529735	0.6159972	0.6160848	0.6153319	0.00114	0.00128	5.62E−05
0.8	0.45694829	0.4575440	0.4575418	0.4568811	0.0013	0.001299	0.00015
0.9	0.25428579	0.2547669	0.2545918	0.2541129	0.00189	0.001203	0.00068
1.0	0.000	1×10^{-10}	1×10^{-10}	3×10^{-10}	0.000	0.000	0.000

8.1.5 CONCLUSION

In this section, the WRMs, namely, CM, GM, and LSM, have been successfully applied to find the solution to MHD Jeffery-Hamel nanofluid flow. The following points can be concluded from the present study:

- The obtained solutions revealed that WRMs can be simple, powerful and efficient techniques for finding analytical solutions in science and engineering nonlinear differential equations.

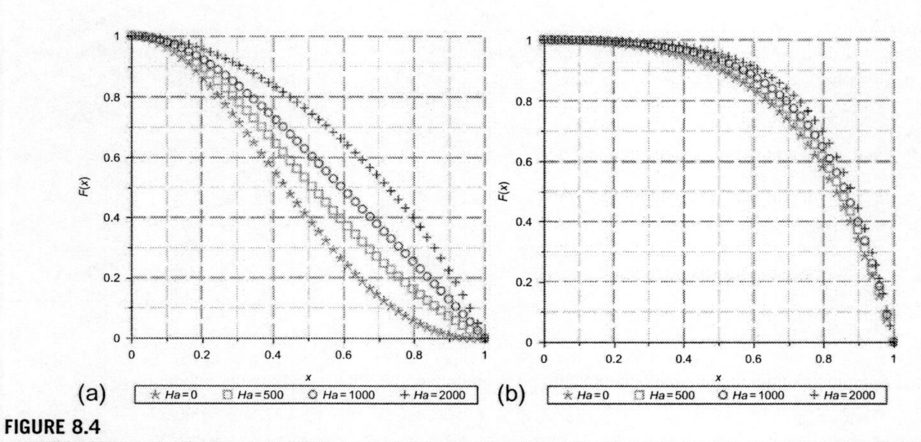

FIGURE 8.4

Effect of *Ha* number on Cu-water velocity profile when $Re=100$, $\phi=0.05$ for (a) divergent channel ($\alpha=5°$) and (b) convergent channel ($\alpha=-5°$).

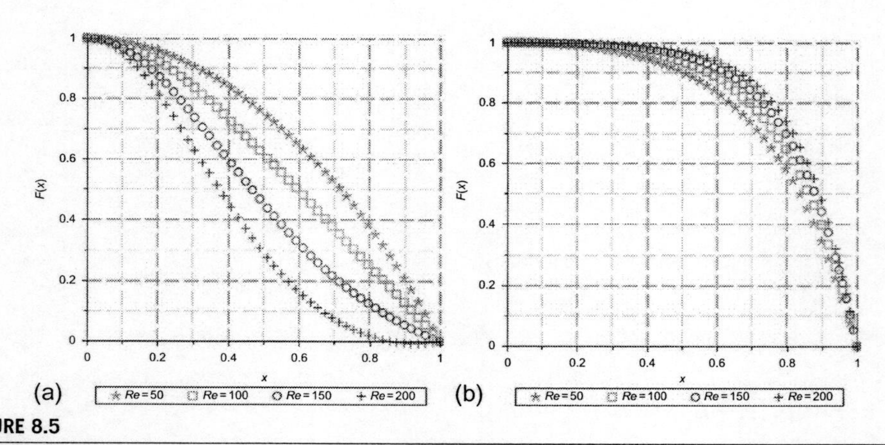

FIGURE 8.5

Effect of *Re* number on Cu-water velocity profile when $Ha=1000$, $\phi=0.05$ for (a) divergent channel ($\alpha=5°$) and (b) convergent channel ($\alpha=-5°$).

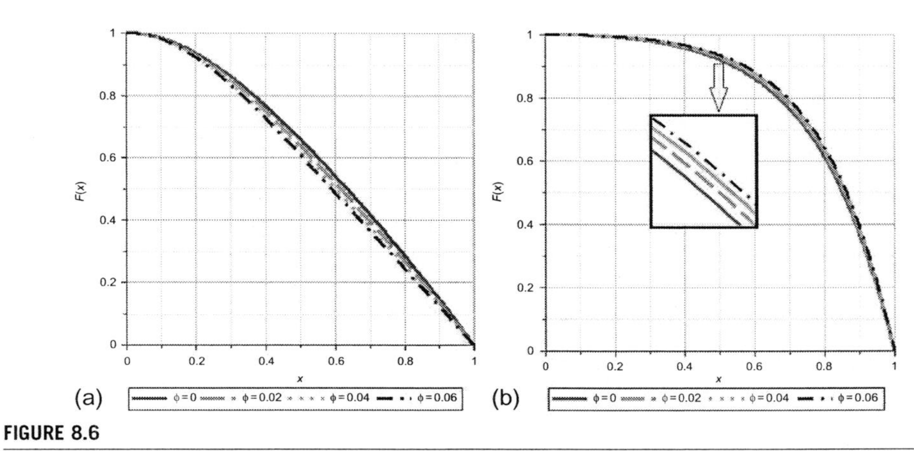

FIGURE 8.6

Effect of nanoparticle volume fraction, ϕ, on velocity profile when $Re = 100$, $Ha = 1000$ for (a) divergent channel ($\alpha = 5°$) and (b) convergent channel ($\alpha = -5°$).

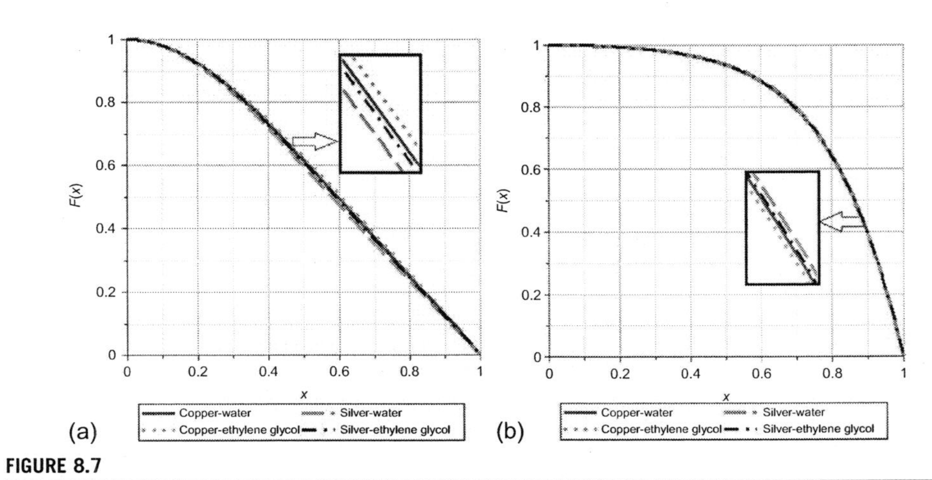

FIGURE 8.7

Effect of nanofluid structure on dimensionless velocity profile when $Re = 100$, $Ha = 1000$, $\phi = 0.05$ for (a) divergent channel ($\alpha = 5°$) and (b) convergent channel ($\alpha = -5°$).

- Calculated errors revealed that LSM led to more appropriate results when compared with numerical solution.
- Velocity boundary layer thickness decreases with increase in Reynolds number and nanoparticle volume friction and it increases as Hartmann number increases.
- For divergent channels, increase in slope angle and Reynolds number makes a decrease in velocity profile; so, for large Reynolds number, a region of back flow was observed.
- For convergent channels, increase in slope angle and Reynolds number makes an increase in velocity profile; so, for large Reynolds numbers, velocity profile is approximately constant near the centerline and suddenly reaches to zero near the wall.

8.2 MHD STAGNATION-POINT FLOW OF A NANOFLUID AND HEAT FLUX

This section presents an analytical solution for MHD stagnation-point flow of a nanofluid past a heated porous stretching sheet with suction or blowing conditions and prescribed surface heat flux. The effects of Brownian motion and thermophoresis occur in the transport equations. The analytic solutions for this special problem are developed by homotopy perturbation method (HPM) employing Padé technique. The results of this analytical solution for the velocity, temperature and concentration distribution, as well as the local Nusselt number and Sherwood number are presented graphically and discussed. A comparative study between the present results in a limiting sense with the already published results is found in an excellent agreement (*this section has been worked by B. Jalilpour, D.D. Ganji, S. Jafarmadar, A.B. Shotorban, S.H. Hashemi Kachapi in nonlinear dynamics team in Mechanical Engineering Department, 2012-2013*).

8.2.1 INTRODUCTION

Momentum, heat, and mass transfer in the laminar boundary layer flow over a stretching sheet with suction or blowing and prescribed surface heat flux is an important type of flow due to wide range of applications in engineering and several technological purposes. Mahapatra and Gupta (2002) numerically analyzed two-dimensional boundary layer flow, stagnation-point flow, and heat transfer over a stretching sheet. The heat transfer over a stretching surface with variable surface heat flux has been considered by Elbashbeshy (1998). Hayat et al. (2009) discussed the MHD flow of a micropolar fluid near a stagnation point toward a nonlinear stretching surface. Boundary layer and MHDs stagnation-point flow toward a stretching sheet studied numerically by Mahapatra and Gupta (2001) and Ishak et al. (2009).

Up to date, no investigation is made which characterizes MHD boundary layer flow, heat and mass transfer of a nanofluid past a heated porous stretching sheet with suction or blowing and surface heat flux. Therefore, current investigation deals with the MHD boundary layer flow of a nanofluid toward a stretching surface with Suction or blowing and surface heat flux. Combined effects of heat and mass transfer in the presence of Brownian motion and thermophoresis are also taken into account. The analytical solution of the resulting problem is derived by HPM employing Padé technique (He, 1999, 2003).

8.2.2 MATHEMATICAL MODEL

Consider steady two-dimensional flow of a viscous incompressible electrically conducting nanofluid near the stagnation point on a stretching sheet kept at a constant temperature T_w and concentration C_w, as shown in Figure 8.8. The ambient temperature and concentration are T_∞ and C_∞, respectively. Fluid saturates is a porous medium in the presence of transverse magnetic field. It is also assumed that the surface of the sheet is subjected to a prescribed heat flux $q_w(x) = cx^n$, where c and n are constants with $c > 0$. Further, a uniform magnetic field of strength B_0 is applied in the positive direction of y-axis. The magnetic Reynolds number is assumed to be small so that the induced magnetic field is negligible. It is further assumed that the base fluid and the suspended nanoparticles are in thermal equilibrium. Under these assumptions, the simplified equations governing the conservation of mass, momentum, energy, and nanoparticle fraction in the presence of magnetic field pasta stretching sheet can be expressed as

$$\frac{\partial u}{\partial x} + \frac{\partial \nu}{\partial y} = 0 \tag{8.33}$$

$$u\frac{\partial u}{\partial x} + \nu\frac{\partial u}{\partial y} = u_e\frac{du_e}{dx} + \upsilon\frac{\partial^2 u}{\partial y^2} + \frac{\sigma {\beta_0}^2}{\rho}(u_e - u) + \frac{\upsilon}{k}(u_e - u) \tag{8.34}$$

$$u\frac{\partial T}{\partial x} + \nu\frac{\partial T}{\partial y} = \alpha\frac{\partial^2 T}{\partial y^2} + \tau\left\{D_B\frac{\partial C}{\partial y}\frac{\partial T}{\partial y} + \frac{D_T}{T_\infty}\left(\frac{\partial T}{\partial y}\right)^2\right\} \tag{8.35}$$

$$u\frac{\partial C}{\partial x} + \nu\frac{\partial C}{\partial y} = D_B\frac{\partial^2 C}{\partial y^2} + \frac{D_T}{T_\infty}\frac{\partial^2 T}{\partial y^2} \tag{8.36}$$

where u and ν are the velocity components in x and y directions, T is the local temperature of the fluid, C is the nanoparticle fraction, υ is the kinematic viscosity, ρ is the density, σ is the electrical conductivity, k is the permeability of saturated porous medium, $\alpha = K/(\rho c)_f$ is the thermal diffusivity, D_B is the

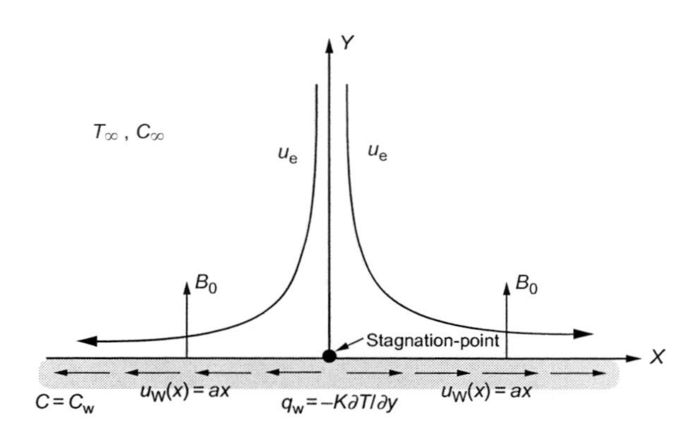

FIGURE 8.8

Physical model and coordinate system.

Brownian motion coefficient, D_T is the thermophoretic diffusion coefficient, $\tau=(\rho c)_f/(\rho c)_p$ is the ratio of effective heat capacity of the nanoparticle material to heat capacity of the fluid, and ρ_f is the density of nanofluid. At constant pressure, the boundary conditions are

$$\begin{aligned} &u=u_w(x)=ax,\ \ v=v_w(x),\ \ q_w=-K\frac{\partial T}{\partial y},\ \ C=C_w \ \text{ at } y=0 \\ &u=u_e(x)=bx,\ \ T=T_\infty,\ \ C=C_\infty,\ \text{ as } y\to\infty \end{aligned} \tag{8.37}$$

where a and b are constants with $a>0$ and $b\geq 0$ and K is the thermal conductivity. We look for a similarity solution of Equations (8.33)–(8.36) and boundary conditions (8.37) as the following:

$$\psi=x(a\nu)^{1/2}f(\eta),\ \ \theta(\eta)=\frac{K(T-T_\infty)}{cx^n}(a/\nu)^{1/2},\ \ \phi(\eta)=\frac{C-C_\infty}{C_w-C_\infty},\ \ \eta=(a/\nu)^{1/2}y \tag{8.38}$$

where ψ is the stream function, which is defined as:

$$u=\frac{\partial\psi}{\partial y}\ \text{ and }\ u=-\frac{\partial\psi}{\partial x} \tag{8.39}$$

Substituting (8.6) into Equations (8.33)–(8.36), we obtain the following ordinary differential equations:

$$f'''+ff''-f'^2-(N+M^2)(f'-\lambda)+\lambda^2=0 \tag{8.40}$$

$$\theta''+Pr\left(f\theta'-n\theta f'+\mathrm{Nb}\phi'\theta'+\mathrm{Nt}\theta'^2\right)=0 \tag{8.41}$$

$$\phi''+Lef\phi'+(\mathrm{Nt}/\mathrm{Nb})\theta''=0 \tag{8.42}$$

And the boundary conditions (8.5) become

$$\begin{aligned} &f(0)=s,\ \ f'(0)=1,\ \ \theta'(0)=-1,\ \ \phi(0)=1 \\ &f'(\eta)\to\lambda,\ \ \theta(\eta)\to 0\ \ \phi(\eta)\to 0\ \text{ as } \eta\to\infty \end{aligned} \tag{8.43}$$

where primes denote differentiation with respect to η, $M^2={\beta_0}^2(\sigma/\rho a)^{1/2}$ is the magnetic number, $N=\upsilon/ak$ is the porosity parameter, $\lambda=b/a$ is the velocity parameter, $Pr=\upsilon/\alpha$ is the Prandtl number, n is the surface heat flux, $\mathrm{Nb}=\tau D_B(\phi_w-\phi_\infty)/\upsilon$ is the Brownian motion parameter, $\mathrm{Nt}=\tau D_B(T_w-T)/T_\infty\upsilon$ is the thermophoresis parameter, $Le=\upsilon/D_B$ is the Lewis number, and $s=\nu_w/(a\upsilon)^{1/2}$ is the suction or injection parameter.

The quantities of physical interest in this problem are the local skin-friction coefficient, the local Nusselt number, and the Sherwood number which can be expressed as:

$$Nu=\frac{xq_w}{K(T_w-T_\infty)},\ \ Sh=\frac{xq_m}{D_B(C_w-C_\infty)} \tag{8.44}$$

where q_w and q_m are the wall heat and mass fluxes, respectively. Using variables in Equation (8.38), we obtain local Reynolds number

$$Re_x^{1/2}c_f=f''(0),\ \ Re_x^{-1/2}Nu_x=1/\theta(0),\ \ Re_x^{-1/2}Sh=-\phi'(0) \tag{8.45}$$

8.2.3 METHODS OF SOLUTION

8.2.3.1 Review of HPM

To illustrate the basic idea of this method, we consider the following nonlinear differential equation:

$$A(u)-f(r)=0 \quad r\in\Omega \tag{8.46}$$

Considering the boundary conditions of:

$$B\left(u,\frac{\partial u}{\partial n}\right)=0 \quad r\in\Gamma \tag{8.47}$$

where A is the general differential operator, B is the boundary operator, $f(r)$ is the known analytical function, and Γ is the boundary of the domain Ω. The operator A can be divided into two parts L and N, where L is the linear part while N is a nonlinear one. Therefore, Equation (8.46) can be rewritten as follows:

$$L(u)+N(u)-f(r)=0 \quad r\in\Omega \tag{8.48}$$

The homotopy perturbation structure is shown as follows:

$$H(\nu,p)=(1-p)[L(\nu)-L(u_0)]+p[A(\nu)-f(r)]=0 \tag{8.49}$$

where

$$\nu(r,p):\Omega\times[0,1]\rightarrow R \tag{8.50}$$

We can assume that the solution of Equation (8.50) can be written as a power series in p as follows:

$$\nu=\nu_0+\varepsilon\nu_1+\varepsilon^2\nu_2+\cdots \tag{8.51}$$

Setting $p=1$, result in the approximate solution is:

$$V=\lim_{p\rightarrow1}\nu=\nu_0+\nu_1+\nu_2+\cdots \tag{8.52}$$

The combination of the perturbation method and the homotopy method is called the HPM, which lacks the limitations of the traditional perturbation methods, although this technique has full advantages of the traditional perturbation techniques.

8.2.3.2 Padé approximants

It is well known that Padé approximations have the advantage of manipulating the polynomial approximation into a rational function of polynomials (see Baker, 1975). This manipulation provides us with more information about the mathematical behavior of the solution. Besides that, power series are not useful for large values of η, say $\eta=\infty$. Boyd (1997) and others have formally shown that power series in isolation are not useful for handling boundary value problems (BVPs). Therefore, the combination of the HPM with the Padé approximation provides an effective tool for handling BVPs on infinite or semi-infinite domains. It is a known fact that Padé approximation converges on the entire real axis if $f(\eta)$, $\theta(\eta)$, and $\phi(\eta)$ are free of singularities on the real axis. More importantly, the diagonal approximants are most accurate approximants; therefore, we have to construct only diagonal approximants. Using the boundary conditions, $f'(\infty)=\lambda, \theta(\infty)=0$, and $\phi(\infty)=0$, the diagonal approximant $[M/M]$ vanishes if the coefficient of η with the highest power in the numerator vanishes. Choosing the coefficients of the highest power of η equal to zero, we get polynomial equations in α, β, and γ which can be solved very easily by using the built-in utilities in the most manipulation languages such as Maple and Mathematica.

8.2.4 ANALYTICAL SOLUTION STATEMENT

In this section, we apply HPM to nonlinear ordinary differential equations (Equations (8.40)–(8.42)). According to HPM, we can construct a homotopy of Equations (8.40)–(8.42) as

$$H(f,p)=(1-p)(f'''-f_0''')+p\left(f'''+ff''-f'^2-(N+M^2)(f'-\lambda)+\lambda^2\right)=0 \tag{8.53a}$$

$$H(\theta,p)=(1-p)(\theta'''-\theta_0''')+p\left(\theta''+Prf\theta'-nPr\theta f'+\mathrm{Nb}\phi'\theta'+\mathrm{Nt}\theta'^2\right)=0 \tag{8.53b}$$

$$H(\theta,p)=(1-p)(\phi'''-\phi_0''')+p(\phi''+Lef\phi'+(\mathrm{Nt}/\mathrm{Nb})\theta'')=0 \tag{8.53c}$$

We consider

$$f(\eta)=f_0(\eta)+pf_1(\eta)+p^2f_2(\eta)+p^3f_3(\eta)+\cdots \tag{8.54a}$$

$$\theta(\eta)=\theta_0(\eta)+p\theta_1(\eta)+p^2\theta_2(\eta)+p^3\theta_3(\eta)+\cdots \tag{8.54b}$$

$$\phi(\eta)=\phi_0(\eta)+p\phi_1(\eta)+p^2\phi_2(\eta)+p^3\phi_3(\eta)+\cdots \tag{8.54c}$$

Substituting f and h from Equations (8.54a)–(8.54c) into Equations (8.53a)–(8.53c) and some simplification and rearranging based on powers of p-terms, we have

$$p^0: f_0'''=0,\quad \theta_0''=0,\quad \phi_0''=0 \tag{8.55}$$

And boundary conditions are:

$$\begin{aligned} &f_0(0)=s,\ f_0'(0)=1,\ f_0''(0)=\alpha \\ &\theta_0(0)=\beta,\ \theta_0'(0)=-1 \\ &\phi_0(0)=1,\ \phi_0'(0)=\gamma \end{aligned} \tag{8.56}$$

$$\begin{aligned} &p^1: \\ &f_1'''+f_0f_0''-f'^2_0-(N+M^2)f_0'+(N+M^2)\lambda+\lambda^2=0 \\ &\theta_1''+Pr\cdot f_0\cdot\theta_0'-nPr\theta_0f_0'+\mathrm{Nb}Pr\phi_0'\theta_0'+\mathrm{Nt}Pr\theta'^2_0=0 \\ &\phi_1''+Le\cdot f_0.\phi_0'f_0'+\frac{\mathrm{Nt}\theta_0''}{\mathrm{Nb}}=0 \end{aligned} \tag{8.57}$$

Boundary conditions are:

$$\begin{aligned} &f_1(0)=0,\ f_1'(0)=0,\ f_1''(0)=0 \\ &\theta_1(0)=0,\ \theta_1'(0)=0 \\ &\phi_1(0)=0,\ \phi_1'(0)=0 \end{aligned} \tag{8.58}$$

Solving Equations (8.55) and (8.57) with boundary conditions Equations (8.56) and (8.58)

$$f_0(\eta)=\frac{1}{2}\alpha\eta^2+\eta+s \tag{8.59}$$

$$\theta_0(\eta)=-\eta+\beta \tag{8.60}$$

$$\phi_0(\eta)=\gamma\eta+1 \tag{8.61}$$

$$\begin{aligned}f_1(\eta)=&\frac{1}{120}\alpha^2\eta^5+\left(\frac{1}{24}\alpha+\frac{1}{24}\alpha(N+M^2)\right)\eta^4\\&+\left(-\frac{1}{6}\lambda^2-\frac{1}{6}\lambda(N+M^2)+\frac{1}{6}(N+M^2)+\frac{1}{6}-\frac{1}{6}\alpha s\right)\eta^3\end{aligned} \tag{8.62}$$

$$\begin{aligned}\theta_1(\eta)=&\left(-\frac{1}{12}nPr\alpha+\frac{1}{24}\alpha Pr\right)\eta^4+\left(\frac{1}{6}Pr-\frac{1}{6}nPr+\frac{1}{6}nPr\beta\alpha\right)\eta^3\\&+\left(\frac{1}{2}sPr+\frac{1}{2}nPr\beta+\frac{1}{2}Pr\mathrm{Nb}\gamma-\frac{1}{2}Pr\mathrm{Nt}\right)\eta^2\end{aligned} \tag{8.63}$$

$$\phi_1(\eta)=\left(-\frac{1}{24}Le\alpha\gamma\right)\eta^4-\left(\frac{1}{6}Le\gamma\right)\eta^3+\left(\frac{1}{2}sLe\gamma\right)\eta^2 \tag{8.64}$$

We avoid listing the other components. However, according to Equations (8.54a)–(8.54c), we obtain $f(\eta)$ up to $0(\eta^5)$ and $\theta(\eta)$ up to $0(\eta^4)$ and $\phi(\eta)$ up to $0(\eta^4)$ as follows:

$$\begin{aligned}f(\eta)=&s+\eta+\frac{1}{2}\alpha\eta^2+\left(-\frac{1}{6}\lambda^2-\frac{1}{6}\lambda(N+M^2)+\frac{1}{6}(N+M^2)+\frac{1}{6}-\frac{1}{6}\alpha s\right)\eta^3\\&+\left(\frac{1}{24}\alpha+\frac{1}{24}\alpha(N+M^2)\right)\eta^4+\frac{1}{120}\alpha^2\eta^5+\cdots\end{aligned} \tag{8.65}$$

$$\begin{aligned}\theta(\eta)=&\beta-\eta+\left(\frac{1}{2}sPr+\frac{1}{2}nPr\beta+\frac{1}{2}Pr\mathrm{Nb}\gamma-\frac{1}{2}Pr\mathrm{Nt}\right)\eta^2\\&+\left(\frac{1}{6}Pr-\frac{1}{6}nPr+\frac{1}{6}nPr\beta\alpha\right)\eta^3+\left(-\frac{1}{12}nPr\alpha+\frac{1}{24}\alpha Pr\right)\eta^4+\cdots\end{aligned} \tag{8.66}$$

$$\phi(\eta)=1+\gamma\eta+\left(\frac{1}{2}sLe\gamma\right)\eta^2-\left(\frac{1}{6}Le\gamma\right)\eta^3+\left(-\frac{1}{24}Le\alpha\gamma\right)\eta^4+\cdots \tag{8.67}$$

8.2.5 RESULTS AND DISCUSSION

Equations (8.40)–(8.42) subject to the boundary conditions (8.43) are solved analytically using HPM employing Padé technique. A comparison of our results for the local skin-friction number $f''(0)$ and local Nusselt number with the already published results are shown in Tables 8.4 and 8.5, respectively.

It is clear that the comparison shows good agreement and therefore we are confident that our results are highly accurate. Figures 8.9–8.11 elucidate the exact behaviors of velocity and temperature profiles that have been obtained by HPM-Padé in comparison to HPM. The results of HPM-Padé are in good agreement with numerical results. Also, these figures show that the increase in the number of HPM

Table 8.4 Comparison of the Results for $f''(0)$ When $\gamma = 0$, $N = M = 0$ at Different Values of λ

	$f''(0)$		
λ	Present Result	Hayat et al. (2009)	Mahapatra and Gupta (2002)
0.1	−0.9693875	−0.969388	−0.9694
0.2	−0.9181645	−0.91811	−0.9181
0.5	−0.6672699	−0.667271	−0.6673
2.0	2.0175911	2.017615	2.0175
3.0	4.7275219	4. 4.729694	4.7293

Table 8.5 Comparison of Values of $\theta(0)$ with Previous Result at Nt = Nb = 0 for Different Values of *Pr* and *n*

		$\theta(0)$		
n	Pr	Present	Elbashbeshy (1998)	Ishak et al. (2009)
0	0.72	2.1570683	2.13767	2.1591531
	1.0	1.7181812	1.71792	1.7182818
	10	0.4332753	0.43341	0.4332748
1.0	0.72	1.2363473	1.2253	1.2366575
	1.0	1.0	1.0	1.0
	10	0.2687687	–	0.2687685

terms does not affect the convergence and these series diverge around infinity. Figure 8.12 shows values of $f(\eta), f'(\eta)$, and $f''(\eta)$ for fixed values of $N = M = 1, s = 0$ and various values of λ and the comparison of them with numerical results.

Figures 8.13 and 8.14 show that the difference between numerical and HPM solutions depends on the order of Padé approximation. It can be concluded that higher degree of the Padé approximations provides a highly accurate solution in comparison to the numerical solution.

Figures 8.15–8.28 plot the variations of the porosity parameter N, thermophoresis parameter Nt, the Prandtl number Pr, the Lewis number Le, suction or injection parameter S, the velocity ratio parameter λ, heat flux index n, the Brownian motion parameter Nb, and the magnetic parameter M, respectively on the temperature and mass fraction profiles.

The influence of porosity parameter N on temperature and concentration distributions is displayed in Figure 8.15. It is noticed that as porosity parameter increases, the temperature and concentration distributions increase. Figure 8.16 depicts typical profiles for temperature distributions and concentration for various values of the thermophoretic parameter (Nt).

It is observed that an increase in the thermophoretic parameter Nt leads to a decrease in the fluid temperature and concentration. Effects of Prandtl number on the temperature and concentration distributions are shown in Figure 8.17. It is observed that the temperature and concentration distributions decrease with increase in Prandtl number. Figure 8.18 shows typical profiles for concentration distribution for various values of the Lewis number Le.

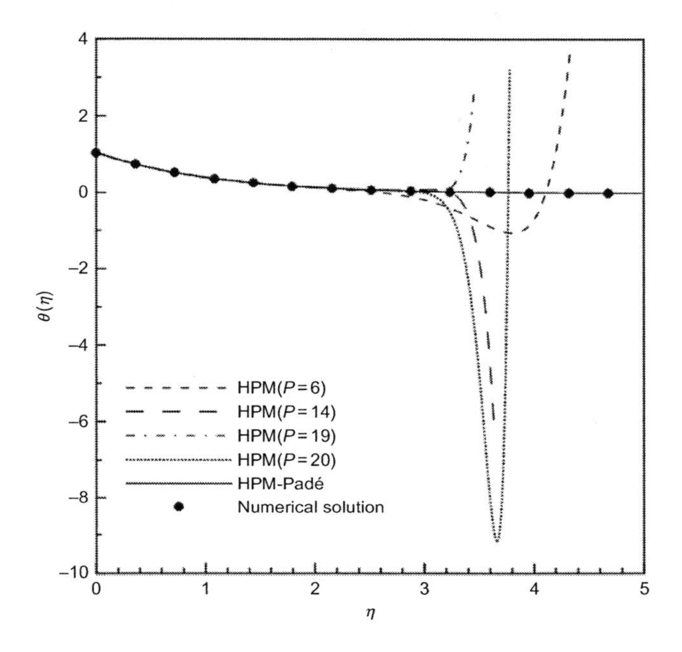

FIGURE 8.9

The effect of increasing the number of HPM terms in comparison with HPM-Padé [8/8] and numerical solution for temperature distribution when $s=0$, $\lambda=N=\mathrm{Nt}=\mathrm{Nb}=0.1$, $Pr=Le=n=1$, and $M=0.25$.

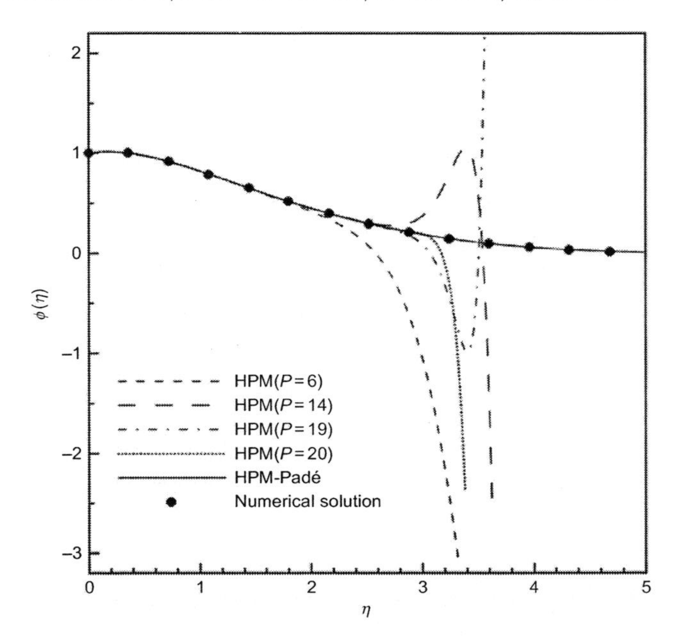

FIGURE 8.10

The effect of increasing the number of HPM terms in comparison with HPM-Padé [10/10] and numerical solution for concentration distribution when $s=0$, $\lambda=N=\mathrm{Nt}=\mathrm{Nb}=0.1$, $Pr=Le=n=1$, and $M=0.25$.

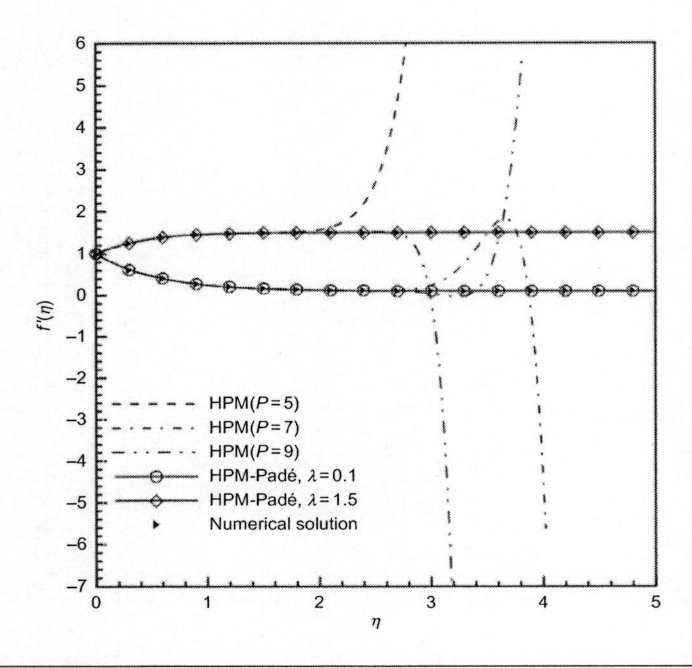

FIGURE 8.11

The effect of increasing the number of HPM terms in comparison with HPM-Padé [14/14] and numerical solution for velocity profile when $s=0$, $N=M=1$, and various values of λ.

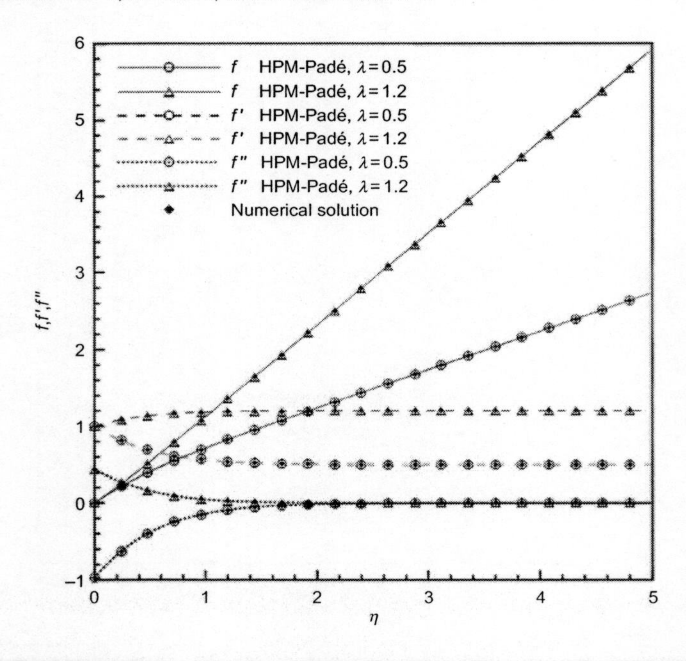

FIGURE 8.12

Comparison of HPM-Padé solution with numerical solution when $N=M=1$, $s=0$, and various values of λ.

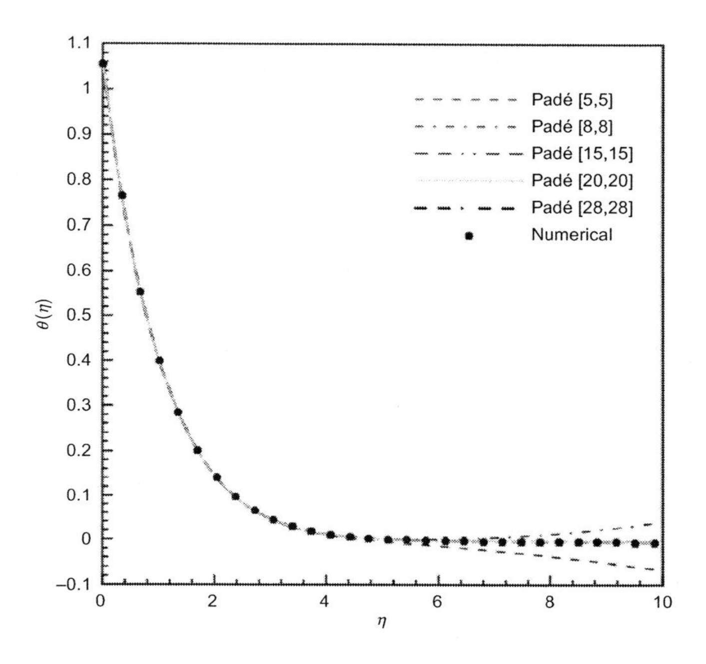

FIGURE 8.13

The effect of increasing the order of Padé approximation on temperature distribution when $s=0$, $\lambda=N=\mathrm{Nt}=\mathrm{Nb}=0.1$, $Pr=Le=n=1$ and $M=0.25$.

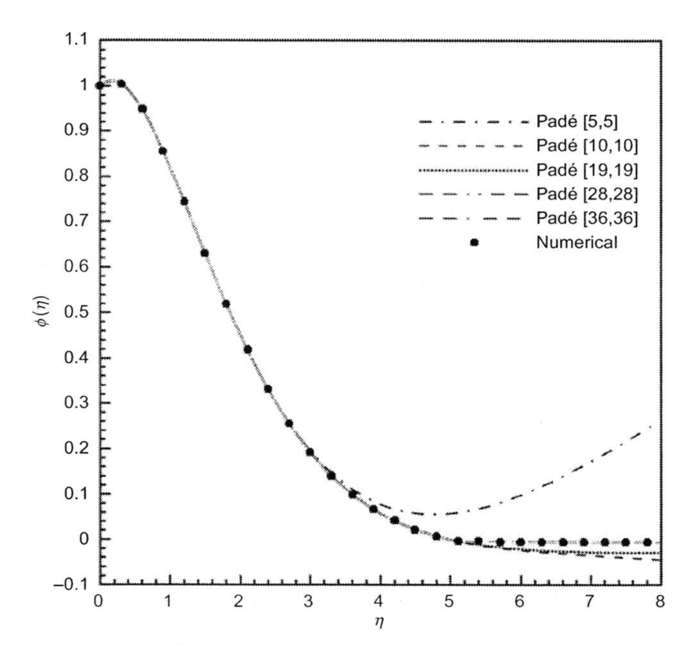

FIGURE 8.14

The effect of increasing the order of Padé approximation on concentration distribution when $s=0$, $\lambda=N=\mathrm{Nt}=\mathrm{Nb}=0.1$, $Pr=Le=n=1$, and $M=0.25$.

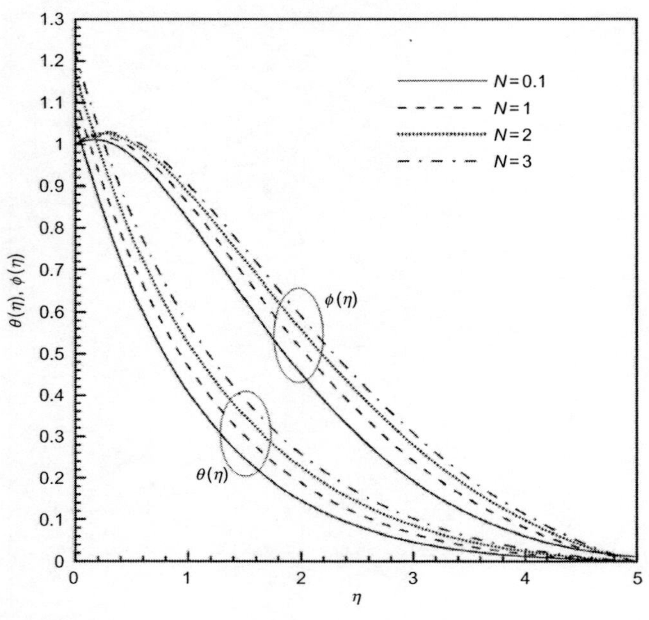

FIGURE 8.15

Effect N on temperature and concentration distribution when $s=0$, $\lambda=\mathrm{Nt}=\mathrm{Nb}=0.1$, $Pr=n=1$ and $M=0.25$.

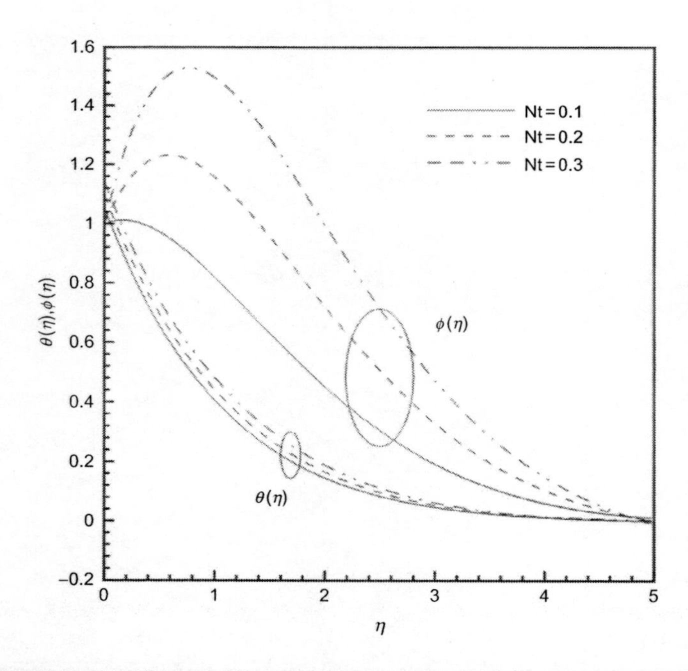

FIGURE 8.16

Effect Nt on temperature and concentration distribution when $s=0$, $\lambda=N=\mathrm{Nb}=0.1$, $Pr=n=1$ and $M=0.25$.

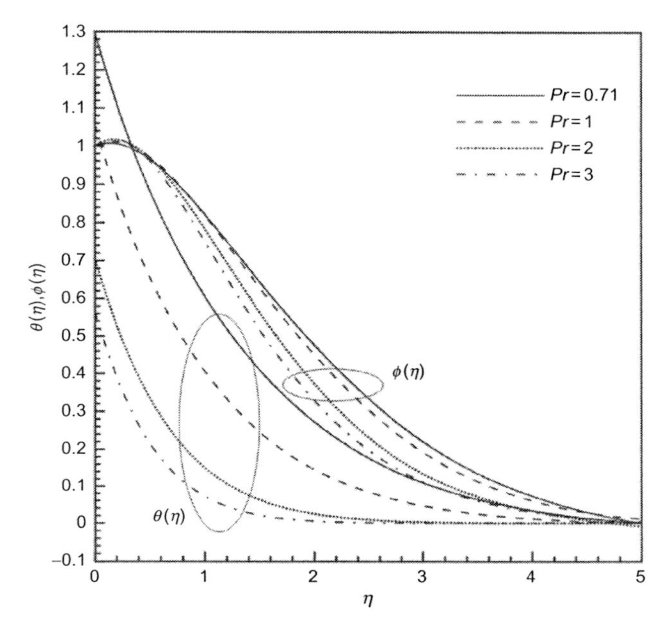

FIGURE 8.17

Effect *Pr* on temperature and concentration distribution when $s=0$, $\lambda=N=\mathrm{Nt}=\mathrm{Nb}=0.1$, $n=1$, and $M=0.25$.

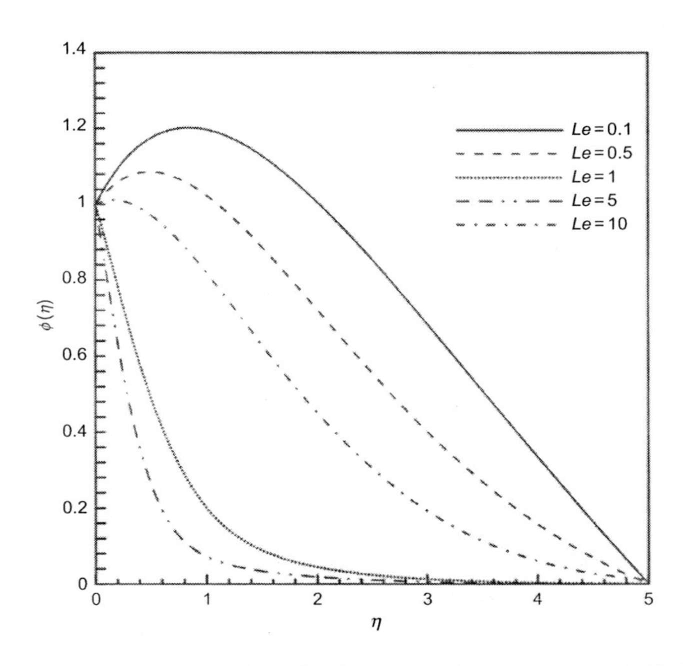

FIGURE 8.18

Effect *Le* on concentration distribution when $s=0$, $\lambda=N=\mathrm{Nt}=\mathrm{Nb}=0.1$, $Pr=n=1$, and $M=0.25$.

It is found that an increase in Le results in reduction of concentration and the concentration boundary layer thickness. Figures 8.19 and 8.20 are plotted to analyze the effects of blowing ($S<0$) and suction ($S>0$) on the temperature and concentration distributions, respectively.

It is noted that blowing ($S<0$) causes an increase in the temperature and concentration distributions, whereas by increasing suction ($S>0$), the temperature and concentration distributions decrease rapidly near the boundary. Figures 8.21–8.24 show the effects of the velocity ratio parameter λ and heat flux index n on the temperature and concentration distributions, respectively.

As shown, the temperature and concentration distributions are decreasing with increasing λ and n. The effects of the Brownian motion parameter Nb on temperature and concentration distributions are depicted in Figures 8.25 and 8.26.

As expected, the boundary layer profiles for the temperature are of the same form as in the case of regular heat transfer fluids. The temperature in the boundary layer increases with increasing Brownian motion parameter Nb. However, the concentration profile decreases with increasing Brownian motion parameter Nb. Figures 8.27 and 8.28 show the influence of magnetic field parameter M on the temperature and volume fraction of nanoparticles, respectively. It is depicted that the temperature and concentration decrease when the values of magnetic field parameter M increases at a fixed value of η.

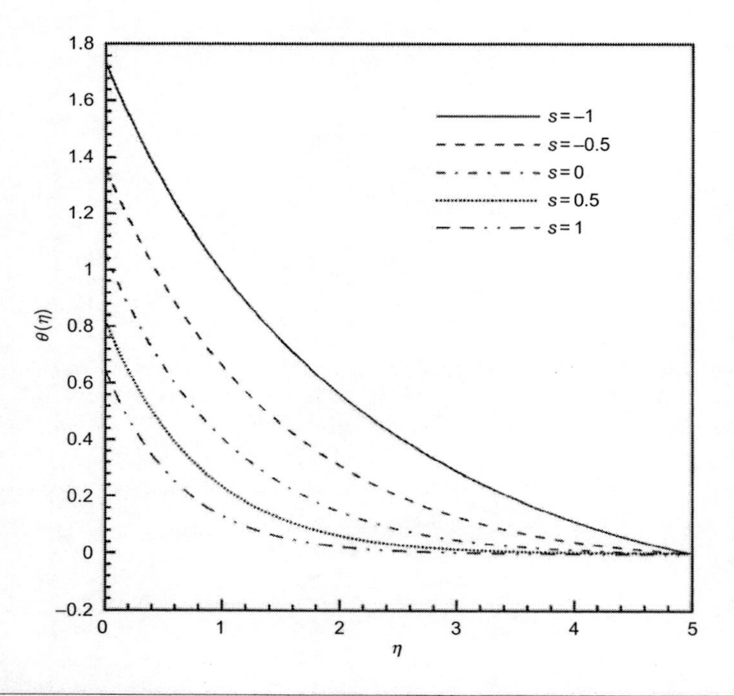

FIGURE 8.19

Effect s on temperature distribution when $\lambda = N = \mathrm{Nt} = \mathrm{Nb} = 0.1$, $Pr = Le = n = 1$, and $M = 0.25$.

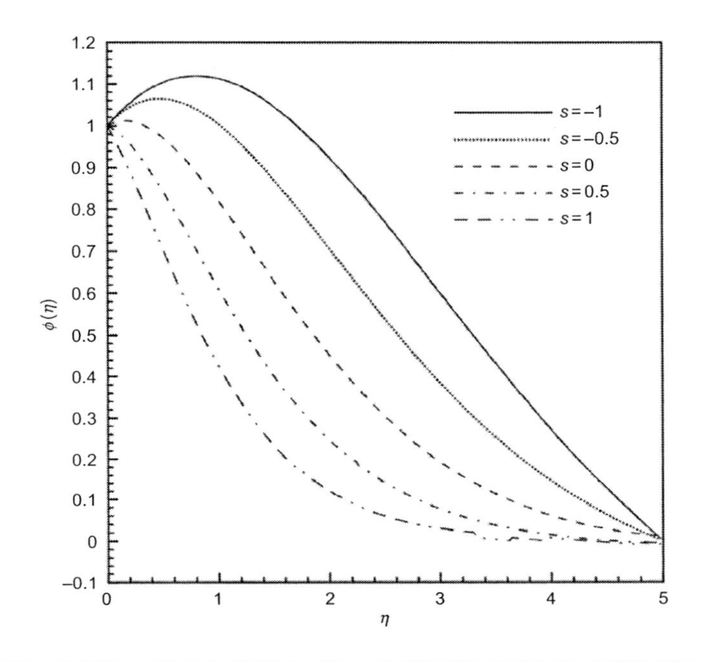

FIGURE 8.20

Effect s on concentration distribution when $\lambda=N=\mathrm{Nt}=\mathrm{Nb}=0.1$, $Pr=Le=n=1$, and $M=0.25$.

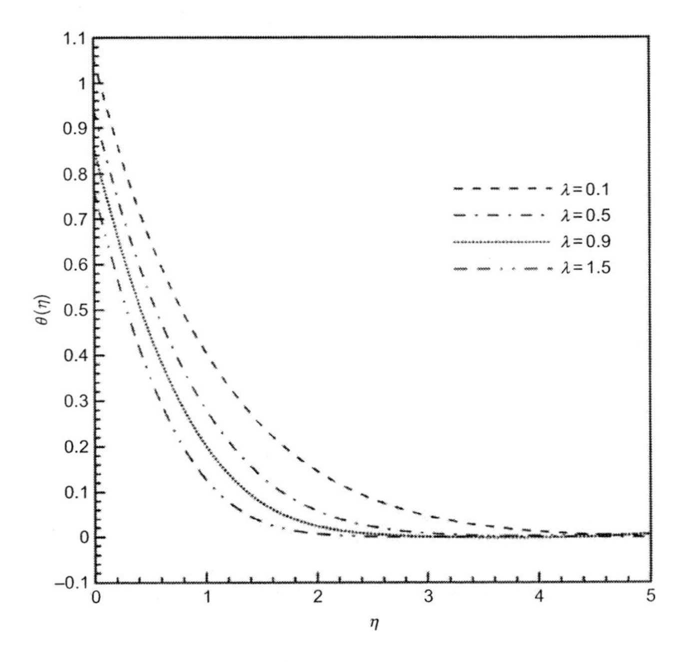

FIGURE 8.21

Effect λ on temperature distribution when $s=0$, $N=\mathrm{Nt}=\mathrm{Nb}=0.1$, $Pr=Le=n=1$, and $M=0.25$.

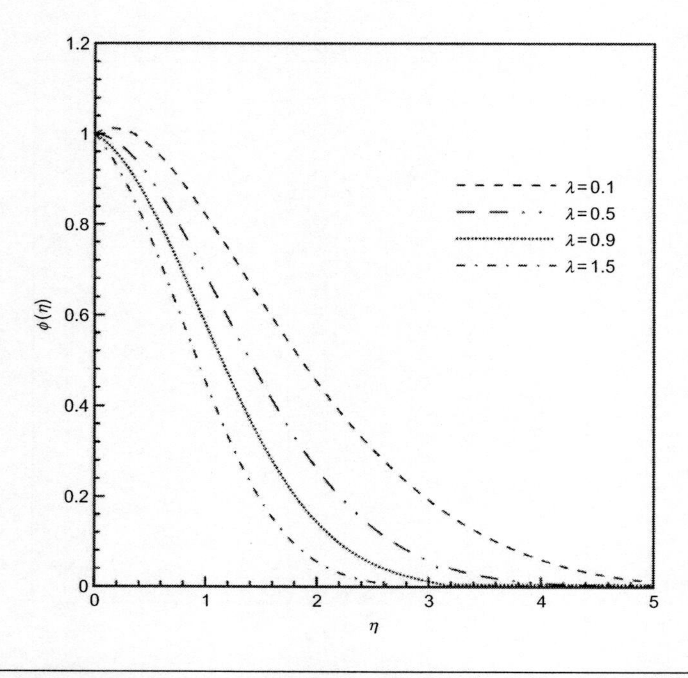

FIGURE 8.22

Effect λ on concentration distribution when $s=0$, $N=\text{Nt}=\text{Nb}=0.1$, $Pr=Le=n=1$, and $M=0.25$.

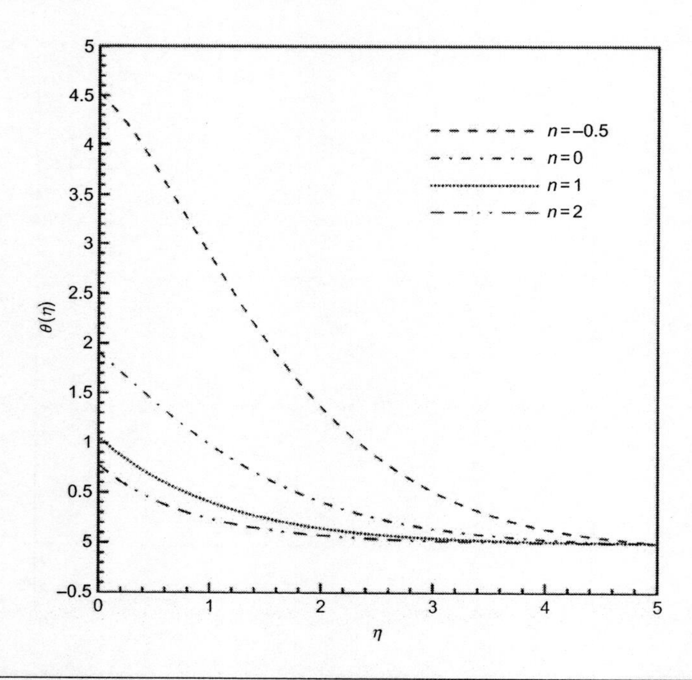

FIGURE 8.23

Effect n on temperature distribution when $s=0$, $\lambda=N=\text{Nt}=\text{Nb}=0.1$, $Pr=Le=1$, and $M=0.25$.

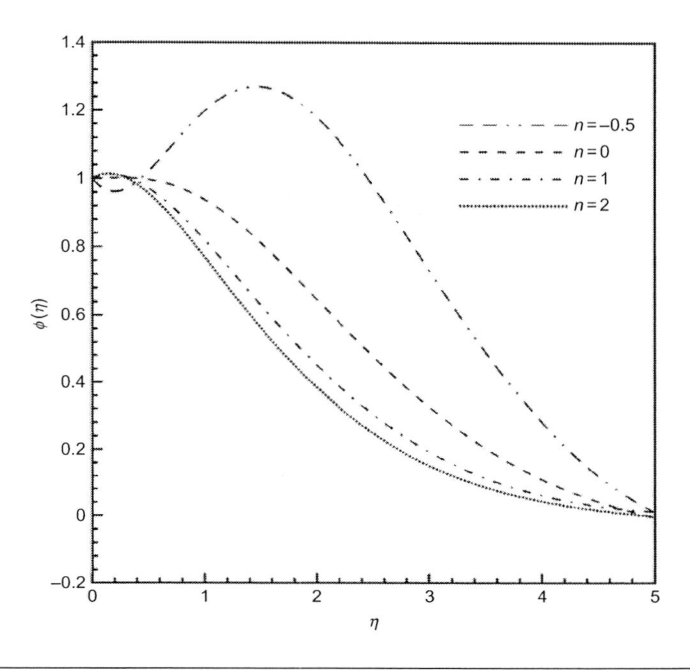

FIGURE 8.24

Effect n on concentration distribution when $s=0$, $\lambda=N=\text{Nt}=\text{Nb}=0.1$, $Pr=Le=1$, and $M=0.25$.

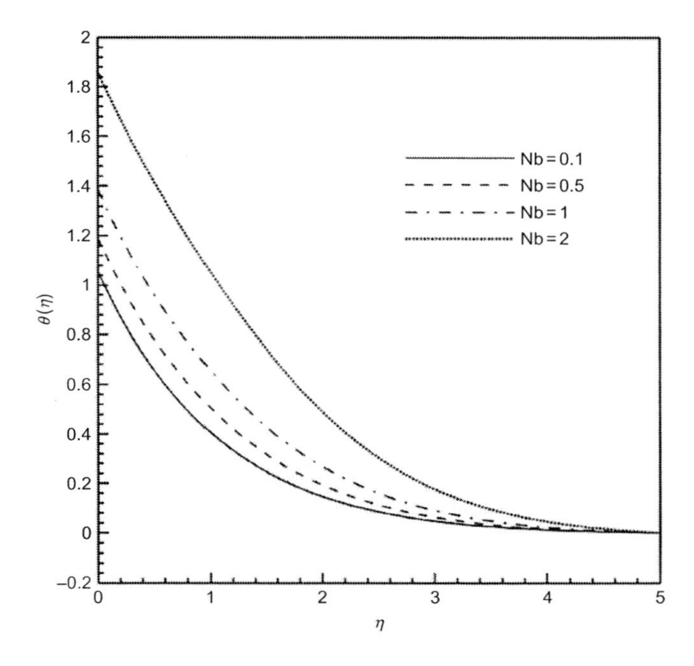

FIGURE 8.25

Effect Nb on temperature distribution when $s=0$, $\lambda=N=\text{Nt}=0.1$, $Pr=Le=n=1$, and $M=0.25$.

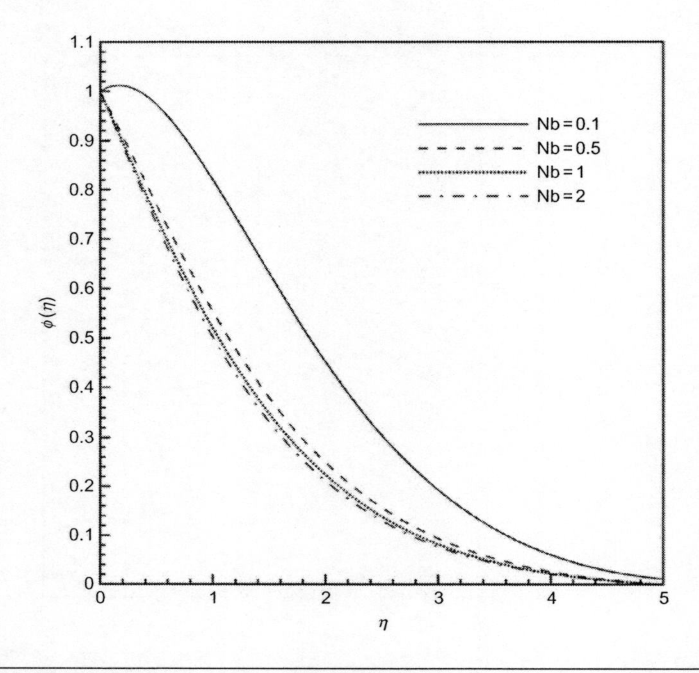

FIGURE 8.26

Effect Nb on concentration distribution when $s=0$, $\lambda=N=\mathrm{Nt}=0.1$, $Pr=Le=n=1$, and $M=0.25$.

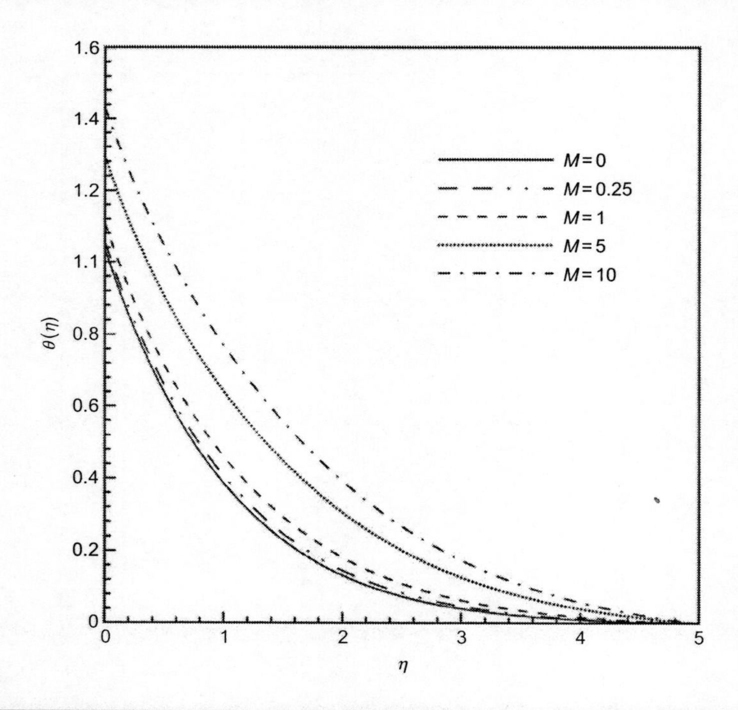

FIGURE 8.27

Effect M on temperature distribution when $\lambda=N=\mathrm{Nt}=\mathrm{Nb}=0.1$, $Pr=Le=n=1$, and $s=0$.

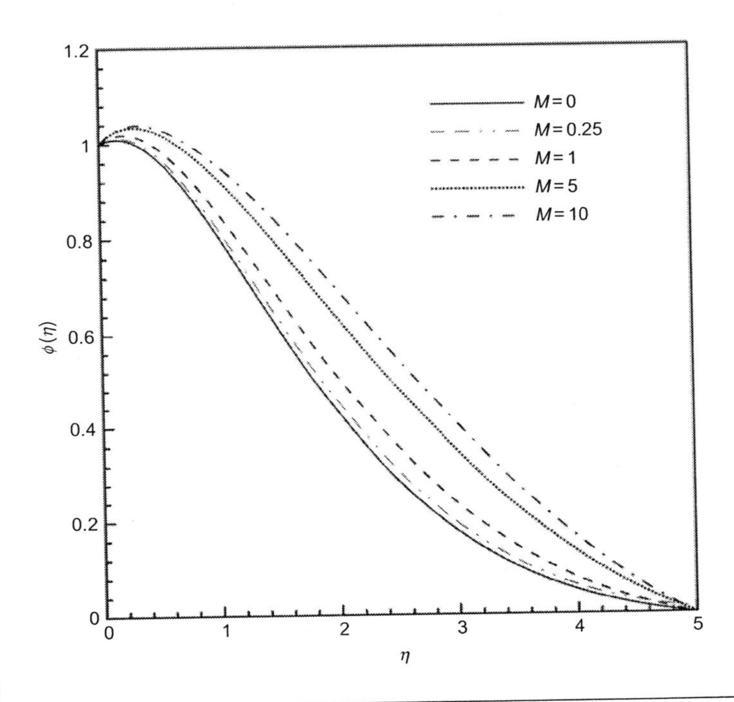

FIGURE 8.28

Effect M on concentration distribution when $\lambda = N = \mathrm{Nt} = \mathrm{Nb} = 0.1$, $Pr = Le = n = 1$, and $s = 0$.

Figures 8.29 and 8.30 show the variation of the dimensionless local Nusselt numbers and Sherwood numbers as a function of the magnetic field parameter M for parametric values of thermophoresis parameter Nt and suction or injection parameter S.

The dimensionless heat transfer and concentration rates decrease as M and Nt increase; also, the dimensionless heat transfer and concentration rates are higher when suction ($S > 0$) is induced. The effects of magnetic field parameter M, Lewis number Le, and suction or injection parameter S on the dimensionless local Nusselt numbers and Sherwood numbers are plotted in Figures 8.31 and 8.32. Figure 8.31 indicates that the dimensionless heat transfer rate decreases as M and Le increase. On the other hand, Figure 8.32 shows that dimensionless concentration rates increase as the suction and Lewis number Le increase but decrease when magnetic field parameter M increases.

8.2.6 CONCLUSIONS

The problem of MHD stagnation-point flow of a nanofluid past a heated porous stretching sheet with suction or blowing conditions and prescribed surface heat flux has been investigated analytically by using the HPM employing Padé technique. The effects of Brownian motion and thermophoresis occur in the transport equations. The solution for velocity, temperature, and nanoparticle concentration

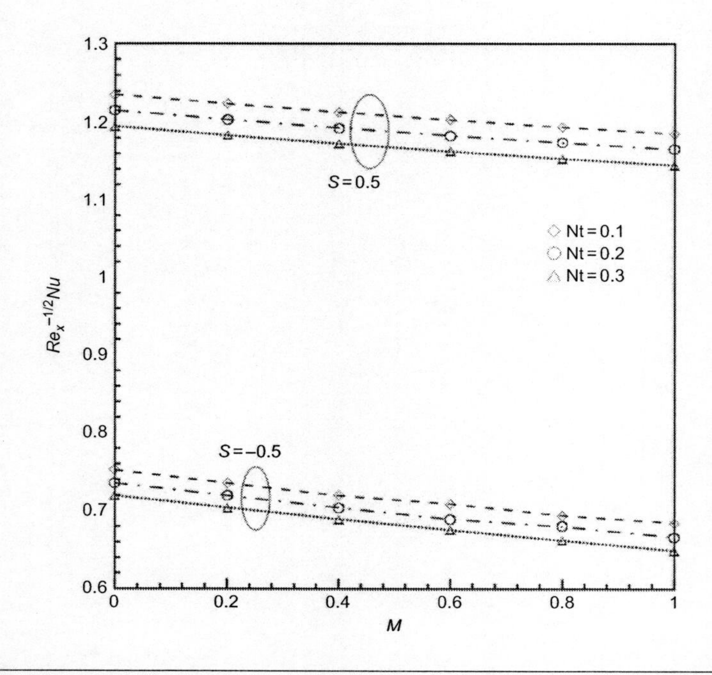

FIGURE 8.29

Effects of Nt on dimensionless heat transfer rates for $S=-0.5<0$ (injection) and $S=0.5>0$ (suction) when $\lambda=N=\mathrm{Nb}=0.1$, $Pr=Le=n=1$, and $M=0.25$.

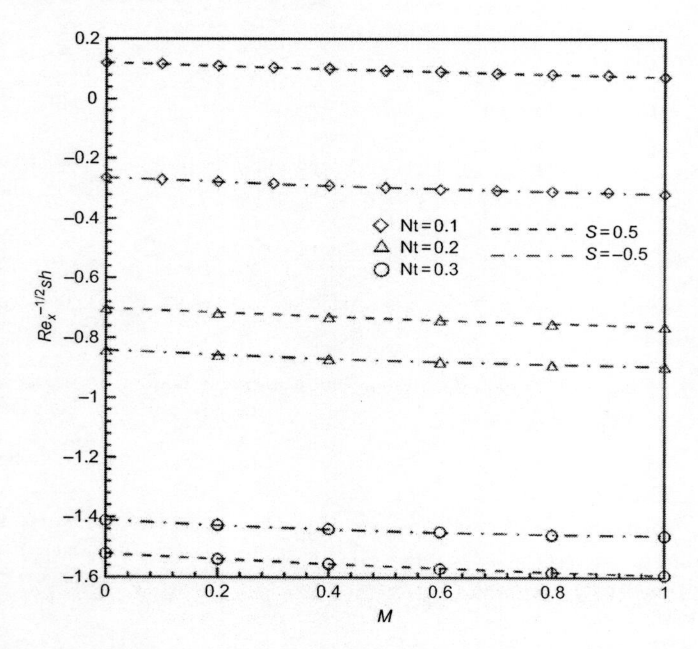

FIGURE 8.30

Effects of Nt on dimensionless concentration rates for $S=-0.5<0$ (injection) and $S=0.5>0$ (suction) when $\lambda=N=\mathrm{Nb}=0.1$, $Pr=Le=n=1$, and $M=0.25$.

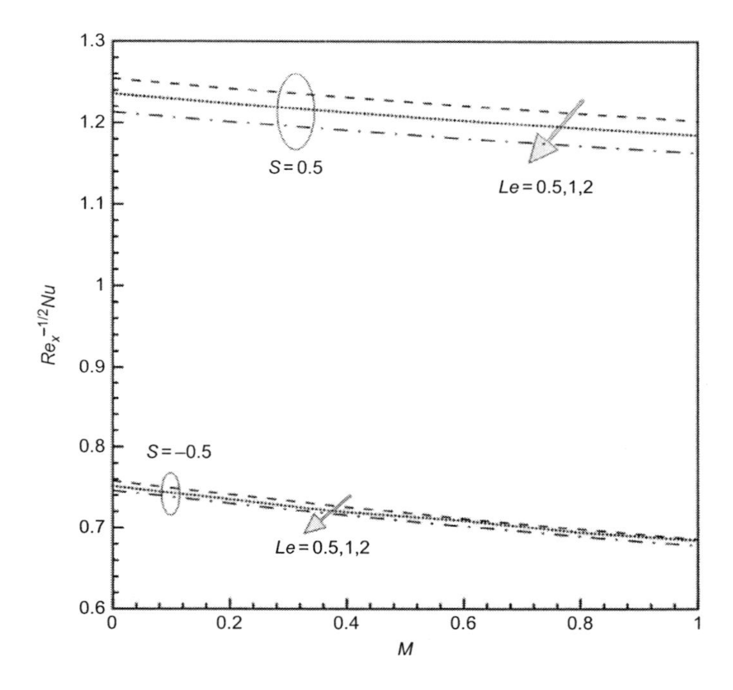

FIGURE 8.31

Effects of *Le* on dimensionless heat transfer rates for $S=-0.5<0$ (injection) and $S=0.5>0$ (suction) when $\lambda=N=\mathrm{Nt}=\mathrm{Nb}=0.1$, $Pr=n=1$ and $M=0.25$.

depends on parameters, viz., magnetic number M, porosity parameter N, velocity parameter λ, Prandtl number Pr, surface heat flux n, the Brownian motion parameter Nb, the thermophoresis parameter Nt, the Lewis number Le, and the suction $(S>0)$or injection parameter $(S<0)$.

It is found that the magnitude of the reduced Nusselt number decreases with an increase in magnetic number M, thermophoresis parameter Nt, and Lewis number Le. The reduced Sherwood number decreases with increasing magnetic number M and thermophoresis parameter Nt and increases with increasing Lewis number Le. Furthermore, the reduced Nusselt number and reduced Sherwood number are higher when suction $(S>0)$ is induced.

8.3 JEFFERY-HAMEL FLOW WITH HIGH MAGNETIC FIELD AND NANOPARTICLE

In this study, the effect of magnetic field and nanoparticle on the Jeffery-Hamel flow is studied by a powerful analytical method called adomian decomposition method (ADM). The traditional Navier-Stokes equation of fluid mechanics and Maxwell's electromagnetism governing equations are reduced

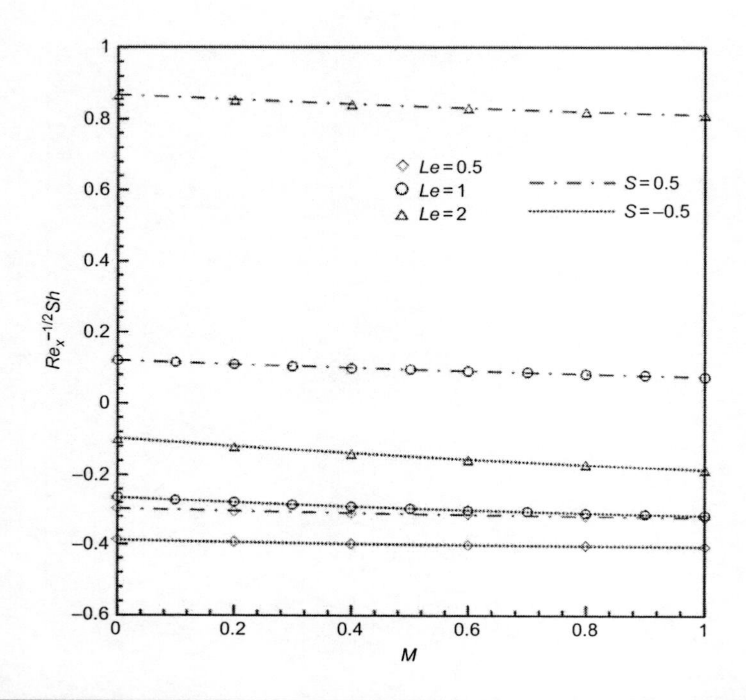

FIGURE 8.32

Effects of *Le* on dimensionless concentration rates for $S=-0.5<0$ (injection) and $S=0.5>0$ (suction) when $\lambda=N=\text{Nt}=\text{Nb}=0.1$, $Pr=n=1$ and $M=0.25$.

to nonlinear ordinary differential equations to model this problem. The obtained results by this method are well agreed with the numerical (Runge-Kutta method) results and tabulated in a table. The plots confirm that the used method is in high accuracy for different α, Ha, and Re numbers. First, the flow field inside the divergent channel is studied for various values of Hartmann number and angle of channel, and at last, the effect of nanoparticle volume fraction in absence of magnetic field is investigated (*this section has been worked by M. Sheikholeslami, D.D. Ganji, H.R. Ashorynejad, Houman B. Rokni in nonlinear dynamics team in Mechanical Engineering Department, 2012-2013*).

8.3.1 INTRODUCTION

After introducing the problem of the flow of fluid through a divergent channel by Jeffery (1915) and Hamel (1916), respectively, it is called Jeffery-Hamel flow. On the other hand, the term of MHD was first introduced by Alfvén (1942). The theory of MHD states that inducing current in a moving conductive fluid in the presence of magnetic field exerts force on the ions of the conductive fluid. The theoretical study of MHD channel has been a subject of great interest due to its extensive applications in designing cooling systems with liquid metals, MHD generators, accelerators, pumps, and flow meters (see Cha et al., 2002; Tendler, 1983).

The small disturbance stability of MHD stability of plane Poiseuille flow has been investigated by Makinde and Motsa (2001, 2003) for generalized plane Couette flow. Their results show that magnetic field has stabilizing effects on the flow. Considerable efforts have been made to study the MHD theory for technological application of fluid pumping system in which electrical energy forces the working conductive fluid. Damping and controlling of electrically conducting fluid can be achieved by means of an electromagnetic body force (Lorentz force) produced by the interaction of an applied magnetic field and an electric current that usually is externally supplied. Anwari et al. (2005) studied the fundamental characteristics of linear Faraday MHD theoretically and numerically, for various loading configurations. Homsy et al. (2005) emphasized on the idea that in such problems, the moving ions drag the bulk fluid with themselves and such MHD system induces continuous pumping of conductive fluid without any moving part. Taking into account the rising demands of modern technology, including chemical production, power station, and microelectronics, there is a need to develop new types of fluids that will be more effective in terms of heat exchange performance.

In this chapter, we have applied ADM (see Adomian, 1988) to find the approximate solutions of nonlinear differential equations governing the MHD Jeffery-Hamel flow, and a comparison between the results and the numerical solution has been provided. The numerical results of this problem are obtained by using Maple 12.

8.3.2 GOVERNING EQUATIONS

Consider a system of cylindrical polar coordinates (r,θ,z) which steady two-dimensional flow of an incompressible conducting viscous fluid from a source or sink at channel walls, lie in planes, and intersect in the axis of z (see Figure 8.33).

Assuming purely radial motion which means that there is no change in the flow parameter along the z-direction. The flow depends on r and θ and further assume that there is no magnetic field in the z-direction.

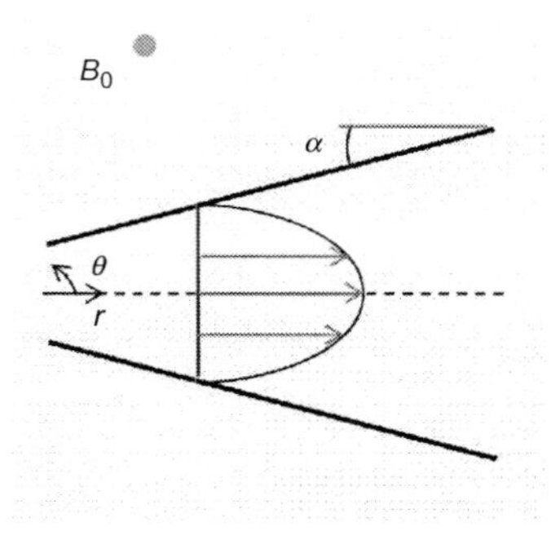

FIGURE 8.33

Geometry of the MHD Jeffery-Hamel flow.

The reduced form of continuity, Navier-Stokes, and Maxwell's equations are

$$\frac{\rho_{\mathrm{nf}}}{r}\frac{\partial(ru(r,\theta))}{\partial r}(ru(r,\theta))=0 \tag{8.68}$$

$$u(r,\theta)\frac{\partial u(r,\theta)}{\partial r}=-\frac{1}{\rho_{\mathrm{nf}}}\frac{\partial P}{\partial r}+\upsilon_{\mathrm{nf}}\left[\frac{\partial^2 u(r,\theta)}{\partial r^2}+\frac{1}{r}\frac{\partial u(r,\theta)}{\partial r}+\frac{1}{r^2}\frac{\partial^2 u(r,\theta)}{\partial\theta^2}-\frac{u(r,\theta)}{r^2}\right]-\frac{\sigma B_0^2}{\rho_{\mathrm{nf}} r^2}u(r,\theta) \tag{8.69}$$

$$\frac{1}{\rho_{\mathrm{nf}} r}\frac{\partial P}{\partial\theta}-\frac{2\upsilon_{\mathrm{nf}}}{r^2}\frac{\partial u(r,\theta)}{\partial\theta}=0 \tag{8.70}$$

where B_0 is the electromagnetic induction, σ is the conductivity of the fluid, $u(r)$ is the velocity along radial direction, P is the fluid pressure, υ_{nf} is the coefficient of kinematic viscosity, and ρ_{nf} is the fluid density. The effective density ρ_{nf}, the effective dynamic viscosity μ_{nf}, and kinematic viscosity υ_{nf} of the nanofluid are given as (Aminossadati and Ghasemi, 2009; Ghasemi and Aminossadati, 2010)

$$\rho_{\mathrm{nf}}=\rho_{\mathrm{f}}(1-\phi)+\rho_{\mathrm{s}}\phi,\quad \mu_{\mathrm{nf}}=\frac{\mu_{\mathrm{f}}}{(1-\phi)^{2.5}},\quad \upsilon_{\mathrm{nf}}=\frac{\mu_{\mathrm{f}}}{\rho_{\mathrm{nf}}} \tag{8.71}$$

Here, ϕ is the solid volume fraction.

Considering $u_\theta=0$ for purely radial flow, one can define the velocity parameter as

$$f(\theta)=ru(r) \tag{8.72}$$

Introducing $\eta=\dfrac{\theta}{\alpha}$ as the dimensionless degree, the dimensionless form of the velocity parameter can be obtained by dividing that to its maximum value as

$$f(\eta)=\frac{f(\theta)}{f_{\max}} \tag{8.73}$$

Substituting Equation (8.72) into Equations (8.69) and (8.70), and eliminating P, one can obtain the ordinary differential equation for the normalized function profile as (see Anwari et al., 2005)

$$f'''(\eta)+2\alpha Re\cdot A^*(1-\phi)^{2.5}f(\eta)f'(\eta)+\left(4-(1-\phi)^{1.25}Ha\right)\alpha^2 f'(\eta)=0 \tag{8.74}$$

where A^* is a parameter. Reynolds number and Hartmann number based on the electromagnetic parameter are introduced as following:

$$A^*=(1-\phi)+\frac{\rho_{\mathrm{s}}}{\rho_{\mathrm{f}}}\phi \tag{8.75}$$

$$Re=\frac{f_{\max}\alpha}{\upsilon_{\mathrm{f}}}=\frac{U_{\max}r\alpha}{\upsilon_{\mathrm{f}}}\begin{pmatrix}\text{divergent-channel}: \alpha>0,\ f_{\max}>0\\ \text{convergent-channel}: \alpha<0,\ f_{\max}<0\end{pmatrix} \tag{8.76}$$

$$Ha=\sqrt{\frac{\sigma B_0^2}{\rho_{\mathrm{f}}\upsilon_{\mathrm{f}}}} \tag{8.77}$$

With the following reduced form of boundary conditions

$$f(0)=1,\ f'(0)=0,\ f(1)=0 \tag{8.78}$$

Physically, these boundary conditions mean that maximum values of velocity are observed at centerline ($\eta=0$) as shown in Figure 8.33, and we consider fully developed velocity profile; thus, rate of velocity is zero at ($\eta=0$). Also, in fluid dynamics, the no-slip condition for fluid states that at a solid boundary, the fluid will have zero velocity relative to the boundary. The fluid velocity at all fluid-solid boundaries is equal to that of the solid boundary, so we can see that value of velocity is zero at ($\eta=1$).

8.3.3 FUNDAMENTALS OF ADM

Consider equation $Fu(t)=g(t)$, where F represents a general nonlinear ordinary or partial differential operator including both linear and nonlinear terms. The linear terms are decomposed into $L+R$, where L is easily invertible (usually the highest order derivative) and R is the remainder of the linear operator. Thus, the equation can be written as (Adomian, 1988)

$$Lu+Nu+Ru=g \tag{8.79}$$

where Nu indicates the nonlinear terms. By solving this equation for Lu, since L is invertible, we can write

$$L^{-1}Lu=L^{-1}g-L^{-1}Ru-L^{-1}Nu \tag{8.80}$$

If L is a second-order operator, L^{-1} is a twofold indefinite integral. By solving Equation (8.79), we have

$$u=A+Bt+L^{-1}g-L^{-1}Ru-L^{-1}Nu \tag{8.81}$$

where A and B are constants of integration and can be found from the boundary or initial conditions. Adomian method assumes that the solution u can be expanded into infinite series as

$$u=\sum_{n=0}^{\infty}u_n \tag{8.82}$$

Also, the nonlinear term Nu will be written as

$$Nu=\sum_{n=0}^{\infty}A_n \tag{8.83}$$

where A_n is the special Adomian polynomial. By specified A_n, next component of u can be determined as

$$u_{n+1}=L^{-1}\sum_{n=0}^{n}A_n \tag{8.84}$$

Finally, after some iteration and getting sufficient accuracy, the solution can be expressed by Equation (8.81).

In Equation (8.82), the Adomian polynomials can be generated by several means. Here, we used the following recursive formulation

$$A_n = \frac{1}{n!}\left[\frac{d^n}{d\lambda^n}\left[N\left(\sum_{i=0}^{n}\lambda^i u_i\right)\right]\right]_{\lambda=0}, \quad n=0,1,2,3,\ldots \tag{8.85}$$

Since the method does not resort to linearization or assumption of weak nonlinearity, the solution generated is, in general, more realistic than those achieved by simplifying the model of the physical problem.

8.3.4 APPLICATION

According to Equation (8.12), Equation (8.7) must be written as following:

$$Lf = -2Re\cdot\alpha\cdot A^*(1-\phi)^{2.5}ff' - \left(4-(1-\phi)^{1.25}Ha\right)\alpha^2 f' \tag{8.86}$$

where the differential operator L is given by $L=\frac{d^3}{d\eta^3}$. Assume the inverse of the operator L exists and it can be integrated from 0 to η, i.e. $L^{-1} = \int_0^\eta\int_0^\eta\int_0^\eta (\bullet)\,d\eta\, d\eta\, d\eta$.

Operating with L^{-1} on Equation (8.86) and after exerting boundary condition on it, we have

$$f(\eta) = f(0) + f'(0)\eta + f''(0)\frac{\eta^2}{2} + L^{-1}(Nu) \tag{8.87}$$

where

$$Nu = -2Re\alpha\cdot A^*(1-\phi)^{2.5}f(\eta)f'(\eta) - \left(4-(1-\phi)^{1.25}Ha\right)f'(\eta) \tag{8.88}$$

ADM introduced the following expression

$$f(\eta) = \sum_{m=0}^{\infty} f_m(\eta) \tag{8.89}$$

$$f(\eta) = \sum_{m=0}^{\infty} f_m = f_0 + L^{-1}(Nu) \tag{8.90}$$

To determine the components of $f_m(\eta)$, $f_0(\eta)$ is defined by applying the boundary condition of Equation (8.74) and by assuming $f''(0)=\beta$.

$$f_0(\eta) = 1 + \beta\frac{\eta^2}{2} \tag{8.91}$$

$$\begin{aligned} f_1(\eta) = {} & \left(\frac{-1}{120}\right)\alpha Re\cdot A^*(1-\phi)^{2.5}\beta^2\eta^6 - \frac{1}{4}\left(\frac{1}{3}\alpha Re\cdot A^*(1-\phi)^{2.5}\beta \right. \\ & \left. + \frac{1}{6}\left(4-(1-\phi)^{1.25}Ha\right)\alpha^2\beta\right)\eta^4 - \frac{1}{120}\left(4-(1-\phi)^{1.25}Ha\right)\alpha^2\beta\eta^5 \\ & - \frac{1}{6}(4-Ha)\alpha^2\eta^3 \end{aligned} \tag{8.92}$$

$$
\begin{aligned}
f_2(\eta) = {} & \left(\frac{1}{10800}\right)\alpha^2\left(Re\cdot A^*(1-\phi)^{2.5}\right)^2\beta^3\eta^{10} + \frac{1}{7560}\alpha^3 Re\cdot A^*(1-\phi)^{2.5}\left(4-(1-\phi)^{1.25}Ha\right)\beta^2\eta \\
& -\frac{1}{8}\left(\frac{1}{21}\alpha Re\cdot A^*(1-\phi)^{2.5}\left(-\frac{1}{12}\alpha Re\cdot A^*(1-\phi)^{2.5}\beta - \frac{1}{24}\left(4-(1-\phi)^{1.25}Ha\right)\alpha^2\beta\right)\beta\right. \\
& -\frac{1}{420}\alpha^2\left(Re\cdot A^*(1-\phi)^{2.5}\right)^2 + \frac{1}{42}\alpha^2 Re\cdot A^*(1-\phi)^{2.5}\beta\left(-\frac{1}{3}\alpha Re\cdot A^*(1-\phi)^{2.5}\beta\right. \\
& \left.\left.-\frac{1}{6}\left(4-(1-\phi)^{1.25}Ha\right)\alpha^2\beta\right) - \frac{1}{840}\alpha^3 Re\cdot A^*(1-\phi)^{2.5}\left(4-(1-\phi)^{1.25}Ha\right)\beta^2\right)\eta^8 \\
& -\frac{1}{7}\left(-\frac{11}{360}\alpha^3 Re\cdot A^*(1-\phi)^{2.5}\left(4-(1-\phi)^{1.25}Ha\right)\beta - \frac{1}{720}\left(4-(1-\phi)^{1.25}Ha\right)^2\alpha^4\beta\right)\eta^7 \\
& -\frac{1}{6}\left(\frac{1}{10}\alpha Re\cdot A^*(1-\phi)^{2.5}\left(-\frac{1}{3}\alpha Re\cdot A^*(1-\phi)^{2.5}\beta - \frac{1}{6}\left(4-(1-\phi)^{1.25}Ha\right)\alpha^2\beta\right)\right. \\
& \left.+\frac{1}{20}\left(4-(1-\phi)^{1.25}Ha\right)\alpha^2\left(-\frac{1}{3}\alpha Re\cdot A^*(1-\phi)^{2.5}\beta - \frac{1}{6}\left(4-(1-\phi)^{1.25}Ha\right)\alpha^2\beta\right)\right)\eta^6 \\
& -\frac{1}{5}\left(-\frac{1}{12}\alpha^3 Re\cdot A^*(1-\phi)^{2.5}\left(4-(1-\phi)^{1.25}Ha\right) - \frac{1}{720}\left(4-(1-\phi)^{1.25}Ha\right)^2\alpha^4\right)\eta^5 \\
& +\frac{1}{40320}\left(4-(1-\phi)^{1.25}Ha\right)^2\alpha^4\beta\eta^8 - \frac{1}{120}\left(4-(1-\phi)^{1.25}Ha\right)\alpha^2\left(-\frac{1}{12}\alpha Re\beta\right. \\
& \left.-\frac{1}{24}\left(4-(1-\phi)^{1.25}Ha\right)\alpha^2\beta\right)\eta^7 + \frac{1}{720}\left(4-(1-\phi)^{1.25}Ha\right)^2\alpha^4\eta^6
\end{aligned}
\tag{8.93}
$$

$f_3(\eta), f_4(\eta), \ldots$ can be determined in similar way from Equation (8.90).

Using $f(\eta) = \sum_{m=0}^{\infty} f_m(\eta) = f_0(\eta) + f_1(\eta) + f_2(\eta) + f_3(\eta) + \cdots$, thus

$$
\begin{aligned}
f(\eta) = {} & 1 + \beta\frac{\eta^2}{2} + \left(\frac{-1}{120}\right)\alpha Re\cdot A^*(1-\phi)^{2.5}\beta^2\eta^6 - \frac{1}{4}\left(\frac{1}{3}\alpha Re\cdot A^*(1-\phi)^{2.5}\beta\right. \\
& \left.+\frac{1}{6}\left(4-(1-\phi)^{1.25}Ha\right)\alpha^2\beta\right)\eta^4 - \frac{1}{120}\left(4-(1-\phi)^{1.25}Ha\right)\alpha^2\beta\eta^5 \\
& -\frac{1}{6}(4-Ha)\alpha^2\eta^3 + \cdots
\end{aligned}
\tag{8.94}
$$

According to Equation (8.94), the accuracy of ADM solution increases by increasing the number of solution terms (m). For the complete solution of Equation (8.94), β should be determined with boundary condition of $f''(0) = \beta$.

8.3.5 RESULTS AND DISCUSSION

In this study, the objective was to apply ADM to obtain an explicit analytic solution of the MHD Jeffery-Hamel problem. At first, viscous fluid (ϕ) with magnetic field effect is considered and finally nanofluid flow without magnetic field ($Ha = 0$) is investigated. The magnetic field acts as a control parameter, such as the flow Reynolds number and the angle of the walls, in MHD Jeffery-Hamel problem. There is an additional nondimensional parameter that determines the solutions, namely, the

Table 8.6 Value of $f''(0)=\beta$ at Various Re, Ha, α when $\phi=0$

		$Ha=100$	$Ha=200$	$Ha=300$
Re	α	β	β	β
25	2.5	−2.42483	−2.54507	−2.66789
	5	−3.09259	−3.57685	−4.09816
	7.5	−3.98607	−5.14061	−6.44257
50	2.5	−2.77059	−2.87965	−2.99167
	5	−3.85619	−4.25077	−4.69303
	7.5	−5.16973	−6.04706	−7.13483

Hartmann number. Table 8.6 shows the value of constant β for different α, Ha, and Reynolds number at the divergent channel.

For comparison, a few limited cases of the ADM solutions are compared with the numerical results. The comparison between the numerical results and ADM solution for velocity when $Re=25$ and $\alpha=5°$ is shown in Table 8.7. The error bar shows an acceptable agreement between the results observed, which confirms the validity of the ADM. In these tables, error is introduced as follows: $\text{Error}=\left|f(\eta)_{\text{NM}}-f(\eta)_{\text{ADM}}\right|$.

Figures 8.34 and 8.35 show the magnetic field effect on the velocity profiles for divergent channels. There are good agreements between the numerical solution obtained by the fourth-order Runge-Kutta method and ADM. The velocity curves show that the rate of transport is considerably reduced with increase of Hartmann number. This clearly indicates that the transverse magnetic field opposes the transport phenomena. This is due to the fact that variation of Ha leads to the variation of the Lorentz force due to magnetic field and the Lorentz force produces more resistance to transport phenomena.

Under magnetic field, the Lorentz force affect in opposite of the momentum's direction that stabilize the velocity profile.

The results show moderate increases in the velocity with increasing Hartmann numbers at small angle ($\alpha=2.5°$), and difference between velocity profiles are more noticeable at greater angles. Backflow is excluded in converging channels, but it may occur for large Reynolds numbers in diverging channels. For specified opening angle, after a critical Reynolds number, we observe that separation and backflow are started (Figure 8.34).

Figure 8.35 shows the magnetic field effects at constant α and different Reynolds numbers. At $\alpha=5°$, $Re=75$ with increasing Hartmann number, the velocity profile becomes flat and thickness of boundary layer decreases, but at this Reynolds number, no back flow is observed as it is shown in Figure 8.35a. It can be seen in Figure 8.35b that without magnetic field at $\alpha=5°$, $Re=150$ back flow starts and that with Hartmann number increasing this phenomenon is eliminated. By increasing Reynolds number, the back flow expands; so, greater magnetic field is needed in order to eliminate it. As it is shown in Figure 8.35c and d, $\alpha=5°$, $Re=225$ that back flow is eliminated at $Ha=1000$ while at $\alpha=5°$, $Re=300$ this occurs in $Ha=2000$.

Finally, we consider a Cu-water nanofluid flow and investigate effect of nanoparticle volume fraction. It is assumed that the base fluid and the nanoparticles are in thermal equilibrium and no slip occurs between them. The density of water is $\rho_f=997.1$ and density of Cu is $\rho_s=8933$. It can be seen from

Table 8.7 Comparison Between the Numerical Results and ADM Solution for Velocity When $\phi=0$, Re = 25, and $\alpha=5°$

	$Ha=0$			$Ha=250$			$Ha=500$		
η	Numerical	ADM	Error	Numerical	ADM	Error	Numerical	ADM	Error
0	1	1	0	1	1	0	1	1	0
0.1	0.986671	0.986637	0.000035	0.988606	0.990196	0.001591	0.99022	0.992695	0.002475
0.2	0.947258	0.947127	0.000136	0.9547	0.960841	0.006141	0.960933	0.970544	0.009611
0.3	0.883419	0.883146	0.000273	0.899076	0.912079	0.013003	0.912273	0.912273	0.006304
0.4	0.797697	0.797259	0.000438	0.822925	0.811225	0.011700	0.844383	0.832683	0.009146
0.5	0.693233	0.692638	0.000595	0.727664	0.713799	0.013865	0.757286	0.743421	0.010234
0.6	0.573424	0.572716	0.000709	0.614709	0.604866	0.009843	0.650719	0.643816	0.006902
0.7	0.441593	0.44085	0.000743	0.485232	0.476625	0.008606	0.523909	0.515303	0.018548
0.8	0.300674	0.300013	0.000662	0.33989	0.325834	0.014056	0.37529	0.361234	0.013709
0.9	0.152979	0.152552	0.000427	0.178555	0.17116	0.007395	0.202125	0.19473	0.009368
1	0	0	0	0	0	0	0	0	0

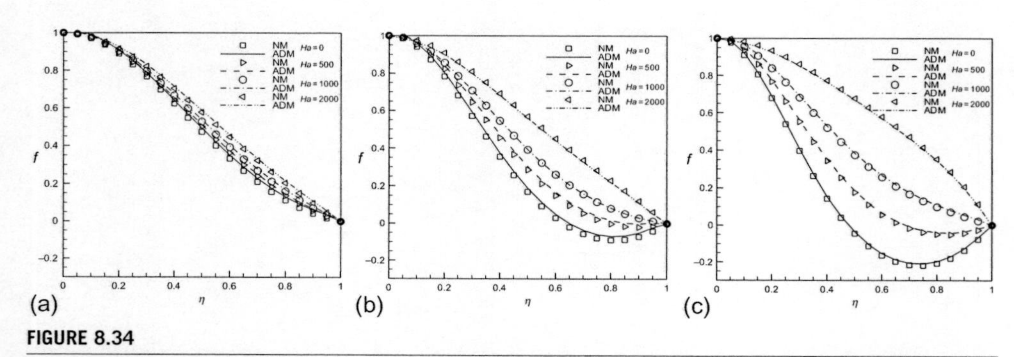

FIGURE 8.34

The ADM solution for velocity in divergent channel for (a) $\alpha=2.5°$, $Re=200$, and $\phi=0$, (b) $\alpha=5°$, $Re=200$, and $\phi=0$ and (c) $\alpha=7.5°$, $Re=200$, and $\phi=0$.

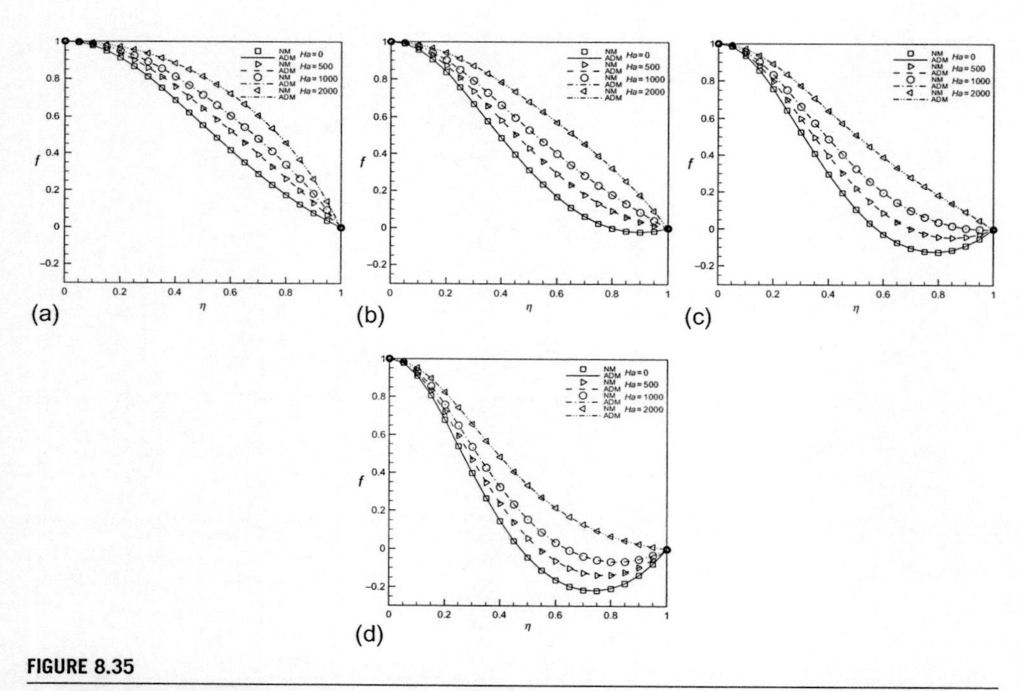

FIGURE 8.35

The ADM solution for velocity in divergent channel for (a) $\alpha=5°$, $Re=75$, and $\phi=0$, (b) $\alpha=5°$, $Re=150$, and $\phi=0$, (c) $\alpha=5°$, $Re=225$, and $\phi=0$, and (d) $\alpha=5°$, $Re=300$, and $\phi=0$.

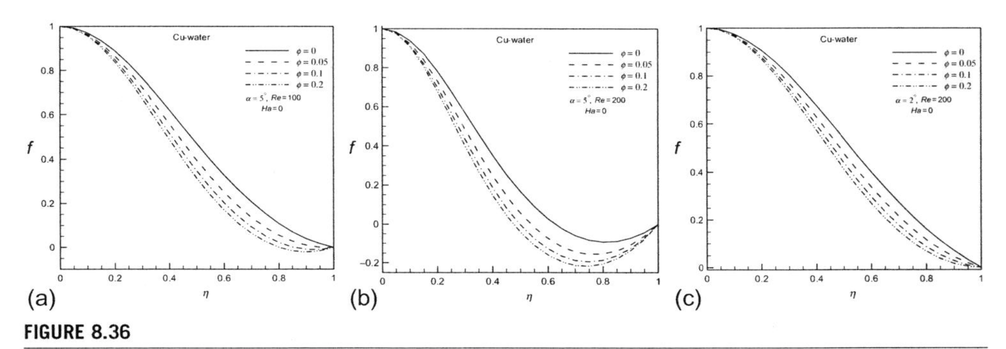

FIGURE 8.36

Effect of nanoparticle volume fraction on velocity profiles for Cu-water.

Figure 8.35 that increasing in nanoparticle volume fraction leads to increase in momentum boundary layer thickness as a result of an increase in the energy transportation, the fluid with the increasing of volume fraction. Also, it can be seen that increasing nanoparticle volume fraction as high values of Reynolds number and angle of the channel leads to start back flow (Figure 8.36).

8.3.6 CONCLUSION

In this section, MHD Jeffery-Hamel flow has been solved via a sort of analytical method, ADM. Also, this problem is solved by a numerical method (the Runge-Kutta method of order 4) and at last, effect of adding nanoparticle in absence of magnetic field is considered. It can be found that ADM is a powerful approach for solving this problem. Also, it is observed that there is a good agreement between this and the numerical result. Increasing Reynolds numbers leads to adverse pressure gradient which causes velocity reduction near the walls. Increasing Hartmann number will lead to backflow reduction. In greater angles or Reynolds numbers, high Hartmann number is needed to reduce backflow. Also, the results show that momentum boundary layer thickness increases as nanoparticle volume fraction increases.

8.4 THE TRANSVERSE MAGNETIC FIELD ON JEFFERY-HAMEL PROBLEM WITH CU-WATER NANOFLUID

This chapter studies the problem of an incompressible and viscous fluid between two rigid plane walls. In the presence of external magnetic field, commonly known as the MHD, Jeffery-Hamel flow with nanoparticles has been investigated. Copper as nanoparticle with water as its base fluid has been considered. The governing equations, continuity, and momentum for this problem are reduced to an ordinary form and are solved by HAM. The accuracy of HAM is authenticated by comparing with numerical solution as BVP. It has been attempted to show the capabilities and wide-range applications of the HAM in comparison with the numerical method. This method led to high accurate appropriate results for nonlinear problems in comparison with numerical solution (*this section has been worked by D.D. Ganji, I. Rahimi Petroudi, M. Khazayi Nejad, J. Rahimi, E. Rahimi, A. Rahimifar in nonlinear dynamics team in Mechanical Engineering Department, 2012-2013*).

8.4.1 INTRODUCTION

During the past few years, the incompressible viscous fluid flow through convergent-divergent channels is one of the most applicable cases in fluid mechanics, civil, environmental, mechanical, and biomechanical engineering. As mentioned in the previous section, the mathematical investigations of this problem were pioneered by Jeffery and Hamel and so, it is known as Jeffery-Hamel problem, too. One of the most significant examples of Jeffery-Hamel problems is that subjected to an applied magnetic field. As everyone knows, most scientific problems such as Jeffery-Hamel flows and other fluid mechanic problems are inherently nonlinear. In most cases, these problems do not admit analytical solution, so these equations should be solved using special techniques. In recent decades, much attention has been devoted to the newly developed methods to construct an analytic solution of equation (*to see various applied examples of these methods*, see Ganji and Hashemi Kachapi, 2011a,b; Hashemi Kachapi and Ganji, 2013; Hashemi Kachapi and Ganji, 2014a,b).

The main purpose of this study is to apply HAM to find approximate solutions of the velocity profile on MHD Jeffery-Hamel flow with nanoparticles (see Liao, 2003). A clear conclusion can be drawn from the numerical method's (NUM) results that the HAM provides highly accurate solutions for nonlinear differential equations.

8.4.2 PROBLEM STATEMENT AND MATHEMATICAL FORMULATION

We consider the boundary layer flow of an electrically conducting viscous fluid with nanoparticle. A magnetic field $B(x)$ acts transversely to the flow. As it can be seen in Figure 8.37, the steady two-dimensional flow of an incompressible conducting viscous fluid from a source or sink at the intersection between two nonparallel plane walls is considered.

We assume that the velocity is purely radial and depends on r and θ only. The governing equations in polar coordinates are (Ganji et al., 2009; Ganji and Azimi, 2013):

$$\frac{\rho_{\mathrm{nf}}\partial}{r\partial r}(ru(r,\theta))=0 \tag{8.95}$$

$$\begin{aligned} u(r,\theta)\frac{\partial u(r,\theta)}{\partial r} = & -\frac{1}{\rho_{\mathrm{nf}}}\frac{\partial p}{\partial r}+\upsilon_{\mathrm{nf}}\left[\frac{\partial^2 u(r,\theta)}{\partial r^2}+\frac{1}{r}\frac{\partial u(r,\theta)}{\partial r}+\frac{1}{r^2}\frac{\partial^2 u(r,\theta)}{\partial \theta^2}-\frac{u(r,\theta)}{r^2}\right] \\ & -\frac{\sigma B_0^2}{\rho_{\mathrm{nf}} r^2}u(r,\theta) \end{aligned} \tag{8.96}$$

$$\frac{1}{\rho_{\mathrm{nf}} r}\frac{\partial p}{\partial \theta}-\frac{2\upsilon_{\mathrm{nf}}}{r^2}\frac{\partial u(r,\theta)}{\partial \theta}=0 \tag{8.97}$$

Here B_0 is the electromagnetic induction, $u(r)$ is the velocity along radial direction, P is the fluid pressure, σ is the conductivity of the fluid, ρ_{nf} is the density of fluid, and υ_{nf} is the coefficient of kinematic viscosity. By introducing φ as a solid volume fraction, fluid density, dynamic viscosity, and the kinematic viscosity of nanofluid can be written as follows:

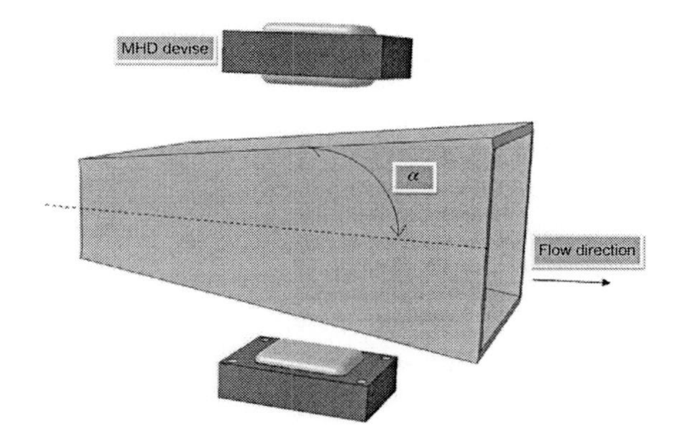

FIGURE 8.37

Schematic diagram of the physical system.

$$\rho_{nf} = \rho_f(1-\phi) + \rho_s\phi_s \tag{8.98}$$

$$\mu_{nf} = \frac{\mu_f}{(1-\phi)^{2.5}} \tag{8.99}$$

$$\upsilon_{nf} = \frac{\mu_f}{\rho_{nf}} \tag{8.100}$$

From Equation (8.95) and using dimensionless parameters, we get

$$f(\theta) = ru(r,\theta) \tag{8.101}$$

$$f(\eta) = \frac{f(\theta)}{f_{max}}, \quad \eta = \frac{\theta}{\alpha} \tag{8.102}$$

Substituting Equation (8.99) into Equations (8.96) and (8.97) and eliminating P, we obtain an ordinary differential equation for the normalized function profile $f(\eta)$ (He, 2006)

$$f'''(\eta) + 2\alpha ReA^*(1-\phi)^{2.5}f(\eta)f'(\eta) + \left(4-(1-\phi)^{1.25}H\right)\alpha^2 f'(\eta) = 0 \tag{8.103}$$

The relevant boundary conditions are

$$f(0) = 1, \; f'(0) = 0, \; f(1) = 0 \tag{8.104}$$

The Reynolds number is

$$Re = \frac{f_{max}\alpha}{\upsilon} = \frac{U_{max}r\alpha}{\upsilon}\begin{pmatrix} \text{divergevt-channel}: \alpha \succ 0, \; f_{max} \succ 0 \\ \text{convergevt-channel}: \alpha \prec 0, \; f_{max} \prec 0 \end{pmatrix} \tag{8.105}$$

The Hartmann number is

$$H=\sqrt{\frac{\sigma B_0^2}{\rho \upsilon}} \tag{8.106}$$

The particle parameter is

$$A^*=(1-\phi)+\frac{\rho_s}{\rho_f}\phi \tag{8.107}$$

8.4.3 APPLICATION OF HAM ON MHD JEFFERY-HAMEL FLOW

For HAM solutions, we choose the initial guess and auxiliary linear operator in the following form:

$$f_0(\eta)=-\eta^2+1 \tag{8.108}$$

$$L(f)=f''' \tag{8.109}$$

$$L\left(\frac{1}{2}c_1\eta^2+c_2\eta+c_3\right)=0 \tag{8.110}$$

where c_i $(i=1,2,3)$ are constants. Let $P\in[0,1]$ denote the embedding parameter and $\hbar$ indicate non-zero auxiliary parameters. We then construct the following equations:

8.4.3.1 Zeroth-order deformation equations

$$(1-P)L[F(\eta;p)-f_0(\eta)]=p\hbar H(\eta)N[F(\eta;p)] \tag{8.111}$$

$$F(0;p)=1;\ F'(0;p)=0,\ F(1;p)=0 \tag{8.112}$$

$$\begin{aligned} N[F(y;p)]=&\frac{d^3F(\eta;p)}{d\eta^3}+\left(4-(1-\phi)^{1.25}H\right)\alpha^2\frac{dF(\eta;p)}{d\eta}\\ &+2\alpha ReA^*(1-\phi)^{2.5}F(\eta;p)\frac{dF(\eta;p)}{d\eta} \end{aligned} \tag{8.113}$$

For $p=0$ and $p=1$, we have

$$F(\eta;0)=f_0(\eta)\ \ F(\eta;1)=f(\eta) \tag{8.114}$$

When p increases from 0 to 1 then $F(\eta;p)$ varies from $f_0(\eta)$ to $f(\eta)$. By Taylor's theorem and using Equation (8.113), $F(\eta;p)$ can be expanded in a power series of p as follows:

$$F(\eta;p)=f_0(\eta)+\sum_{m-1}^{\infty}f_m(\eta)p^m,\qquad f_m(\eta)=\frac{1}{m!}\frac{\partial^m(F(\eta;p))}{\partial p^m}\bigg|_{p=0} \tag{8.115}$$

In which, $\hbar$ is chosen in such a way that this series is convergent at $p=1$; therefore, we have through Equation (8.115) that

$$f(\eta)=f_0(\eta)+\sum_{m-1}^{\infty}f_m(\eta), \tag{8.116}$$

8.4.3.2 *m*th-Order deformation equations

$$L[f_m(\eta)-\chi_m f_{m-1}(\eta)]=\hbar H(\eta)R_m(\eta) \tag{8.117}$$

$$F(0;p)=0;\ \ F'(0;p)=0,\ \ F(1;p)=0 \tag{8.118}$$

$$R_m(\eta)=f'''_{m-1}+\left(4-(1-\phi)^{1.25}H\right)\alpha^2 f'_{m-1}+\sum_{k=0}^{m-1}\left[2\alpha ReA^*(1-\phi)^{2.5}f_{m-1-k}f'_k\right] \tag{8.119}$$

Now, we determine the convergency of the result, the differential equation, and the auxiliary function according to the solution expression. So, let us assume $H(\eta)=1$.

We have found the answer by Maple analytic solution device. For first deformation, the solution are presented below:

$$\begin{aligned}f_1(\eta)=&-\frac{1}{15}\hbar\alpha\eta^6 ReA^*\sqrt{1-Q}Q+\frac{1}{30}\hbar\alpha\eta^6 ReA^*\sqrt{1-Q}Q^2+\frac{1}{30}\hbar\alpha\eta^6 ReA^*\sqrt{1-Q}\\&+\frac{1}{12}\hbar\alpha^2\eta^4(1-Q)^{1/4}H-\frac{1}{12}\hbar\alpha^2\eta^4(1-Q)^{1/4}HQ-\frac{1}{6}\hbar\alpha\eta^4 ReA^*\sqrt{1-Q}-\frac{1}{3}\hbar\alpha^2\eta^4\\&-\frac{1}{6}\hbar\alpha\eta^4 ReA^*\sqrt{1-Q}Q^2+\frac{1}{3}\hbar\alpha\eta^4 ReA^*\sqrt{1-Q}Q+\frac{1}{2}\left(-\frac{2}{15}\hbar\alpha ReA^*\sqrt{1-Q}Q\right.\\&+\frac{1}{15}\hbar\alpha ReA^*\sqrt{1-Q}Q^2+\frac{1}{15}\hbar\alpha ReA^*\sqrt{1-Q}-\frac{1}{24}(1-Q)^{1/4}\hbar\alpha^2 H+\frac{1}{6}\hbar\alpha^2\\&\left.+\frac{1}{24}(1-Q)^{1/4}\hbar\alpha^2 HQ\right)\eta^2\end{aligned} \tag{8.120}$$

The solutions $f(\eta)$ were too long to be mentioned here; therefore, they are shown graphically.

8.4.4 CONVERGENCE OF THE HAM SOLUTION

As pointed out by Liao, the convergence region and rate of solution series can be adjusted and controlled by means of the auxiliary parameter $\hbar$ (Liao, 2003, 2012). In general, by means of the so-called $\hbar$ curve, it is straightforward to choose an appropriate range for $\hbar$, which ensures the convergence of the solution series. To influence $\hbar$ on the convergence of solution, we plot the so-called $\hbar$ curve of $f''(0)$ by 7th-order approximation, as shown in Figures 8.38–8.42.

For $\alpha=5$, $Re=100$, $H=0$, and different values of ϕ, the ranges $-1.7<\hbar<0$; for $\alpha=-5$, $\phi=0.05$, $H=250$, and different values of Re, the ranges $-1.4<\hbar<0$; for $\alpha=5$, $\phi=0.05$, $H=250$, and different values of Re, the ranges $-1.4<\hbar<0$; for $\alpha=-5$, $\phi=0.05$, $Re=100$, and different values of H, the ranges $-1.4<\hbar<0.0$; and for $\alpha=5$, $\phi=0.05$, $Re=100$, and different values of H, the ranges $-1.8<\hbar<0.0$ give suitable value of $\hbar$ for convergency. Then, $\hbar=-0.9$ is a suitable value which is used for solution.

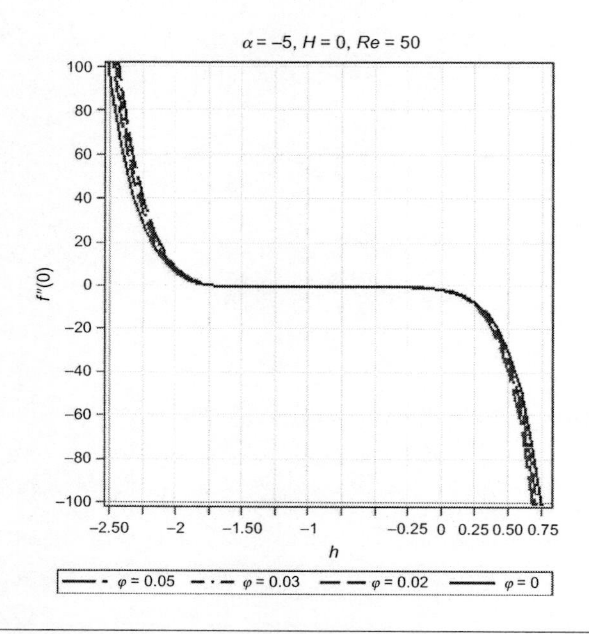

FIGURE 8.38

The $\hbar$ validity for $\alpha = -5$, $Re = 50$, $H = 0$, and different value of ϕ.

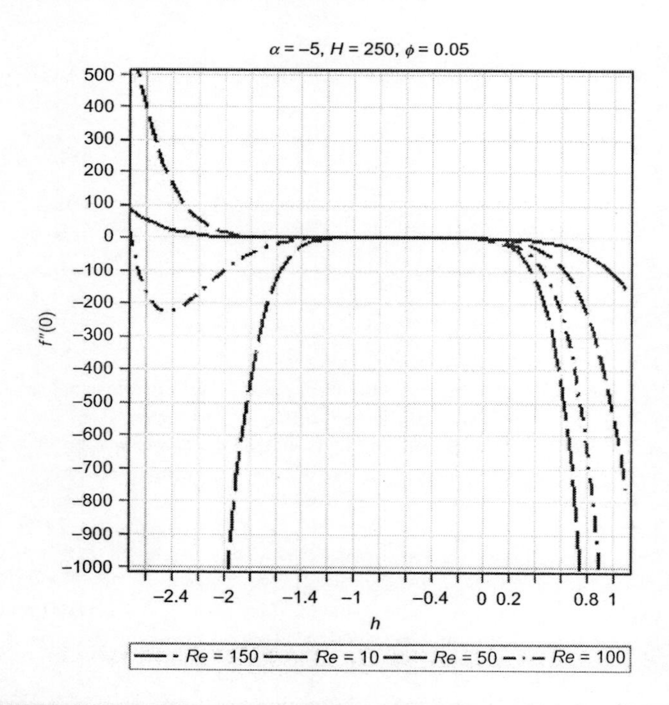

FIGURE 8.39

The $\hbar$ validity for $\alpha = -5$, $H = 250$, $\phi = 0.05$, and different value of Re.

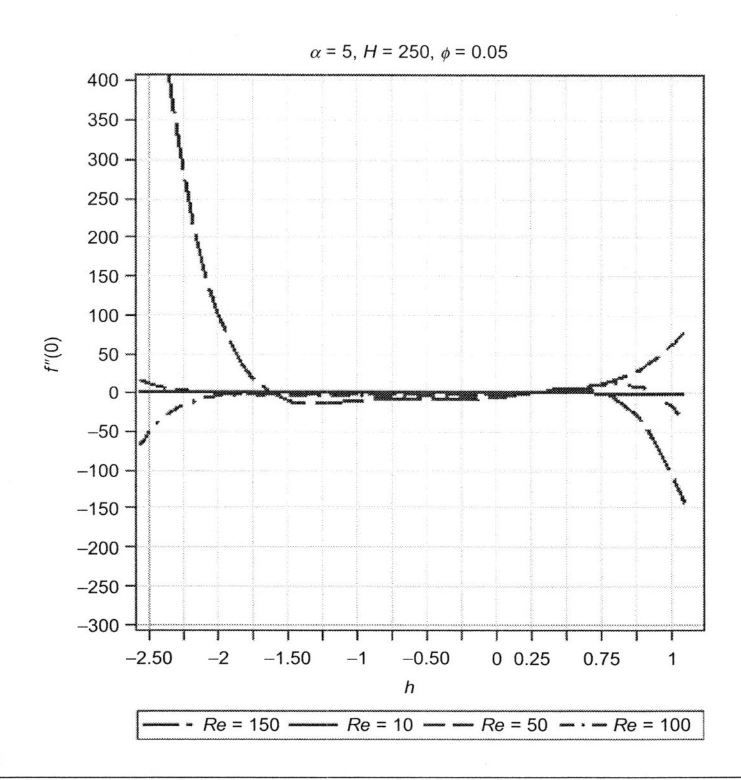

FIGURE 8.40

The $\hbar$ validity for $\alpha=5$, $H=250$, $\phi=0.05$, and different value of *Re*.

8.4.5 RESULTS AND DISCUSSIONS

In this study, the accuracy and validity of this approximate solution on MHD Jeffery-Hamel problem with nanoparticles for various values of Hartmann number, Reynolds number, and particle parameter has been investigated. The results of HAM and numerical solution are compared in Figures 8.43 and 8.48.

Figures 8.43 and 8.44 show the comparison of the nondimensional velocity $f(\eta)$ and $f'(\eta)$ for known values of the parameters $\alpha=\pm5$, $Re=100$, $\phi=0.05$, $\hbar=-0.9$, and different value of H.

Figure 8.44 shows comparison between numerical solution and HAM solution for $f(\eta)$ and $f'(\eta)$ when $\alpha=+5$, $H=250$, $Re=120$, $\hbar=-0.9$, and different value of ϕ, respectively. Figures 8.45 and 8.46 show comparison between the numerical solution and HAM solution for $f(\eta)$ and $f'(\eta)$, when $\alpha=\pm5$, $H=0$, $\phi=0.05$, $\hbar=-0.9$, and different value of Re.

By drawing the 2-D Figures 8.42–8.47, of numerical solution and HAM solution for $f(\eta)$ and $f'(\eta)$ with different values of α,Re, H, and ϕ. As can be seen in graphical results, the obtained analytical solution in comparison with the numerical ones represents a remarkable accuracy.

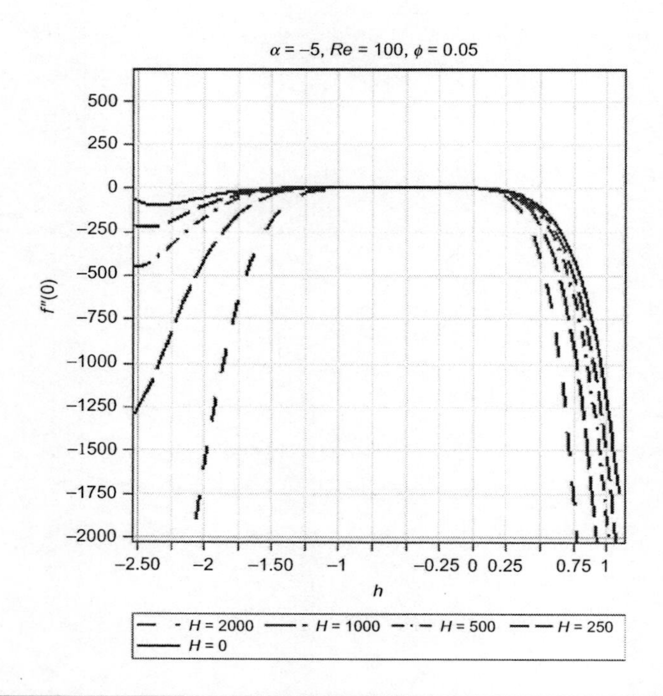

FIGURE 8.41

The $\hbar$ validity for $\alpha=-5$, $Re=100$, $\phi=0.05$, and different value of H.

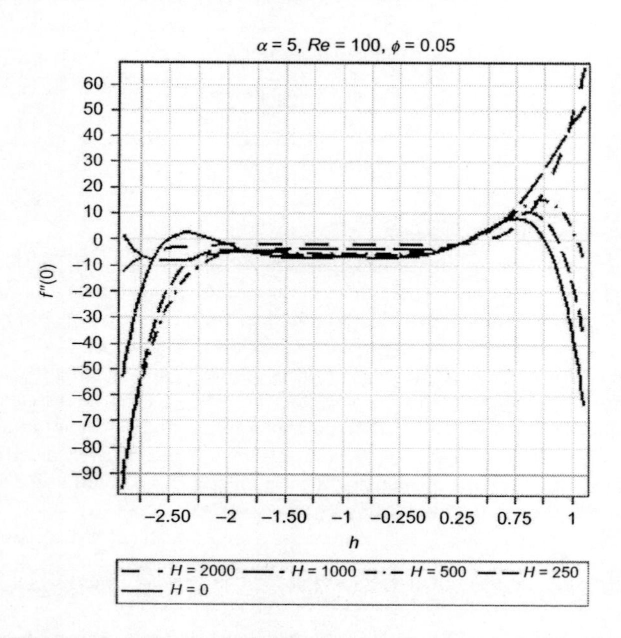

FIGURE 8.42

The $\hbar$ validity for $\alpha=5$, $Re=100$, $\phi=0.05$, and different value of H.

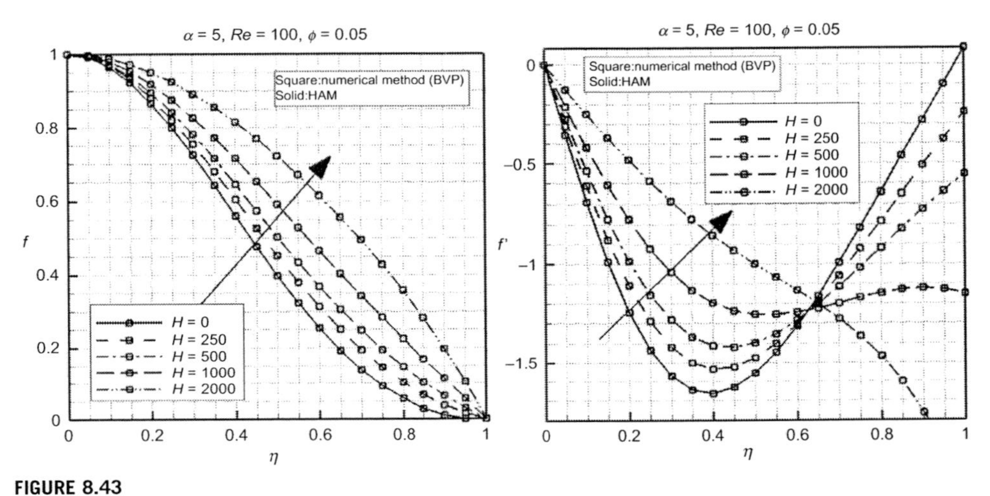

FIGURE 8.43

The comparison between the numerical and HAM solution for $f(\eta)$ and $f'(\eta)$ when $\alpha=5$, $Re=100$, $\phi=0.05$, and $H=0, 250, 500, 1000, 2000$.

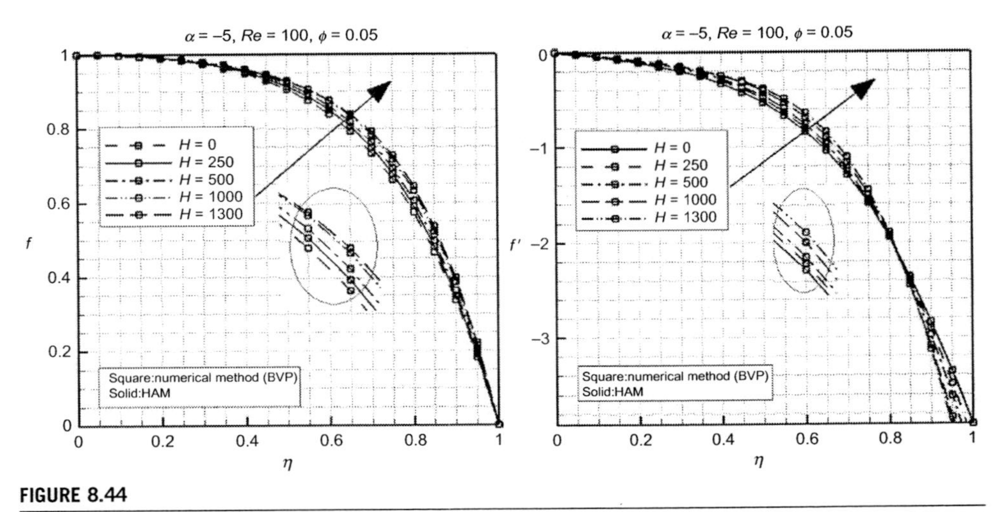

FIGURE 8.44

The comparison between the numerical and HAM solution for $f(\eta)$ and $f'(\eta)$ when $\alpha=5$, $Re=100$, $\phi=0.05$, and $H=0, 250, 500, 1000, 1300$.

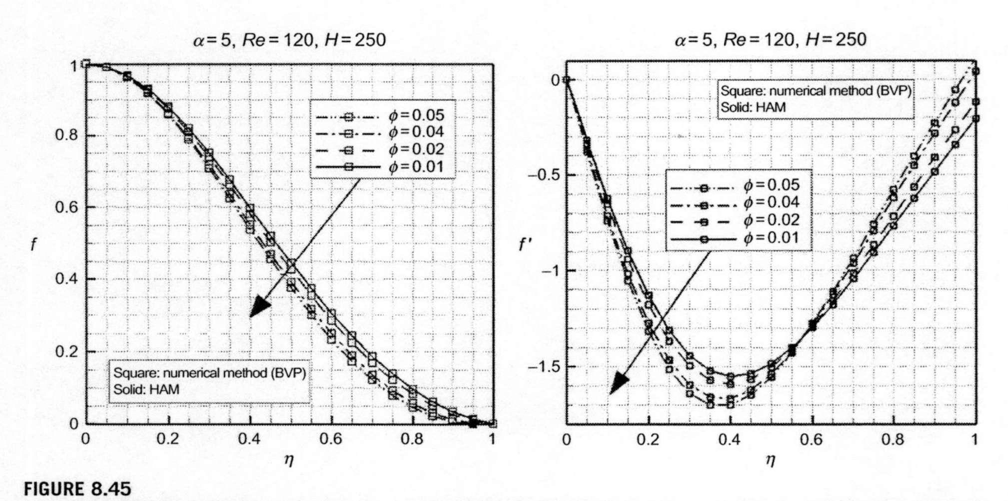

FIGURE 8.45

The comparison between the numerical and HAM solution for $f(\eta)$ and $f'(\eta)$ when $\alpha=5$, $H=250$, $Re=120$, and $\phi=0.01, 0.02, 0.04, 0.05$.

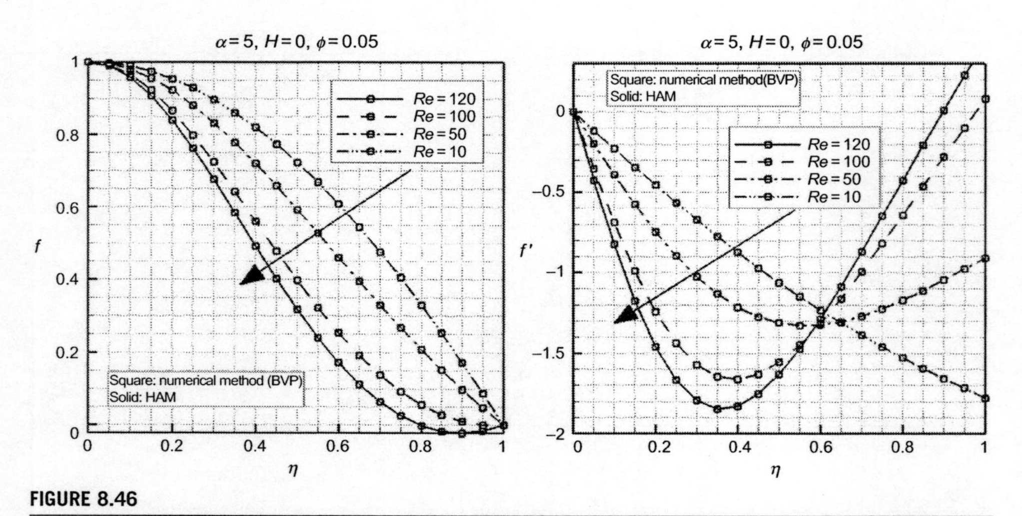

FIGURE 8.46

The comparison between the numerical and HAM solution for $f(\eta)$ and $f'(\eta)$ when $\alpha=5$, $H=0$, $\phi=0.05$, and $Re=10, 50, 100, 120$.

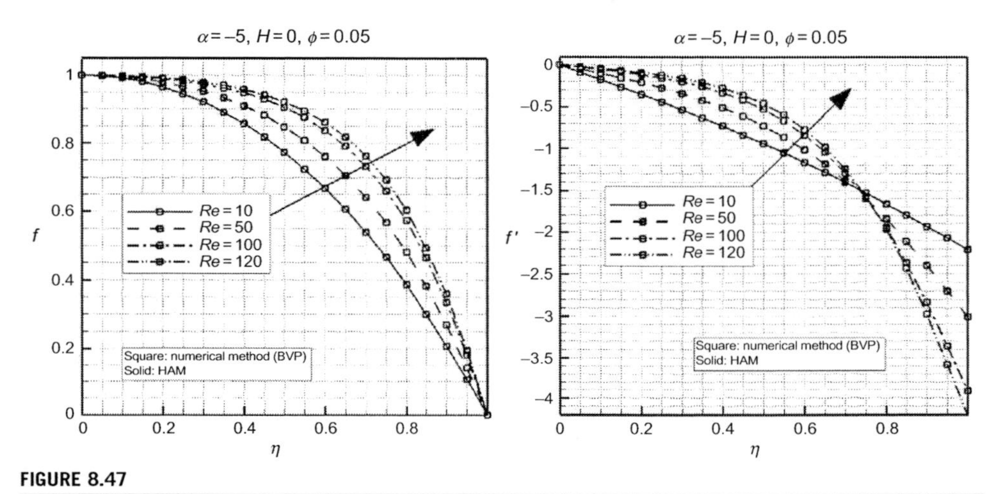

FIGURE 8.47

The comparison between the numerical and HAM solution for $f(\eta)$ and $f'(\eta)$ when $\alpha = -5$, $H = 250$, and $Re = 10, 50, 100, 150, 200$.

8.4.6 CONCLUSION

In this section, we implemented the HAM for finding solutions of MHD Jeffery-Hamel problem with nanoparticles. The governing equations, continuity, and momentum for this problem are reduced to an ordinary single third form by using a similarity transformation. Furthermore, the obtained solutions by HAM are compared with numerical solution. The good agreement between numerical solution and analytical results proves the accuracy and validity of method.

8.5 INVESTIGATION OF MHD NANOFLUID FLOW IN A SEMIPOROUS CHANNEL

In this section, LSM and GM are used to solve the problem of laminar nanofluid flow in a semiporous channel in the presence of transverse magnetic field. It has been attempted to show the capabilities and wide-range applications of the GM in comparison with the CM in this problem. The effective thermal conductivity and viscosity of nanofluid are calculated by the MG and Brinkman models, respectively. The influence of the three dimensionless numbers, the nanofluid volume friction, Hartmann number, and Reynolds number, on nondimensional velocity profile is considered. The results show that velocity boundary layer thickness decreases with increase of Reynolds number and nanoparticle volume friction and it increases as Hartmann number increases (*this section has been worked by M. Sheikholeslami, M. Hatami, D.D. Ganji in nonlinear dynamics team in Mechanical Engineering Department, 2012-2013*).

8.5.1 INTRODUCTION

There are some simple and accurate approximation techniques for solving differential equations called the WRMs. Collocation, Galerkin, and least square are examples of the WRMs. Stern and Rasmussen (1996) used CM for solving a third-order linear differential equation. Vaferi et al. (2012) have studied the feasibility of applying orthogonal collocation method to solve diffusivity equation in the radial transient flow system. Hendi and Albugami (2010) used CM and GM for solving Fredholm-Volterra integral equation. Recently, LSM is introduced by Bouaziz and Aziz (2010) and is applied for predicting the performance of a longitudinal fin. They found that LSM is simple compared with other analytical methods. Shaoqin and Huoyuan (2008) developed and analyzed least-squares approximations for the incompressible MHD equations.

The flow problem in porous tubes or channels has been under considerable attention in recent years because of its various applications in biomedical engineering, for example, in the dialysis of blood in artificial kidney, in the flow of blood in the capillaries, in the flow in blood oxygenators, as well as in many other engineering areas such as the design of filters, in transpiration cooling boundary layer control (Andoh and Lips, 2003), and gaseous diffusion (Runstedtler, 2006). Berman (1953) described an exact solution for the Navier-Stokes equation for steady two-dimensional laminar flow of a viscous, incompressible fluid in a channel with parallel, rigid, porous walls driven by uniform, steady suction, or injection at the walls. This mass transfer is paramount in some industrial processes. More recently, Chandran et al. (1996) analyzed the effects of a magnetic field on the thermodynamic flow past a continuously moving porous plate.

The main aim of this chapter is to investigate the problem of laminar nanofluid flow in a semiporous channel in the presence of transverse magnetic field using LSM and GM. The effects of the nanofluid volume friction, Hartmann number, and Reynolds number on velocity profile are considered. Furthermore, velocity profiles for different structures of nanofluid (copper and silver nanoparticles in water or ethylene glycol) are depicted.

8.5.2 PROBLEM DESCRIPTION

Consider the laminar two-dimensional stationary flow of an electrically conducting incompressible viscous fluid in a semiporous channel made by a long rectangular plate with length L_x in uniform translation in x^* direction and an infinite porous plate.

The distance between the two plates is h. We observe a normal velocity q on the porous wall. A uniform magnetic field B is assumed to be applied toward direction y^* (Figure 8.48).

In the case of a short circuit to neglect the electrical field and perturbations to the basic normal field and without any gravity forces, the governing equations are:

$$\frac{\partial u^*}{\partial x^*}+\frac{\partial v^*}{\partial y^*}=0 \tag{8.121}$$

$$u^*\frac{\partial u^*}{\partial x^*}+v^*\frac{\partial u^*}{\partial y^*}=-\frac{1}{\rho_{\text{nf}}}\frac{\partial P^*}{\partial x^*}+\frac{\mu_{\text{nf}}}{\rho_{\text{nf}}}\left(\frac{\partial^2 u^*}{\partial x^{*2}}+\frac{\partial^2 u^*}{\partial y^{*2}}\right)-u^*\frac{\sigma_{\text{nf}}B^2}{\rho_{\text{nf}}} \tag{8.122}$$

$$u^*\frac{\partial v^*}{\partial x^*}+v^*\frac{\partial v^*}{\partial y^*}=-\frac{1}{\rho_{\text{nf}}}\frac{\partial P^*}{\partial y^*}+\frac{\mu_{\text{nf}}}{\rho_{\text{nf}}}\left(\frac{\partial^2 v^*}{\partial x^{*2}}+\frac{\partial^2 v^*}{\partial y^{*2}}\right) \tag{8.123}$$

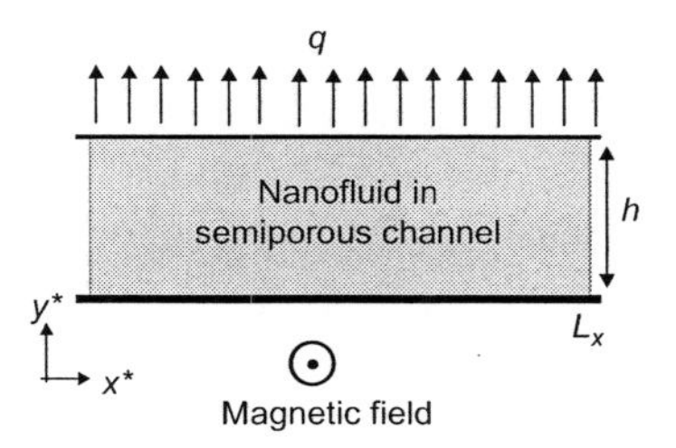

FIGURE 8.48

Schematic of the problem (nanofluid in a porous media between parallel plates and magnetic field).

The appropriate boundary conditions for the velocity are

$$y^* = 0 : u^* = u_0^*, \quad v^* = 0 \tag{8.124}$$

$$y^* = h : u^* = 0, \quad v^* = -q \tag{8.125}$$

Calculating a mean velocity U by the relation

$$y^* = 0 : u^* = u_0^*, \quad v^* = 0 \tag{8.126}$$

we consider the following transformations

$$x = \frac{x^*}{L_x}; \quad y = \frac{y^*}{h} \tag{8.127}$$

$$u = \frac{u^*}{U}; \quad v = \frac{v^*}{q}, \quad P_y = \frac{P^*}{\rho_f \cdot q^2} \tag{8.128}$$

Then, we can consider two dimensionless numbers: the Hartman number Ha for the description of magnetic forces (Desseaux, 1999) and the Reynolds number Re for dynamic forces

$$Ha = Bh\sqrt{\frac{\sigma_f}{\rho_f \cdot \upsilon_f}} \tag{8.129}$$

$$Re = \frac{hq}{\mu_{nf}}\rho_{nf} \tag{8.130}$$

where the effective density(ρ_{nf}) is defined as

$$\rho_{nf} = \rho_f(1-\phi) + \rho_s\phi \tag{8.131}$$

where ϕ is the solid volume fraction of nanoparticles. The dynamic viscosity of the nanofluids given by Brinkman (1952) is

$$\mu_{nf} = \frac{\mu_f}{(1-\phi)^{2.5}} \tag{8.132}$$

The effective thermal conductivity of the nanofluid can be approximated by the MG model as (see Maxwell, 1904)

$$\frac{k_{\mathrm{nf}}}{k_{\mathrm{f}}}=\frac{k_{\mathrm{s}}+2k_{\mathrm{f}}-2\phi(k_{\mathrm{f}}-k_{\mathrm{s}})}{k_{\mathrm{s}}+2k_{\mathrm{f}}+\phi(k_{\mathrm{f}}-k_{\mathrm{s}})} \tag{8.133}$$

The effective electrical conductivity of nanofluid was presented by Maxwell as below:

$$\frac{\sigma_{\mathrm{nf}}}{\sigma_{\mathrm{f}}}=1+\frac{3\left(\frac{\sigma_{\mathrm{s}}}{\sigma_{\mathrm{f}}}-1\right)\phi}{\left(\frac{\sigma_{\mathrm{s}}}{\sigma_{\mathrm{f}}}+2\right)-\left(\frac{\sigma_{\mathrm{s}}}{\sigma_{\mathrm{f}}}-1\right)\phi} \tag{8.134}$$

The thermophysical properties of the nanofluid are given in Table 8.8.

Introducing Equations (8.126) and (8.130) into Equations (8.121) and (8.123) leads to the dimensionless equations:

$$\frac{\partial u}{\partial x}+\frac{\partial v}{\partial y}=0 \tag{8.135}$$

$$u\frac{\partial u}{\partial x}+v\frac{\partial u}{\partial y}=-\varepsilon^2\frac{\partial P_y}{\partial x}+\frac{\mu_{\mathrm{nf}}}{\rho_{\mathrm{nf}}}\frac{1}{hq}\left(\varepsilon^2\frac{\partial^2 u}{\partial x^2}+\frac{\partial^2 u}{\partial y^2}\right)-u\frac{Ha^2 B^*}{Re\ A^*} \tag{8.136}$$

$$u\frac{\partial v}{\partial x}+v\frac{\partial v}{\partial y}=-\frac{\partial P_y}{\partial x}+\frac{\mu_{\mathrm{nf}}}{\rho_{\mathrm{nf}}}\frac{1}{hq}\left(\varepsilon^2\frac{\partial^2 v}{\partial x^2}+\frac{\partial^2 v}{\partial y^2}\right) \tag{8.137}$$

where A^* and B^* are constant parameters.

$$A^*=(1-\phi)+\frac{\rho_{\mathrm{s}}}{\rho_{\mathrm{f}}}\phi,\quad B^*=1+\frac{3\left(\frac{\sigma_{\mathrm{s}}}{\sigma_{\mathrm{f}}}-1\right)\phi}{\left(\frac{\sigma_{\mathrm{s}}}{\sigma_{\mathrm{f}}}+2\right)-\left(\frac{\sigma_{\mathrm{s}}}{\sigma_{\mathrm{f}}}-1\right)\phi} \tag{8.138}$$

Quantity of ε is defined as the aspect ratio between distance h and the characteristic length L_x of the slider. This ratio is normally small. Berman's similarity transformation is used to be free from the aspect ratio of ε.

Table 8.8 Thermo Physical Properties of Nanofluids and Nanoparticles

Material	Density (kg/m^3)	Electrical Conductivity, σ ((Ω m)$^{-1}$)
Silver	10,500	6.30×10^7
Copper	8933	5.96×10^7
Ethylene glycol	1113.2	1.07×10^{-4}
Drinking water	997.1	0.05

$$v = -V(y);\; u = \frac{u^*}{U} = u_0 U(y) + x\frac{\mathrm{d}V}{\mathrm{d}y} \tag{8.139}$$

Introducing Equation (8.139) in the second momentum equation (8.137) shows that quantity $\partial P_y/\partial y$ does not depend on the longitudinal variable x. With the first momentum equation, we also observe that $\partial^2 P_y/\partial x^2$ is independent of x. We omit asterisks for simplicity. Then a separation of variables leads to (Desseaux, 1999)

$$V'^2 - VV'' - \frac{1}{Re}\frac{1}{A^*(1-\phi)^{2.5}}V''' + \frac{Ha^2 B^*}{Re\ A^*}V' = \varepsilon^2\frac{\partial^2 P_y}{\partial x^2} = \varepsilon^2\frac{1}{x}\frac{\partial P_y}{\partial x} \tag{8.140}$$

$$UV' - VU' = \frac{1}{Re}\frac{1}{A^*(1-\phi)^{2.5}}\left[U'' - Ha^2 B^*(1-\phi)^{2.5}U\right] \tag{8.141}$$

The right-hand side of Equation (8.140) is constant. So, we derive this equation with respect to x. This gives

$$V'''' = Ha^2 B^*(1-\phi)^{2.5}V'' + ReA^*(1-\phi)^{2.5}[V'V'' - VV'''] \tag{8.142}$$

where primes denote differentiation with respect to y and asterisks have been omitted for simplicity. The dynamic boundary conditions are

$$\begin{cases} y=0: U=1;\; V=0;\; V'=0 \\ y=1: U=0;\; V=1;\; V'=0 \end{cases} \tag{8.143}$$

8.5.3 WEIGHTED RESIDUAL METHODS

There existed an approximation technique for solving differential equations called the WRMs. Suppose a differential operator D is acted on a function u to produce a function p

$$D(u(x)) = p(x) \tag{8.144}$$

it is considered that u is approximated by a function $\tilde{u}$, which is a linear combination of basic functions chosen from a linearly independent set. That is,

$$u \cong \tilde{u} = \sum_{i=1}^{n} c_i \varphi_i \tag{8.145}$$

Now, when substituted into the differential operator, D, the result of the operations generally is not $p(x)$. Hence, an error or residual will exist

$$R(x) = D(\tilde{u}(x)) - p(x) \neq 0 \tag{8.146}$$

The notion in WRMs is to force the residual to zero in some average sense over the domain. That is

$$\int_X R(x)\,W_i(x) = 0 \quad i = 1,2,\ldots,n \tag{8.147}$$

where the number of weight functions W_i is exactly equal to the number of unknown constants c_i in $\tilde{u}$. The result is a set of n algebraic equations for the unknown constants c_i. Three methods of WRMs are explained in the following subsections.

8.5.3.1 Least square method

8.5.3.1.1 Mathematical formulation

If the continuous summation of all the squared residuals is minimized, the rationale behind the name can be seen. In other words, a minimum of

$$S = \int_X R(x)R(x)\,\mathrm{d}x = \int_X R^2(x)\,\mathrm{d}x \tag{8.148}$$

In order to achieve a minimum of this scalar function, the derivatives of S with respect to all the unknown parameters must be zero. That is,

$$\frac{\partial S}{\partial c_i} = 2\int_X R(x)\frac{\partial R}{\partial c_i}\,\mathrm{d}x = 0 \tag{8.149}$$

Comparing with Equation (8.147), the weight functions are seen to be

$$W_i = 2\frac{\partial R}{\partial c_i} \tag{8.150}$$

However, the "2" coefficient can be dropped, since it cancels out in the equation. Therefore, the weight functions for the LSM are just the derivatives of the residual with respect to the unknown constants

$$W_i = \frac{\partial R}{\partial c_i} \tag{8.151}$$

8.5.3.1.2 Application

Because trial functions must satisfy the boundary conditions in Equation (8.143), they will be considered as,

$$\begin{cases} U(y) = 1 - y + c_1(y - y^2) + c_2(y - y^3) \\ V(y) = c_3\left(\dfrac{y^2}{2} - \dfrac{y^3}{3}\right) + c_4\left(\dfrac{y^2}{2} - \dfrac{y^4}{4}\right) + c_5\left(\dfrac{y^2}{2} - \dfrac{y^5}{5}\right) \end{cases} \tag{8.152}$$

In this problem, we have two coupled equations (Equations (8.141) and (8.142)); so, two residual functions will appear as

$$\begin{cases} R_1(c_1,c_2,c_3,c_4,c_5,y)=(1-y+c_1(y-y^2)+c_2(y-y^3))(c_3(y-y^2)+c_4(y-y^3)+c_5(y-y^4)) \\ -\left(c_3\left(\frac{y^2}{2}-\frac{y^3}{3}\right)+c_4\left(\frac{y^2}{2}-\frac{y^4}{4}\right)+c_5\left(\frac{y^2}{2}-\frac{y^5}{5}\right)\right)(-1+c_1(1-2y)+c_2(1-3y^2)-2c_1-6c_2y) \\ \dfrac{-Ha^2\left(1+\dfrac{3\left(\frac{\sigma_s}{\sigma_f}-1\right)\phi}{\left(\frac{\sigma_s}{\sigma_f}+2\right)-\left(\frac{\sigma_s}{\sigma_f}-1\right)\phi}\right)(1-\phi)^{2.5}(1-y+c_1(y-y^2)+c_2(y-y^3))}{Re\left(1-\phi+\frac{\rho_s\phi}{\rho_f}\right)(1-\phi)^{2.5}} \\ R_2(c_1,c_2,c_3,c_4,c_5,y)=-6c_4-24c_5y-Ha^2\left(1+\dfrac{3\left(\frac{\sigma_s}{\sigma_f}-1\right)\phi}{\left(\frac{\sigma_s}{\sigma_f}+2\right)-\left(\frac{\sigma_s}{\sigma_f}-1\right)\phi}\right)(1-\phi)^{2.5}(c_3(1-2y) \\ +c_4(1-3y^2)+c_5(1-4y^3))+\mathrm{Re}\left(1-\phi+\frac{\rho_s\phi}{\rho_f}\right)(1-\phi)^{2.5}(c_3(y-y^2)+c_4(y-y^3)+c_5(y-y^4))(c_3(1-2y) \\ +c_4(1-3y^2)+c_5(1-4y^3))-\left(c_3\left(\frac{y^2}{2}-\frac{y^3}{3}\right)+c_4\left(\frac{y^2}{2}-\frac{y^4}{4}\right)+c_5\left(\frac{y^2}{2}-\frac{y^5}{5}\right)\right)(-2c_3-6c_4y-12c_5y^2) \end{cases} \tag{8.153}$$

By substituting the residual functions, $R_1(c_1,c_2,c_3,c_4,c_5y)$ and $R_2(c_1,c_2,c_3,c_4,c_5y)$, into Equation (8.149), a set of equation with five equations will appear, and by solving this system of equations, coefficients c_1-c_5 will be determined. For example, Using LSM for a water-copper nanofluid with $Re=0.5$, $Ha=0.5$, and $\phi=0.05$. $U(y)$ and $V(y)$ are as follows:

$$\begin{cases} U(y)=1-1.334953917y+0.3461783819y^2-0.01122446534y^3 \\ V(y)=1.8703229y^2+3.1584693y^3-6.9279074y^4+2.8991152y^5 \end{cases} \tag{8.154}$$

8.5.3.2 Galerkin method

8.5.3.2.1 Mathematical formulation

This method may be viewed as a modification of the LSM. Rather than using the derivative of the residual with respect to the unknown c_i, the derivative of the approximating function or trial function is used. In this method, weight functions are

$$W_i=\frac{\partial \tilde{u}}{\partial c_i} \quad i=1,2,\ldots,n \tag{8.155}$$

8.5.3.2.2 Application

Now, we apply GM for solving the $U(y)$ and $V(y)$ functions as nondimensional velocities equation for nanofluid flow. First, as already described, consider the trial functions as Equation (8.152) which satisfies described boundary condition in Equation (8.143). Using Equation (8.155), weight functions will be obtained as

$$W_1 = y - y^2,\ W_2 = y - y^3,\ W_3 = \frac{y^2}{2} - \frac{y^3}{3},\ W_4 = \frac{y^2}{2} - \frac{y^4}{4} \tag{8.156}$$

Applying Equation (8.147), a set of algebraic equations can be defined and solving this set of equations, c_1-c_5 coefficients and finally $U(y)$ and $V(y)$ functions for a for water-copper nanofluid with $Re=0.5$, $Ha=0.5$, and $\phi=0.05$ will be calculated as following:

$$\begin{cases} U(y) = 1 - 1.362453681y + 0.3918602208y^2 - 0.02940653969y^3 \\ V(y) = 2.88352y^2 - 1.8470776y^3 + 0.043577119y^4 - 0.0800255y^5 \end{cases} \tag{8.157}$$

8.5.4 RESULTS AND DISCUSSIONS

In this chapter, LSM and GM methods are applied to obtain an explicit analytic solution of the laminar nanofluid flow in a semiporous channel in the presence of uniform magnetic field (Figure 8.48). First, a comparison between the applied methods, LSM and GM, is investigated. For this aim, Equations (8.141) and (8.142) are solved for different nanofluid structures (see Table 8.8) and comparison between described methods is demonstrated in Figure 8.49.

All these figures confirm that GM is an accurate and convenient method for solving these kinds of problems. The values for one case from Figure 8.49 are presented in Tables 8.9 and 8.10 for $U(y)$ and $V(y)$, respectively. It can be concluded from the tables that GM has better compatibility with numerical solution than LSM.

Figure 8.50 shows the effect of nanoparticle volume fraction on $U(y)$ and $V(y)$ for water with copper nanoparticles when $Re=1$ and $Ha=1$. Velocity boundary layer thickness decreases with increase of nanoparticle volume fraction.

Effect of Hartman number (Ha) on dimensionless velocities for water with copper is shown in Figures 8.51 and 8.52.

Generally, when the magnetic field is imposed on the enclosure, the velocity field is suppressed owing to the retarding effect of the Lorenz force. For low Reynolds number, as Hartmann number increases, $V(y)$ decreases for $y > y_m$ but opposite trend is observed for $y < y_m$, y_m is a meeting point that all curves joint together at this point. When Reynolds number increases, this meeting point shifts to the solid wall and it can be seen that $V(y)$ decreases with increase of Hartmann number. Effect of Reynolds number (Re) on dimensionless velocities is shown in Figures 8.53 and 8.54.

It is worth to mention that the Reynolds number indicates the relative significance of the inertia effect compared to the viscous effect. Thus, velocity profile decreases as Re increases and, in turn, increasing Re leads to increase in the magnitude of the skin-friction coefficient.

The effects of the nanoparticle and liquid phase material on velocity's profiles are shown in Tables 8.11 and 8.12 for $U(y)$ and $V(y)$, respectively. These tables reveal that when nanofluid includes copper (as nanoparticles) or ethylene glycol (as fluid phase) in its structure, the $U(y)$ and $V(y)$ values are greater than the other structures.

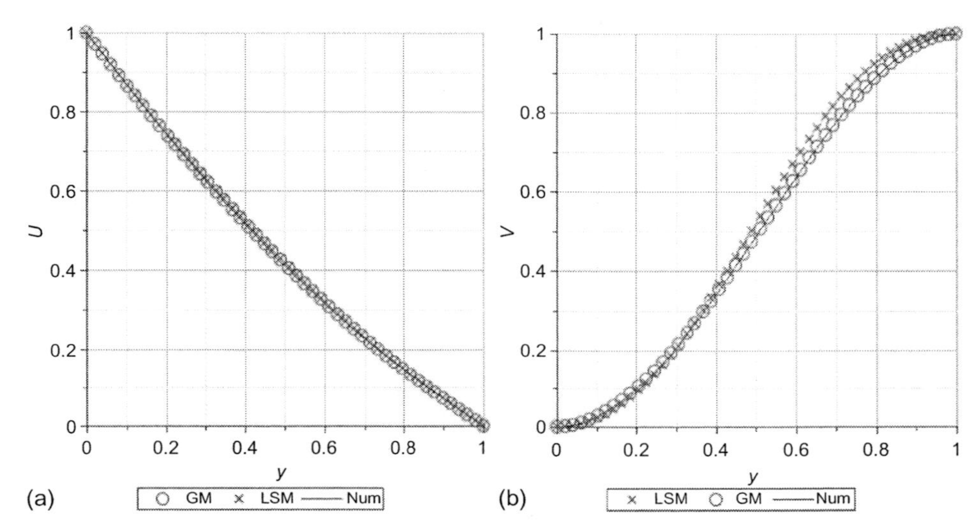

Re = 0.5, *Ha* = 0.5, and $\phi = 0.05$ nanofluid's (water with copper nanoparticles)

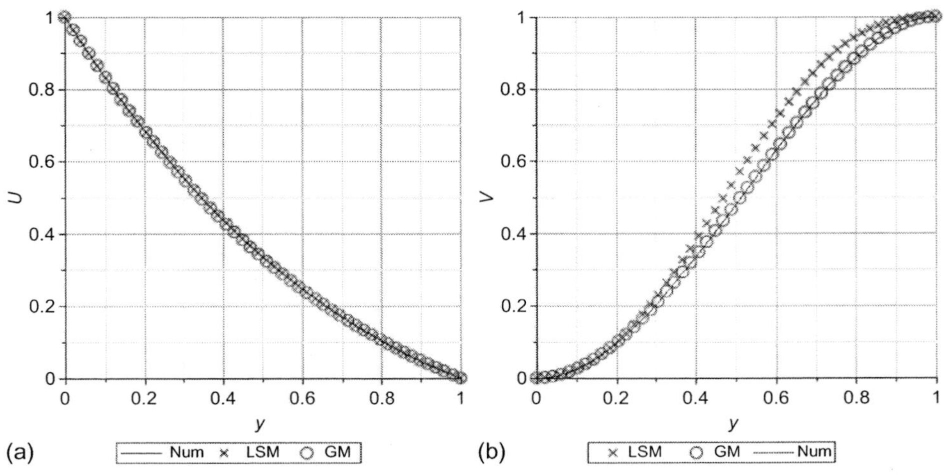

Re = 1, *Ha* = 1, and $\phi = 0.04$ (ethylene glycol with copper nanoparticles)

FIGURE 8.49

Comparison of LSM, GM, and numerical results for dimensionless velocities (a) $U(y)$ and (b) $V(y)$.

Continued

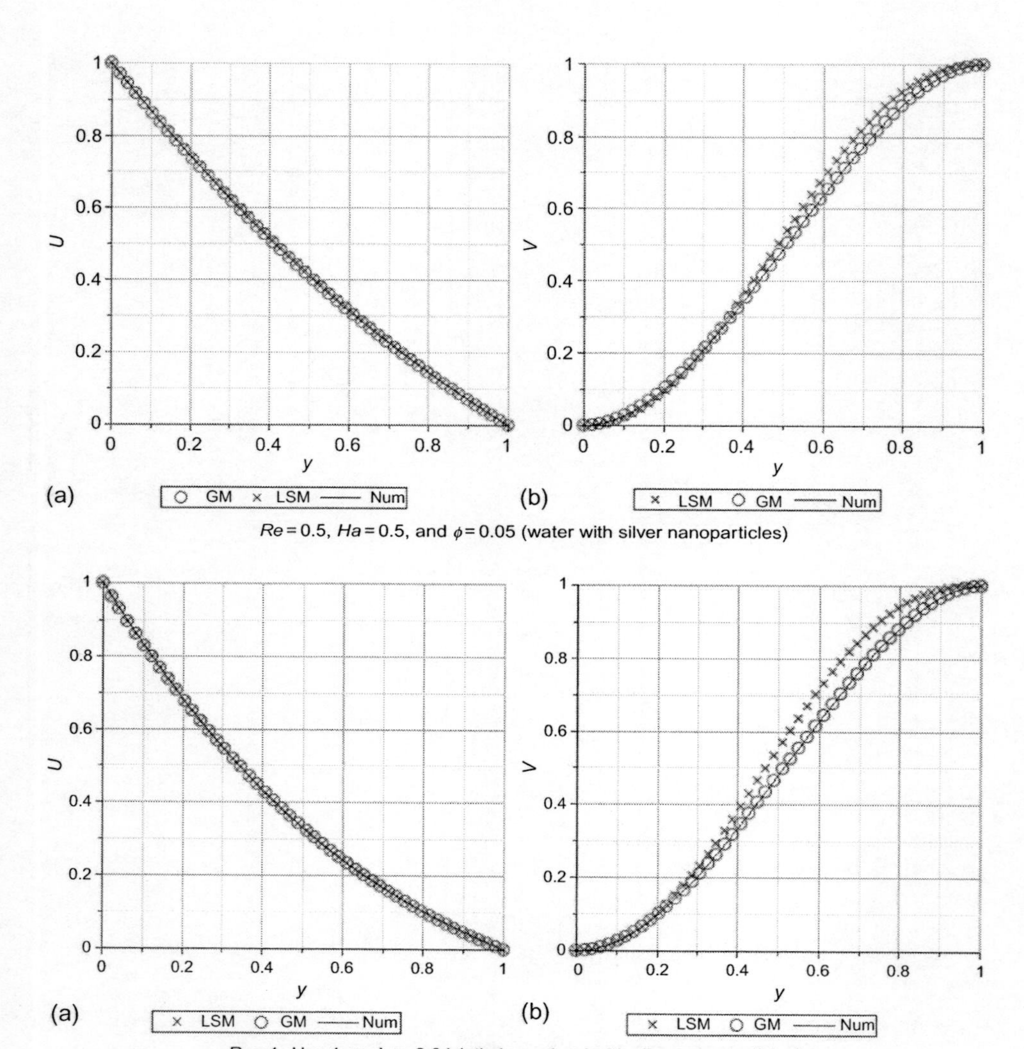

FIGURE 8.49, cont'd

Table 8.9 Comparison Between $U(y)$ Results from Applied Methods for $Re = 0.5$, $Ha = 0.5$, and $\phi = 0.05$

y	Numerical	LSM	GM	% Error LSM	% Error GM
0.0	1.0	1.0	1.0	0.0	0.0
0.1	0.86936870	0.86995516	0.8684172	6.75E−06	1.09E−05
0.2	0.74386446	0.74676655	0.7443189	3.9E−05	6.1E−06
0.3	0.62520345	0.63036681	0.6275303	8.26E−05	3.7E−05
0.4	0.51430050	0.52068860	0.5178766	0.000124	7E−05
0.5	0.41141966	0.41766457	0.4151830	0.000152	9.1E−05
0.6	0.31632490	0.32122738	0.3192748	0.000155	9.3E−05
0.7	0.22843209	0.23130967	0.2299772	0.000126	6.8E−05
0.8	0.14695892	0.14784410	0.1471153	6.02E−05	1.1E−05
0.9	0.071068179	0.07076332	0.0705145	4.3E−05	7.79E−05
1.0	0.0	0.0	0.0	0.0	0.0

Table 8.10 Comparison Between $V(y)$ Results from Applied Methods for $Re = 0.5$, $Ha = 0.5$, and $\phi = 0.05$

y	Numerical	LSM	GM	% Error LSM	% Error GM
0.0	0.0	0.0	0.0	0.0	0.0
0.1	0.02691402	0.02119789	0.02702730	0.00212	4.2E−05
0.2	0.10041111	0.08992373	0.10072712	0.00104	3.1E−05
0.3	0.20956921	0.20453653	0.21001968	0.00024	2.1E−05
0.4	0.34329260	0.35372621	0.34374192	0.000304	1.3E−05
0.5	0.49022237	0.51999253	0.49055433	0.000607	6.8E−06
0.6	0.63867700	0.68312401	0.63884782	0.000696	2.7E−06
0.7	0.77660672	0.82367692	0.77665058	0.000606	5.6E−07
0.8	0.89154404	0.92645414	0.89153498	0.000392	1.02E−07
0.9	0.97053097	0.98398416	0.97052437	0.000139	6.8E−08
1.0	1.0	1.0	1.0	0.0	0.0

8.5.6 CONCLUSION

In this section, LSM and GM are used to solve the problem of laminar nanofluid flow in a semiporous channel in the presence of uniform magnetic field. The comparison between these two methods revealed that GM is a more accurate and powerful approach than LS for solving this problem. The results indicate that velocity boundary layer thickness decreases with increase of Reynolds number and nanoparticle volume friction and it increases as Hartmann number increases. Choosing copper (as nanoparticles) or ethylene glycol (as fluid phase) leads to maximum increment in velocity.

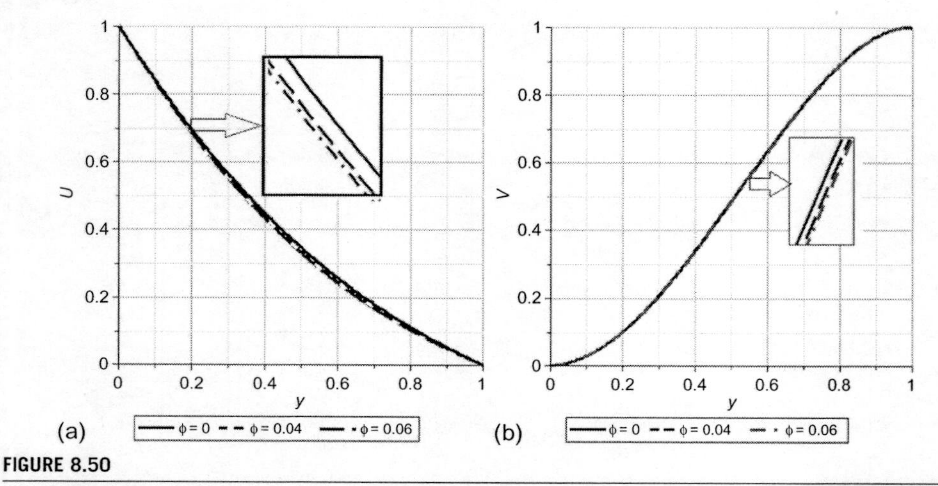

FIGURE 8.50

Effect of nanoparticle volume fraction, ϕ, on (a) $U(y)$ and (b) $V(y)$, for water with copper nanoparticles when $Re=1$ and $Ha=1$.

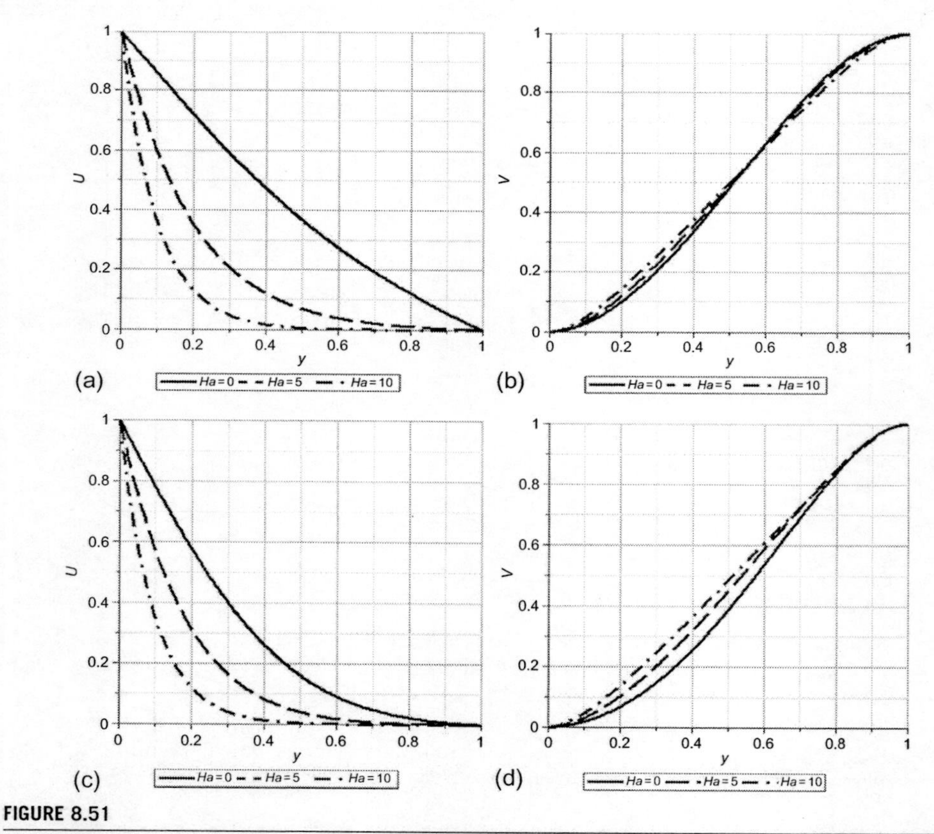

FIGURE 8.51

Effect of Hartman number (Ha) on dimensionless velocities for water with copper nanoparticles, $\phi=0.04$, (a) $U(y)$, $Re=1$, (b) $V(y)$, $Re=1$, (c) $U(y)$, $Re=5$, and (d) $V(y)$, $Re=5$.

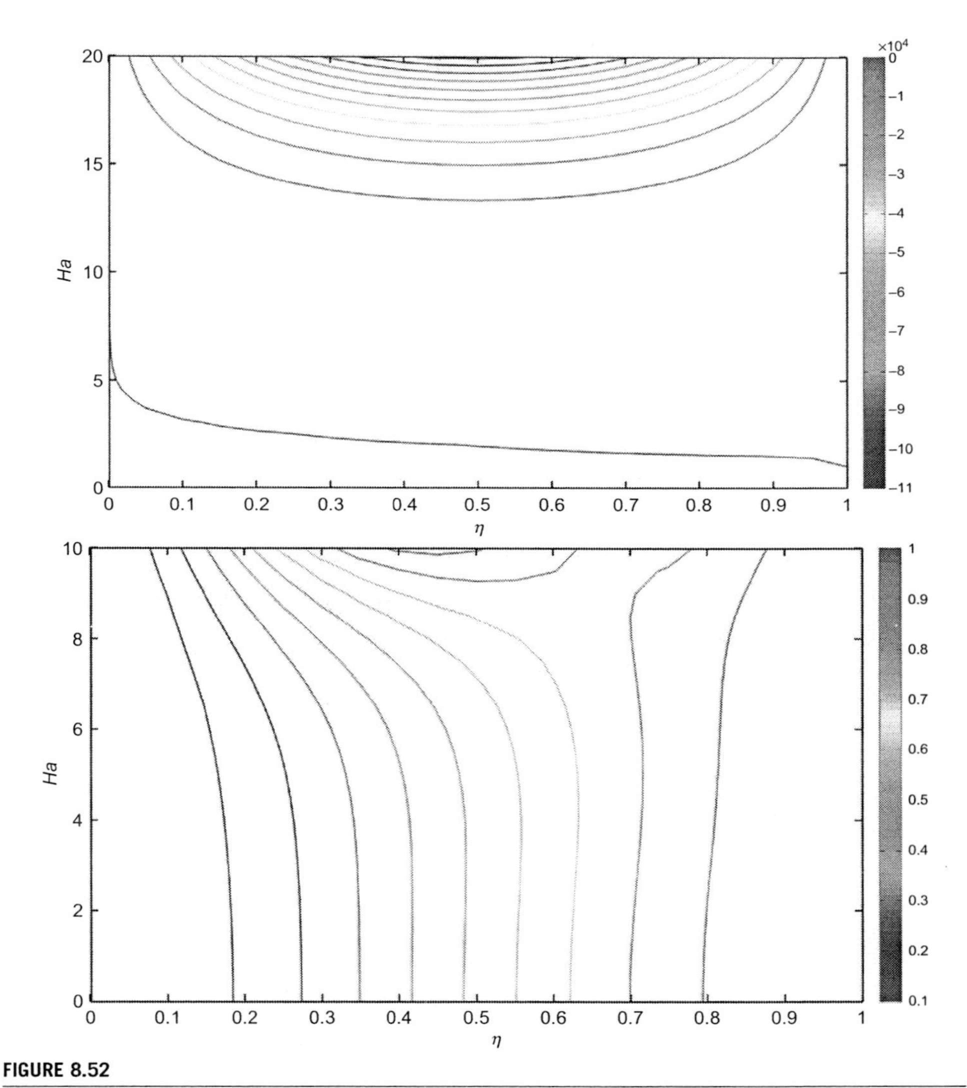

FIGURE 8.52

Contour plots of V (left) and U (right) when $\phi = 0.06$, $Re = 1$ for Cu-water case.

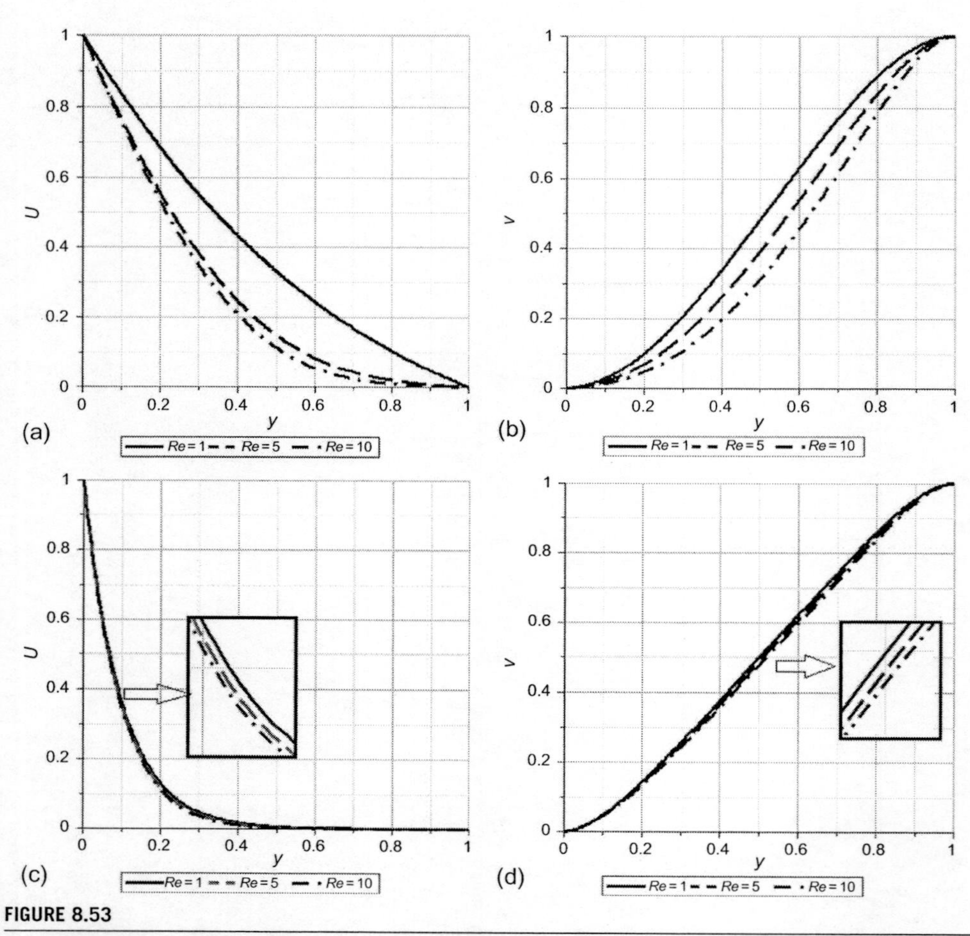

FIGURE 8.53

Effect of Reynolds number (*Re*) on dimensionless velocities for water with copper nanoparticles, $\phi = 0.04$, (a) $U(y)$, $Ha = 1$, (b) $V(y)$, $Ha = 1$, (c) $U(y)$, $Ha = 10$, and (d) $V(y)$, $Ha = 10$.

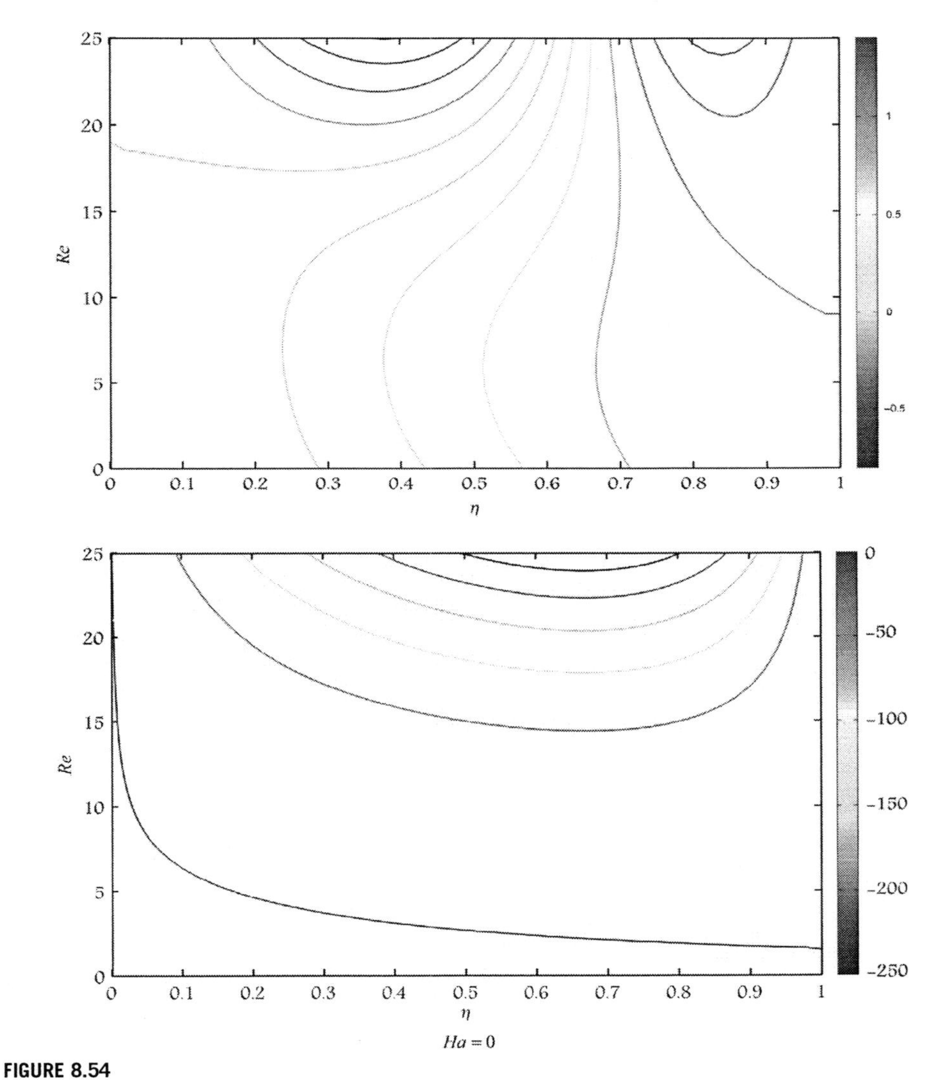

FIGURE 8.54

Contour plots of V (left) and U (right) when $\phi = 0.06$ for Cu-water case.

Continued

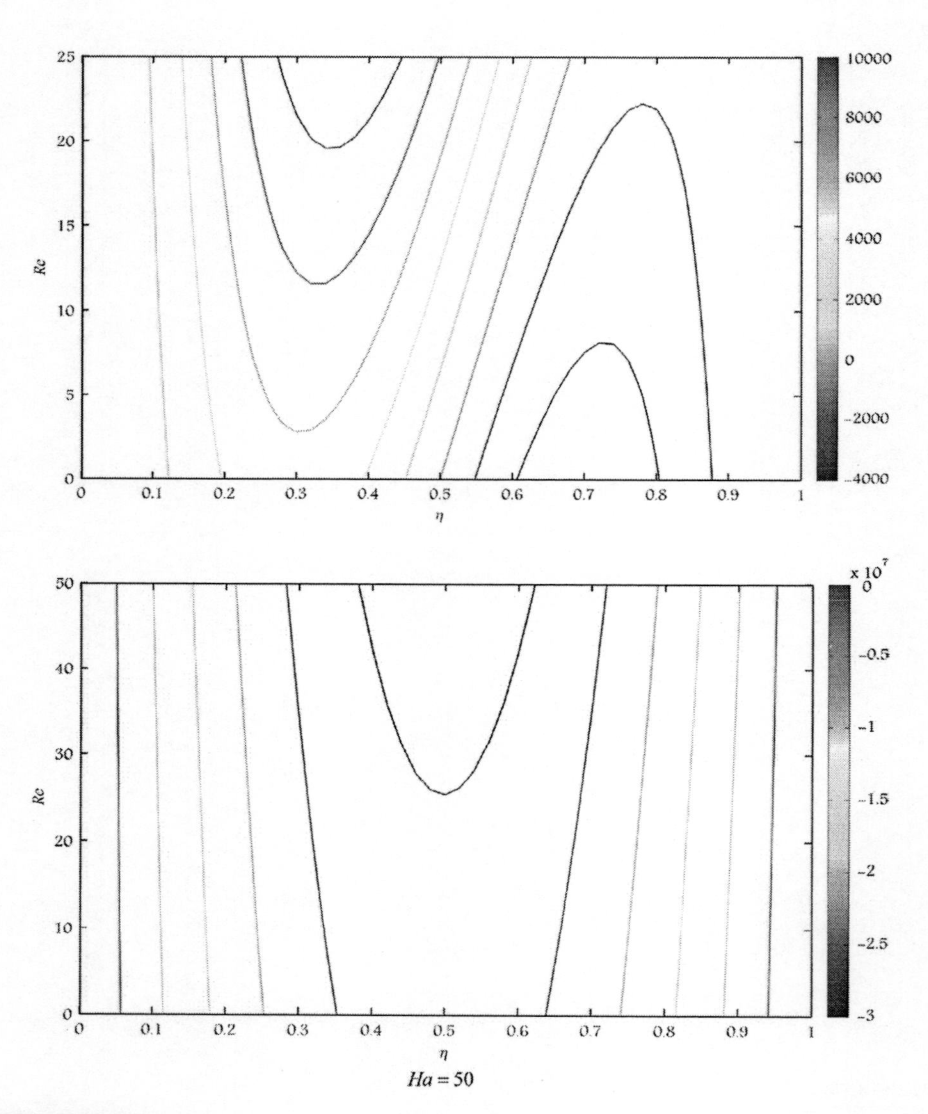

FIGURE 8.54, cont'd

Table 8.11 ***U*(*y*) Variations for Different Types of Nanofluids and Nanoparticles, $Re = 1$, $Ha = 1$, and $\phi = 0.04$**

y	Water-Copper	Water-Silver	Ethylene Glycol-Copper	Ethylene Glycol-Silver
0.0	1.0	1.0	1.0	1.0
0.1	0.83570944	0.83422691	0.8366021	0.8352604
0.2	0.68526977	0.68251844	0.6869270	0.6844364
0.3	0.55072934	0.54706012	0.5529409	0.5496176
0.4	0.43267586	0.42850606	0.4351915	0.4314120
0.5	0.33056756	0.32632363	0.3331308	0.3292806
0.6	0.24305045	0.23912555	0.2454244	0.2418595
0.7	0.16826652	0.16499404	0.1702490	0.1672728
0.8	0.10414284	0.10178554	0.1055735	0.1034265
0.9	0.04864722	0.04739956	0.0494059	0.0482677
1.0	0.0	0.0	0.0	0.0

Table 8.12 ***V*(*y*) Variations for Different Types of Nanofluids and Nanoparticles, $Re=1$, $Ha=1$, *and* $\phi=0.04$**

y	Water-Copper	Water-Silver	Ethylene Glycol-Copper	Ethylene Glycol-Silver
0.0	0.0	0.0	0.0	0.0
0.1	0.02610404	0.025997188	0.026167546	0.026071871
0.2	0.097561353	0.097211791	0.097768963	0.097456128
0.3	0.204165243	0.203542825	0.204534660	0.203977939
0.4	0.335581162	0.334740930	0.336079488	0.335328393
0.5	0.48112359	0.480180063	0.481682731	0.480839846
0.6	0.62958117	0.628678803	0.630115463	0.629309909
0.7	0.769057186	0.768337156	0.769483116	0.768840824
0.8	0.88678803	0.886349813	0.887046994	0.886656410
0.9	0.96889373	0.96874723	0.968980208	0.968849751
1.0	1.0	1.0	1.0	1.0

REFERENCES

Adomian, G., 1988. A review of the decomposition method in applied mathematics. J. Math. Anal. Appl. 135 (2), 501–544.

Alfvén, H., 1942. Existence of electromagnetic-hydrodynamic waves. Nature 150, 405.

Aminossadati, S.M., Ghasemi, B., 2009. Natural convection cooling of a localized heat source at the bottom of a nanofluid-filled enclosure. Eur. J. Mech. B Fluids 28, 630–640.

Andoh, Y.H., Lips, B., 2003. Prediction of porous walls thermal protection by effusion or transpiration cooling: an analytical approach. Appl. Therm. Eng. 23, 1947–1958.

Anwari, M., Harada, N., Takahashi, S., 2005. Performance of a magnetohydrodynamic accelerator using air-plasma as working gas. Energy Convers. Manag. 4, 2605–2613.
Baker, G.A., 1975. Essentials of Padé Approximants. Academic Press, London.
Berman, A.S., 1953. Laminar flow in channels with porous walls. J. Appl. Phys. 24, 1232.
Bouaziz, M.N., Aziz, A., 2010. Simple and accurate solution for convective-radiative fin with temperature dependent thermal conductivity using double optimal linearization. Energy Convers. Manag. 51, 76–82.
Boyd, J., 1997. Padé approximant algorithm for solving nonlinear ordinary differential equation boundary value problems on an unbounded domain. Comput. Phys. 11 (3), 299–303.
Brinkman, H.C., 1952. The viscosity of concentrated suspensions and solution. J. Chem. Phys. 20, 571.
Cha, J.E., Ahn, Y.C., Kim, M.-H., 2002. Flow measurement with an electromagnetic flow meter in two-phase bubbly and slug flow regimes. Flow Meas. Instrum. 12 (5–6), 329–339.
Chandran, P., Sacheti, N.C., Singh, A.K., 1996. Hydromagnetic flow and heat transfer past a continuously moving porous boundary. Int. Comm. Heat Mass Tran. 23, 889–898.
Desseaux, A., 1999. Influence of a magnetic field over a laminar viscous flow in a semi-porous channel. Int. J. Eng. Sci. 37, 1781–1794.
Elbashbeshy, E.M.A., 1998. Heat transfer over a stretching surface with variable surface heat flux. J. Phys. D. Appl. Phys. 31, 1951–1954.
Ganji, D.D., Azimi, M., 2013. application of DTM on MHD Jeffery Hamel problem with nanoparticles. UPB Sci. Bull. Ser. A 75, 1.
Ganji, D.D., Hashemi Kachapi, S.H., 2011a. Analytical and numerical method in engineering and applied science. Prog. Nonlinear Sci. 3, 1–579.
Ganji, D.D., Hashemi Kachapi, S.H., 2011b. Analysis of nonlinear equations in fluids. Prog. Nonlinear Sci. 3, 1–294.
Ganji, Z.Z., Ganji, D.D., Esmaeilpour, M., 2009. Study on nonlinear Jeffery-Hamel flow by He's semi-analytical methods and comparison with numerical results. Comput. Math. Appl. 58, 2107–2116.
Ghasemi, B., Aminossadati, S.M., 2010. Mixed convection in a lid-driven triangular enclosure filled with nanofluids. Int. Commun. Heat Mass Transf. 37, 1142–1148.
Hamel, G., 1916. Spiralförmige Bewgungen Zäher Flüssigkeiten. Jahresber. Deutsch. Math.-Verein. 25, 34–60.
Hashemi Kachapi, S.H., Ganji, D.D., 2013. Dynamics and Vibrations: Progress in Nonlinear Analysis, 2014 ed. Solid Mechanics and Its Applications. Springer, Netherlands.
Hashemi Kachapi, S.H., Ganji, D.D., 2014a. Nonlinear Differential Equations: Analytical Methods and Application, 2015 ed. Cambridge International Science Publishing, Cambridge, UK.
Hashemi Kachapi, S.H., Ganji, D.D., 2014b. Nonlinear Analysis in Science and Engineering, Cambridge International Science Publishing, Cambridge, UK, 450 p.
Hayat, T., Javed, T., Abbas, Z., 2009. MHD flow of a micropolar fluid near a stagnation point towards a non-linear stretching surface. Nonlinear Anal. Real World Appl. 10, 1514–1526.
He, J.H., 1999. Homotopy perturbation technique. Comput. Meth. Appl. Mech. Eng. 178, 257–262.
He, J.H., 2003. Homotopy perturbation method: a new nonlinear analytical technique. Appl. Math. Comput. 135, 73–79.
He, J.H., 2006. Homotopy perturbation method for solving boundary value problems. Phys. Lett. A 350, 87–88.
Hendi, F.A., Albugami, A.M., 2010. Numerical solution for Fredholm–Volterra integral equation of the second kind by using collocation and Galerkin methods. J. King Saudi Univ. (Sci.) 22, 37–40.
Homsy, A., Koster, S., Eijkel, J.C.T., Ven der Berg, A., Lucklum, F., Verpoorte, E., de Rooij, N.F., 2005. A high current density DC magnetohydrodynamic (MHD) micropump. Lab Chip 5, 466–471.
Ishak, A., Jafar, K., Nazar, R., Pop, I., 2009. MHD stagnation point flow towards a stretching sheet. Physica A 388, 3377–3383.
Jeffery, G.B., 1915. The two-dimensional steady motion of a viscous fluid. Philos. Mag. 6, 455–465.

Liao, S.J., 2003. Beyond Perturbation: Introduction to the Homotopy Analysis Method. Chapman & Hall, CRC Press, Boca Raton.

Liao, S.J., 2012. Homotopy Analysis Method in Nonlinear Differential Equations. Springer & Higher Education Press, Heidelberg.

Mahapatra, T.R., Gupta, A.S., 2001. Magneto hydrodynamics stagnation-point flow towards a stretching sheet. Acta Mech. 152, 191–196.

Mahapatra, T.R., Gupta, A.G., 2002. Heat transfer in stagnation point flow towards a stretching sheet. Heat Mass Transf. 38, 517–521.

Makinde, O.D., Motsa, S.S., 2001. Hydromagnetic stability of plane Poiseuille flow using Chebyshev spectral collocation method. J. Ins. Math. Comput. Sci. 12 (2), 175–183.

Makinde, O.D., Motsa, S.S., 2003. Magneto-hydrodynamic stability of plane-Poiseuille flow using multi-deck asymptotic technique. Math. Comput. Model. 37 (3–4), 251–259.

Maxwell, J.C., 1904. A Treatise on Electricity and Magnetism, second ed. Oxford University Press, Cambridge, 435–441.

Runstedtler, A., 2006. On the modified Stefan–Maxwell equation for isothermal multicomponent gaseous diffusion. Chem. Eng. Sci. 61, 5021–5029.

Shaoqin, G., Huoyuan, D., 2008. negative norm least-squares methods for the incompressible magneto-hydrodynamic equations. Acta Math. Sci. 28 B (3), 675–684.

Sheikholeslami, M., Ganji, D.D., Ashorynejad, H.R., Rokni, H.B., 2012a. Analytical investigation of Jeffery-Hamel flow with high magnetic field and nanoparticle by adomian decomposition method. Appl. Math. Mech. (English Ed.) 33 (1), 25–36.

Sheikholeslami, M., Soleimani, S., Gorji-Bandpy, M., Ganji, D.D., Seyyedi, S.M., 2012b. Natural convection of nanofluids in an enclosure between a circular and a sinusoidal cylinder in the presence of magnetic field. Int. Commun. Heat Mass Transf. 39, 1435–1443.

Stern, R.H., Rasmussen, H., 1996. Left ventricular ejection: model solution by collocation, an approximate analytical method. Comput. Biol. Med. 26, 255–261.

Tendler, M., 1983. Confinement and related transport in extrap geometry. Nucl. Instrum. Methods Phys. Res. 207 (1–2), 233–240.

Vaferi, B., Salimi, V., Dehghan Baniani, D., Jahanmiri, A., Khedri, S., 2012. Prediction of transient pressure response in the petroleum reservoirs using orthogonal collocation. J. Pet. Sci. Eng. 98-99, 156–163. http://dx.doi.org/10.1016/j.petrol.2012.04.023.

This page intentionally left blank

Index

Note: Page numbers followed by *f* indicate figures and *t* indicate tables.

A

Adomian decomposition method (ADM)
  definition, 349–350
  Eckert number, 138–139, 147, 148*f*
  fundamentals, 141, 353–354
  kinematic viscosity, 145–147
  lubrication system, 138
  vs. NM, 143, 145*t*
  numerical method, 143, 357*t*
  resolution, 142–143
  skin-friction coefficient, 138–139, 147
  squeeze number, 143–144
  temperature profile, 146*f*, 148*f*
  thermo physical properties, 138–139, 139*t*
  velocity, divergent channel, 358*f*
  volume fraction, 146*f*
Average Nusselt number, 71–72, 77–78, 81, 82*f*, 211, 212, 217–218, 218*f*

B

Bejan number (*Be*), 116, 150, 153–154, 156–161, 286, 290, 291, 300
Boubaker Polynomials Expansion Scheme (BPES)
  first derivatives properties, 199
  linear equations, 199–200
  nonnull set of coefficients, 200
  resolution protocol, 199
Boundary-layer flow
  mathematical model, 111–113
  Sherwood number, 110
Boundary value problem (BVP), 246–247, 260, 333–334, 359–360
BPES. *See* Boubaker Polynomials Expansion Scheme (BPES)
Brinkman models, 83, 138, 206, 211, 287, 318, 369
Brownian motion number/parameter (*Nb*), 110–111, 113, 115–116, 127, 219–220, 222, 332, 336, 342, 347–349
Bucky tubes. *See* Carbon nanotubes (CNTs)
BVP. *See* Boundary value problem (BVP)

C

Cantilevered SWCNT
  basic bending vibration and resonant frequencies, 31
  CNT-based biosensors, 28–29
  continuum mechanics approach and equations, 30–31
  Euler–Bernoulli theory, 28–29
  mass detection, resonators, 30
  mass sensor mode comparison, 33–35, 35*t*
  resonant frequency, 31–33, 32*f*
  vibration mode analysis, 33, 33*t*, 34*f*
Carbon nanotubes (CNTs)
  conductive materials, transport properties, 16, 16*t*
  DWNT, 15–17
  engineering fibers, mechanical properties, 16, 16*t*
  MWNT, 15
  with rippling deformations, 54–68
  SWNT, 15
Cauchy-Born rule, 8
CNTs. *See* Carbon nanotubes (CNTs)
Collocation method (CM)
  application, 321–322
  mathematical formulation, 321
Conservation of energy (first law of thermodynamics), 195–196
Continuum mechanics approach and equations, 30–31
Curved SWCNT
  amplitude frequency response curves, 49, 50*f*
  Chebyshev collocation approach, 45–46
  CNT's waviness on nonlinear frequency, 49–53, 51*f*, 52*f*, 53*f*
  continuum modeling approach, 45
  Hamilton's energy principle, 45–46
  midplane stretching nonlinearity, 46
  Pasternak model, 45–46, 51, 52*f*
  solution methodology, 48–49
  and stretching nonlinearity, combination, 49, 50*f*
  vibrational model, 46–48
  Winker model, 45–46
Cu/water nanofluid. *See also* Transverse magnetic field
  heat transfer, 83–91
  number, *Ec*, 86
  turbulent natural convection, 71–83

D

Differential transformation method (DTM)
  combustion of iron particles, 198–199, 198*t*
  fundamental operations, 197, 198*t*
  Maclaurin series, 197
  and numerical method, comparison, 200, 200*f*
  Taylor series expansion function, 197
  thermophysical properties, 197–198, 198*t*
Double-wall nanotubes (DWNT), 15–17
DTM. *See* Differential transformation method (DTM)
DWNT. *See* Double-wall nanotubes (DWNT)

E

Eckert number, 138–139, 147, 148*f*
Elastic foundation, 46–48, 47*f*
Entropy generation minimization (EGM)
  Be number, 156, 159*f*, 160*f*
  Blasius equation, 149
  definition, 150, 286
  governing equation, 150–153
  horizontal flat plate equations, 153
  solid volume fraction, 154–156
  temperature profile and gradients, 156, 157*f*, 158*f*, 159*f*
  velocity profile and gradient, 155*f*, 156*f*, 157*f*
  volumetric entropy generation, 156–161, 160*f*, 161*f*
Euler–Bernoulli beam model, 17–18, 19, 28–29, 31, 32

F

Falkner–Skan boundary-layer equation
  *m*th-order deformation equations, 188
  for nanofluid, 181, 182
  rule of solution expression, 186–187
  zeroth-order deformation equations, 187–188
Finite element method (FEM)
  hermite interpolation polynomials, 23
  Newton–Raphson method, 22–23
Forced convection analysis
  basic EG water fluid, 238, 239*f*
  continuity equation, 231
  CuO-EG water, 235, 235*f*, 236*f*, 237*f*, 238*f*
  CuO nanoparticle, 238, 240*f*, 241*f*
  density of nanofluid, 231
  dimensionless groups, 232
  energy equation, 231
  HAM and numerical solution, comparison, 238, 240*t*
  heat capacitance and thermal expansion coefficient, 231–232
  heat transfer, 229–241
  MG model, 230, 232–233
  for MHD AL_2O_3-water nanofluid
    flow (*see* Magnetohydrodynamics (MHD))
  momentum equation, 231
  physical configuration, 230, 230*f*
  temperature profiles, 235, 235*f*, 236*f*
  thermal conductivity, 232
  thermophysical properties, 230, 231, 231*t*
Fourth-order Runge–Kutta–Fehlberg method, 260
Fourth-order Runge–Kutta method, 89, 97, 253, 260, 272, 277, 293, 308

G

Galerkin method (GM)
  application, 323, 324, 325*f*, 326*f*, 327*t*, 375–376
  mathematical formulation, 323, 375
  and numerical comparison, 376, 377*f*, 379*t*
Generalized differential quadrature (GDQ) method, 17, 20–21
GM. *See* Galerkin method (GM)
Grashof number, 232

H

HAM. *See* Homotopy analysis method (HAM)
Hartman number (*Ha*), 324, 328*f*, 376, 380*f*, 381*f*
Heat transfer and flow, porous medium
  applications, 286
  Bejan number, 286
  boundary-layer approximations, 288
  Brinkman model, 287
  fourth-order Runge–Kutta method, 293
  HAM (*see* Homotopy analysis method (HAM))
  HTI and FFI, 300
  hydrodynamics, 288–289
  Nusselt number, 295, 298*f*, 299*t*, 300*t*
  power-law temperature, 288
  power-law velocity variation, 286
  shear stress and rate of heat transfer, 300, 301*f*, 302*f*
  skin-friction coefficients, 295, 297*f*, 299*t*
  steady boundary-layer flow, 286
  temperature profiles, nanoparticle volume fraction, 293–295, 294*f*, 296*f*
  thermal analysis, 289–291
  thermodynamics second law, 288
  thermo physical properties, 287, 287*t*
  velocity, nanoparticle volume fraction, 293–295, 294*f*, 296*f*
  volumetric entropy generation rate, 288
  water-based nanofluid, 287
Heat transfer, Cu-water nanofluid flow
  Brinkman models, 83
  differential equations, 85
  equations for momentum and energy, 85
  fluid heating and cooling, 83
  fourth-order Runge–Kutta method, 89
  heat transfer analysis, 84, 84*f*
  HPM, 83–84, 86–87, 88–89, 88*f*, 89*t*
  implementation, method, 87–88
  Maxwell–Garnett (MG), 83
  parameters, 85
  skin fraction coefficient and Nusselt number, 86
  squeeze number, effects, 90, 91*f*
  squeezing flow in lubrication system, 83
  thermal boundary layer thickness, 90, 93*f*
  thermo physical properties, 84, 84*t*
Heat transfer, slip-flow boundary condition
  BVP, 246–247
  continuity and momentum equations, 244
  Cu nanoparticle concentration, 248, 248*f*, 249*f*

density of nanofluid, 244
dimensionless boundary conditions, 246
forced convection, 229–241
HAM solution, 233–235
heat capacitance and thermal expansion coefficient, 244
LBM, 242
local Nusselt number, 247, 248, 250, 250*f*, 251*f*, 253*f*
MCHS, 242
nondimensional parameters, 245
physical configuration of microchannel, 243, 243*f*
Rung–Kutta and shooting method, 241–242, 246
shear stress distributions, 250, 252*f*
slip coefficient, 241–253, 251*f*
temperature profiles, 247, 247*f*, 248, 249*f*
thermal conductivity, 244, 245, 245*t*
thermophysical properties, 243–244, 247*f*
Heat transfer, unsteady stretching surface
dimensionless temperature, 128–131, 136*f*
film thickness, 133, 133*f*
flow field and heat transfer, 127
formulation and governing equation, 128–131
free surface temperature, 133, 135*f*
numerical procedure and validation, 131–132
Nusselt number, 133, 135*f*
ODEs, 132
Rung–Kutta shooting method, 128
skin-friction coefficient, 133, 134*f*, 137–138, 137*t*
surface velocity, 133, 134*f*
He's energy balance method (HEBM), 36
Homotopy analysis method (HAM)
advantages, 257
analytical and numerical results, 116, 117*f*
auxiliary linear operators, 291–292
Brownian motion number, 115–116
convergence and rate of approximation, 115, 261, 293, 307–308, 363–364
Falkner–Skan problem, 186–188
fourth-order Runge–Kutta procedure, 260
and HPM results, comparison, 262, 264*t*
magnetic parameter, effect, 265, 266*f*
MAPLE software, 293
MHD Jeffery–Hamel flow, 362–363
*m*th-order deformation equation, 114–115, 186, 234–235, 257, 258, 259–260, 292–293, 307
nanoparticle volume fraction, 124, 125*f*, 126*f*, 262, 265, 265*f*
nonlinear differential equation, 181, 182, 185, 257
numerical method's (NUM), 262, 263*f*, 263*t*, 360, 365, 367*f*, 368*f*
Nusselt number, 116–120, 117*f*, 119*f*
physical properties, 262, 262*t*
power-law heat flux, 293
power-law temperature, 293
Pr numbers, 120, 120*f*
Sherwood number, 116–120, 118*f*
skin-friction coefficient, 116*f*
solution, convergence, 188–189, 189*f*, 190*f*, 191*f*
Taylor series expansion, 185–186, 257, 258, 292
temperature profiles, 124, 125*f*, 126*f*
transverse magnetic field, 362–364
values of auxiliary parameters, 261, 261*f*
velocity profiles, 120, 123*f*
zeroth-order deformation, 113–114, 185, 233–234, 257, 259, 292–293, 306–307
Homotopy method (zero-order deformation equation), 185
Homotopy perturbation method (HPM), 83–84, 86–87, 163–164, 166–169, 219–220, 223–224, 333, 334, 335–336. *See also* Nonlinearly stretching sheet

I

Interatomic potentials, 6

J

Jeffery–Hamel flow
ADM, 349–350, 353–354
application, 354–355
constant value, 355–356, 356*t*
Cu-water nanofluid flow, 356–359, 359*f*
cylindrical polar coordinates, 351
definition, 350–351
electromagnetic body force, 351, 352
Lorentz force, 356
magnetic field effects, 355–356
Navier–Stokes and Maxwell's equations, 349–350, 352
numerical and ADM solution, comparison, 356, 357*t*
Poiseuille flow, 351
Reynolds number, 355–356, 359
transverse magnetic field *(see* Transverse magnetic field)
velocity profiles, divergent channels, 356, 358*f*

K

Kirchhoff's law of radiation, 196

L

Large-eddy simulations (LESs) method
nanofluid-filled enclosure, 72
numerical analysis, 72
turbulent flows, 71–72
Lattice Boltzmann equation (LBE), 208
Lattice Boltzmann method (LBM)
advantages, 206
based on LES model, 75–76
Boussinesq approximation, 75, 209

Lattice Boltzmann method (LBM) *(Continued)*
buoyancy forces, 209
Chapman–Enskog expansion, 74–75
effective thermal conductivity, 211
flow and temperature distribution function, 73–74
LBE, 208
macroscopic variables, 75, 209
nanofluid flow, 76–77, 207, 210–211
natural convection flow, 206, 207, 209
Navier–Stokes equations, 73
Nusselt number, 211
square grid and D2Q9 model, 73, 74, 74*f*, 208, 208*f*
temperature field, 74
thermophysical properties, 211
third-order Gauss–Hermite quadrature, 74
types of nanoparticles, 211
viscosity, 211
LBE. *See* Lattice Boltzmann equation (LBE)
LBM. *See* Lattice Boltzmann method (LBM)
Least square method (LSM)
application, 323, 374–375
comparison, GM and numerical, 324, 327*t*, 376, 377*f*, 379*t*
mathematical formulation, 322, 374–375
LESs. *See* Large-eddy simulations (LESs) method
Lewis number (*Le*), 110–111, 113, 115–116, 127, 141, 219–220, 222, 332, 336, 347, 349
Local Nusselt number, 71–72, 76, 77, 78, 80–81, 80*f*, 96, 110, 131, 211, 276, 301–302
LSM. *See* Least square method (LSM)

M

Maclaurin series, 197
Magnetohydrodynamics (MHD)
AL_2O_3-water nanofluid flow
dimensionless parameters, 256
fourth-order Runge–Kutta numerical method, 253
HAM, 253, 254, 257–260
HPM, 254
nanofluid boundary layer, 254, 254*f*
numerical method, 260
physical boundary conditions, 255–256
stream function and temperature, 256
transformation parameters, 256
transformed boundary conditions, 256
in divergent and convergent channels
cylindrical polar coordinates, 319
fourth-order Runge–Kutta procedure, 324
Jeffery–Hamel nanofluid flow, 318
nanofluids and nanoparticles, 324, 326*t*
Navier–Stokes and Maxwell's equations, 319
nondimensional velocity profiles, 324, 329*f*
numerical method, comparison, 324, 325*f*, 326*f*
Reynolds number and Hartmann number, 320, 324, 328*f*
skin-friction coefficient, 324, 328*f*
solid volume fraction, 320
stagnation-point flow *(see* Stagnation-point flow)
values, 324, 327*t*
volume fraction, 324, 329*f*
WRMs *(see* Weighted residual methods (WRMs))
Maxwell–Garnett (MG) model, 83, 206, 301–302, 303, 318
MCHS. *See* Microchannel heat sinks (MCHS)
MG. *See* Maxwell–Garnett (MG) model
Microchannel heat sinks (MCHS), 242
Micromechanics, 6–8
Midplane stretching
curved SWCNT, 45, 46
embedded CNT, 37, 39, 47–48, 49–53, 50*f*
nonlinear amplitude frequency response, 36, 37, 41–42, 41*f*, 43, 43*f*, 44*f*
rippling deformation, 45
Mixed convection flow
dimensionless variables, 222
geometry and coordinate system, 220, 221*f*
heat and mass transfer, 220
homotopy perturbation method, 219–220, 223–224
in horizontal channel, 219–229
HPM and exact solution, 224, 224*f*, 225*t*
mass flow rate, 221
mathematical model, 221
Newtonian fluids, 222
numerical procedure and validation, 246–247
temperature distribution, 224–225, 225*f*, 226–229, 226*f*, 227*f*
temperature profiles, 226–229, 228*f*
Molecular dynamics (MD) method, 6, 36
Molecular nanotechnology, 1–2
*m*th-order deformation equations, 114–115, 186, 187–188, 234–235, 259–260, 292–293, 307, 363
Multiwalled carbon nanotubes (MWCNTs), 17, 31, 54
Multiwall nanotubes (MWNT), 15
MWCNTs. *See* Multiwalled carbon nanotubes (MWCNTs)
MWNT. *See* Multiwall nanotubes (MWNT)

N

Nanobeam model, 18*f*, 19–20
Nanofluid
applications, 8–9, 10
computational fluid dynamics, 9
definition, 8, 72
nanoparticles, definition, 8
sensing applications, 9–10
smart cooling, 9
synthesis, 9

Nanomechanics, 5–6
Nanotechnology
  description, 1–2
  fundamental concepts, 4
  future implications, 3
  nanomaterials, 4–5
  origins, 3–4
  size comparison, 2–3, 2*f*
Natural convection flow
  in concentric annulus
    Cu-water case, 214, 215*f*, 217–218, 218*f*
    grid testing and code validation, 211–212
    heat transfer enhancement, 213–214, 213*f*, 214*f*, 218, 219, 219*f*
    in horizontal annuli, 206
    LBM (*see* Lattice Boltzmann method (LBM))
    Nusselt number, 213–214, 216*f*, 217–218, 217*f*
    Rayleigh number, 214–217
    temperature, curved boundary treatment, 210
    thermophysical characteristics, 213
    types of nanoparticles, 212–213
    velocity, curved boundary treatment, 209–210
  non-Darcy porous medium
    Al_2O_3-water and Cu-water nanofluids, 303
    comparison, HAM solution and numerical solutions, 308
    engineering applications, 303
    HAM, 306–307
    heat capacitance, 304–305
    heat transfer rate, 311, 313*f*, 314*f*
    local Nusselt number, 311–313, 313*f*, 314*f*
    Maxwell–Garnett model, 301–302, 303
    nondimensional heat transfer coefficient, 305–306
    physical configuration, 303, 303*f*
    Rayleigh number, 305
    temperature profiles, 308, 309*f*, 310*f*
    thermal conductivity, 305
    thermophysical properties, 304, 304*t*
    two-dimensional Cartesian coordinates, 304
    variation, nondimensional velocity, 311, 311*f*, 312*f*
    velocity profiles, 311, 312*f*
    wall surface, heat transfer rate, 301–302
Newton–Raphson method, 22–23, 94
Nonlinearly stretching sheet
  comparison, 169
  formulation, 164–166
  heat transfer rate, 178*f*
  HPM (*see* Homotopy perturbation method (HPM))
  HPM-Padé, 172*f*
  industrial application, 162
  laminar boundary layer flow, 162–163
  Nusselt number, 169
  skin friction coefficient, 173*t*, 177*f*
  temperature profile, 173, 175*f*
  velocity distribution, 172–173, 174*f*
  viscous dissipation parameter, 173, 176*f*, 177*f*
Nonlinear vibration model, CNTs
  beam bending theory, 38
  classical analysis of beam, 37
  deflection deformation, 38
  differential equation, 39
  free vibration equation, 37, 41
  frequency ratio, 41–42
  midplane stretching, 42–43, 44*f*
  rippling deformation, effect, 42
  rippling formation, 37, 38, 43, 43*f*
  solution methodology, 40–41
  vibration equation, 39
Nusselt number (Nu)
  average, 71–72, 77–78, 81, 82*f*, 211, 212, 217–218, 218*f*
  heat transfer, 276, 311
  local, 71–72, 76, 77, 78, 80–81, 80*f*, 96, 110, 131, 276, 301–302
  nanofluids types, 284, 285*t*, 299
  nanoparticle volume fraction, 299*t*, 300*t*
  power-law temperature and heat flux, 290
  reduced, 116–120, 119*f*, 121*f*

P

Padé approximation, 163, 164–166, 170*t*, 171*t*, 172*f*, 333–334, 336, 337*f*, 338*f*, 339*f*
Pasternak elastic foundation, 45–48, 47*f*, 49, 51, 52*f*, 53
Perturbation method (PM), 54, 57–58, 83–84, 164, 333
Pore space, 272
Poromechanics, 185
Prandtl number (*Pr*), 75, 76, 86, 96, 110, 113, 115–116, 127, 140, 166, 212–213, 222, 232, 248, 336

Q

Quasicontinuum method, 8

R

Rayleigh number (*Ra*), 71–72, 76, 78, 97, 195, 214–217, 241–242, 246, 250–253, 273, 305
Resonant frequency, 28–29, 30, 31–33, 32*f*, 62
Reynolds number (*Re*)
  defined, 256, 289, 320, 352, 361–362, 371
  divergent and convergent channel, 324, 355–356
  high/large, 324, 356
  increasing function, 318, 324, 356, 359, 369, 379
  local, 96, 113, 131, 166, 276, 332–333
  low, 376
  magnetic field, 331–332
Rippling, 36, 54

Rippling deformations
CNTs on nonlinear frequency, 62, 64, 64*f*, 65*f*
nonlinear elastic beam model, 54–55
vibration model (*see* Vibrational model)
Winkler stiffness, effects, 62, 63*f*, 66*f*, 67*f*
Young modulus, 54
Rippling formation, 37, 38, 43, 43*f*
Rippling mode, 54

S

Semi nonlinear analysis
CNTs (*see* Carbon nanotubes (CNTs))
SWCNT (*see* Single-walled carbon nanotube (SWCNT))
Semiporous channel, MHD nanofluid flow
Berman's similarity transformation, 372–373
biomedical engineering applications, 370
boundary conditions, velocity, 371
dimensionless equations, 372
electrical conductivity, Maxwell, 372
flow problem, 370
governing equations, 370–371
Hartman number, 376, 380*f*, 381*f*
laminar two-dimensional stationary flow, 370
LSM, GM and numerical, 376, 377*f*, 379*t*
mean velocity calculation, 371
MG and Brinkman models, 369
momentum equation, 373
nanoparticle volume fraction, 376, 380*f*
Reynolds number, 376, 382*f*, 383*f*
thermal conductivity, MG model, 372
thermophysical properties, 372, 372*t*
variations, nanofluids and nanoparticles, 376, 385*t*
WRMs, 373–376
Sherwood number, 116–120, 118*f*, 122*f*
Single SWCNT
beam, 19–20
differential quadrature and solution procedure, 20–22
FEM, 22–23
length effect, 25, 26*f*, 27*f*, 28*f*
mesh point number effect, 24–25, 24*f*, 25*f*
nanobeam model, 18*f*
schematic of problem, 18, 19*f*
validation of GDQ approach, 25–27, 29*f*, 30*f*
Single-walled carbon nanotube (SWCNT)
cantilevered, 28–36
curved, 45–54
single, 17–28
Single-wall nanotubes (SWNT), 15
Square cavity, 71–72, 73, 130, 273
Squeeze number (*S*), 83, 86, 90, 91, 91*f*, 93*f*, 138, 140, 143, 144, 145–147, 145*f*, 147*f*
Squeezing flow, 143–144
Stagnation-point flow
MHD
boundary layer flow, 330
Brownian motion parameter, 342
concentration distributions, 336, 340*f*, 341*f*, 343*f*, 344*f*, 345*f*, 346*f*, 347*f*
dimensionless Nusselt numbers, 347, 348*f*, 349
heat transfer and concentration rates, 347, 349*f*, 350*f*
heat transfer and mass, 330
HPM review, 333
mathematical model, 331–332
nonlinear ordinary differential equations, 334
Padé approximants, 333–334, 339*f*
Prandtl number, 336, 341*f*
Sherwood numbers, 347, 348*f*, 349
skin-friction number and Nusselt number, comparison, 335, 336*t*
temperature distributions, 336, 340*f*, 341*f*, 342*f*, 343*f*, 344*f*, 345*f*, 346*f*
thermophoresis and Brownian motion, 347–349
velocity and temperature profiles, 335–336, 337*f*, 338*f*
in porous medium
boundary-layer structure, 279–284
Cu-water, Al_2O_3-water and TiO_2-water, 272
fourth-order Runge–Kutta method, 277, 284–286
heat transfer, 272–273
hydraulic boundary layer, 284
mathematical formulation, 273–276
nanoparticle volume fraction, 278, 278*f*, 279*f*, 280*f*, 281*f*, 282*f*
natural convection, 273
numerical procedure and validation, 276–277, 277*f*
Nusselt number, 284, 285*t*
skin-friction coefficients, 272, 284, 285*t*
thermal behavior, 273
unsteady (*see* Unsteady stagnation point flow)
Stretching and slip parameter, 102, 104*f*
SWCNT. *See* Single-walled carbon nanotube (SWCNT)
SWNT. *See* Single-wall nanotubes (SWNT)

T

Taylor series expansion function, 185–186, 197, 292, 306–307
Temperature variation analysis
Boubaker polynomials expansion scheme, 194–195
BPES, 199–200
conservation of energy (first law of thermodynamics), 195–196
DTM, 195, 197–199
homotopy perturbation method, 194
ignition temperature, 196
Kirchhoff's law of radiation, 196
micro and nanoparticles, 201, 201*f*, 202*f*

nonlinear differential equation, 196
particle combustion, 201
polynomial expansion methods, 194–195
Runge–Kutta method, 195
Thermal properties
CuO, temperature profile, 192, 193*f*
Falkner–Skan problem, 186–188
fluids, temperature profile, 192, 193*f*
governing equations, 182–185
HAM *(see* Homotopy analysis method (HAM))
temperature curve variation, 192, 192*f*
velocity profile of fluid, 189, 191*f*
Thermophoresis number (*Nt*), 110–111, 115–116, 127
Thermophysical properties
ADM, 138–139, 139*t*
Cu-water nanofluid flow, 84, 84*t*
DTM, 197–198, 198*t*
forced convection analysis, 230, 231, 231*t*
heat transfer and flow, 243–244, 247*f*, 287, 287*t*
LBM, 211
MHD nanofluid flow, 372, 372*t*
turbulent natural convection, 73, 73*t*
unsteady stagnation point flow, 97, 97*t*
Transverse magnetic field
convergence region, HAM, 363–364, 364*f*, 365*f*, 366*f*
Jeffery–Hamel problem, 360, 362–363
mathematical formulation, 360–362
numerical and HAM solution, comparison, 365–368, 367*f*, 368*f*, 369*f*, 371*f*
Tubular carbon structures, 17
Turbulent natural convection
average Nusselt number, 81, 82*f*
code validation, 77–78, 78*f*, 79*f*
flow, 77
heat transfer, 72, 80, 83–91
LBM *(see* Lattice Boltzmann method (LBM))
LES method, 75
local Nusselt number, 80–81, 80*f*
nanofluids, 72
pure fluid and *Ra* numbers, comparison, 78–80, 79*f*
Rayleigh number, 72
temperatures, 73, 77
thermophysical properties, 73, 73*t*
turbulent flows, 72
vertical and horizontal velocity values, 81, 81*f*

U

Unsteady stagnation point flow
base fluid and nanoparticles, thermophysical properties, 97, 97*t*
comparison of values, 97, 97*t*
Cu-water nanofluid, 98, 99*f*, 100*f*, 101*f*, 102*f*, 102*t*
for different nanoparticles, 98, 98*f*, 99*f*
field boundary conditions, 97
fourth-order Runge–Kutta, 97
governing equations, 94–96
Navier's slip condition, 92–94
physical model and coordinate system, 94, 95*f*
stretching and slip parameter, effects, 102, 104*f*

V

Van der Waals interaction, 17–18, 19–20, 27–28
Variational iteration method (VIM), 83–84
Vibrational model
CNTs
damping forced Duffing's equation, 58
load-to-damping ratio, 62, 63*f*
nonlinear analysis, 55–62, 56*f*, 58*t*
curved SWCNT
linear vibration version, 47
Pasternak elastic foundation, 46–48, 47*f*

W

Water/copper nanofluid. *See* Cu/water nanofluid
Weighted residual methods (WRMs)
collocation method, 321–322
Galerkin method, 323, 375–376
least square method, 322–323, 374–375
solving differential equations, 320–321
Winker model, 45–46, 62, 63*f*, 66*f*, 67*f*
WRMs. *See* Weighted residual methods (WRMs)

Y

Young's modulus, 17, 31, 35, 36, 37, 46–47, 54

Z

Zeroth-order deformation equations, 113–114, 185, 186, 187–188, 233–234, 257, 258, 259, 292–293, 306–307, 362–363